ORCHIDS OF THE SOLOMON ISLANDS AND BOUGAINVILLE

ORCHIDS of the SOLOMON ISLANDS and BOUGAINVILLE

by B.A. Lewis & P.J. Cribb

Illustrations by
S. Wickison

ROYAL BOTANIC GARDENS
KEW

Dendrobium spectabile

First published 1991

General Editor of Series J.M. Lock. Special Editors for this volume M. Clements & T.A. Cope. Cover design and title and dedication pages by Sue Wickison. The orchid on the front cover is *Dendrobium gouldii*. Typeset by Electronic Images, Cambridge.

ISBN 0 947643 27 3

Printed in Great Britain by Whitstable Litho Printers Ltd., Whitstable, Kent.

For Geoffrey Dennis

and

G. Hermon Slade

Bulbophyllum dennisii

FOREWARD

As a resident naturalist and amateur field botanist in the Solomon Islands for the past 44 years it gives me much pleasure and satisfaction to write this forward to the 'Orchids of the Solomon Islands and Bougainville', particularly because those who have been involved in its compilation are good friends of mine. This scientifically valuable and much awaited publication records the knowledge we have gained to the present date of the rich and fascinating orchid flora of these predominantly jungle-clad oceanic islands, some of which are still botanically unexplored. Indeed, it is possible that these islands – which appear merely as oversized dots on a world map – are home to more than twice the number of orchid species that are native to Europe.

In such heavily forested and wet regions as this, epiphytic orchids usually predominate over terrestrial species, especially in the lowland and lower montane or foot-hill zones. With increasing elevation, resulting in lower temperatures (especially at night) and a lower and more open tree canopy, a much greater variety of terrestrial orchids thrives in the leaf mould and mossy humus of the forest floor, usually in association with small ferns, mosses, gesneriads, *Hoya* species etc. The truly epiphytic orchids of these higher elevations (above 900 m) usually differ from their lowland relatives in having a smaller vegetative habit with relatively large flowers which are often brightly coloured, for example, *Dendrobium laevifolium*. The few larger-growing orchids of the montane forest, such as *Coelogyne* species, are epiphytes in the coastal and lowland habitats but usually become entirely or partly terrestrial in habit at high elevations.

Along the coastal areas *Dendrobium gouldii* is outstanding in numbers and massed displays of blooms, the flowers being long-lasting and resistant to strong winds and salt spray. Likewise, in open scrub or grassland along coastal areas and on lowland ridge slopes, the showy deep mauve to almost pure white blooms of *Spathoglottis plicata* are delightfully displayed on their tall-growing flower heads. It soon becomes apparent to visitors to these tropical islands that orchids are the most obvious and decorative of our wildflowers.

Showy orchids such as *Dendrobium gouldii*, *Grammatophyllum scriptum* var. *boweri* and *G. speciosum* have long been cultivated on coconut palms, citrus and other host trees by lowland village communities and individuals. Traditionally the extremely strong and durable fibers of certain species of *Dendrobium* and *Diplocaulobium* have also been used for weaving decorative arm-bands, small baskets of ceremonial significance, and for providing highly decorative woven embellishments on ceremonial spears, clubs and other weapons, custom chief's batons of office, and on certain kinds of combs worn for show and as 'good luck' talismans.

Many of our orchids are of recognised or potential horticultural value. Fortunately there has never been much commercial collecting and exporting of Solomon Islands orchids, as has occurred in most countries in South-east Asia, resulting in the loss of wild-growing populations. However, Bougainville has not been spared and commercial growers have acquired substantial numbers of two rare *Paphiopedilum* species, and possibly other orchid species, during the past 20 years or so. Orchid and other plant (and animal) exports from the Solomon Islands are now strictly controlled. Such controls are particularly relevant to such rare orchids as the terrestrial *Paphiopedilum wentworthianum*, a much sought-after slipper orchid that has been found in only one or two locations on Bougainville and a single location in central Guadalcana

Increasing loss of orchid and other plant habitats is the most serious threat to the preservation of our species' diversity. Ever since the first humans stepped ashore in these islands, thousands of years ago, destruction of rain forest environments has taken place due to clearing for village settlements and subsistence food gardens with the subsequent development of fire-maintained grasslands. This has resulted in serious climatic change due to the lack of daily forest-generated (from evaporation and transpiration) rainfall. European and other settlers have converted large sections of coastal forest into coconut plantations and cattle runs and some of the smaller islands are completely denuded of natural forest. In more recent years, especially since World War II, greatly accelerated and widespread destruction of the lowland forests has resulted from further expansive agricultural and mining (especially on Bougainville and Guadalcanal) projects, town, road and airfield developments and widespread environmentally ruthless logging followed by subsistence food gardeners with increasing pressures from a rapidly growing population. It is therefore essential for us all to recognise and protect the richness and variety of our natural resources and to enjoy them and to exploit them, if we wish, in a wise and sustainable way, so that future generations will be able to experience the same qualities of our unique island lifestyle as we do today.

Geoffrey Dennis
Honiara

CONTENTS

INTRODUCTION

The Solomon Islands form a chain some 1450 km long, lying to the southeast of the Bismarck Archipelago and New Guinea and about 1900 km to the east of Queensland across the Coral Sea, between latitudes 5° to 12° south of the equator and longitudes 154° and 162° (map 1). The archipelago is made up of a double chain of 993 scattered islands, atolls and cays, which vary from large islands which are rugged, heavily forested and mountainous, to tiny, low-lying coral atolls. At the northwest end of the island chain lies the largest island, Bougainville (map 2), with its satellite island Buka. Although politically Bougainville forms a part of the state of Papua New Guinea it is floristically similar to the Solomon Islands and is therefore included in this account.

The Solomon Islands were a Protectorate of Great Britain from 1893 until July 1978 when they became a fully independent nation.

The Solomon archipelago is divided into seven provinces: Western, Isabel, Malaita, Guadalcanal, Central, Makira and Timotu (see map 3). (See appendix I for a list of alternative or previous island names).

Western Province

The main islands in the Western Province are New Georgia, Vanguna, Vella Lavella, Kolombangara, Rendova, Choiseul, the Shortlands and the Treasury Islands, the two largest islands being New Georgia and Choiseul.

New Georgia is nearly 85 km long and 41 km wide at its broadest point and is 2145 km^2 in extent. The inshore coast is mainly swampy, whilst inland there are several breached volcanoes (Mt Mase, 843 m and Mt Mahimba, 821 m). The volcanic island Vanguna lies at the southwest tip of New Georgia.

Vanguna is 520 km^2 and has two distinct mountains, Mt Vangunu (1123 m) in the south and Mt Reku (520 m) in the northeast. The swampy Nggevala Valley forms a natural division between the island's two parts.

Vella Lava, or Vella, is 670 km^2. The island is wooded and mountainous and the highest mountain, Tambisila Hill (790 m), lies in the northeast.

Kolombangara is almost circular. The island is 685 km^2 and rises from low, flat-topped ridges through increasingly steep escarpments to the rugged crater rim of Mt Veve (1770 m). Logging has been Kolombangara's main industry for the past ten years and, at its height, 75% of the Solomons, logging production came from there. Consequently much of the original rain forest below 450 m has gone.

Rendova lies southwest of New Georgia and is 400 km^2. Rendova Peak (1063 m) dominates the island.

Choiseul and its adjacent islands together are 179 km long, 27 km wide and 3294 km^2 in area. Choiseul is a long, narrow, densely wooded island. Its highest point, near the centre, is Mt Maetambe (1060 m), a volcanic cone.

The scattered islands of the Shortland group lie at the Solomons' northwestern tip, only 9 km from Bougainville and 340 km^2 in extent. The main island is Shortland, or Alu, which is low-lying, the highest point being Balo Hill (185 m). Its principal neighbour is Fauro, a hilly island with a highest point of 600 m. The Treasury Islands lie 29 km southwest of the Shortlands with a combined area of 80 km^2.

Map 1: Position of the Solomon Islands in the south-west Pacific

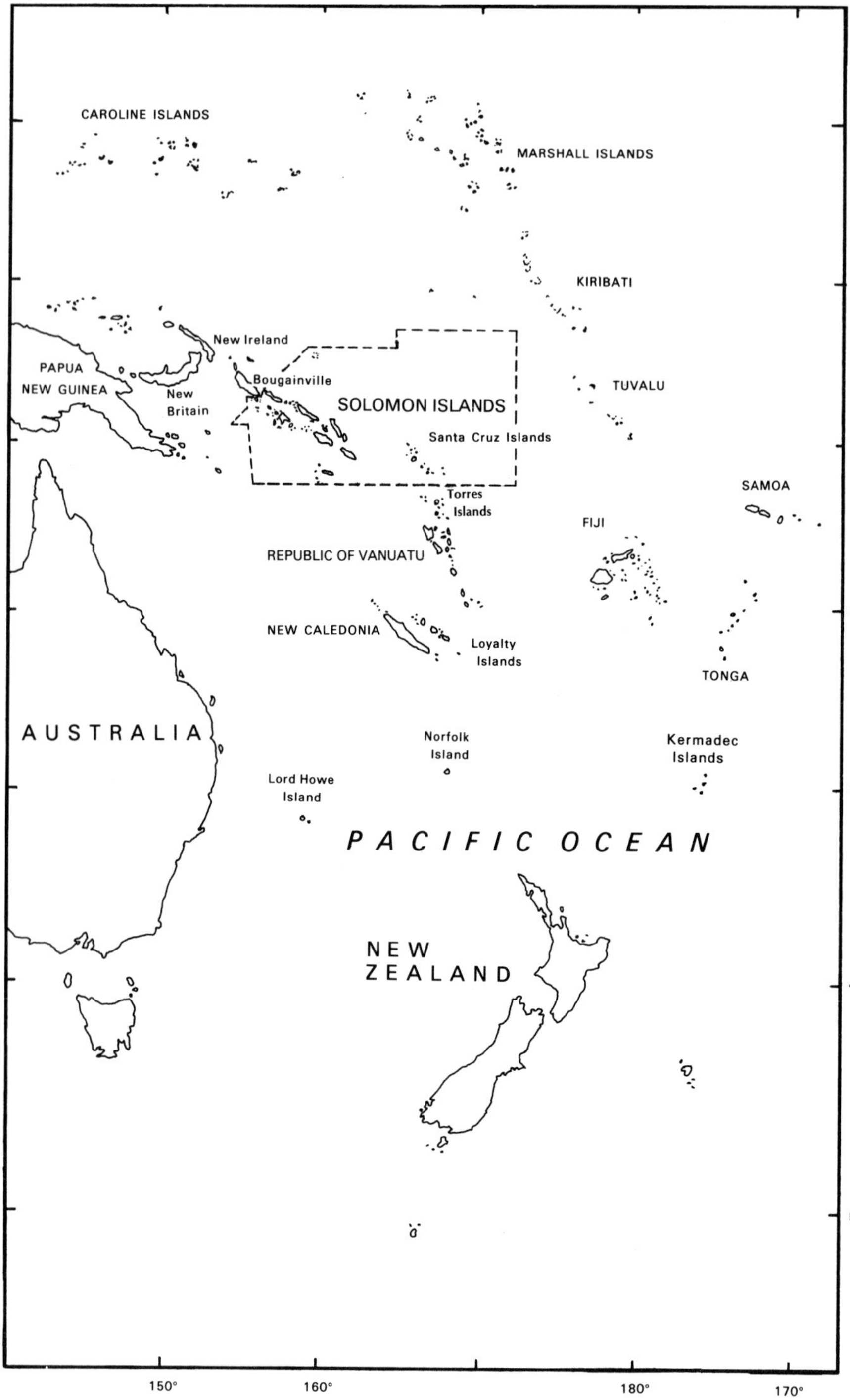

Map 2: Solomon Islands and Bougainville

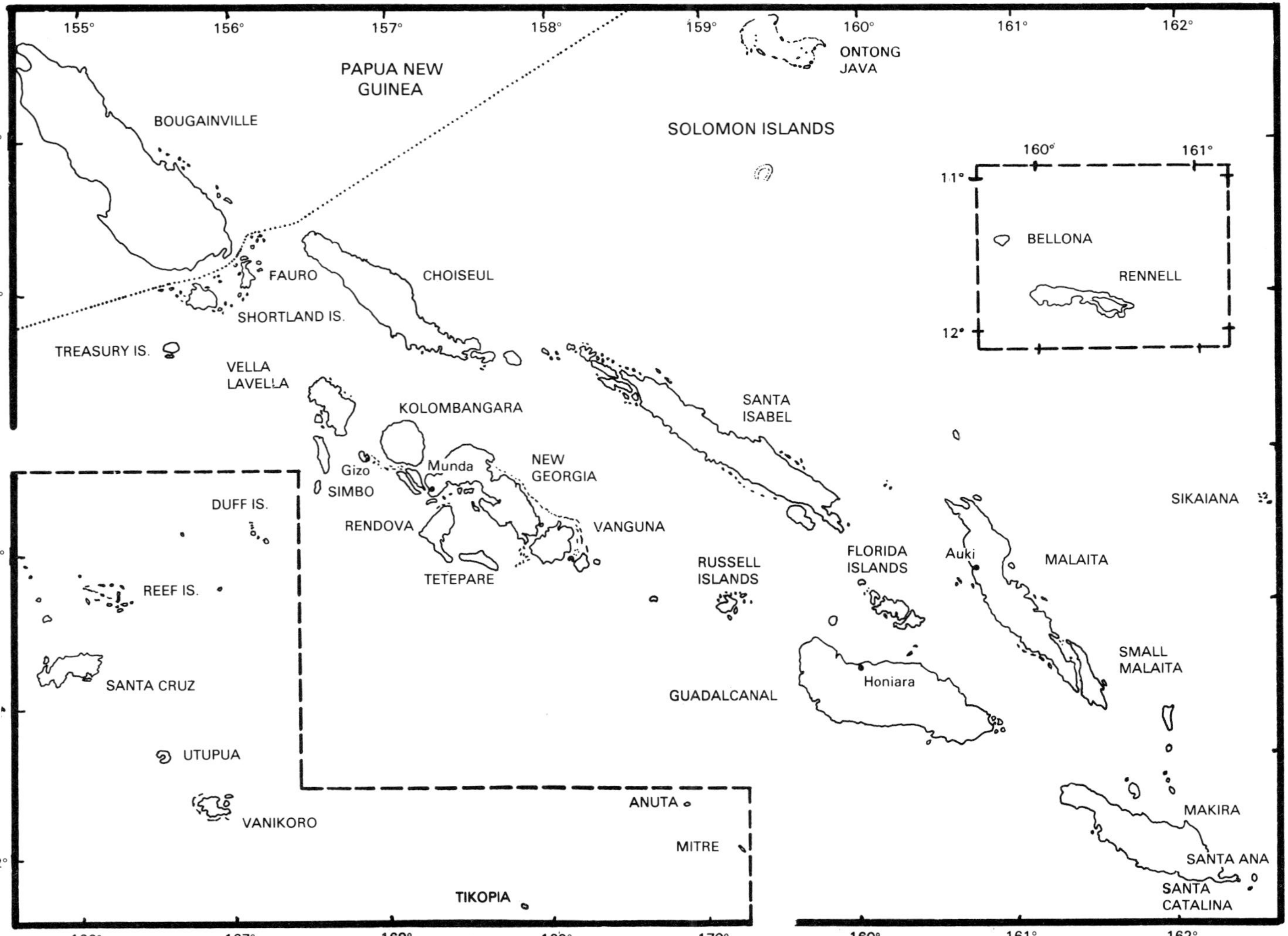

Map 3: Provinces of the Solomon Islands

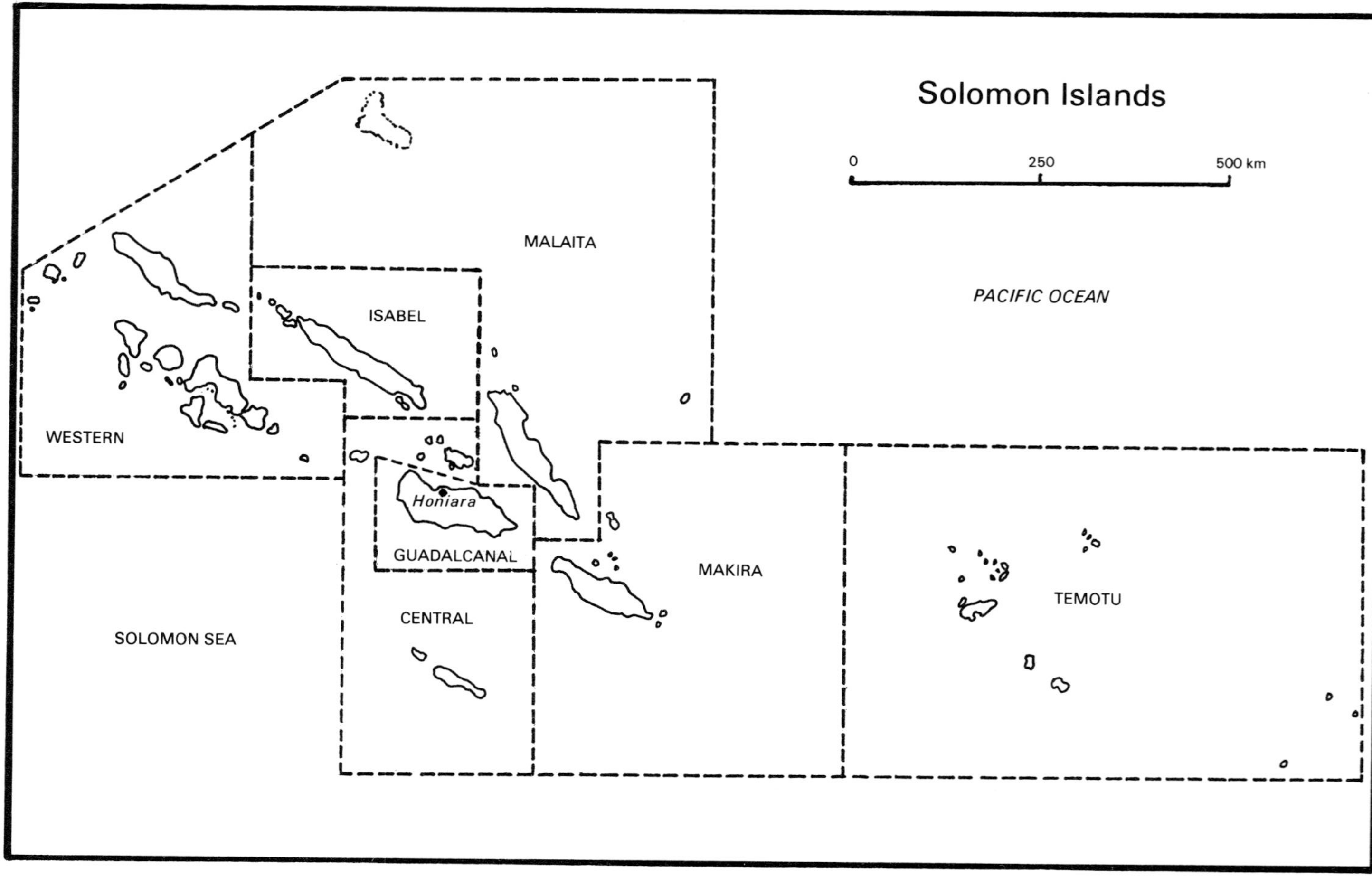

Isabel Province

Isabel Province comprises Santa Isabel and the smaller island of San Jorge.

Santa Isabel is 171 km long, in the shape of a cigar. The island consists of steep and sheer-sided mountain ranges dissected by river valleys, the highest mountain being Mt Sasari (1120 m) in the southeast. Santa Isabel is one of the few islands in the country that still has a large population living inland in long-established villages.

San Jorge lies off the southwest coast of Santa Isabel and is 173 km^2. Much of the island is swampy.

Malaita Province

Malaita and its immediate neighbour Maramasike, or Small Malaita, comprise the majority of the province, amounting to about 4900 km^2, and together are 180 km long by 43 km wide. The inland highlands are eroded into deep valleys and narrow ridges which rise to a peak of 1303 m at Mt Kolovrat. Maramasike is separated from Malaita by the 20 km long Maramasike Passage, a narrow waterway in places less than 400 m wide and only 4 m deep.

The Solomon Islands northernmost point, Ontong Java, is also included in this province. Lying just south of the Equator and 270 km north of Santa Isabel, it is about 60 km from north to south and 59 km wide. This immense lagoon is the largest in the Solomon Sea and exceeds 1400 km^2. The total land area however, is only 12 km^2.

Guadalcanal Province

This province consists of Guadalcanal, the largest of the Solomon Islands. It is 160 km long and 48 km wide with an area of 6475 km^2 with Honiara, the capital, lying on the north coast. It also has the country's highest peaks, Mt Makarakombu (2447 m) and Mt Popomanaseu (2440 m) which lie in a mountain range along the length of the island. Mountains with steep-sided narrow valleys characterise the highland areas. On the northeast coast the mountains descend into wide alluvial plains whilst in the northwest the island is composed of quiescent volcanic cones.

Central Province

This comprises the Florida and the Russell Islands plus Rennell and Bellona in the south. The largest island Rennell, 210 km south of Guadalcanal, is nearly 80 km long and almost 14 km wide. It is an uplifted atoll and its surface of deeply eroded, jagged, pitted limestone carries high forest. At its southeastern end is Lake Te'Nggano, the South Pacific's largest expanse of fresh water.

The Florida Islands, or Nggela Islands, lie between Guadalcanal and Malaita and cover an area of 391 km^2. The Russell Islands lie northwest of Guadalcanal and are 210 km^2 in extent.

Makira Province

Makira (San Cristobal), the main island, is 136 km long by 39 km wide, with an area of 3090 km^2. A backbone of mountains lies in the centre of the island with its highest point reaching 1040 m.

Off the eastern tip of Makira are 2 smaller islands: Santa Ana, 8 km from Makira, is a raised coral atoll and Santa Catalina, about 9 km from Makira, is a low raised coral platform.

Temotu Province

Formerly called the Santa Cruz Islands, Temotu Province lies at the extreme southeastern end of the Solomons. The main islands are Santa Cruz (also known as Nendo), Vanikoro, the Reef Islands and the Duff Islands. The islands of Temotu have a total land area of only 836 km^2, yet they are scattered over 200,000 km^2 of ocean. Their nearest neighbours are the Torres and Banks groups at the northern tip of Vanuatu, 270 km to the south, while Makira lies 400 km to the west.

Bougainville

Bougainville is about 200 km long and has a maximum breadth of 100 km. There are two major mountain ranges: the Crown Prince Range in the south and the higher Emperor Range, whose highest peak is Mt Balbi at 2745 m, in the north.

Buka, in the north, is separated from Bougainville by a channel only 300 m wide and 1 km long. Buka is generally low-lying apart from a hilly region in the south. There are many coral islands off the south and west coasts of Bougainville.

GEOLOGY

The Solomon Islands and Bougainville are geologically relatively young. The larger islands are almost entirely of volcanic origin and consist of lavas ranging from picrite basalts to basaltic andesites. Because of their recent emergence, some areas are covered with level-topped terraces of coral. Ultrabasic rocks, such as serpentine, occur principally on southern Santa Isabel, adjacent San Jorge and southern Choiseul, with additional smaller areas on Guadalcanal, Makira and the Florida Islands (Coleman, 1966).

The Solomon Islands and Bougainville form part of the Pacific 'Ring of Fire'. This highly volcanic area is at the intersection of the Indo-Australasian and Oceanic tectonic plates. At this plate junction there is constant seismic activity, including earthquakes and volcanoes, though less than in neighbouring Papua New Guinea and Vanuatu.

There are three land and at least two submarine volcanoes in the Solomon Islands, plus a large thermal area. The most active volcano in the last 40 years has been Tinakula, 40 km from Santa Cruz (Temotu Province). Its last violent eruption in 1971 forced almost all of its small population to evacuate the island. Savo Island, north of Guadalcanal, has a number of hot springs and, in the New Georgia Group, Simbo's crater emits sulphurous fumes. Further north a large area of boiling mud can be found in northeastern Vella Lavella. The two areas of submarine volcanoes are both in Western Province and eruptions sometimes cause the seas above them to boil.

Bougainville possesses two active volcanoes, Mt Balbi (2745 m) in the north and Mt Bagana (1999 m) in the centre.

A few of the outer Solomon Islands are former reefs lifted high out of the water by volcanic activity. In the case of Rennell this has happened several times creating the finest example of an uplifted coral atoll in the Pacific (see Taylor, 1973).

Coleman (1966) notes that the Solomon and Bougainville block is not sharply demarcated from New Ireland, but is clearly separated from the Louisiade Archipelago and the Rennell group by the Solomon trough. As a geological entity the Solomons include Bougainville in the north but do not include Temotu Province to the east which, geologically, is part of the Vanuatu archipelago.

CLIMATE

The Solomon Islands and Bougainville lie within the humid maritime tropical zone, characterised by high ambient temperatures and a high annual rainfall. However, the climate is tempered by the surrounding sea. The trade winds roughly divide the climate into a 'dry' and 'wet' season depending on wind direction. However, rainfall does not always follow a regular pattern and in some years there is little seasonal variation, while in the interior of the larger islands rain falls throughout the year.

The southeast trade winds usually blow from April to November and this tends to be the 'dry' season with steady winds, fine days and occasional rain. The northwest trade winds generally blow from November to April and this is usually the 'wet' season. Tropical downpours are common and cyclones may build up in the surrounding seas but tend to move south, only occasionally hitting the Solomon Islands.

The mean annual temperature is 27.2 °C (Kieta, Bougainville) and 27.5 °C (Tulagi, Florida Islands) with an average daily maximum of 31.7°C (Kieta) (Walter & Lieth, 1967). Montane temperatures are cooler.

Rainfall averages 3,035 mm per year (Kieta) and 3,133 mm year (Tulagi) but can be considerably heavier in inland areas and, on the windward side of the main islands, precipitation in excess of 8,000 mm per year occurs in the mountain areas on the high islands such as Guadalcanal and Makira. Honiara is in a rain shadow and has a lower than average annual rainfall of about 2000 mm year (see Pendleton, 1949).

Throughout the year humidity is highest in the morning and regularly reaches 90%. In the cooler months of the year it is usually about 60%.

VEGETATION OF THE ISLANDS

The Solomon Islands and Bougainville are amongst the wettest places in the tropics. This has a profound effect on the vegetation which is also strongly influenced by the young unstable landscape and natural catastrophes such as earthquakes, landslips, volcanic eruptions and cyclones (see plate 1C) as well as centuries of disturbance by man.

The predominant vegetation of the islands is tropical rain forest. This is floristically related to that of the Malay Archipelago, particularly New Guinea, but with fewer families, genera and species represented (Whitmore, 1966). Mangroves form extensive stands by the sea and on these and other strand trees, such as *Calophyllum inophyllum* and *Barringtonia asiatica*, orchids are found in abundance. *Robiquetia mooreana*, *Dendrobium gouldii* and *Thrixspermum spp.* are all common in such habitats. In lowland forest, terrestrial orchids such as

Corymborkis veratrifolia and *Hetaeria oblongifolia* are common, whilst common epiphytes include *Grammatophyllum scriptum* var. *boweri*, *Sarcanthopsis nagarensis* and *Ascoglossum calopterum.*

The montane and lower montane forests are much richer in orchid species with the epiphytic genera *Dendrobium* and *Bulbophyllum* particularly well represented. Terrestrial orchids such as *Calanthe*, *Corybas* and *Malaxis* are also found here. None of the Solomon or Bougainville mountains extends above the tree line but strong zonation of the forest is evident with changes in altitude.

The summits of most of the higher peaks are covered by moss forest, the canopy of which can be quite low, and here the epiphytic orchids *Dendrobium laevifolium*, *Bulbophyllum spp.*, *Phreatia spp.* and the cherry orchids, *Mediocalcar spp.*, are found. More detailed surveys of the vegetation of the Solomon Islands is given by Whitmore (1969) and Hancock & Henderson (1988). Whitmore emphasises the effect that man's activities have had on the landscape. The grasslands behind Honiara, for example, are man-made, the result of shifting cultivation, fire and the widespread use of timber for building purposes. The clearing of forest throughout the islands has progressed rapidly in recent years. Extensive clear felling of forest (see plate 1, D & E) by logging companies and for shifting agriculture and mining operations such as that at Panguna, in Bougainville, has destroyed many hectares of botanically unexplored forest. Few orchids can survive in such conditions but a spectacular exception is the terrestrial *Spathoglottis plicata* which can become abundant in such places.

BOTANICAL EXPLORATION OF THE SOLOMON ISLANDS AND BOUGAINVILLE

The first botanical collections of any note in the Solomon Islands were made in the late nineteenth century. Some of the most significant of these were made by H.B. Guppy, a surgeon on the naval vessel H.M.S. Lark, which visited the Bougainville Straits in 1884 (see Whitmore, 1966). His collections included the type of *Eurycentrum salomonense* from the Shortland Islands. Other early collectors included Rev. R.B. Comins, a Melanesian Mission priest, who collected the type of *Macodes cominsii* from Makira, and Sir Everard Im Thurn, the first British High Commissioner for the Western Pacific, who collected the type of *Arachnis beccarii* var. *imthurnii* from Malaita.

Commercial orchid collections were made in the Solomon Islands by John Gould Veitch, a collector for the Veitch nursery in Chelsea (see Kent, 1900). Veitch collected the type of *Dendrobium gouldii* in the 1860's and Heinrich Reichenbach named the plant in his honour in 1867. Wilhelm Micholitz, the bold and adventurous plant collector for the nurseryman Frederick Sander, collected in Guadalcanal and Makira in October and November 1899. Micholitz collected 1500 *Dendrobium spectabile* plants which he shipped back to the Sander nursery in St. Albans.

From 1929 until 1932, S. Frank Kajewski, financed by the Arnold Arboretum (see Ames 1932, 1933 and Kajewski 1946) collected from the Santa Cruz Islands, Bougainville, Malaita, Makira, Santa Isabel and Guadalcanal. Many of Kajewski's collections from the Santa Cruz Islands were selected as types of species named by Oakes Ames, including *Dendrobium ruginosum*, *Mediocalar vanikorense* (a synonym of *M. paradoxum*) and *Appendicula vanikorense* (a synonym of *A. reflexa*).

J.H.L. Waterhouse, a Methodist missionary, collected orchids from the islands of New Georgia and Bougainville between 1929 and 1936 and recorded many of the vernacular names which are cited in this account. His specimens were studied by the New Zealand botanist Cedric Errol Carr and include the types of *Dendrobium solomonense* and *D. waterhousei* (a synonym of *D. punamense*) from Bougainville (see Carr 1934).

More recently, noteworthy collections have been made by the Solomon Islands Forestry Department, particularly by Timothy C. Whitmore, who was a forest botanist in the Solomon Islands from 1962 to 1964, Geoffrey Dennis and Graham Chaplin. Geoffrey Dennis, former curator of the Herbarium and Botanic Gardens in Honiara, and a resident of the Solomon Islands, has been active for over 40 years collecting and cultivating native Solomon Island orchids, and the unusual endemic orchid, *Bulbophyllum dennisii*, has been named after him.

Many orchid collections were made in 1965 by Peter Francis Hunt, of the Royal Botanic Gardens, Kew, on the Royal Society's expedition to the Solomons. This expedition resulted in the first detailed scientific survey of the islands (see Corner, 1969).

The Danish Noona Dan Expedition visited Rennell Island in 1961–1962 and 1965 (see Hansen & Dissing, 1980), and collected the type specimen of the endemic *Dendrobium rennellii*. More recently Phillip Cribb collected on Rennell and recorded 22 species of Orchidaceae from the island (see Cribb, 1990a).

Significant collections have been made on Bougainville by Karl Rechinger senior and the Australian CSIRO expedition to Bougainville in 1961. P.S. Lavarack, C. Ridsdale and Andrée Millar have also made noteworthy collections there for the Papua New Guinea Herbarium in Lae. In the 1950's a ten acre orchid collection growing both native and exotic orchids was established at Arawa by Kip McKillop (see Slade, 1965). Some of the exotic orchids established here have since become naturalised, for example, *Bulbophyllum medusae*, *Arundina graminifolia* and *Dendrobium crumenatum*. However, Bougainville still remains little explored.

In recent years several expeditions from the Royal Botanic Gardens, Kew, have studied the orchids of the Solomon Islands and Bougainville and have greatly increased our knowledge of the flora of these islands. Since 1980, these have included trips made by Phillip Cribb, John Campbell, Geoffrey Dennis, Alasdair Morrison, Robert and Caron Mitchell and Sue Wickison. The islands visited include Bougainville, New Georgia, Choiseul, Rendova, Rennell, Gizo, Guadalcanal and Malaita. Other collections now at the Royal Botanic Gardens, Kew, have been made by Margaret Dickson and members of Operation Raleigh who visited New Georgia and Rendova in 1986.

THE ORCHIDS

The Orchidaceae form the major component of the epiphytic flora of the Solomon Islands and Bougainville. 285 species are described in this account, 8 of these being introduced species, namely *Arundina graminifolia*, *Bulbophyllum medusae*, *Dendrobium crumenatum*, *D. insigne*, *D. rutriferum*, *D. vandoides*, *Robiquetia wassellii* and *Vanilla planifolia*, which have become naturalised. The species are distributed in 87 genera although very unevenly, with such genera as *Dendrobium* and *Bulbophyllum* representing a considerable proportion of the total number of species. Conversely, many of the genera are represented by one

Table 1. Survey of the epiphytic, climbing and terrestrial native genera and species in the Solomon Islands and Bougainville

Epiphytic or climbing		Terrestrial	
Genera	**No. of species**	**Genera**	**No. of species**
Acriopsis	1	*Acanthephippium*	1
Aglossorhyncha	2	*Acianthus*	1
Agrostophyllum	7	*Calanthe*	5
Appendicula	7	*Chrysoglossum*	1
Arachnis	1	*Corybas*	4
Ascoglossum	1	*Corymborkis*	1
Bulbophyllum	40	*Cryptostylis*	1
Cadetia	1	(*Dendrobium*	1)
Calymmanthera	1	*Didymoplexis*	1
Ceratostylis	2	*Epipogium*	1
Chilopogon	1	*Eulophia*	1
Cleisostoma	2	*Eurycentrum*	1
Coelogyne	4	*Geodorum*	1
(*Dendrobium*	51)	*Goodyera*	2
Diplocaulobium	7	*Hetaeria*	1
Dipodium	1	*Hylophila*	1
Epiblastus	1	(*Liparis*	2)
Eria	7	*Macodes*	3
Flickingeria	2	*Malaxis*	6
Glomera	4	*Mischobulbon*	1
Glossorhyncha	1	*Nervilia*	1
Grammatophyllum	2	*Neuwiedia*	1
Gunnarella	1	*Oeceoclades*	1
Hippeophyllum	1	*Peristylus*	3

[continued on next page]

species only. Some of the genera reach the eastern limit of their range in the Solomon Islands, for example the slipper orchids in the genus *Paphiopedilum*.

The majority of the native species (76%) are epiphytic or climbing (see table 1). Of the terrestrial species (24%), three are saprophytic: *Epipogium roseum*, *Stereosandra javanica* and *Didymoplexis micradenia*.

Table 2 shows the number of orchid species recorded from the main islands of the Solomon Islands and Bougainville. These figures almost certainly reflect the amount of collecting done on each island as much as the species diversity of the islands. Guadalcanal is the best explored island in the Solomon Islands because it is the most accessible. Several of the more remote larger islands such as Choiseul, Santa Isabel and Malaita are botanically under-explored and might be expected to have richer orchid floras than suggested here. The outlying islands such as the Santa Cruz Islands and Rennell are also under-explored and can be expected to yield more species when better studied.

Table 1 (contd.)

Genus	Species	Genus	Species
(*Liparis*	5)	*Phaius*	2
Luisia	1	*Paphiopedilum*	2
Malleola	2	*Pilophyllum*	1
Mediocalcar	3	(*Plocoglottis*	2)
Micropera	1	*Pristiglottis*	1
Oberonia	7	*Spathoglottis*	2
Octarrhena	1	*Spiranthes*	1
Pedilochilus	1	*Stereosandra*	1
Pholidota	2	*Tainia*	1
Phreatia	9	*Tropidia*	1
(*Plocoglottis*	1)	*Vrydagzynea*	3
Podochilus	1	*Zeuxine*	3
Pomatocalpa	2		
Pseuderia	1		
Pseudovanilla	2		
Renanthera	1		
Rhinerrhiza	1		
Robiquetia	3		
Sarcanthopsis	1		
Schoenorchis	1		
Taeniophyllum	4		
Thelasis	1		
Thrixspermum	4		
Trichoglottis	2		
Trichotosia	3		
Tuberolabium	1		
Vanda	1		
GENERA= 50 (59 % of total) SPECIES = 210 (76 % of total)		GENERA = 35 (41 % of total) SPECIES = 67 (24 % of total)	

TOTAL NO. OF NATIVE GENERA = 85
TOTAL NO. OF NATIVE SPECIES = 277

Genera with both epiphytic and terrestrial species appear in parentheses.

It is perhaps surprising that only 130 species of orchid have been recorded from Bougainville when New Guinea is estimated to have between 1500 and 2000 species. A major reason is the lack of collecting on the island which in recent times has been beset by political unrest. Large areas of forest have also been cleared for the open-cast copper mines at Panguna. However, both Bougainville and the Solomon Islands have a restricted variety of ecological niches for orchids in comparison with New Guinea, and much of the rain forest has been replaced by thickets of small trees and climbers due to the influence of cyclones, earthquakes, landslips and man. Furthermore, the islands have only a limited altitudinal range with everywhere relatively close to the sea.

Table 2. Number of species of orchids recorded from the major islands in the Solomon Islands and Bougainville.

Island	No. of species
Bougainville	130
Choiseul	36
Guadalcanal	175
Kolombangara	82
Makira	74
Malaita	45
New Georgia	97
Rendova	32
Rennell	24
The Santa Cruz Islands	28
Santa Isabel	36
The Shortland Islands	7

Although we do not consider this account to be a comprehensive list of the species in the Solomon Islands and Bougainville, and we envisage many more orchids being added to this flora with further exploration, it is interesting to note the affinities of the orchid species in the Solomon Islands and Bougainville. Almost three-quarters (74%) of the orchid species occuring in the Solomon Islands and Bougainville also occur in New Guinea (see table 3). This strong relationship is to be expected from the proximity of New Guinea and the rich orchid flora that it possesses.

Table 3: Number of Solomon Islands and Bougainville species recorded from other areas.

Area	No. of orchid species in common	As a % of the Solomon Is. & Bougainville orchid flora
New Guinea (incl. New Ireland and New Britain)	205	74 %
Pacific Islands (Vanuatu, New Caledonia, Fiji, Samoa etc.)	113	41 %
Australia	41	15 %
Malay Peninsula and Archipelago	55	20 %
Endemic	28	10 %

A significant portion (41 %) of the species are also found in other Pacific Islands, especially in Vanuatu. This is also to be expected as the Santa Cruz Islands are only 270 km from the Torres and Banks Islands in Vanuatu.

A small number of species (15 %) also occur in Australia. However, the species held in common with Australia are all of New Guinea or Malay Archipelago origin and not Australian orchids in the strict sense. This relationship is perhaps weaker than expected for two reasons. First, the land masses of the Solomon Islands and Bougainville, which are recent volcanic islands, have never been attached to the Australian continent. Secondly, the predominant winds of the area are the southeast trades which blow from the northwest to the southeast and therefore the Solomon Islands and Bougainville are more likely to receive seed of Asiatic and New Guinea than of Australian orchids. Some 20 % of the Solomon Island and Bougainville orchids, such as *Calanthe triplicata* and *Bulbophyllum longiflorum*, are widely distributed throughout southeast Asia across to the Pacific.

The number of endemic orchids (see table 4) recorded is lower than that predicted by Hunt (1969), who suggested that about 90 species were endemic, and is considerably lower than is estimated for New Guinea. It suggests strongly that the islands have a young immigrant flora and that the colonisation of the archipelago has taken place very rapidly in recent times. The flora has been insufficiently isolated in space and time to permit speciation to develop to any significant degree. The endemic orchids all have close affinities with New Guinea species. Some of them, such as *Paphiopedilum bougainvilleanum*, *P. wentworthianum* and *Dendrobium rennellii* appear to have a very restricted distribution and are under threat from plant hunters and development in the islands.

Table 4. Orchid taxa endemic to the Solomon Islands and Bougainville

Acianthus vulcanicus
Arachnis beccarii var. *imthurnii*
Bulbophyllum dennisii
B. fractiflexum subsp. *solomonense*
B. melanoxanthum
Coelogyne susanae
Corybas longipedunculatus
C. gemmata
C. solomonensis
Dendrobium gnomus
D. rechingerorum
D. rennellii
D. ruginosum
D. salomonense
D. sancristobalense
D. vanikorense
D. campbellii
Diplocaulobium bicolor
D. magnilabre
D. solomonense
D. subintegrum
Habenaria bougainvilleae
Macodes cominsii
Paphiopedilum bougainvilleanum
P. wentworthianum
Robiquetia woodfordii
Trichotosia vanikorensis
Zeuxine elatior var. *angustata*

THE CATALOGUE

The orchids of the Solomon Islands and Bougainville are amongst the least known in the world. Orchids are the largest family of flowering plants in the Solomon Islands and Bougainville and are found in all habitats from seashore to the tops of mountains. The increasing awareness of the need for plant conservation has emphasised the dearth of knowledge of the orchids of these islands. Before it is possible to pinpoint areas or islands needing special care or protection it is necessary to know what grows there. Orchids can be used to pinpoint areas of biodiversity which can be useful to forestry and conservation in planning which areas might need protection.

The islands are relatively rich in numbers of orchid species and it is hoped that this account will act as a basis for future research. We anticipate that many species will be added to those listed here as the flora becomes better known and studied.

The identification and naming of the orchids of the region is far from straightforward. Many of the orchids of the Solomon Islands and Bougainville are also found in New Guinea. Schlechter and Kraenzlin named many of the orchid species from New Guinea but, unfortunately, many of their type specimens were destroyed in Berlin in the Second World War. However, drawings of many of Schlechter's types can be found in 'Die Orchidaceen von Deutsch-Neu-Guinea' published in Feddes, Repertorium Specierum Novarum Regni, Beihefte 1: 1–1079 (available in English translation from the Australian Orchid Foundation) and other publications. Duplicates of only about 40% of Schlechter's New Guinea types have been traced in other herbaria (Cribb, Wood & Stirton, 1988; Cribb & Robbins, 1991).

In the absence of type material of some of the species described by Schlechter and Kraenzlin we have sometimes had to base identifications on Schlechter's published illustrations and on herbarium material at Kew annotated by V.S. Summerhayes who had access to the Berlin herbarium prior to 1939.

Since the publication of 'Orchids of Vanuatu' by Lewis & Cribb (1989), two other significant publications have been produced by Kores (1989) and Clements (1989) on the orchids of Fiji and Australia respectively. These have increased our knowledge of orchids of this area and we have incorporated relevant findings in this account.

LIST OF NEW TAXA

NEW COMBINATIONS

Trichotosia vanikorensis (Ames) Cribb & B. Lewis **comb. nov.**

NEW TAXA

Bulbophyllum melanoxanthum J. Vermeulen & B. Lewis **sp. nov.**

Corybas gemmatus Cribb & B. Lewis **sp. nov.**

Dendrobium campbellii Cribb & B. Lewis **sp. nov.**

Bulbophyllum fractiflexum J.J. Smith subsp. *solomonense* J. Vermeulen & B. Lewis **subsp. nov.**

Zeuxine elatior Schltr. var. *angustata* Cribb & B. Lewis **var. nov.**

CLASSIFICATION OF GENERA

(After Dressler, 1981)

Genera marked with an asterisk (*) are not native to the Solomon Islands or Bougainville but have become naturalised.

SUBFAMILY APOSTASIOIDEAE
1. *Neuwiedia*

SUBFAMILY CYPRIPEDIOIDEAE
2. *Paphiopedilum*

SUBFAMILY SPIRANTHOIDEAE

Tribe Erythrodeae
3. *Corymborkis*
4. *Tropidia*
5. *Eurycentrum*
6. *Goodyera*
7. *Hylophila*
8. *Macodes*
9. *Pristiglottis*
10. *Hetaeria*
11. *Vrydagzynea*
12. *Zeuxine*

Tribe Cranichideae
13. *Spiranthes*
14. *Cryptostylis*

SUBFAMILY ORCHIDOIDEAE

Tribe Diurideae
15. *Acianthus*
16. *Corybas*

Tribe Orchideae
17. *Peristylus*
18. *Habenaria*

SUBFAMILY EPIDENDROIDEAE

Tribe Vanilleae
19. *Pseudovanilla*
20. **Vanilla*

Tribe Gastrodieae
21. *Nervilia*
22. *Didymoplexis*

Tribe Epipogieae
23. *Epipogium*
24. *Stereosandra*

Tribe Arethuseae
25. **Arundina*
26. *Acanthephippium*
27. *Calanthe*
28. *Phaius*
29. *Spathoglottis*
30. *Plocoglottis*
31. *Mischobulbon*
32. *Tainia*

Tribe Coelogyneae
33. *Coelogyne*
34. *Pholidota*

Tribe Malaxideae
35. *Hippeophyllum*
36. *Liparis*
37. *Malaxis*
38. *Oberonia*

Tribe Epidendreae
39. *Eria*
40. *Trichotosia*
41. *Mediocalcar*
42. *Ceratostylis*
43. *Epiblastus*
44. *Agrostophyllum*
45. *Appendicula*
46. *Chilopogon*
47. *Podochilus*
48. *Octarrhena*
49. *Phreatia*
50. *Thelasis*
51. *Agrossorhyncha*
52. *Glomera*
53. *Glossorhyncha*
54. *Cadetia*
55. *Dendrobium*
56. *Diplocaulobium*
57. *Flickingeria*
58. *Pseuderia*
59. *Bulbophyllum*
60. *Pedilochilus*

SUBFAMILY VANDOIDEAE

Tribe Vandeae

61. *Calymmanthera*
62. *Gunnarella*
63. *Rhinerrhiza*
64. *Sarcanthopsis*
65. *Thrixspermum*
66. *Ascoglossum*
67. *Arachnis*
68. *Schoenorchis*
69. *Luisia*
70. *Renanthera*
71. *Vanda*
72. *Cleisostoma*
73. *Malleola*
74. *Micropera*
75. *Pomatocalpa*
76. *Robiquetia*
77. *Taeniophyllum*
78. *Tuberolabium*
79. *Trichoglottis*

Tribe Cymbidieae

80. *Chrysoglossum*
81. *Pilophyllum*
82. *Eulophia*
83. *Geodorum*
84. *Oeceoclades*
85. *Grammatophyllum*
86. *Dipodium*
87. *Acriopsis*

ARTIFICIAL KEY TO THE GENERA

Genera marked with an asterisk (*) are not native to the Solomon Islands or Bougainville but have become naturalised and are therefore included in this key.

1. Plant terrestrial 2
   Plant epiphytic, lithophytic or climbing 41

2. Plant saprophytic, lacking green leaves and stems 3
   Plant with green leaves 5

3. Lip spurred .... **23. Epipogium**
   Lip not spurred 4

4. Lip boat-shaped, obtuse, with 2 fleshy, round calli at the base .... **24. Stereosandra**
   Lip obscurely trilobed, with the central disc of the mid-lobe raised, fleshy, with 3 verrucose calli .... **22. Didymoplexis**

5. Lateral sepals united to form a synsepalum; lip slipper-shaped, deeply saccate .... **2. Paphiopedilum**
   Plants without the above combination of features 6

6. Style and anthers showing partial fusion only at the base; anthers 3 .... **1. Neuwiedia**
   Style and anthers fused to form a column; anther solitary 7

7. Leaf solitary, broadly heart-shaped 8
   Leaves not as above 10

8. Inflorescence produced before the leaf appears .... **21. Nervilia**
   Inflorescence produced with the leaf 9

9. Inflorescence few-flowered; dorsal sepal lanceolate; lip with a jagged raised callus at the base .... **15. Acianthus**
   Inflorescence bearing a single flower; dorsal sepal large, helmet-shaped; lip without a raised callus .... **16. Corybas**

10. Flowers sessile, many, arranged in a tight spiral on the inflorescence .... **13. Spiranthes**
    Flowers on a short to long peduncle, few to many, not arranged spirally, or only laxly so 11

11. Inflorescence or inflorescences lateral, from nodes of stem or pseudobulb 12
    Inflorescence terminal or arising from the base of the pseudobulb or the rhizome 18

12. Inflorescence branched .... **3. Corymborkis**
    Inflorescence simple 13

13. Flowers urn-shaped; sepals connate, with free apices .... **26. Acanthephippium**
Flowers not urn-shaped; sepals free or lateral sepals joined for part of their length to form a mentum 14

14. Lateral lobes of lip inrolled to form a tube which encompasses the column .... **28. Phaius** (in part, *P. amboinensis* var. *papuanus*)
Lip not as above 15

15. Lip without a spur 16
Lip with a spur 17

16. Flowers resupinate; lip trilobed with a glabrous or hairy callus at the base of the mid-lobe .... **29. Spathoglottis**
Flowers not resupinate; lip trilobed with a callus of 3 ridges, the outer two warty with a warty apex .... **55. Dendrobium** (in part, *D. lobbii*)

17. Lip entire; sepals 2 cm long or more .... **82. Eulophia**
Lip 4-lobed; sepals 1 cm long or less .... **84. Oeceoclades**

18. Inflorescence apparently arising from the rhizome or the base of the pseudobulb 19
Inflorescence terminal, from apex of stem or pseudobulb 27

19. Stem or pseudobulb with several leaves 20
Pseudobulb bearing a single leaf 23

20. Flowers turning blue when damaged 21
Flowers not turning blue when damaged 22

21. Lateral lobes of lip inrolled to form a tube which encompasses the column .... **28. Phaius** (in part, *P. tankervilleae*)
Lip not as above .... **27. Calanthe**

22. Rhachis recurved; flowers pink, mauve or white; lip without an elastic joint .... **83. Geodorum**
Rhachis straight; flowers yellow or orange and spotted with maroon or red with a white or yellow lip; lip with an elastic joint that springs when the lip is touched .... **30. Plocoglottis** (in part)

23. Leaves pale green mottled with dark green; petiole mottled with purple .... **14. Cryptostylis**
Leaves and petiole not as above 24

24. Leaf covered with brown hairs .... **81. Pilophyllum**
Leaf not hairy 25

25. Lip ovate, acute .... **31. Mischobulbum**
Lip trilobed 26

26. Flowers yellowish-green with a yellowish lip; the column with small wings on its front edge .... **80. Chrysoglossum**
Flowers olive-yellow, the petals with a broad crimson streak, the lip yellow suffused with purple; the column without small wings on its front edge .... **32. Tainia**

27. Lip without a spur and not saccate at base 28
Lip with a spur, or saccate at base 31

28. Leaves tough, pleated 29
Leaves thin-textured 30

29. Leaves narrowly linear, acute, less than 1.5 cm wide .... ***25. Arundina**
Leaves lanceolate, more than 3 cm wide .... **33. Coelogyne**

30. Column very short, embraced by basal auricles of lip; lip flat .... **37. Malaxis**
Column relatively long, not embraced by base of lip; lip porrect to recurved, never flat .... **36. Liparis** (in part)

31. Bracts large and imbricate .... **4. Tropidia**
Bracts not as above 32

32. Lip without basal glands or papillae 33
Lip with basal glands or papillae 35

33. Lip triangular, acute .... **7. Hylophila**
Lip trilobed 34

34. Stigma entire or only partly divided, sessile or very shortly stalked .... **17. Peristylus**
Stigma lobes 2, distinctly stalked. .... **18. Habenaria**

35. Leaves almost black to very dark green or purple with a pink central vein above, purple below .... **5. Eurycentrum**
Leaves not as above 36

36. Leaves rather fleshy, ovate, elliptic to subcircular, with coloured venation; flowers asymmetrical .... **8. Macodes**
Leaves thin-textured, without coloured veins but may be variegated; flowers symmetrical 37

37. Lip tripartite, with a strong median constriction, the margins of the mesochile sinuous .... **9. Pristiglottis**
Lip not as above 38

38. Lip apex widened into a transversely oblong blade .... **12. Zeuxine**
Lip apex not widened into an oblong blade 39

39. Flowers non-resupinate .... **10. Hetaeria**
Flowers resupinate 40

40. Lip with stalked basal glands .... **11. Vrydagzynea**
Lip without glands but often hairy or papillose within at base
.... **6. Goodyera**

****** ALL SUBSEQUENT GENERA ARE EPIPHYTIC ******

41. Plant covered in red-brown hairs on the stem and usually on the leaves
....**40. Trichotosia**
Plant glabrous or finely pubescent, but the hairs never red-brown 42

42. Plant leafless; roots green, fasciculate, assuming the function of leaves
.... **77. Taeniophyllum**
Plant with leaves 43

43. Leaves bilaterally compressed 44
Leaves dorsiventrally flattened or almost terete 47

44. Inflorescence lateral; lip hooked .... **48. Octarrhena**
Inflorescence terminal 45

45. Lateral sepals more than 7 mm long
.... **55. Dendrobium** (section Rhopalanthe)
Lateral sepals less than 2 mm long 46

46. Stems clustered or solitary, not on a creeping rhizome; ovary glabrous
....**38. Oberonia**
Stems arising from a creeping, rigid rhizome; ovary pubescent
.... **35. Hippeophyllum**

47. Leaf or leaves almost terete 48
Leaves dorsiventrally flattened 50

48. Leaf solitary; inflorescence terminal .... **42. Ceratostylis**
Leaves several; inflorescence lateral 49

49. Flowers not bell-shaped; lateral sepals 7–8 mm long; lip divided into two distinct portions, lacking a spur .... **69. Luisia**
Flowers bell-shaped; lateral sepals 2 mm long; lip trilobed with a globular spur .... **68. Schoenorchis**

50. Leaves 1 or 2 at the apex of the stem or pseudobulb 51
Leaves several, arranged distichously along the stem, in a fan on a short stem or from upper part of stem or pseudobulb 64

51. Flowers orange to cherry-red with white, yellow or pale green tips; sepals connate in basal half 52
Plants without the above combination of features 53

52. Plants large and sturdy; pseudobulbs superposed; flowers in clusters of up to 8 .... **43. Epiblastus**
Plants small and creeping; pseudobulbs borne on a creeping rhizome; 1– or 2–flowered at the apex .... **41. Mediocalcar**

53. Pseudobulb bearing 2 unequal sized leaves at the apex .... **50. Thelasis**
Pseudobulb bearing a solitary leaf or, if 2-leaved, leaves of equal size 54

54. Flowers ephemeral, lasting less than a day 55
Flowers lasting for more than a day 56

55. Sepals and petals ovate to linear; stems bulbous towards apex, superposed .... **57. Flickingeria**
Sepals and petals filiform above; stems not branching, bulbous at the base .... **56. Diplocaulobium**

56. Inflorescence with one or two flowers 57
Inflorescence with three or more flowers 59

57. Pedicel and ovary more or less warty .... **54. Cadetia**
Pedicel and ovary smooth 58

58. Lip saccate, immovably attached to the column-foot .... **60. Pedilochilus**
Lip not saccate, hinged to the column-foot or column footless with the lip immovably adnate to it .... **59. Bulbophyllum** (in part)

59. Inflorescence from the base of the pseudobulb 60
Inflorescence terminal or apparently terminal 61

60. Lip immovably attached to the column-foot .... **49. Phreatia** (in part)
Lip hinged to the column-foot .... **59. Bulbophyllum** (in part)

61. Flowers medium-sized to large, not concealed by bracts; lip usually with longitudinal papillose or verrucose keels 62
Flowers small, often concealed by bracts; lip without papillose keels 63

62. Sepals free; column long, winged around the top, lacking a column-foot .... **33. Coelogyne**
Lateral sepals forming a more or less conical mentum; column not winged around the top, with a pronounced column-foot .... **55. Dendrobium** (in part)

63. Inflorescence pendent, bearing flowers alternately in 2 ranks; bracts large, concave, persistent, almost hiding the flowers; lip saccate at base .... **34. Pholidota**
Inflorescence usually erect, with all or only a few flowers opening together; bracts small, persistent or deciduous; lip not saccate at base .... **36. Liparis** (in part)

64. Inflorescence terminal, apparently terminal or arising from the base of the pseudobulb 65
Inflorescence or inflorescences lateral 74

65. Bracts larger than the flowers, purple with white, serrate margins
.... **46. Chilopogon**
Bracts not as above 66

66. Plant small, moss-like; sepals less than 2 mm long .... **47. Podochilus**
Plant larger, not moss-like, erect or pendent; sepals more than 3 mm long
.... 67

67. Lip with an elastic joint that springs when the lip is touched
.... **30. Plocoglottis** (in part)
Lip without an elastic joint 68

68. Leaf sheaths with fimbriate margins; flower solitary
.... **53. Glossorhyncha**
Leaf sheaths without fimbriate margins; inflorescence one- to many-flowered 69

69. Column with a column-foot .... **55. Dendrobium** (in part)
Column lacking a column-foot 70

70. Leaves (2-)3(-4) from apex of pseudobulb; base of plant surrounded by erect tapering roots .... **87. Acriopsis**
Leaves distichous; base of plant not surrounded by erect roots 71

71. Inflorescence 1- to 2-flowered .... **51. Aglossorhyncha**
Inflorescence 3- to many-flowered 72

72. Inflorescence simple or branched basally; pollinia 6
.... **45. Appendicula** (in part)
Inflorescence with the flowers in a globose head, in clusters on an elongated inflorescence, or in a panicle; pollinia 4 or 8 73

73. Flowers always in a globose head; pollinia 4 .... **52. Glomera**
Flowers in a globose head, in clusters on an elongated inflorescence, or in a panicle; pollinia 8 .... **44. Agrostophyllum**

74. Plants climbing 75
Plants epiphytic or lithophytic 78

75. Fruits like a bean-pod, up to 20 cm long 76
Fruits not like a bean-pod, less than 5 cm long 77

76. Stems brown; leaves bract-like, ovate-orbicular, brown; inflorescence branching; fruits not fragrant .... **19. Pseudovanilla**
Stems green; leaves elliptic-oblong, green; inflorescence simple; fruits fragrant .... ***20. Vanilla**

77. Leaves well-spaced on stem; lip entire, obovate, with 2 raised callus ridges in the basal half converging in the middle of the lip .... **58. Pseuderia**
Leaves imbricate; lip trilobed, the mid-lobe pubescent and fleshy
.... **86. Dipodium**

78. Lip with entrance to spur blocked with a tongue or callus 79
Lip lacking a spur, or if spur present then not blocked by a tongue or callus 82

79. A large epiphyte forming coarse, straggling clumps; stems usually over 1 m long; flowers non-resupinate .. .. **74. Micropera**
An epiphyte with stems up to 50 cm long; flowers resupinate 80

80. Inflorescence simple, less than 2 cm long, with one to few flowers .. .. **79. Trichoglottis**
Inflorescence branched or unbranched, 2–80 cm long, densely many-flowered 81

81. Spur with a tongue or callus on front wall .. .. **72. Cleisostoma**
Spur with a tongue or callus on back wall .. .. **75. Pomatocalpa**

82. Inflorescence branched 83
Inflorescence simple 87

83. Flowers small, not opening widely; sepals lanceolate, thin-textured .. .. **61. Calymmanthera**
Flowers relatively large; sepals spathulate, fleshy 84

84. Flowers vivid purple, or reddish-orange with darker or red spots and a bright red lip 85
Flowers yellow spotted with reddish-brown, or white becoming yellow towards the apices; lip yellow marked with violet-blue 86

85. Flowers vivid purple occasionally with an orange or pink lip and spur; lip mid-lobe ligulate, porrect .. .. **66. Ascoglossum**
Flowers reddish-orange with darker or red spots and a bright red lip; lip mid-lobe oblong, reflexed, with 2 ovoid, fleshy calli at the base .. .. **70. Renanthera**

86. Lip immovably attached to the column-foot; flowers yellow spotted with reddish-brown; sepals less than 1 cm long .. .. **64. Sarcanthopsis**
Lip articulated to the column-foot by a short strap of flexible tissue; flowers white becoming yellow towards the apices with a yellow lip marked with violet-blue; sepals up to 3 cm long .. .. **67. Arachnis**

87. Plants pseudobulbous 88
Plants lacking pseudobulbs 90

88. Plants forming clumps; leaves up to 1 m long; flowers large; plant with erect tapering roots at base of pseudobulb .. .. **85. Grammatophyllum**
Plants not forming clumps; leaves less than 20 cm long; flowers small to medium-size; plant lacking erect roots at base of stems 89

89. Pollinia 4; plant form extremely variable .. .. **55. Dendrobium** (in part)
Pollinia 8; pseudobulbs slender or thick and fleshy, basal part always covered by sheaths .. .. **39. Eria**

90. Inflorescence rather like a bottle-brush; flowers non-resupinate .... **76. Robiquetia**
Inflorescence not as above; flowers resupinate 91

91. Flowers medium-sized, fairly showy, the sepals shiny red-brown with yellow margins; lip trilobed with the mid-lobe projecting as a landing platform .... **71. Vanda**
Flowers small to medium-sized; flower colour and lip not as above 92

92. Growth sympodial (i.e. stems when branching producing branches with limited growth potential) 93
Growth monopodial (i.e. all branches of the stem having unlimited growth potential) 95

93. Column without a foot .... **45. Appendicula** (in part, *A. reflexa*)
Column with a foot 94

94. Inflorescence with one to many flowers; lateral sepals 7 mm long or more .... **55. Dendrobium** (in part)
Inflorescence with many tiny white flowers; lateral sepals up to 2 mm long .... **49. Phreatia** (in part)

95. Bracts persistent, in two distinct ranks or arranged spirally .... **65. Thrixspermum**
Bracts deciduous, distichous 96

96. Flowers thin-textured, white 97
Flowers fleshy, pale yellow spotted with red 98

97. Sepals 4–5 mm long; lip lacking a spur .... **62. Gunnarella**
Sepals 2 mm long or less; lip with an elongate, tapering spur 1.5–2.5 mm long .... **73. Malleola**

98. Inflorescence short, up to 3 cm long, few-flowered; lateral sepals 4–5 mm long .... **78. Tuberolabium**
Inflorescence long, up to 60 cm long, many-flowered; lateral sepals more than 1 cm long .... **63. Rhinerrhiza**

CATALOGUE OF THE GENERA AND SPECIES

Genera and species marked with an asterisk (*) are not native in the Solomon Islands or Bougainville but have become naturalised.

1. NEUWIEDIA Blume

Terrestrial herbs with short rhizomes. *Stems* erect, leafy. *Leaves* plicate, lanceolate to linear, acuminate, the upper ones hairy, sheathing at base. *Inflorescence* terminal, erect, unbranched. *Flowers* nodding, not opening widely; sepals and petals subsimilar; lip entire, similar to the sepals and petals with a central longitudinal but obscure callus; column showing partial fusion of style and anthers; anthers 3. *Fruit* a thin-walled capsule.

A small genus of about eight species from tropical SE. Asia to the SW. Pacific Islands, a single species being recorded.

N. veratrifolia *Blume* in van d. Hoeven & de Vriese, Tijd. Nat. Gesch. Phys. 1: 140 (1834). Type: Java, *Blume* s.n. (lectotype L!).

For full synonymy see de Vogel (1969).

A terrestrial plant 70–120 cm tall, growing in small colonies. *Rhizome* 4 mm in diameter. *Stem* erect, leafy. *Leaves* distichous, erect, lanceolate, acuminate, up to 78 cm long, 6.7 cm wide, sheathing at base. *Inflorescence* terminal, erect, unbranched, many-flowered, hairy; bracts lanceolate, finely pubescent, 2–3.5 cm long, 2–4 mm wide. *Flowers* yellow-orange to bright orange, not opening widely, fragrant; pedicel and ovary finely pubescent, triangular in cross-section, c. 1 cm long; sepals narrowly elliptic-lanceolate, shortly apiculate, finely pubescent on outer surface, 1.4–1.6 cm long, 4.5 mm wide; petals obovate, apiculate, with a central fleshy ridge on the outer surface, 1.6 cm long, 7.5 mm wide; lip obovate, acute, with a central fleshy ridge, 1.5 cm long, 0.7 cm wide; column with three anther filaments united at base only to a 1.2 cm long style. *Fruit* triangular in cross-section, 1–1.4 cm long, pale green. (See fig. 1 and plate 2F).

DISTRIBUTION: Guadalcanal. Also in the Malay Peninsula and Archipelago, New Guinea and Vanuatu.

HABITAT: Well drained primary forest in deep leaf litter overlaying calcareous soil and limestone outcrops, 250–400 m.

COLLECTIONS: *Cribb* 6002 (K); *Dennis* s.n. & 12 (K); *Fa'arodo et al.* in BSIP 12256 (K).

2. PAPHIOPEDILUM Pfitzer

Terrestrial or rarely epiphytic herbs. *Rhizome* present but usually extremely abbreviated. *Stems* usually short but plants may produce elongated, ascending stems with the leaves spaced along them. *Leaves* arranged in a fan, linear-ligulate to elliptic-oblong, rather fleshy or leathery, mottled on upper surface in some species. *Inflorescence* erect or arching, usually hairy, 1– to many-flowered; bracts lanceolate and acute to elliptic and rounded. *Flowers* large and usually showy; lateral sepals united to form a synsepalum; petals deflexed and/or reflexed and often twisted; lip slipper- or helmet-shaped, deeply saccate; column bearing two ventral anthers and an apical, usually shield-shaped staminode.

A genus of about sixty-five species distributed from India, southern China and SE. Asia to the Philippines, the Malay Archipelago, New Guinea and the Solomon Islands; two species being recorded.

Petals falcate-decurved and borne at an angle of c. 45° to horizontal, not glossy, white or greenish with green veins, ± faintly flushed purple towards apex; lip green; staminode lateral teeth obtuse .. .. **1. P. bougainvilleanum**
Petals spreading, borne at an angle of c. 20° to the horizontal, glossy, purple; lip ochre-coloured; staminode lateral teeth acute .. .. **2. P. wentworthianum**

1. P. bougainvilleanum *Fowlie* in Orchid Dig. 35: 122 (1971). Type: Bougainville, *Wentworth*, cult. Los Angeles Arb. R69 BI (holotype UCLA).

See Cribb (1987b) for more information.

A terrestrial plant. *Stems* may rarely be elongated. *Leaves* 6–7, arranged in a fan, narrowly elliptic, obtuse and minutely tricuspidate at apex, 14–22 cm long, 2.5–4.2 cm wide, pale green faintly lined and mottled dark green, margins erose at apex. *Inflorescence* 1-flowered; peduncle 10–23 cm long, pale green, purple- and white-pubescent; bract elliptic, rounded, 2–2.6 cm long, pale green, pubescent. *Flower* c. 5 cm across; sepals white with green veins, purple-pubescent on outside; petals white, veined with green, purple-flushed on apical margins; lip and staminode green with darker veins; pedicel and ovary 4 cm long, green and purple-pubescent; dorsal sepal ovate, acuminate or acute, 3.2–3.8 cm long, 3.3–4 cm wide; synsepalum concave, ovate, acute, 2.5–2.7 cm long, 1.7 cm wide; petals deflexed, falcate, narrowly elliptic, obtuse, 4.5–5 cm long, 1.8–2.6 cm wide; staminode lunate, 8–10 mm long, 12–14 mm wide, minutely pubescent, apical lateral teeth obtuse. (See plate 2B).

DISTRIBUTION: Bougainville, very rare.
HABITAT: Wet montane forest at foot of large exposed rocky outcrops growing amongst sedges etc. and in leaf litter, 1100–1850 m.
COLLECTIONS: cult. *Brybaker* 21.11.69 (K); *Cribb* S65 (K); *Cribb & Morrison* 1984 (K); *Mason* s.n. (K); *Robinson* s.n. & cult. Kew (K).

2. P. wentworthianum *Schoser & Fowlie* in Orchid Dig. 32: 285 (1968). Type: Bougainville, cult. Los Angeles Arb. (holotype TUB).

Fig. 1. Neuwiedia veratrifolia. A, habit reduced; **B**, leaf x 2/3; **C**, fruit x 2/3; **D**, inflorescence x 2/3; **E**, dorsal sepal, anthers and column x 2; **F**, lateral sepal x 2; **G**, petal x 2; **H**, lip x 2. **A** & **C** drawn from *Fa'arodo et al.* in BSIP 12256; **B** & **D** from *Streimann* in NGF 24478; **E-H** from *Dennis* s.n. (Kew spirit no. 45741). All drawn by Sue Wickison.

E
F
G
H
C
D
B
A
SUE
WICKISON

See Cribb (1987b) and Cribb, Campbell & Dennis (1985) for more information.

A terrestrial plant. *Stem* short. *Leaves* c. 5, arranged in a fan, narrowly oblong to elliptic-oblong, tricuspidate at subacute apex, 12–25 cm long, 3.5–4.5 cm wide, tessellated light and dark green. *Inflorescence* 1-flowered; scape up to 35 cm tall, purple-pubescent; bract elliptic-ovate, obtuse, 1.5–2.4 cm long, pubescent. *Flower* c. 8 cm across; dorsal sepal cream with a green centre and veined with green; petals glossy, purple above merging to brown and green towards the base; lip yellow-green, flushed with brown; staminode green; pedicel and ovary 4–5.5 cm long, densely pubescent; dorsal sepal concave, broadly ovate, obtuse, 2.8–3.2 cm long, 2.5–3.5 cm wide; synsepalum ovate, acute, 1.75–2.5 cm long, 2.5–2.6 cm wide; petals spreading at c. 20 ° below horizontal, oblong-elliptic, obtuse, 4.3–4.5 cm long, 1.5–2.5 cm wide, ciliate, undulate on upper margin; lip 4–4.5 cm long, 2.2 cm wide; staminode lunate, 8–10 mm long, 11–12 mm wide, shortly pubescent, lateral teeth incurved, subacute or acute. (See plate 2A).

DISTRIBUTION: Bougainville and Guadalcanal, very rare.
HABITAT: In deep leaf litter in montane forest on ridges, 900–1800 m.
COLLECTIONS: *Cribb, Dennis & Campbell* 5051 (K); *Dennis* s.n. (K, LAE).

3. CORYMBORKIS Thouars

Terrestrial herbs with subterranean rhizomes and persistent roots. *Stems* erect, slender but woody, unbranched. *Leaves* scattered in the upper part of the stem, plicate, usually rather broad, thin but tough, sheathing at base. *Inflorescences* lateral, paniculate. *Flowers* with long, slender, spathulate sepals and petals of about equal length; lip about the same length as the sepals and petals, being widest near the apex, the disc with 2 faint longitudinal keels; column straight, long and slender, dilated at the apex, rostellum erect; pollinia 2.

A pantropical genus of five species, a single species being recorded.

C. veratrifolia *(Reinw.) Blume* in Orch. Arch. Ind.: 125 (1855). Type: Java, *Lobb* 162 (neotype K!).

var. **veratrifolia**

For full synonymy see Rasmussen (1977).

A terrestrial herb. *Stems* up to 2 m high, erect. *Leaves* 6–15, scattered in the upper part of the stem, narrowly elliptic or ovate, acute to acuminate, 18–35 cm long, 5–10 cm wide, ribbed and fluted, dark green, sheathing at base. *Inflorescences* 1–4, lateral, arising near the apex of the stem, erect to drooping, corymbose-paniculate, with few to 60 flowers. *Flowers* white to greenish-white, fragrant; dorsal sepal linear-spathulate, acute, 2–5.3 cm long, 2.3–5.3 mm wide;

Fig. 2. Corymborkis veratrifolia. A, inflorescence and habit x 2/3; **B**, habit x 1/12; **C**, lip x 1; **D**, column, ovary and pedicel x 1; **E**, dorsal sepal x 1; **F**, lateral sepal x 1; **G**, petal x 1; **H**, column apex and anther-cap x 4; **J**, pollinia x 4. **A** drawn from living plant in G. Dennis's garden, Honiara; **B-J** from *Wickison* s.n. All drawn by Sue Wickison.

C
D
E
F
G
H
B
A
J
SUE WICKISON

lateral sepals obliquely linear-spathulate, acute, 2–5.1 cm long, 2.3–5.1 mm wide; petals obliquely linear-spathulate, acute, margins often undulate, 2.3–5.2 mm long, 2.3–5.1 mm wide; lip about the same length as the sepals, basal portion narrow, embracing the column, with 2 barely discernible, long, raised lines which diverge on the lamina and terminate in small calli, the apical lamina round and reflexed with irregularly undulate margins; column slightly shorter than the lip. (See fig. 2 and plate 2C).

DISTRIBUTION: Bellona, Guadalcanal, Kolombangara, New Georgia, the Reef Islands (Ngamanie), Rennell, the Santa Cruz Islands and Santa Isabel. Widely distributed from SE. Asia and the Malay Archipelago to New Guinea, Vanuatu, Fiji, Samoa, Tonga and Australia.

VERNACULAR NAMES: Sisiduru (New Georgia), Laulau (Kwarae'e name).

USES: Waterhouse notes that the leaves are used in native medicine on New Georgia.

HABITAT: Shaded lowland forest, often on calcareous soil or raised limestone, 15–100 m.

COLLECTIONS: *Cribb* 6012 (K); *Danish Noona Dan Expedition* 2827 (K); *Dennis* s.n. (K); *Hadley* 125 (BM); *Hunt* in RSS 2673 (K); *Inimua* in BSIP 6583 (K, LAE); *Mauriasi et al.* in BSIP 7574 (K, LAE); *Sirute et al.* in BSIP 9611 (K, LAE); *Waterhouse* 225 (K); *Wickison* s.n. & 217 (K).

The Solomon Island material is variable in both the size and number of flowers. Rasmussen (1977) commented on this and the apparently continuous nature of the variation and has taken a broad species' view. However, G. Dennis (pers. comm.) considers that there are two distinct taxa readily distinguishable in flower size and habit and a specimen grown by him (*Dennis* s.n.) does indeed have many significantly larger flowers than is typical (this plant is figured in plate 2C).

4. TROPIDIA Lindley

Terrestrial herbs. *Stems* erect, leafy. *Leaves* distichous, tough, pleated, sheathing at base. *Inflorescences* terminal, simple, with 1 to few flowers opening together; bracts large, imbricate and sheathing. *Flowers* not opening widely; sepals and petals free with the lateral sepals enclosing the base of the lip; lip entire, concave; column short, with a long, erect rostellum; pollinia 2. *Fruit* dehiscing along 6 lines of weakness.

A pantropical genus of about twenty species, a single species being recorded.

T. disticha *Schltr.* in K. Schum. & Laut., Nachtr. Fl. Deutsch. Südsee: 94 (1905). Types: New Guinea, *Schlechter* 13890, 14090 & 14585 (syntypes B†), 14298 (syntype B†, isosyntype K!); New Ireland, *Schlechter* 14701 (syntype B†, isosyntype K!).

Fig. 3. Tropidia disticha. A, habit reduced; **B**, flower x 4; **C**, column x 4; **D**, dorsal view of lip x 4; **E**, ventral view of lip x 4; **F**, anther-cap x 6; **G**, pollinia x 6; **H**, dorsal sepal x 4; **J**, lateral sepal x 4; **K**, petal x 4. **A-B** drawn from *Hunt* 2041; **C-K** from *Mitchell* 28. All drawn by Sue Wickison.

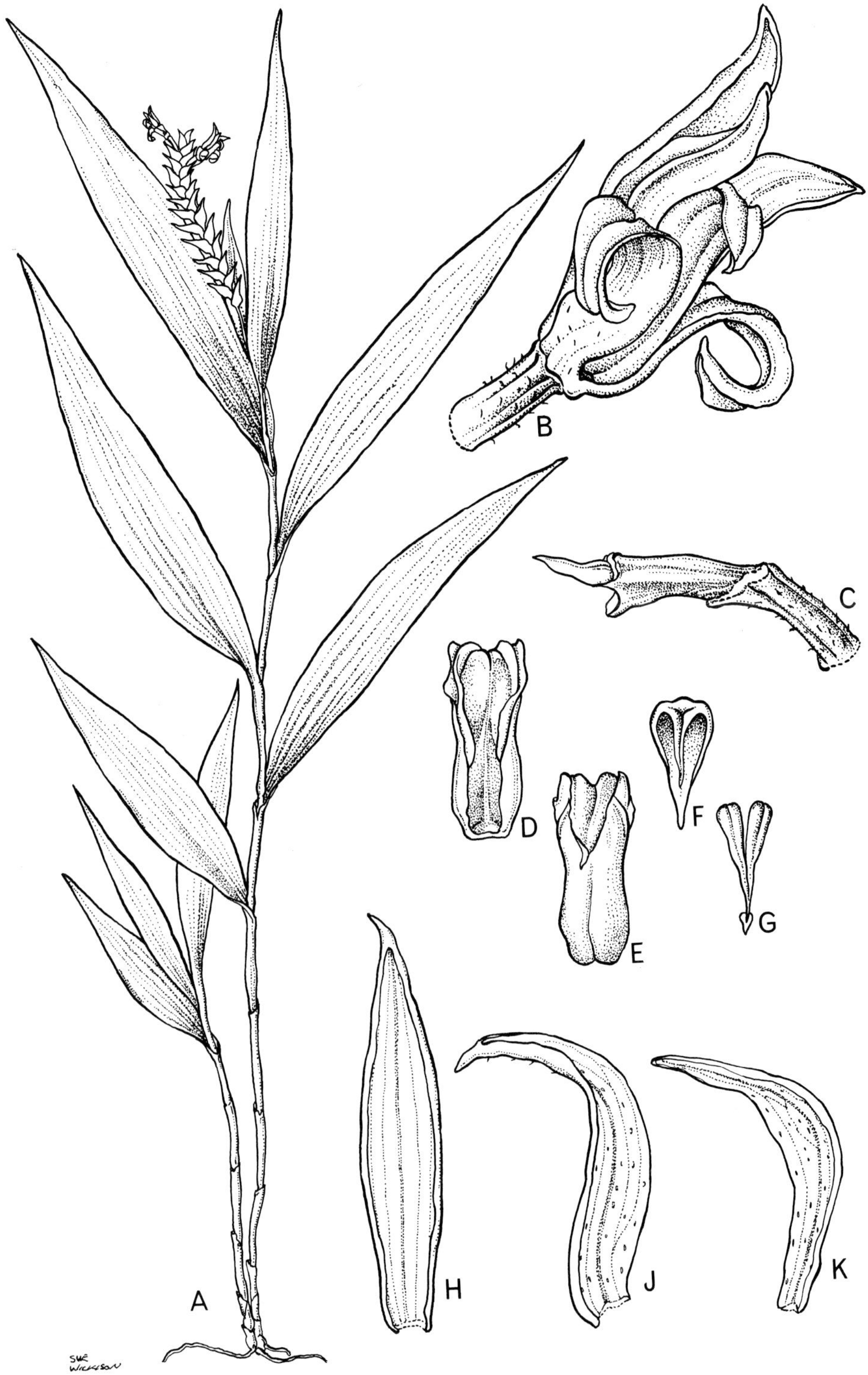
A
B
C
D
E
F
G
H
J
K

A terrestrial herb. *Stems* erect, 20–40 cm high, leafy. *Leaves* 5–8, scattered towards apex of stem, distichous, lanceolate, acute, up to 30 cm long, 6.5 cm wide, sheathing at base. *Inflorescences* 1–3, terminal, 3–18 cm long; rhachis with 1–2 flowers opening together; bracts imbricate and sheathing, lanceolate, 0.5–1.5 cm long, 3–6 mm wide. *Flowers* pale yellow or green with a pale orange disc on the lip; pedicel and ovary sparsely hairy; dorsal sepal ovate, 1–1.2 cm long, 2 mm wide, sparsely pubescent; lateral sepals falcate, 1–1.1 cm long, 2 mm wide, sparsely pubescent; petals falcate, 8.5–10 mm long, 2 mm wide; lip oblong, with a saccate base, 6.5 mm long, 3 mm wide, with a triangular, acute, reflexed, 3 mm long apex, the lateral lobes being inrolled; column 6 mm long. (See fig. 3).

DISTRIBUTION: Bougainville, Choiseul, Duke Island (off New Georgia), Guadalcanal, Makira, Santa Isabel. Also in New Guinea.and New Ireland.

VERNACULAR NAMES: Tukurutukuru (Bougainville); Puni (Duke Island); Oi'oi (Kwara'ae name).

USES: Waterhouse noted that the local people of Duke Island mix the leaves with lime to stain native cloth.

HABITAT: Rain forest, in leaf litter, 30–650 m.

COLLECTIONS: *Hunt* in RSS 2041 & 2193 (K); *Mitchell* 28 (K); *Runikera et al.* in BSIP 12750 (K); *Susui* in BSIP 8258 (K); *Waterhouse* 287 & 557 (K); *Wickison* 118 (K).

5. EURYCENTRUM Schlechter

Terrestrial herbs. *Rhizomes* fleshy, creeping, with a few, short, fat, lateral roots. *Stems* fleshy, erect and leafy. *Leaves* several, ovate, very dark green above, purple below. *Inflorescence* terminal, laxly many-flowered, pubescent. *Flowers* small, sessile, resupinate, usually pubescent on the outer surface; dorsal sepal forming a hood over the column; lateral sepals free, spreading; petals smaller than the sepals, adnate to the dorsal sepal; lip bipartite or obscurely trilobed, somewhat pandurate, the basal part concave, the apical part small and ovate to transversely oblong, prominently spurred at the base, the spur with a broad mouth and 2 sessile glands at the apex; column porrect, short, the rostellum truncate; pollinia 4.

A small genus of about five species from New Guinea to the Pacific Islands, a single species being recorded.

E. salomonense *Schltr.* in K. Schum. & Laut., Nachtr. Fl. Deutsch. Südsee: 90, t. 5, fig. B (1905). Type: Solomon Islands, Shortland Islands, *Guppy* 78 (holotype B†, isotypes K!, MEL).

A terrestrial plant 20–35 cm tall, rooting at the nodes of the rhizome. *Leaves* 4–7, scattered, obliquely ovate, acute, 4–7.5 cm long, 1.5–2.8 cm wide, satiny and almost black to very dark green or purple with a pink central vein above, purple beneath, sometimes with undulate margins. *Inflorescence* terminal, 18–23 cm long, laxly many-flowered; peduncle pubescent, slender, pink; bracts lanceolate, setose, 4–9 mm long, red. *Flowers* off-white with a red dorsal sepal; pedicel and ovary 4–8 mm long, pubescent, pinkish-red; sepals pubescent on outer surface;

dorsal sepal concave, ovate, obtuse to bluntly apiculate, 2–3 mm long, 2–2.5 mm wide; lateral sepals obliquely ovate, acuminate, 3.5–5 mm long, 2–2.5 mm wide; petals obliquely ovate, acute, 3 mm long, 1 mm wide; lip obscurely trilobed, concave, 3–5 mm long, 1.5–2 mm wide, the lateral lobes rounded, triangular, the mid-lobe tiny, shortly clawed, ovate, obtuse, 1 mm long, 1 mm wide, the spur compressed, cylindrical-conical, 4 mm long, 2 mm wide; column 1 mm long. (See fig. 4 A–J).

DISTRIBUTION: Duke Island (near New Georgia), Guadalcanal, Kolombangara, Makira, New Georgia, the Santa Cruz and the Shortland Islands. Also in the Banks Islands (Vanuatu).

VERNACULAR NAME: Buburu Zimiri (Duke Island).

HABITAT: Rain forest to montane moss forest, sea level to 1150 m.

COLLECTIONS: *Chaplin* 823 & 842 (K); *Cribb, Dennis & Campbell* 5045 & 5077 (K); *Dennis* in RSS 8170 (K); *Guppy* 78 (K); *Hunt* in RSS 2364 (K); *Mauriasi et al.* in BSIP 17208 (K); *Officers of HMS Penguin* s.n. (K); *Waterhouse* 288 (K); *Wickison* 62 & 113C (K).

6. GOODYERA R. Brown

Terrestrial or rarely epiphytic herbs with a horizontal, creeping, fleshy rhizome about the same thickness as the stem; roots rather thick. *Stems* erect, short or long, leafy. *Leaves* in a rosette or scattered, ovate, often asymmetric; petioles sheathing the stem. *Inflorescence* terminal, erect; peduncle finely pubescent, 1– to many-flowered. *Flowers* usually not opening widely, off-white to pale green, light pink to reddish-brown; pedicel and ovary pubescent; dorsal sepal and petals adnate, hooded; lateral sepals free, spreading, deflexed or porrect; lip saccate, glabrous or papillose; column long with a long porrect rostellum; pollinia 2.

A cosmopolitan genus of about fifty species, two species being recorded.

Leaves dark glossy green with a paler or white mid-vein; exterior of sepals pubescent; apex of lip obtuse .... **1. G. erythrodoides**

Leaves midgreen with a darker green venation; exterior of sepals glabrous; apex of lip acute .... **2. G. viridiflora**

1. G. erythrodoides *Schltr.* in Fedde, Repert. Spec. Nov. Regni Veg., Beih. 1: 49 (1911). Types: New Guinea, *Schlechter* 16293, 17748 & 17981 (syntypes B†).

A terrestrial herb. *Stems* up to 30 cm high. *Leaves* 5–7, scattered on the stem, obliquely ovate, up to 12 cm long, 3.5 cm wide, dark glossy green with a paler or white mid-vein; petiole 2–3 cm wide, sheathing at base. *Inflorescence* terminal, erect, up to 30 cm long, many-flowered; bracts lanceolate, 1–1.5 cm long, 4–5 mm wide, pubescent, reddish-green. *Flowers* reddish-green with off-white petals and lip; pedicel and ovary pubescent, 1 cm long; exterior of sepals pubescent; dorsal sepal ovate, 6.5 mm long, 2 mm wide; lateral sepals 8–8.5 mm long, 4 mm wide; petals spathulate, 6.5 mm long, 2 mm wide; lip oblong-rotund, saccate, 7 mm long, 3–5.5 mm wide, papillose, with an obtuse deflexed apex. (See fig. 4 K–S).

DISTRIBUTION: Bougainville, Guadalcanal, Kolombangara and Ranongga. Also in New Guinea.
HABITAT: Montane forest, 450–1000 m.
COLLECTIONS: *Craven & Schodde* 174 (LAE); *Cribb* 5087 (K); *Dennis* in RSS 2057 (K, LAE); *Hunt* in RSS 2496 & 2545 (K).

2. G. viridiflora *(Blume) Blume*, Fl. Java n. ser. 1: 34 (1858). Type: Java, *Blume* s.n. (holotype L – probably lost, isotype P).
Neottia viridiflora Blume, Bijdr. Fl. Ned. Ind.: 408 (1825).

For full synonymy see Clements (1989).

A terrestrial herb. *Stems* up to 10 cm high, pale to silver-green. *Leaves* 4–5, scattered on stem, obliquely ovate, 5.5–7 cm long, 2.5–3.5 cm wide, mid-green with slightly darker venation; petiole 2–4 cm long, sheathing at base. *Inflorescence* terminal, erect, 12–20 cm long, 3- to 13-flowered. *Flowers* pale green turning orange with age, the lip white; pedicel and ovary 1–2 cm long, pubescent; exterior of sepals glabrous; dorsal sepal ovate, 9 mm long, 3 mm wide; lateral sepals ovate, 8 mm long, 3 mm wide; petals rhombic, 7 mm long, 3 mm wide; lip oblong, saccate, 5.5 mm long, 3.5 mm wide, papillose at base, with an acute, deflexed apex. (See plate 3E).

DISTRIBUTION: Bougainville and Kolombangara. Widely distributed from SE. Asia and the Malay Archipelago to New Guinea, New Britain, Vanuatu, New Caledonia and Australia.
HABITAT: Primary rain forest, 800–1750 m.
COLLECTIONS: *Hunt* in RSS 2485 (K); *Lavarack & Ridsdale* in NGF 31440 (LAE).

This species is similar to *G. brachiorrhynchos* Schltr., from New Guinea, and the two species should be critically compared.

7. HYLOPHILA Lindley

Terrestrial herbs with creeping, fleshy rhizomes. *Stems* erect, leafy. *Leaves* distichous, ovate, often asymmetric; petioles sheathing stem. *Inflorescence* erect, terminal, with many small flowers. *Flowers* usually not opening widely; dorsal sepal and petals forming a hood; lateral sepals oblique, enclosing whole of lip; lip saccate, hairy or glabrous with a narrow reflexed blade; column short with a long, divided rostellum; pollinia 2.
A small genus of about six species from SE. Asia to the Solomon Islands, a single species being recorded.

Fig. 4. Eurycentrum salomonense A-J. A, habit x 2/3; **B**, flower x 4; **C**, lip x 6; **D**, column x 6; **E**, column, pedicel and ovary x 6; **F**, anther-cap x 12; **G**, dorsal sepal x 6; **H**, lateral sepal x 6; **J**, petals x 6. **Goodyera erythrodoides K-S. K**, flower x 4; **L**, column x 6; **M**, column, pedicel, ovary and bract x 6; **N**, lip x 6; **O**, lateral sepal x 6; **P**, dorsal sepal x 6; **Q**, petal x 6; **R**, anther- cap x 6; **S**, pollinia x 6. **A-J** drawn from *Cribb, Dennis & Campbell* 5045; **K-S** from *Cribb* 5087. All drawn by Sue Wickison.

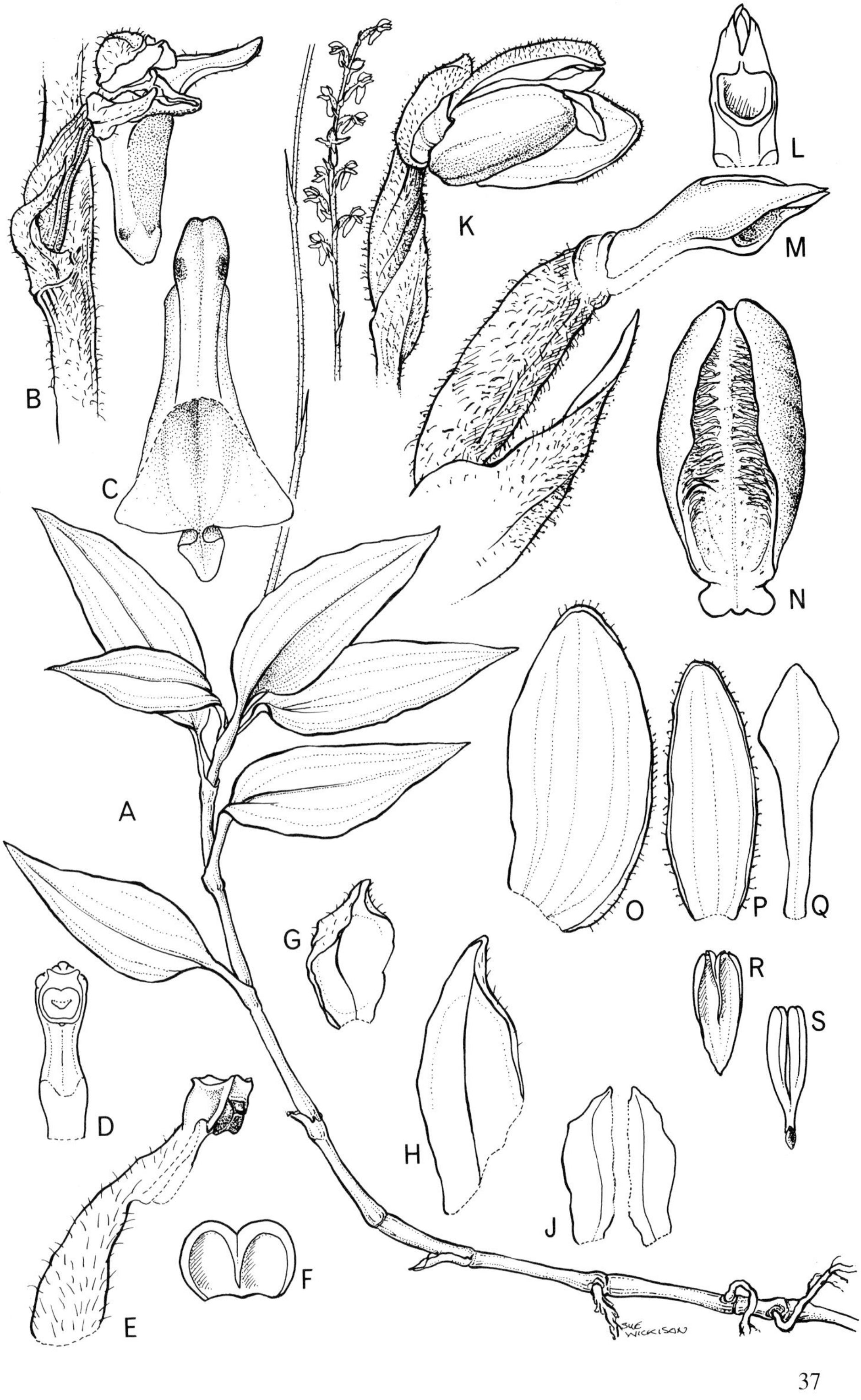
A
B
C
D
E
F
G
H
J
K
L
M
N
O
P
Q
R
S
SUE WICKISON

H. gracilis *Schltr.* in Fedde, Repert. Spec. Nov. Regni Veg., Beih. 1: 57 (1911). Type: New Guinea, *Schlechter* 20005 (holotype B†, isotype L).

A terrestrial herb. *Stems* erect, up to 18 cm long, leafy. *Leaves* distichous, obliquely ovate, acute, 6–9 cm long, 1.5–2.2 cm wide, shortly petiolate, the petiole up to 2 cm long, sheathing at base. *Inflorescence* erect, terminal, up to 40 cm long, finely pubescent; rhachis cylindrical, densely many-flowered; bracts lanceolate, up to 8 mm long, 1–2 mm wide. *Flowers* not opening widely, translucent white with green mid-veins, with a pale yellow-green lip and a yellow column; sepals pubescent on exterior; dorsal sepal and petals forming a hood; dorsal sepal concave, ovate, 2.5–3 mm long, 2 mm wide; lateral sepals obliquely ovate, 3–3.5 mm long, 3 mm wide, enclosing the lip; petals obliquely ovate, 2.5 mm long, 2 mm wide; lip triangular, acute, 2–2.5 mm long, 2 mm wide, with a reflexed apex, roundly saccate at base; column 1–2 mm long, with a long, divided rostellum.

DISTRIBUTION: Makira. Also in New Guinea and the Admiralty Islands.
HABITAT: *Casuarina* forest, 500–550 m.
COLLECTION: *Dennis* in RSS 2353 (K).

8. MACODES Blume

Terrestrial herbs with creeping, fleshy rhizomes. *Stems* erect, short or long. *Leaves* apical, rather fleshy, broad, with coloured veins, sheathing at base. *Inflorescence* terminal, laxly few- to many-flowered, the peduncle and bracts pubescent. *Flowers* rather small, with lip uppermost, asymmetric owing to twisting of lip and column; exterior of sepals pubescent; lateral sepals enclosing base of lip; lip bipartite or trilobed, with the epichile twisted to one side and a saccate hypochile containing 2 glands; column short; pollinia 2.

A genus of about ten species from the Malay Archipelago to the Pacific Islands, three species being recorded.

1. Leaves more than twice as long as broad .... **2. M. dendrophila**
   Leaves less than twice as long as broad 2
2. Leaves elliptic-subcircular, olive-green with pale green veins; flowers reddish-bronze with white petals, lip and column and a dull-orange spur .... **1. M. cominsii**
   Leaves oblong-subcircular, olive-green with silver, greenish-yellow or pinkish veins; flowers green with pale green petals and lip with a white apex to the lip and column .... **3. M. sanderiana**

1. M. cominsii *(Rolfe) Rolfe* in Orchid Review 19: 240 (1911). Type: Solomon Islands, Makira, *Comins* 116 (holotype K!).
Pseudomacodes cominsii Rolfe in Bull. Misc. Inf., Kew 1892: 127 (1892).

A terrestrial herb. *Stems* erect, up to 75 cm high, succulent. *Leaves* 3–4, apical, elliptic-subcircular, obtuse to apiculate, 5–13 cm long, 4–8.3 cm wide, with undulate margins, olive-green with pale green veins, the petiole 7–9 cm long, sheathing at base. *Inflorescence* terminal, laxly 10- to 30-flowered; peduncle 8–47 cm long, pubescent with red-brown hairs; bracts lanceolate, acuminate, 6–9 mm long, 1–2 mm wide, red. *Flowers* non-resupinate, asymmetric, opening widely, reddish-bronze with white petals, lip and column and a dull-orange spur; pedicel and ovary 1.4–1.7 cm long, pubescent; exterior of sepals pubescent; dorsal sepal ovate, obtuse, concave, 7 mm long, 4 mm wide; lateral sepals obliquely ovate-elliptic, 8 mm long, 4.5 mm wide; petals sigmoid-linear, rounded, 8 mm long, 1.5 mm wide; lip 6 mm long, 4 mm wide, the apex spathulate, rounded, slightly bilobed, twisted to one side, the base saccate, globose and fleshy, containing 2 large glands; column beak-like, 4 mm long. (See plate 2D).

DISTRIBUTION: Choiseul, Guadalcanal and Makira.
HABITAT: Rain forest, 140–530 m.
COLLECTIONS: *Comins* 116 (K); *Hunt* in RSS 2130 (K, LAE); *Wickison* 113E (K); *Woodford* 7 (K).

2. M. dendrophila *Schltr.* in Fedde, Repert. Spec. Nov. Regni Veg., Beih. 1: 73 (1911). Types: New Guinea, *Schlechter* 17088, 18019 & 18210 (syntypes B†; *Schlechter* 18210 & 19165 (syntypes B†, isosyntypes G, K!, L, NSW).

A terrestrial herb. *Stems* erect, 30–40 cm long, succulent. *Leaves* 3–5, apical, ovate, acute, 8–10 cm long, 2.9–4.3 cm wide, slightly tessellated, pale velvety green, the petiole 3–6 cm long, sheathing at base. *Inflorescence* terminal, subdensely many-flowered; peduncle 18–20 cm long, sparsely pubescent; bracts lanceolate, 9 mm long, 3 mm wide, pubescent. *Flowers* non-resupinate, asymmetric, not opening widely, green with pale green petals and lip, the apex of the lip and the column white; pedicel and ovary glandular pubescent, 8–14 mm long; exterior of sepals glandular pubescent; dorsal sepal ovate, obtuse, concave, 6.5 mm long, 3.5 mm wide; lateral sepals obliquely ovate, obtuse, concave, 7.5 mm long, 4.5 mm wide; petals linear, rounded at apex, 5 mm long, 1.2 mm wide; lip 6.5 mm long, 3 mm wide, the epichile linear-spathulate, obtuse, twisted to one side, 2.5 mm long, the base saccate with 2 large glands within; column 3 mm long with 2 broad wings.

DISTRIBUTION: Makira. Also in New Guinea.
HABITAT: Old garden on ridge, 600 m.
COLLECTION: *Dennis* in RSS 2281 (K).

3. M. sanderiana *(Kraenzl.) Rolfe* in Bull. Misc. Inf., Kew 1896: 47 (1896). Type: Sunda Islands, *Sander* s.n. (holotype B†).
Anoectochilus sanderianus Kraenzl. in Gard. Chron. ser. 3, 18: 484 (1895).
Anoectochilus spec. ex. aff. *A. roxburghii* Kraenzl. in K. Schum. & Laut., Nachtr. Fl. Deutsch. Südsee: 240 (1900).

See Rolfe in Curtis's Bot. Mag. 144: t. 8744 (1918) for further details.

A terrestrial herb. *Stems* erect, 2–4 cm high. *Leaves* 2–4, apical, oblong-subcircular, 4.5–8.5 cm long, 3–5 cm wide, dark olive-green with silver, greenish-yellow or pinkish veins. *Inflorescence* terminal, the raceme 14 cm long, laxly many-flowered; peduncle 32 cm long, pubescent, pinkish fawn. *Flowers* non-resupinate, asymmetric, pinkish white, yellow or pale green lightly suffused with brown; pedicel and ovary 1 cm long, pubescent; sepals hairy on the exterior; dorsal sepal ovate, concave, 7 mm long, 3 mm wide; lateral sepals obliquely ovate, concave, 7 mm long, 3 mm wide; petals linear, 7 mm long, 1.5 mm wide; lip trilobed, 6 mm long, 4 mm wide, twisted to one side, saccate at the base, containing 2 glands, the lateral lobes erect, short, rounded, the mid-lobe with a narrow base and short spreading blade; column 4 mm long.

DISTRIBUTION: Bougainville. Also in the Malay Archipelago, Sumatra, New Guinea and Vanuatu.
HABITAT: Rain forest, 600 m.
COLLECTION: *Ridsdale & Lavarack* in NGF 31130 (LAE).

9. PRISTIGLOTTIS Cretz. & J.J. Smith

Small terrestrial or rarely epiphytic herbs. *Stems* erect, leafy, arising from a creeping, fleshy rhizome. *Leaves* distichous, ovate; petioles sheathing. *Inflorescence* terminal, short, of few flowers. *Flowers* with sepals and petals not spreading; lateral sepals enclosing the saccate base of the lip; lip saccate at the base with 2 enclosed glands, with a long, narrow claw with sinuous margins, widened to a blade at the apex; column with a long porrect rostellum; pollinia 2.

A small genus of about ten species from India and SE. Asia to New Guinea and the Pacific Islands, a single species being recorded.

P. longiflora *(Reichb. f.) Kores* in Allertonia 5 (1): 28 (1989). Types: Fiji, *Seemann* 601 & *Graeffe* s.n. (syntypes W, isosyntypes BM!, K!).
Anoectochilus longiflorus Reichb. f. in Seem., Fl. Vit.: 294 (1868).

For full synonymy see Kores (1989).

A terrestrial herb up to 20 cm high. *Stems* erect, leafy. *Leaves* 3–6, distichous, ovate, acute, 2.5–4 cm long, 1.5–2.8 cm wide, midgreen with light green mid-veins. *Inflorescence* terminal, usually 2-(to 3-) flowered; bracts pubescent. *Flowers* white; pedicel and ovary pubescent; exterior of sepals pubescent; dorsal sepal oblong, obtuse, 9–10 mm long, 2 mm wide; lateral sepals oblong, obtuse, 10–11 mm long, 2–3 mm wide; petals oblong, obtuse, 9–10 mm long, 3–3.5 mm wide; lip 11–17 mm long, saccate at base with 2 glands, with sinuous margins in the oblong middle part, widened to a reniform, apiculate blade at apex, 2–2.5 mm long, 5 mm wide; column 2 mm long, 2 mm wide.

DISTRIBUTION: Guadalcanal and Rendova. Also in New Guinea, Vanuatu, Fiji and Samoa.
HABITAT: Montane forest, 800–1150 m.
COLLECTIONS: *Dennis* in RSS 2038 & 4628 (K); *Dickson* MD36 (K).

Having recently been able to examine living plants from Vanuatu (see *Lewis & McDonagh* 180) it is obvious that we confused two distinct species in our treatment of *Pristiglottis* in Vanuatu (see Lewis & Cribb, 1989). Both *P. longiflora* and *P. montana* (Schltr.) Gretz. & J.J. Smith occur in Vanuatu. *P. longiflora* can be distinguished by its larger flowers and midgreen leaves veined with light green. *P. montana* appears to be restricted to Vanuatu and New Caledonia whilst *P. longiflora* is a relatively widespread species in the Pacific.

Pristiglottis longiflora is similar to *P. fimbriata* (J.J. Smith) Cretz. & J.J. Smith, from New Guinea, and the two species should be critically compared.

10. HETAERIA Blume

Terrestrial herbs with fleshy, creeping rhizomes about the same thickness as the stem. *Stems* erect, leafy. *Leaves* distichous, broad, asymmetric, with sheathing petioles. *Inflorescence* terminal, pubescent, many-flowered. *Flowers* non-resupinate, not opening widely; exterior of sepals pubescent; dorsal sepal and petals forming a hood; lateral sepals enclosing the saccate base of the lip; lip uppermost in flower, concave, narrowed towards the apex, with 2 groups of basal glands in the pouched lip-base; column short with large lateral wings; pollinia 2.

A genus of about twenty species in the Old World Tropics, a single species being recorded.

H. oblongifolia *(Blume) Blume*, Coll. Orch. Arch. Ind.: 102, t.32 (1858) Type: Java, *Blume* s.n. (holotype L).
Etaeria oblongifolia Blume, Bijdr. Fl. Ned. Ind.: 410, t.14 (1825).
Goodyera erimae Schltr. in K. Schum. & Laut., Nachtr. Fl. Deutsch. Südsee: 93 (1905). Type: New Guinea, *Schlechter* 13676 (holotype B†).
Hetaeria erimae (Schltr.) Schltr. in Fedde, Repert. Spec. Nov. Regni Veg. 9: 88 (1910).

For full synonymy see Clements (1989).

A terrestrial herb. *Stems* erect, up to 30 cm long, leafy. *Leaves* 3–9, of which 3–4 may be dead at flowering time and the remainder scattered towards the apex of the stem, obliquely ovate, 7–10 cm long, 3–4 cm wide; petioles 2–3 cm long, sheathing at base. *Inflorescence* up to 30 cm long, pubescent; rhachis densely many-flowered; bracts lanceolate, 6–10 mm long, 0.7–1 mm wide. *Flowers* yellow, or white with a yellow lip; pedicel and ovary pubescent; exterior of sepals pubescent; dorsal sepal oblong to ovate, 3–3.6 mm long, 1.5 mm wide; lateral sepals ovate, 3–3.5 mm long, 1.5–1.8 mm wide; petals linear, acute, 3 mm long, 0.8 mm wide; lip ovate, 2.5–3.5 mm long, 2.5 mm wide, saccate, with inrolled apical margins and 4–6 basal glands in 2 groups in the pouched lip base.

DISTRIBUTION: Rennell, the Santa Cruz Islands (Vanikoro) and Santa Isabel. Widely distributed from the Philippines and Indonesia through New Guinea to the Caroline Islands (Palau), Vanuatu, New Caledonia, Fiji, Samoa and Australia.

HABITAT: Rain forest on calcareous soils and raised coral, 40–100 m.

COLLECTIONS: *Cribb* 6015 & 6021 (K); *Hadley* 79 (BM); *Hunt* in RSS 2682 (K).

11. VRYDAGZYNEA Blume

Small terrestrial herbs with fleshy, creeping rhizomes about the same thickness as the stem. *Stems* erect, leafy. *Leaves* distichous, scattered towards the apex of the erect stem, obliquely ovate or lanceolate, with a sheathing petiole. *Inflorescence* terminal, short and densely or laxly few-flowered; rhachis elongating after pollination. *Flowers* usually white or cream; sepals and petals not spreading, often thickened at the apex, similar; lip fleshy, entire, with a long spur with 2 stalked glands inside; column very short; pollinia 2.

A genus of about forty species from SE. Asia to Australia and the Pacific Islands, three species being recorded.

1. Leaves variegated, dark green chequered with light green; flowers few, c. 4; apex of spur bilobed. .. .. **1. V. argyrotaenia**
   Leaves not variegated; flowers many, 8–20; apex of spur entire 2
2. Leaves lanceolate, acute, 4–4.5 cm long, 7–9 mm wide; inflorescence laxly many-flowered; bracts almost as long as the flowers, with an entire margin .. .. **2. V. rivularis**
   Leaves obliquely ovate, acute, 6–8 cm long, 2–3 cm wide; inflorescence densely many-flowered; bracts shorter than the flowers, with a serrate apical margin .. .. **3. V. salomonensis**

1. V. argyrotaenia *Schltr.* in K. Schum. & Laut., Nachtr. Fl. Deutsch. Südsee: 84 (1905). Type: New Guinea, *Schlechter* 14477 (holotype B†).

A terrestrial herb. *Stems* erect, up to 35 cm tall, pinkish to olive-brown. *Leaves* 5–7, obliquely ovate, acute, 3–5.5 cm long, 1–1.9 cm wide, green chequered with light green; petiole 0.5–1 cm long, sheathing at base. *Inflorescence* terminal, up to 15 cm long, few-flowered; bracts lanceolate, 7–9 mm long, 2 mm wide, with an obscurely serrate upper margin, shorter than the flower. *Flowers* white with maroon markings near the apex and with light green margins, the petals and lip white with a pale orange spur; pedicel and ovary 1.1 cm long; dorsal sepal pyriform, 6.5 mm long, 2.5 mm wide; lateral sepals obliquely lanceolate, obtuse, 6 mm long, 2.5 mm wide; petals pyriform, narrowed at base, 6 mm long, 2 mm wide; lip 4–5 mm long, ovate, attenuate and reflexed, apex obtuse, with 2 raised hairy calli, the spur 5 mm long, 3 mm wide, upcurved, with a bilobed apex, and 2 stalked glands inside. (See fig. 5 K–T and plate 2E).

Fig. 5. Vrydagzynea salomonensis A-J. A, habit x 2/3; **B**, flower and bract x 4; **C**, dorsal sepal x 6; **D**, lateral sepal x 6; **E**, petal x 6; **F**, anther-cap x 8; **G**, lip x 6; **H**, longitudinal section of lip x 6; **J**, column x 8. **V. argyrotaenia K-T. K**, flower and bract x 2; **L**, dorsal sepal x 4; **M**, lateral sepal x 4; **N**, petal x 4; **O**, habit x 2/3; **P**, lip x 4; **Q**, longitudinal section of lip x 4; **R**, anther-cap x 6; **S**, column x 6; **T**, pollinia x 6. **A** drawn from *Wickison* 119; **B-J** from *Wickison* 56; **K-T** from *Wickison* 149. All drawn by Sue Wickison.

A
B
C
D
E
F
G
H
J
K
L
M
N
O
P
Q
R
S
T

DISTRIBUTION: Guadalcanal and Makira. Also in New Guinea and Vanuatu.
HABITAT: Primary rain forest, 100–600 m.
COLLECTIONS: *Braithwaite* 2218 (K); *Wickison* 149 (K).

2. V. rivularis *Schltr.* in Fedde, Repert. Spec. Nov. Regni Veg., Beih. 1: 85 (1911). Types: New Guinea, *Schlechter* 14592, 16335, 18126 & *Werner* 44 (syntypes B†), 190043 (syntype B†, isosyntypes E, K!, L, MO, NSW), & 19083 (syntype B†, isosyntypes E, K!, L, NSW).
Vrydagzynea elongata sensu Schltr. in K. Schum. & Laut., Nachtr. Fl. Deutsch. Südsee: 84 (1905), non Blume (1858).

A terrestrial herb. *Stems* erect, up to 20 cm tall, pale green. *Leaves* 6, scattered towards apex of stem, lanceolate, acute, 4–4.5 cm long, 7–9 mm wide; petiole 0.5–1.5 cm long, sheathing at base. *Inflorescence* terminal, 11 cm long, laxly 8- to 10-flowered; bracts lanceolate, almost as long as the flowers, 7–10 mm long, 1–1.3 mm wide. *Flowers* pale green at base, white above with a yellow column; pedicel and ovary 5 mm long; sepals and petals thickened at apex; dorsal sepal ovate, attenuate, obtuse, 4–5 mm long, 1.5 mm wide; lateral sepals ovate, obtuse, 4 mm long, 1.5 mm wide; petals oblong-ovate, acute, 3.5 mm long, 1.3 mm wide; lip ovate with apical margins inrolled, 1–2 mm long, 1.5 mm wide, the spur dilated in the middle, acute, 4–5 mm long, 1.5 mm wide, with 2 stalked glands inside.

DISTRIBUTION: Makira. Also in New Guinea.
HABITAT: Rain forest, growing at the edge of a river, probably periodically flooded, 60–70 m.
COLLECTION: *Hunt* in RSS 2246 (K).

3. V. salomonensis *Schltr.* in K. Schum. & Laut., Nachtr. Fl. Deutsch. Südsee: 86 (1905). Type: Solomon Islands, *Guppy* 75 (holotype B†, isotype K!).
Vrydagzynea cheesemanii Ames in Journ. Arn. Arb. 14: 103 (1933). Type: Vanuatu, Malekula, *Cheesman* in RBG Kew 4 (holotype AMES!, isotype K!).

A terrestrial herb. *Stems* erect, up to 40 cm tall, green to brown. *Leaves* 4–7, scattered towards the apex of the stem, obliquely ovate, acute, 6–8 cm long, 2–3 cm wide, midgreen. *Inflorescence* up to 12 cm long, densely many-flowered; rhachis elongating after pollination; bracts lanceolate, with a serrate apical margin, 7–9 mm long, 1–1.5 mm wide, pinkish-brown. *Flowers* white with pale green-brown mottling on exterior of sepals; pedicel and ovary 5–7 mm long, pinkish; dorsal sepal oblong, obtuse, 4 mm long, 1.5 mm wide; lateral sepals obliquely ovate, 4 mm long, 2 mm wide; petals obliquely ovate, 3 mm long, 2 mm wide; lip 1–2 mm long, 2 mm wide, triangular, with incurved margins, the spur conical-cylindrical, downcurved, nectariferous, 5 mm long, 2 mm wide, with 2 stalked glands inside. (See fig. 5 A–J).

DISTRIBUTION: Bougainville, Guadalcanal, Makira, Malaita, New Georgia, Ranongga, Rendova, Rennell, Santa Isabel and the Shortland Islands. Also in Vanuatu.

HABITAT: Common in rain forest, in fairly open to densely shaded forest, 40–400 m.
VERNACULAR NAME: Waingongi (Kwara'ae name).
COLLECTIONS: *Cribb* 6030 (K); *Danish Noona Dan Expedition* 2747 (K); *Dickson* 37 (K); *Dockrill* in NGF 34180 (K); *Foreman* in NGF 45682 & 45741 (K); *Gafui et al.* in BSIP 14772 & 14756 (K); *Guppy* 75 (K); *Hunt* in RSS 2001, 2020, 2195, 2291 & 2678 (K); *Mauriasi et al.* in BSIP 14497 (K, LAE) & 15582 (K); *Nakisi & Babala* in BSIP 8210 (K); *Runikera et al.* in BSIP 13012 (K, LAE); *Waterhouse* 26 (K); *Wickison* 56, 119 & 216 (K).

12. ZEUXINE Lindley

Terrestrial or occasionally lithophytic herbs. *Stems* erect, leafy, arising from a creeping, fleshy rhizome. *Leaves* obliquely ovate to lanceolate, with a sheathing petiole. *Inflorescence* terminal, pubescent, with few to many small flowers. *Flowers* resupinate, scarcely opening, usually pubescent on exterior of sepals; dorsal sepal and petals forming a hood; lateral sepals enclosing the base of lip; lip bipartite, with a saccate base containing 2 glands, and a transversely widened apical blade, connected to the base by a claw; column short; pollinia 2.

A genus of seventy to eighty species from Africa to tropical and subtropical Asia, Australia and the Pacific Islands. Three species are recorded from the Solomon Islands and Bougainville of which *Z. elatior* var. *angustata* is newly described here.

1. Lip not deeply saccate at base, with the apical blade broader than the base of the lip 2
   Lip deeply saccate at base, with a small apical blade less than or equal to the width of the base of the lip 3
2. Leaves obliquely ovate, (6)7–12 cm long, (2)3–5 cm wide .... **1a. Z. elatior** var. **elatior**
   Leaves obliquely lanceolate, 5–8 cm long, 1–1.8 cm wide .... **1b. Z. elatior** var. **angustata**
3. Leaves deep green with a pale green or white mid-vein; inflorescence laxly many-flowered; sepals 5–6 mm long; apical blade transversely oblong .... **3. Z. polygonoides**
   Leaves light green; inflorescence densely many-flowered; sepals 3–3.5 mm long; apical blade reniform .... **2. Z. erimae**

1. Z. elatior *Schltr.* in Fedde, Repert. Spec. Nov. Regni Veg., Beih. 1: 79 (1911). Type: New Guinea, *Schlechter* 19931 (holotype B†).

a. var. **elatior**

A terrestrial herb. *Rhizome* creeping, rooting from the nodes. *Stems* erect, 30–40 cm long, succulent, slender, leafy, pale olive-brown tinged with dull maroon. *Leaves* 4–6, scattered towards apex of stem, obliquely ovate, acute, (6)7–12 cm long, (2)3–5 cm wide; petiole slender, 1–2 cm long, sheathing at base.

Inflorescence terminal, 15–20 cm long, laxly few- to several-flowered, densely pubescent; bracts lanceolate, acuminate, 6–10 mm long, 3–4 mm wide, pink. *Flowers* pale salmon-pink with creamy-white petals and a white lip with a yellow-orange disc, sometimes faintly scented; pedicel and ovary pubescent, 10–11 mm long; exterior of sepals pubescent; dorsal sepal ovate, obtuse, 7.5 mm long, 3.5 mm wide; lateral sepals ovate, obtuse, 7 mm long, 3.5 mm wide; petals adnate to dorsal sepal, obliquely ovate, obtuse, 7 mm long, 4 mm wide; lip fleshy, 6 mm long, 2–3.5 mm wide with a saccate base containing 2 lobed calli, the lateral lobes erect and incurved, the apical blade broadly transversely reniform-oblong, 2.5 mm long, 8–9 mm wide; column 2–3 mm long.

DISTRIBUTION: Bougainville, Guadalcanal, New Georgia, Rennell and Santa Ana. Also in New Guinea.

HABITAT: In leaf litter in rain forest, 10–200 m.

COLLECTIONS: *Cribb* 6031 (K); *Dennis* 8 (K); *Fa'arodo et al.* in BSIP 12371 (K); *Hunt* in RSS 2018 (K); *Mauriasi et al.* in BSIP 17219 (K); *Wickison* 58 (K); *Womersley* in NGF 24828 (K).

b. var. **angustata** *Cribb & B. Lewis* var. nov. a varietate typica foliis angustioribus oblique lanceolatis, ad altitudionem altiorem crescenti recedit. Typus: Solomon Islands, Guadalcanal, *P.F. Hunt* 2050 (holotypus K!).

A terrestrial herb. *Rhizome* creeping, rooting from the nodes. *Stems* 15–40 cm long, slender, leafy. *Leaves* 5–8, scattered towards apex of stem, obliquely lanceolate, 5–8 cm long, 1–1.8 cm wide, dull maroon to dark glossy green; petiole slender 1–2.5 cm long, sheathing at base. *Inflorescence* terminal, 25 cm long, laxly few- to several-flowered, pubescent; bracts ovate, acute, 1 cm long, 4 mm wide, pink. *Flowers* pink with white petals and a white lip with a yellow disc; pedicel and ovary pubescent, 1 cm long; exterior of sepals pubescent; dorsal sepal concave, ovate, 7 mm long, 3 mm wide; lateral sepals ovate, 7 mm long, 3.5 mm wide; petals obliquely ovate, 7.5 mm long, 3 mm wide; lip fleshy, 6 mm long, 2–2.5 mm wide, the lateral lobes erect and incurved, with a saccate base containing 2 lobed calli, the apical blade broadly transversely reniform, 2.5–3 mm long, 9–9.5 mm wide; column 2–3 mm long. (See fig. 6 L–S).

DISTRIBUTION: Choiseul, Guadalcanal and Makira.

HABITAT: Montane rain forest, 650–750 m.

COLLLECTIONS: *Dennis* in RSS 2289 (K); *Hunt* in RSS 2050 (K); *Mitchell* 44 (K).

This variety differs from var. *elatior* in having narrower, obliquely lanceolate leaves but the flowers are identical. Var. *angustata* grows at higher altitudes than the typical variety.

Fig. 6. Zeuxine polygonoides A-K. A, habit x 2/3, **B**, flower, pedicel and bract x3; **C**, anther-cap x 6; **D**, pollinia x 6; **E**, dorsal sepal x 4; **F**, lateral sepal x 4; **G**, petal x 4; **H**, lip x 9; **J**, longitudinal section of lip x 6; **K**, column x 6. **Z. elatior** var. **angustata L-S. L**, flower, pedicel and bract x3; **M**, dorsal sepal x 4; **N**, lateral sepal x 4; **O**, petal x 4; **P**, lip x 4; **Q**, left side of lip x 4; **R**, anther-cap with pollinia x 6; **S** column x 4. **A** drawn from *Wickison* 43; **B-K** from *Wickison* 142; **L-S** from *Hunt* 2050. All drawn by Sue Wickison.

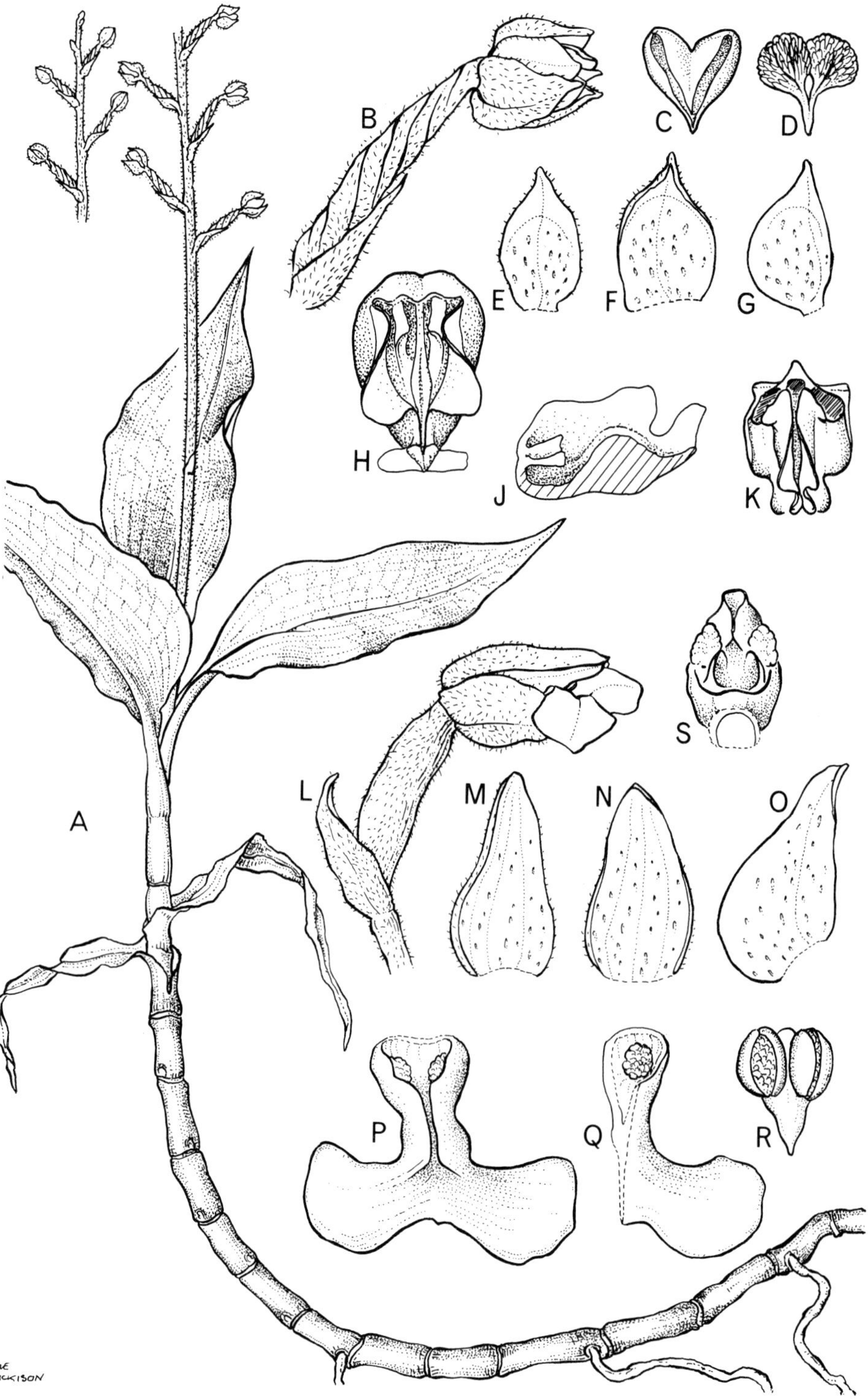

A
B
C
D
E
F
G
H
J
K
L
M
N
O
P
Q
R
S
SUE WICKISON

2. Z. erimae *Schltr.* in K. Schum. & Laut., Nachtr. Fl. Deutsch. Südsee: 90 (1905). Type: New Guinea, *Schlechter* 13677 (holotype B†).

A terrestrial herb. *Rhizome* creeping; roots arising from nodes. *Stems* erect, 25–40 cm tall, leafy, succulent. *Leaves* 3–6, scattered, obliquely ovate, acute, 6–11 cm long, 2–4.8 cm wide, light green; petioles 2–2.5 cm long, sheathing. *Inflorescence* terminal, 25–40 cm long, densely many-flowered, pubescent; raceme 5–15 cm long. *Flowers* creamy green outside, creamy white within, pubescent on outer surface; pedicel and ovary 6–9 mm long, pubescent; dorsal sepal ovate, subacute, 3–3.5 mm long, 1.5 mm wide; lateral sepals oblong, 3 mm long, 2 mm wide; petals adnate to dorsal sepal, oblong-ovate, acute, 3 mm long, 2.5 mm wide; lip fleshy, 3–3.5 mm long, 2 mm wide, with a saccate, ovate base containing 2 lobed calli, and with a clawed, reniform apical blade, 1 mm long, 2 mm wide; column 1.5 mm long.

DISTRIBUTION: Bougainville. Also in New Guinea and Vanuatu (Banks Islands).
HABITAT: Rain forest, 15–1000 m.
COLLECTIONS: *Craven & Schodde* 488 (K); *Ridsdale & Lavarack* in NGF 31603 (K).

3. Z. polygonoides *(F. Muell.) Cribb* in Clements, Cat. Austr. Orchid.: 147 (1989). Type: Australia, *Dallachy* s.n. (holotype MEL).
Goodyera polygonoides F. Muell., Fragm. 8: 29 (1873).
Zeuxine novaehiberniae Schltr. in Fedde, Repert. Spec. Nov. Regni Veg., Beih. 1: 86 (1912). Type: New Ireland, *Schlechter* 14695 (holotype B†, isotype K!).
Hetaeria polygonoides (F. Muell.) Dockr., Austral. Pl. 3: 350 (1966).

A terrestrial herb. *Rhizome* creeping, fleshy, rooting from the nodes. *Stems* erect, 30–50 cm long, succulent. *Leaves* 3–7, clustered towards the top of the stem, lanceolate-ovate, acuminate, 6–12 cm long, 1.5–4 cm wide, dark green with a central pale green or white stripe, sometimes with undulate margins; petiole 1.5–3 cm long, sheathing at base. *Inflorescence* terminal, laxly many-flowered; rhachis up to 16 cm long, pubescent; peduncle 8–14 cm long, pubescent; bracts lanceolate, acuminate, pubescent, 6–9 mm long. *Flowers* pale green to light pink with white petals and a white lip with an orange base; sparsely pubescent on outer surface; pedicel and ovary 9–11 mm long, pubescent; sepals ovate, acute, 5–6 mm long, 2.5–4 mm wide; petals adnate to the dorsal sepal, oblong-ovate, acute, 5–6 mm long, 2.5–4 mm wide; lip fleshy, obovate, 4–5 mm long, 3–3.5 mm wide, deeply saccate, the lateral lobes triangular to elliptic, erect and spreading, the disc with a central fleshy raised keel, the base minutely bilobed with 2 digitate calli on each side, the apical blade transversely oblong, 1 mm long, 2 mm wide; column 2–3 mm long, fleshy. (See fig. 6 A–K).

DISTRIBUTION: Bougainville, Choiseul, Guadalcanal, Kolombangara, Makira and Rendova. Also in New Ireland and Australia.
HABITAT: Montane rain forest, 500–750 m.
COLLECTIONS: *Chaplin* 824 & 834 (K); *Cribb & Campbell* 5131 (K); *Cribb, Dennis & Campbell* 5023 (K); *Dennis* in RSS 2278 (K, LAE); *Dickson* 38 (K);

Hunt in RSS 2278 & 2506 (K); *Mitchell* 43 & 66 (K); *Schodde & Craven* 3829 (K); *Wickison* 43 & 142 (K).

13. SPIRANTHES L.C. Richard

Terrestrial herbs with fleshy, fasciculate roots. *Stems* erect, rather short, leafy. *Leaves* conduplicate, fleshy. *Inflorescence* terminal, more or less densely spicate, the flowers arranged spirally on the rhachis. *Flowers* small, tubular; sepals free; dorsal sepal porrect, connivent with petals to form a hood; lateral sepals oblique, more or less spreading; lip obscurely trilobed at apex; column rather short; pollinia 2.

A cosmopolitan genus of over fifty species, a single species being recorded.

S. sinensis *(Pers.) Ames*, Orch. 2: 53 (1908). Type: China, collector not traced (holotype UPS).
Neottia sinensis Pers., Syn. Pl. 2: 511 (1807).

For full synonymy see Clements (1989).

A terrestrial plant 8–40 cm tall. *Roots* fleshy, tuberous, cylindrical, hairy. *Leaves* mostly basal, varying greatly in size and shape, oblong-elliptic to linear-elliptic, up to 20 cm long, 1 cm wide, often smaller. *Inflorescence* terminal, erect, slender, rather densely or subdensely many-flowered. *Flowers* arranged spirally on the rhachis, probably self-pollinating, scented, semi-open, translucent pearly white tinged on apical half a deep rose-mauve, with a translucent white lip tinged rose-mauve near the apex, the column green; dorsal sepal ovate, 3.2–4 mm long, 1.4 mm wide; lateral sepals oblong-lanceolate, 3.2–4.3 mm long, 1 mm wide; petals forming a hood with the dorsal sepal, oblong-oblanceolate, slightly shorter than the dorsal sepal, 3 mm long, 0.7 mm wide; lip obovate in outline, obscurely trilobed at apex, slightly longer than the sepals, 4.5 mm long, 3 mm wide, the lateral lobes semi-ovate, erect, the mid-lobe subcircular with an undulate-crispate margin, with 2 fleshy, subglobose calli at the base.

DISTRIBUTION: Guadalcanal. A quite remarkable distribution from mainland Asia, SE. Asia and the Malay Archipelago to New Guinea, Vanuatu, New Caledonia, Samoa and Australia.
HABITAT: Edge of grassland, montane forest, 500–600 m.
COLLECTION: *Dennis* s.n. (K).

This specimen was found growing in a window box at the Herbarium (19.9.67) in the Botanic Garden in Honiara, but the label states "believed to have been collected 6/67 by *G. Dennis* from near edge of grassland, Mt. Austen," Guadalcanal.

This is a variable species and M. Clements (pers. comm.) suggests that the current application of this name needs critical reconsideration.

14. CRYPTOSTYLIS R. Brown

Terrestrial herbs with fleshy roots. *Leaf* solitary, erect, arising from the rhizome, ovate, petiolate. *Inflorescence* arising separately on the rhizome; peduncle not sheathed by leaves, erect, long and slender. *Flowers* several, non-resupinate; sepals and petals narrow, spreading, with sepals longer than petals; lip erect, entire, widest near the base and tapering to the apex, strongly concave at the base and surrounding the column; column very short; pollinia 4.

A genus of about twenty species from SE. Asia to Australia and the Pacific Islands, a single species being recorded.

C. arachnites *(Blume) Hassk.* in Cat. Bogor.: 48 (1844). Type: Java, *Blume* s.n. (holotype L!, isotype P).
Zosterostylis arachnites Blume, Bijdr. Fl. Ned. Ind. 1(8): 419 (1825).

See Kores (1989) for full synonymy and discussion.

A terrestrial herb 20–60 cm tall, with several thick hairy roots at the base. *Leaves* erect, arising from the rhizome, ovate, acute, up to 17 cm long, 7.5 cm wide, pale green with a more or less distinct network of darker veins and mottled dark green; petiole slender, 5–15 cm long, mottled with purple. *Inflorescence* arising from the rhizome, 10–30 cm long, 8- to 12-flowered; peduncle covered with several overlapping sheaths. *Flowers* pale green or dull red-flushed, with a purple or orange-red lip which is spotted purple-red and a yellow-green column; dorsal sepal lanceolate, acute, 1.4–1.6 cm long, 2.8 mm wide, with edges inrolled; lateral sepals lanceolate, acute, 1.4–1.6 cm long, 2.8 mm wide; petals lanceolate, acute, 8–10 mm long, 1.7 mm wide; lip ovate, acute, 1.5–2 cm long, 5–7 mm wide at base, with a pubescent upper surface; column short, 1–1.5 mm long.

DISTRIBUTION: Bougainville, Guadalcanal, Kolombangara, Makira, New Georgia and Rendova. Widely distributed from the Malay Peninsula and Archipelago to Vanuatu, New Caledonia, Fiji and Samoa.

HABITAT: In leaf litter in ridge-top forest or in moss in montane forest, 200–2000 m.

COLLECTIONS: *Burn-Murdock et al.* in BSIP 7490 (K); *Chaplin* 835 & 836 (K); *Cribb & Campbell* 5126 (K); *Dennis* in RSS 2355 (K) & 7852 (K, LAE); *Hunt* in RSS 2503 & 2969 (K); *Lavarack & Ridsdale* in NGF 31330 (K) & 31401 (K, LAE); *Mitchell 3* (K); *Schodde & Craven* 3892 (LAE); *Wickison* 28 (K).

15. ACIANTHUS R. Brown

Small terrestrial herbs arising from small tuberoids, often growing in colonies. *Leaf* solitary, subcircular to cordate, rarely lobed, held horizontally on stalk well above the ground. *Inflorescence* terminal, erect, slender, few- to many-flowered. *Flowers* pale yellow, greenish or maroon; sepals and petals free; dorsal sepal narrow, usually acuminate; lateral sepals narrow; lip large, often heart-shaped, with 2 basal glands which secrete nectar.

A small genus of about twenty-five species from the SW. Pacific, Australia and New Zealand, a single species being recorded.

A. vulcanicus *Schodde* in Blumea 15: 403 (1968). Type: Bougainville, *Schodde & Craven* 3889 (holotype CANB, isotype LAE!).

A small terrestrial herb arising from a small tuberoid, growing in colonies. *Stem* erect, up to 5 cm tall, very slender and delicate. *Leaf* solitary, cordate, minutely apiculate, 1–2.5 cm long, 1–2.2 cm wide, light green, held horizontally on the stem well above the ground. *Inflorescence* terminal, erect, slender, 1.5–5 cm long, 2- to 3-flowered; bracts ovate, acuminate, 5 mm long, 4 mm wide. *Flowers* pale greenish-white or greenish with a purple lip; dorsal sepal lanceolate, acute, 8–9 mm long, 1 mm wide; lateral sepals lanceolate, acute, 9 mm long, 1 mm wide; petals lanceolate, acute, 8 mm long, 1 mm wide; lip circular, shortly apiculate, 6 mm long, 6 mm wide, with a jagged raised callus at the base; column 4 mm long, downcurved at apex. (See fig. 7 A–C and plate 3B).

DISTRIBUTION: Bougainville, Guadalcanal and Kolombangara.
HABITAT: Primary montane forest often growing in moss, 1650–2000 m.
COLLECTIONS: *Argent* in NGF 1229 (LAE); *Craven & Schodde* 3889 (LAE); *Cribb & Morrison* 5143 (K); *Dennis* in RSS 7865 (K, LAE); *Lavarack & Ridsdale* in NGF 31327 (K, LAE).

16. CORYBAS Salisbury

Dwarf terrestrial herbs arising from small tuberoids. *Stem* erect. *Leaf* solitary, ovate to orbiculate or cordate, borne horizontally above the ground. *Inflorescence* terminal, erect, 1-flowered. *Flowers* sessile or shortly stalked, helmet-shaped, large for size of plant, resupinate; dorsal sepal large and hooded; lateral sepals and petals small and rudimentary or long and filiform; lip large, recurved, tubular below, with 2 short spurs at base; column small and erect; pollinia 4, mealy.

A genus of about one hundred species from India, S. China, Taiwan, SE. Asia and the Philippines to the SW. Pacific, Australia and New Zealand, four species being recorded.

1. Flower sessile on leaf; peduncle more or less absent; lip pubescent 2
   Flower held well above the leaf; peduncle (0.5)1–5 cm long; lip glabrous 3
2. Lip with erose margins .. .. **1. C. gemmatus**
   Lip with entire margins .. .. **4. C. solomonensis**
3. Dorsal sepal twice as long as lip; lip apical lamina 5 mm long, 5 mm wide .. .. **3. C. mirabilis**
   Dorsal sepal half length of lip; lip apical lamina 18 mm long, 16 mm wide .. .. **2. C. longipedunculatus**

1. C. gemmatus *Cribb & B. Lewis* **sp. nov.**, *C. solomonensis* van Royen affinis sed folio longe apiculato, labello margine valde eroso-laciniato satis differt. Typus: Solomon Islands, Kolombangara, *Cribb & Campbell* 5141 (holotypus K!).

A terrestrial herb. *Stem* erect, 2–3 cm long. *Leaf* solitary, cordate, long-apiculate, 2–2.2 cm long, 1–1.2 cm wide, green veined with crimson. *Inflorescence* 1-flowered, the flower apparently sessile on leaf; bract setose, linear, acute, 1 cm long. *Flower* with a purple dorsal sepal, white lateral sepals and petals and a white and purple lip; pedicel and ovary 3 mm long; dorsal sepal oblanceolate, obtuse, 1.4 cm long, 8 mm wide, pubescent near apex on outer surface; lateral sepals and petals filamentous, 1.6–2 cm long, 1 mm wide; lip strongly recurved in the middle, tubular at base, the apical lamina orbicular, 1 cm long, 1.2 cm wide, pubescent in the middle with erose margin, with 2 basal spurs, the spurs 2 mm long, 1 mm wide at base tapering towards an obtuse apex; column 3–4 mm long with a large oblong stigma. (See fig. 7 D–F and plate 3A).

DISTRIBUTION: Kolombangara.
HABITAT: Growing in moss cushions on ridge-top forest, 500–750 m.
COLLECTION: *Cribb & Campbell* 5141 (K).

This species is similar to *C. solomonensis* but it differs in the leaf which is long-apiculate, and in having a flower with an erose margin to the lip.

2. C. longipedunculatus *van Royen*, The genus *Corybas* in its eastern areas: 124–128 (1983). Type: Bougainville, *Lavarack & Ridsdale* in NGF 31477 (holotype LAE!, isotype K!).

A terrestrial herb. *Stem* erect, 4.5–6 cm long, terete. *Leaf* solitary, cordate, apiculate, 2–2.2 cm long, 1.5–1.7 cm wide. *Inflorescence* 5–18 mm long, 1-flowered; bract ovate, apiculate, 4–12 mm long. *Flower* translucent green with streaks of red or purple; pedicel and ovary 4–5 mm long; dorsal sepal obovate, acute, 1.2–2 cm long, 0.8–1.2 cm wide; lateral sepals and petals linear, 8–12 mm long, 1–1.5 mm wide; lip tubular below, elliptic above, 18 mm long, 16 mm wide, with 2 short spurs at base. (See fig. 7 K–M).

DISTRIBUTION: Bougainville, Guadalcanal and Rendova.
HABITAT: In moss cushions in montane forest, 900–2000 m.
COLLECTIONS: *Dickson* BF1 (K); *Hunt* in RSS 2966 (K); *Lavarack & Ridsdale* in NGF 31477 (K, LAE).

3. C. mirabilis *(Schltr.) Schltr.* in Fedde, Repert. Spec. Nov. Regni Veg. 19: 22 (1923). Type: Vanuatu, Anatom, *Morrison* s.n. (holotype B†, isotype K!).

Fig. 7. Acianthus vulcanicus A-C. A, habit x 2/3; **B**, flower x 4; **C**, lip x 4. **Corybas gemmatus D-F. D**, flower x 4; **E**, habit x 2/3; **F**, lip x 4. **C. mirabilis G-J. G**, Flower x 4; **H**, habit x 2/3; **J**, lip x 4. **C. longipedunculata K-M. K**, flower x 4; **L**, lip x 4; **M**, habit x 2/3. **A-C** drawn from *Cribb* 5143; **D-F** from *Cribb & Campbell* 5141; **G & J** from *Dennis* 2290; **H** from *Hunt* 2504; **K-L** from *Hunt* 2966; **M** from *Dickson* BF1. All drawn by Sue Wickison.

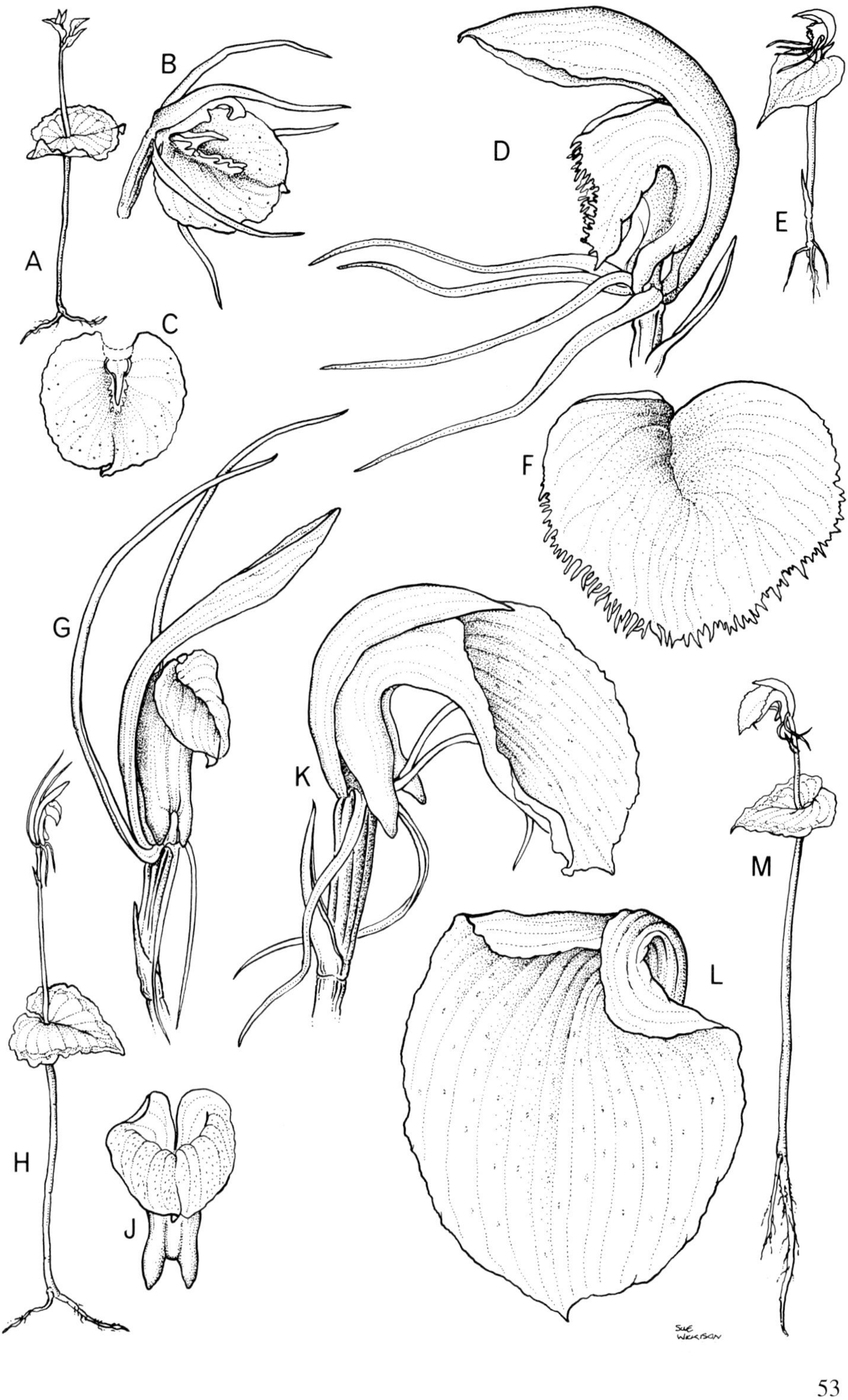
A
B
C
D
E
F
G
H
J
K
L
M
SUE WICKISON

A terrestrial herb. *Stem* erect, 2.5–8 cm long, terete. *Leaf* solitary, cordate, shortly apiculate-acuminate, 2–4 cm long, 1.6–2.5 cm wide, with slightly wavy margins, pale velvety green, tinged with purple below. *Inflorescence* 3–6 cm long, 1-flowered; bract linear, acuminate, 5–10 mm long. *Flower* with a deep maroon dorsal sepal, pink lateral sepals, pale green petals and a maroon- and white-striped lip; dorsal sepal obovate, 1.2–2.4 cm long, 0.5 cm wide; lateral sepals and petals filiform, 1.7–2.8 cm long, c. 1 mm wide; lip spathulate, recurved in apical third, tubular below, 1–1.3 cm long, 5–7 mm wide, with 2 conical spurs. (See fig. 7 G–J).

DISTRIBUTION: Kolombangara and Makira. Also in Vanuatu.
HABITAT: In moss cushions in montane forest, 700–1400 m.
COLLECTIONS: *Dennis* in RSS 2290 (K); *Hunt* in RSS 2481 & 2504 (K).

4. C. solomonensis *van Royen*, The genus *Corybas* in its eastern areas: 123 (1983). Type: Bougainville, *Craven & Schodde* 175 (holotype CANB, isotype LAE!).

A terrestrial herb. *Stem* erect, 1–1.5 cm long. *Leaf* solitary, ovate-cordate, acuminate, 1.3–3.5 cm long, 1–2.4 cm wide, dull purple-green above with purple veins, the underside deep purple with light green veins. *Inflorescence* 1-flowered, the flower apparently sessile on leaf; bract linear, acuminate, 6 mm long. *Flower* with a purplish-red dorsal sepal, faintly striped whitish inside, white lateral sepals and a purplish-red lip with a large mauve-cream patch inside; pedicel and ovary 3 mm long; dorsal sepal obovate, cucullate, obtuse, 1.8 cm long, 1 cm wide, pubescent on hooded part with crested veins on outside; lateral sepals and petals filamentous, 1.8–2.5 cm long, 1 mm wide; lip strongly recurved in middle, tubular in basal half, the apical lamina orbicular, 1.5 cm long, 1.3 cm wide, pubescent, with an entire margin and 2 short, obtuse spurs.

DISTRIBUTION: Bougainville.
HABITAT: In moss cushions in lower montane forest, 780 m.
COLLECTION: *Craven & Schodde* 175 (LAE).

17. PERISTYLUS Blume

Terrestrial or rarely epiphytic herbs. *Stem* erect, arising from a small root-stem tuberoid. *Leaves* scattered or in a basal rosette, the upper bract-like; leaf bases sheathing. *Inflorescence* terminal, unbranched, fairly long, racemose, few- to many-flowered. *Flowers* small, usually greenish; dorsal sepal adhering to the petals and forming a helmet; lateral sepals free; petals entire; lip trilobed, spurred at base; column erect, short; stigma in two parts, sessile; pollinia 2.

A genus of seventy to eighty species from tropical and subtropical Asia to Australia and the Pacific Islands, three species being recorded.

1. Leaves ovate, up to 23 cm long, 10 cm wide .... **3. P. sp. A**
   Leaves lanceolate or elliptic, up to 16 cm long, 5 cm wide 2

2. Lateral lobes of lip c. 4.5 mm long and curled .. .. **1. P. tradescantiifolius**
Lateral lobes of lip c. 2.5 mm long and straight .. .. **2. P. triaena**

1. P. tradescantiifolius *(Reichb. f.) Kores* in Allertonia 5: 12 (1989) (as *P. tradescantifolius*). Type: Fiji, *Seemann* 608 (holotype W, isotype K!).
Habenaria tradescantifolia Reichb. f. in Seem., Fl. Vit.: 293 (1868).
Peristylus remotifolius J.J. Smith in Bull. Dép. Agric. Indes Néerl. 22: 2 (1909). Type: New Guinea, *Lorentz* 380 (holotype BO!).
Habenaria cyrtostigma Schltr. in Fedde, Repert. Spec. Nov. Regni Veg. 9: 83 (1910). Types: Samoa, *Rechinger* 732 (syntype W!, isosyntype G); *Rechinger* 1146 (syntype W!); *Rechinger* 1802 (syntype W!).

For full synonymy and discussion see Kores (1989).

A terrestrial herb up to 60 cm tall, growing in small colonies. *Leaves* 4–7, scattered towards apex of stem, lanceolate, acute, 9–14 cm long, 1.9–2.5 cm wide. *Inflorescence* terminal, laxly many-flowered, up to 28 cm long; bracts lanceolate, 7 mm long. *Flowers* green; ovary and pedicel 8–10 mm long; dorsal sepal ovate, 3.2 mm long, 1.3 mm wide; lateral sepals obliquely ovate, 3 mm long, 1.5 mm wide; petals obliquely rhombic, 3.2 mm long, 1.2 mm wide; lip trilobed, with the mid-lobe shorter than the lateral lobes, the mid-lobe 2 mm long, the lateral lobes 4.5 mm long, curled, the spur fusiform, downcurved, 3.5–5.5 mm long (see fig. 8).

DISTRIBUTION: Makira. Also in Vanuatu, Fiji, Tonga and Samoa.
HABITAT: Montane forest in old garden area, 500 m.
COLLECTION: *Dennis* in RSS 2279 (K).

Kores (1989) notes that this species is similar to *P. papuanus* (Kraenzl.) J.J. Smith and the two species should be critically reviewed.

2. P. triaena *(Schltr.) P.F. Hunt* in Kew Bull. 26(1): 177 (1971). Type: New Guinea, *Schlechter* 18169 (holotype B†).
Habenaria triaena Schltr. in Fedde, Repert. Spec. Nov. Regni Veg., Beih. 1: 7 (1911).

An epiphytic herb. *Leaves* 3 or more, suberect, elliptic, acuminate, 14 cm long, 3–4 cm wide, dark green above, greyish-blue below. *Inflorescence* terminal, 35 cm long, many-flowered; bracts lanceolate, 6–12 mm long. *Flowers* green; pedicel and ovary 8–10 mm long; dorsal sepal ovate, 4 mm long, 2 mm wide; lateral sepals ovate, 4 mm long, 2.5 mm wide; petals obliquely ovate, obtuse, 4 mm long, 2 mm wide; lip trilobed, the lobes linear, obtuse, 2.5 mm long, 0.8 mm wide, the spur cylindrical, slightly inflexed towards apex, 5.5 mm long; column 2 mm long.

DISTRIBUTION: Bougainville. Also in New Guinea.
HABITAT: Lower montane forest, c. 850 m.

COLLECTION: *Lavarack & Ridsdale* in NGF 31361 (K).

The specimen cited here is incomplete having been broken off at the top of the stem, and the flowers are going over. More material of this species is needed to confirm the identification.

3. P. sp. A

Large robust terrestrial plant. *Stem* erect, 60 cm tall, fleshy. *Leaves* 6, scattered on the stem, ovate, acute, up to 23 cm long, 10 cm wide, grey-green. *Inflorescence* terminal, 45 cm long, with many densely-clustered flowers; bracts lanceolate, 1.7–2.5 cm long. *Flowers* not observed.

DISTRIBUTION: Ranongga.
HABITAT: In deep leaf mould in shade, growing with ferns etc. in rain forest, 500 m.
COLLECTION: *Mitchell* 76 (K).

This species is much larger than the other *Peristylus* species from the Solomon Islands and Bougainville but is in fruit and unfortunately we have yet to see it in flower.

18. HABENARIA Willdenow

Terrestrial or rarely epiphytic herbs. *Stem* erect, arising from a root-stem tuberoid. *Leaves* in a basal or apical rosette or scattered along stem, upper ones usually bract-like; leaf bases sheathing. *Inflorescence* terminal, unbranched, erect, long, racemose, few to many-flowered. *Flowers* small to large; dorsal sepal and petals often adnate to form a hood over the column; lateral sepals free, often reflexed; petals entire or bifid; lip trilobed, spurred; column short, erect; stigma lobes distinctively stalked; pollinia 2.

A large genus of six hundred to seven hundred species in the tropics and warm temperate regions of the world, four species being recorded.

1. Leaves in an apical rosette .. .. **2. H. novaehiberniae**
   Leaves scattered in the lower half of the stem or basal 2
2. Lateral lobes of lip and posterior lobe of petal dissected .. .. **3. H. polyschista**
   Lateral lobes of lip and petals entire 3
3. Flowers pale green to white; petals ovate-lanceolate .. .. **1. H. bougainvilleae**
   Flowers green, usually with pale purple spotting, with a pale yellow-green lip; petals linear, obtuse .. .. **4. H. retroflexa**

Fig. 8. Peristylus trandescantiifolius. A, habit x 2/3; **B**, column, dorsal sepal and petal x 12; **C**, flower x 6; **D**, lip x 6; **E**, dorsal sepal x 8; **F**, lateral sepal x 8; **G**, petal x 8. Drawn from *Dennis* 2279 by Sue Wickison.

D
C
B
A
E
F
G
SUE
WILKISON

1. H. bougainvilleae *Renz* in Pl. Syst. Evol. 155: 326 (1987). Type: Bougainville, *Lavarack & Ridsdale* in NGF 31153 (holotype LAE!, isotype K!).

A terrestrial herb up to 80 cm tall. *Stem* erect. *Leaves* 5–8, scattered in lower half of stem, suberect, lanceolate or oblanceolate, acute, 20–30 cm long, 3–5.5 cm wide, narrowed at base, pale green with darker green longitudinal stripes. *Inflorescence* terminal, up to 40 cm long, many-flowered; bracts lanceolate, 1.5–4.5 cm long, pale green. *Flowers* pale green to white with a lime-green lip, strongly scented; pedicel and ovary up to 2 cm long; dorsal sepal falcate, concave, 5 mm long, 3 mm wide, forming a helmet with the petals; lateral sepals obliquely ovate, 7 mm long, 4 mm wide, reflexed; petals ovate-lanceolate, 5 mm long, 0.9 mm wide; lip trilobed, each lobe linear-lanceolate, 5 mm long, 1 mm wide, the spur subclavate, 2 cm long; column 3 mm long, fleshy.

DISTRIBUTION: Bougainville and Guadalcanal.
HABITAT: Primary rain forest, 300–1000 m.
COLLECTIONS: *Corner* 2994 (K); *Lavarack & Ridsdale* in NGF 31053 (K).

2. H. novaehiberniae *Schltr.* in K. Schum. & Laut., Nachtr. Fl. Deutsch. Südsee: 79 (1905). Type: New Ireland, *Schlechter* 14698 (holotype B†).

A terrestrial herb. *Stem* erect, up to 40 cm high, green with a dull purple tinge. *Leaves* 6–7 in an apical rosette, ovate-lanceolate, 7–15 cm long, 1.5–4 cm wide, bluish-green, with bases sheathing stem; leaf sheaths sometimes spotted. *Inflorescence* terminal, up to 40 cm long, 10 to 18-flowered; bracts lanceolate, 1–2.4 cm long, 6 mm wide. *Flowers* pale yellow-green to green; pedicel and ovary 1.5–2.7 cm long; dorsal sepal ovate, 9–10 mm long, 2.5–4 mm wide; lateral sepals obliquely ovate, 9–10 mm long, 2–4 mm wide; petals bipartite, the posterior lobe linear, 8.5–9 mm long, 1.2 mm wide, the anterior lobe filiform, 1.1–2.3 cm long, 0.5–1 mm wide, curled; lip trilobed with linear lobes, the lateral lobes longer than the mid-lobe, 1.5–2.5 cm long, 0.8–1 mm wide, curled, the mid-lobe 7–9 mm long, 1–1.2 mm wide, the spur filiform, 1.8–2.5 cm long; column 3–3.5 mm long. (See fig. 9 A–G).

DISTRIBUTION: Bougainville, Guadalcanal, Kolombangara, Makira, Malaita, Rennell and Santa Isabel. Also in New Ireland and Vanuatu.
HABITAT: Rain forest, often growing on calcareous soil, 30–900 m.
COLLECTIONS: *Braithwaite* 3041 (K); *Cribb* 6033 (K); *Dennis* in RSS 2280 (K); *Hunt* in RSS 2299, 2511 & 2677 (K); *Ridsdale & Lavarack* in NGF 30584 (LAE); *Wickison* 146 (K).

3. H. polyschista *Schltr.* in K. Schum. & Laut., Nachtr. Fl. Deutsch. Südsee: 80 (1905). Type: New Guinea, *Schlechter* 14475 (holotype B†).

Fig. 9. Habenaria novaehiberniae A-G. A, habit x 2/3; **B**, side view of lip, column, ovary, pedicel and bract x 3; **C**, dorsal sepal x 4; **D**, petal x 4; **E**, lateral sepal x 4; **F**, column, lip and spur x 3; **G**, anther-cap x 4. **H. polyschista. H**, flower x 2. **H. retroflexa. J**, flower x 2. **A-G** drawn from *Wickison* 146; **H** from *Corner* 3036; **J** from *Mitchell* 77. All drawn by Sue Wickison.

H
J
B
A
F
C
D
E
G
SUE WICKISON

A terrestrial herb 75–160 cm tall. *Stem* erect. *Leaves* basal, suberect, lanceolate, acute, up to 30 cm long, 4 cm wide, sheathing at base. *Inflorescence* terminal, 60 cm long, many-flowered; bracts lanceolate, 2–3 cm long. *Flowers* pale green with a yellow lip and spur; dorsal sepal ovate, 8 mm long, 2.5 mm wide; lateral sepals obliquely ovate, 9 mm long, 3 mm wide; petals bifid, the posterior lobe finely dissected and longer than the linear, 9 mm long, 0.8 mm wide anterior lobe; lip trilobed, the lateral lobes finely dissected and longer than the mid-lobe, the mid-lobe linear, 1.2 cm long, 0.8 mm wide, the spur subclavate, 1.6–1.8 cm long; column 3 mm long. (See fig. 9 H).

DISTRIBUTION: Malaita. Also in New Guinea.
HABITAT: Primary rain forest, 650 m.
COLLECTION: *Corner* 3036 (K).

4. H. retroflexa *F. Muell. & Kraenzl.* in Engler, Bot. Jahrb. 17: 488 (1893). Type: New Guinea, no collector given (holotype MEL?).

A terrestrial herb up to 70 cm tall. *Stem* erect. *Leaves* basal or scattered along basal half of stem, ovate, acute, narrowed at base, 20–30 cm long, 4.5–7 cm wide; petioles up to 9 cm long, sheathing at base. *Inflorescence* up to 55 cm long, many-flowered; bracts ovate-lanceolate, 1.5–2.2 cm long, 5–8 mm wide. *Flowers* green, usually with pale purple spotting and with a pale yellow-green lip; pedicel and ovary 1.8–2.8 cm long, 3-winged; dorsal sepal ovate, 7–8 mm long, 3.5–4.5 mm wide, forming a helmet with the petals; lateral sepals obliquely ovate, 7.5–8 mm long, 3–3.3 mm wide; petals linear, obtuse, 7–8 mm long, 1–1.8 mm wide; lip trilobed, the lobes linear, the mid-lobe 4.5–7 mm long, 0.8–1 mm wide, with 2 obscure calli at the base, the lateral lobes 6–9 mm long, 1–2 mm wide, the spur filiform, 1.4–2.3 cm long; column 3 mm long. (See fig. 9 J).

DISTRIBUTION: Guadalcanal, Makira and Ranongga. Also in New Guinea.
HABITAT: In deep leaf litter in rain forest, 150–350 m.
COLLECTIONS: *Dennis* in RSS 2002 (K); *Hunt* in RSS 2222 (K); *Mitchell* 77 (K); *Nakisi* in BSIP 8024 (K, LAE).

19. PSEUDOVANILLA Garay

Climbing saprophytes or semi-saprophytes arising from rather thick rhizomes which are both horizontal and vertical. *Stems* usually branching. *Leaves* reduced to bract-like scales at the nodes. *Inflorescences* arising from the axils of the leaves and sometimes terminal, paniculate. *Flowers* with sepals and petals free; petals slightly narrower than the sepals; lip entire or obscurely trilobed, embracing the column, the margins crisped and undulate, the disc with callus ridges; column slender; pollinia mealy. *Fruits* long, ± cylindrical, with winged seeds.

These vine-like orchids are short-lived plants which depend on mycorrhizal fungi in wood at a certain stage of decomposition for their existence.

A genus of eight species from Malaysia, Indonesia, New Guinea, the SW. Pacific and Australia, two species being recorded.

Sepals 2.3–3 cm long; flowers orange to pale yellow with a pale yellow lip marked with orange between the papillae .. .. **1. P. foliata**
Sepals up to 1.7 cm long; flowers cream-coloured with purple markings on the basal half of the lip .. .. **2. P. gracilis**

1. P. foliata *(F. Muell.) Garay* in Bot. Mus. Leafl. Harv. Univ. 30(4): 235 (1986). Type: Australia, *Fitzalan* s.n. (holotype MEL).
Ledgeria foliata F. Muell. in Fragm. 2: 167 (1861).
Galeola foliata (F. Muell.) F. Muell. in Fragm. 8: 31 (1873).
Galeola montigena Schltr. in Fedde, Repert. Spec. Nov. Regni Veg., Beih. 1: 29 (1911). Type: New Guinea, *Schlechter* 20167 (holotype B†).

For full synonymy see Clements (1989).

A large saprophyte or semi-saprophyte with stems up to 15 m long, climbing up dead trees, with a much-branched extensive root system. *Stems* branching towards the apex, pendulous, mid-green. *Leaves* bract-like at nodes, ovate-orbicular, up to 6 cm long, 1.5 cm wide. *Inflorescences* paniculate, many-flowered, providing an impressive floral display. *Flowers* orange to pale yellow with a pale yellow lip, marked with orange between the papillae; dorsal sepal oblong, acute, up to 3 cm long, 7 mm wide; lateral sepals obliquely oblong, up to 3 cm long, 8 mm wide; petals obliquely oblong, up to 3 cm long, 5 mm wide, with a central, fleshy longitudinal ridge; lip entire to obscurely trilobed, up to 2.5 cm long, 1.7 cm wide, almost tubular in basal half and expanded in the apical half into a semi-circular lamina which has crisped and contorted margins, the basal half with raised fleshy transverse calli, the apical half with numerous fleshy papillae, the disc with 2 raised keels; column slender, 1.3 cm long, projected forwards. *Fruit* fleshy, cylindrical, up to 20 cm long, 1 cm wide. (See fig. 10).

DISTRIBUTION: Guadalcanal, Malaita and Santa Ana. Also in New Guinea and Australia.
HABITAT: Primary rain forest, up to 340 m.
COLLECTIONS: *Dennis* in *Cribb* 5 (K); *Cribb, Dennis & Campbell* 5165 (K); *Douglas* K/4 (K); *Wickison* 207 (K).

This species is similar to *P. anomala* (Ames & L.O. Williams) Garay, from Fiji, and the two species should be critically compared.
Dennis (1976) notes that, after the 1971 hurricane in the Solomon Islands, great numbers of *P. foliata* were recorded from Santa Ana growing on the rotting logs and still standing dead trees wherever the sunlight penetrated.

2. P. gracilis *(Schltr.) Garay* in Bot. Mus. Leafl. Harv. Univ. 30(4): 235 (1986). Type: New Guinea, *Schlechter* 19966 (holotype B†, isotypes E, K!, L, NSW).
Galeola gracilis Schltr. in Fedde, Repert. Spec. Nov. Regni Veg., Beih. 1: 28 (1911).

A large saprophyte with stems up to 4 m long, climbing up dead trees. *Stems* branching towards the apex, pendulous. *Leaves* bract-like at nodes, up to 5 cm long, 1.5 cm wide. *Inflorescences* paniculate, many-flowered, providing an impressive floral display. *Flowers* cream-coloured with purple markings on the basal half of the lip; dorsal sepal oblong, acute, 1.7 cm long, 6 mm wide; lateral sepals obliquely oblong, 1.6 cm long, 7 mm wide; petals obliquely oblong, 1.5 cm long, 5 mm wide; lip entire, 1.7 cm long, 1.2 cm wide, almost tubular in basal half and expanded in the apical half into a narrow semi-circular lamina which has crisped and contorted margins, the basal half with small raised fleshy transverse calli, the apical half with numerous fleshy papillae, the disc with 2 raised keels; column slender, 1 cm long, projected forwards.

DISTRIBUTION: Tetepari Island (off New Georgia). Also in New Guinea.
VERNACULAR NAME: Bwagoe (Tetepari Island).
HABITAT: Ridge-top forest, 130 m.
COLLECTION: *Mauriasi et al.* in BSIP 16114 (K).

*20. **VANILLA** Plumier ex Miller

Scandent or climbing plants, leafy (in the Solomon Islands and Bougainville) or leafless herbs. *Stems* long, branched, green. *Leaves* fleshy, sessile or shortly petiolate. *Inflorescences* lateral, arising from the axils of the leaves, short. *Flowers* fugaceous, with subequal, spreading, free sepals; petals similar to sepals; lip tubular, entire or obscurely lobed, with its claw adnate to the column; column elongated, footless; pollinia granular. *Fruit* ('bean') elongated, fleshy, often fragrant when mature.

A genus of about one hundred species widely distributed throughout the tropics, a single species being recorded.

Vanilla planifolia, which is the source of the flavouring vanilla, is economically important and is commercially cultivated for this purpose in tropical regions. In the Solomon Islands and Bougainville *V. planifolia* is cultivated in gardens and has become naturalised in some areas.

***V. planifolia** *Jackson* in Andrews, Bot. Repos. 8: t. 538 (1808). Type: West Indies, *Greville* s.n. (lectotype the original plate).

For full synonymy see Kores (1989).

A climbing herb. *Stems* branched, leafy, up to 20 m long. *Leaves* elliptic-oblong, acute to acuminate, 14–25 cm long, 3.5–7 cm wide; petiole 1.5 cm long. *Inflorescences* arising from the axils of the leaves, up to 13 cm long; few-flowered. *Flowers* creamy-yellow; sepals lanceolate, acute, 3.5–4.8 cm long, 1–1.2 cm wide; petals similar to the sepals, 3.5–4.8 cm long, 1–1.2 cm wide; lip obscurely trilobed in the apical half, 3–4.5 cm long, 2 cm wide, with the claw adnate to the column, the lateral lobes erect, embracing the column, semicircular,

---

Fig. 10. Pseudovanilla foliata. A, habit x 1; **B**, flower x 1; **C**, lip x 2; **D**, column x 4; **E**, ventral and dorsal view of anther-cap x 4; **F**, dorsal sepal x 2; **G**, petal x 2; **H**, lateral sepal x 2. All drawn from *Wickison* 207 by Sue Wickison.

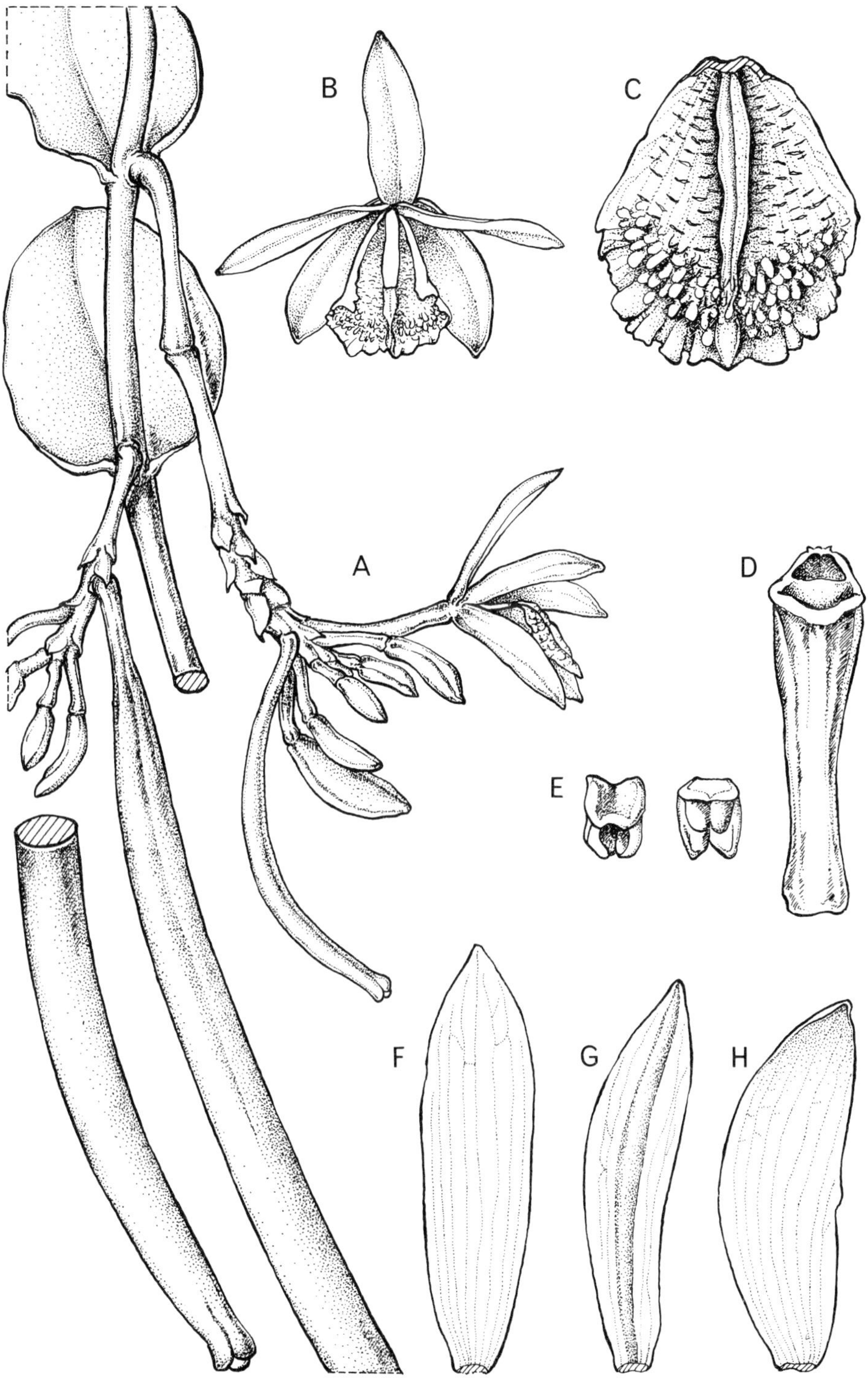
A
B
C
D
E
F
G
H

the disc with longitudinal verrucose lines and a tuft of hairs at the apex of the mid-lobe; column 3 cm long, finely pubescent dorsally. *Fruit* bean-like, up to 20 cm long, 1 cm in diameter, fragrant.

DISTRIBUTION: Naturalised in Bougainville and Guadalcanal. Indigenous in Mexico, Central America and the West Indies and extensively cultivated and frequently naturalised elsewhere.

HABITAT: Lowland and lower montane rain forest.

COLLECTIONS: *Cribb* sight records (Guadalcanal); *Ridsdale and Lavarack* in NGF 31618 (K); *Wickison* s.n. (K).

21. **NERVILIA** Commelin ex Gaudichaud

Terrestrial herbs, arising from a more or less round, reduced corm which has short roots. *Inflorescence* erect, produced before the leaf appears, elongating after fertilization, terminating in a 1– to few-flowered raceme. After the flowers mature a solitary leaf arises on a separate stalk. *Leaf* erect to horizontal, plicate, ovate to cordate, rarely lobed. *Flowers* with sepals and petals similar and free, long and narrow; lip entire or trilobed and usually lacking a spur; column rather long, dilated at apex; pollinia 2.

The whole flowering cycle, from the first emergence of the flowering stem to the dispersal of the seed is brief, occupying only a few weeks, after which the stem collapses and quickly rots away. The leaf, however, persists for several months.

A genus of about eighty species in the tropics and subtropics of the Old World, a single species being recorded.

N. aragoana *Gaud.* in Freyc., Voy. Bot.: 422, t.35 (1826). Type: Mariana Islands, *Gaudichaud* s.n. (holotype P!).

For full synonymy see Seidenfaden (1978).

A terrestrial herb. *Leaf* solitary, erect, cordate, acuminate or acute, 8 cm long, 8 cm wide; petiole 15–20 cm long, produced after the inflorescence. *Inflorescence* erect, 15–30 cm tall, elongating after fertilization, with 3 to 7 flowers; peduncle sheathed by bracts. *Flowers* pale green with a white lip, probably self-pollinating; sepals and petals similar, linear to narrowly ovate, acuminate, 1.5 cm long, 2–3 mm wide; lip pubescent, trilobed, 10.5 mm long, 6.5 mm wide, the lateral lobes erect, obliquely triangular, the mid-lobe broadly ovate; column 7 mm long.

DISTRIBUTION: The Florida Islands and Rennell. Widely distributed from Asia and the Malay Archipelago to the Mariana Islands, New Guinea, Vanuatu, New Caledonia, the Horn Islands, Fiji, Samoa, Niue Island and Australia.

HABITAT: Rain forest, on raised coral, sea level to 100 m.

COLLECTIONS: *Comins* 340 (K); *Cribb* 6013 (K); *Officers of HMS Penguin* s.n. (K).

22. DIDYMOPLEXIS Griffith

Saprophytic herbs lacking leaves. *Rhizomes* small and fleshy. *Inflorescence* a terminal raceme on an erect stem; pedicels elongating rapidly after fruit set. *Flowers* with sepals and petals joined in a short tube at the base but with petals joined to the dorsal sepal for a greater distance than to the lateral ones; lip joined to the column-foot, free from sepals and petals; column long, slightly widened at apex, the foot short; pollinia 4.

A genus of about twenty species in the Old World Tropics, a single species being recorded.

D. micradenia *(Reichb. f.) Hemsley* in Journ. Linn. Soc. 20: 311 (1883). Type: Fiji, *Seemann* 610 (holotype W, isotype K!).
Epiphanes micradenia Reichb. f. in Seem., Fl. Vit.: 295 (1868).
Didymoplexis neocaledonica Schltr. in Engler, Bot. Jahrb. 39: 50 (1906). Type: New Caledonia, *Schlechter* 15748 (holotype B†).
Didymoplexis minor subsp. *samoensis* H. Fleischm. & Rechinger in Denkschr. Akad. Wiss. Wien 85: 251, t. 1, fig. 7 (1910). Type: Samoa, *Rechinger* 1641 (holotype W).
Didymoplexis samoensis (H. Fleischm. & Rechinger) Schltr. in Fedde, Repert. Spec. Nov. Regni Veg. 9: 85 (1910).
Didymoplexis papuana Schltr. in Fedde, Repert. Spec. Nov. Regni Veg., Beih. 1: 43 (1911). Types: New Guinea, *Schlechter* 16578, 18106, 18413, 19974, 20359 (syntypes B†).
Didymoplexis fimbriata Schltr. in Engler, Bot. Jahrb. 56: 449 (1921). Type: Palau Islands, *Ledermann* 14572 (holotype B†).

A saprophytic herb. *Rhizome* fleshy, up to 8 cm long. *Stems* erect, 5–12 cm long. *Inflorescence* terminal, racemose, few-flowered. *Flowers* pale brown, olive or pinkish, with a yellowish-white lip, spotted with pink; dorsal sepal ovate, obtuse, joined to petals, 7–8 mm long, 2 mm wide; lateral sepals ovate, obtuse, connate, 6.5–7.5 mm long, 2–2.5 mm wide; petals ovate, obtuse, shorter than dorsal sepal, 7 mm long, 2 mm wide; lip erect, with margins rolled inwards enclosing the column, weakly trilobed, widest towards apex, 5.5 cm long, 3.2 mm wide, with central area of mid-lobe raised, fleshy and with verrucose calli; column 4.5 mm long. (See fig. 11 M–O).

DISTRIBUTION: Rennell and Santa Ana Island. Widely distributed from Java and New Guinea to Vanuatu, New Caledonia, Fiji, Tonga, Niue and Samoa.
HABITAT: Rain forest and secondary forest on raised coral or on peaty soil, 100 m.
COLLECTIONS: *Cribb* 6011 (K); *Dennis* 9 (K).

Because of the ephemeral nature of the flowers, specimens of *Didymoplexis* are almost always collected in fruit and are therefore difficult to identify. Consequently this species has been confused with the widespread *D. pallens* Griffith, but it differs in having a lip which is longer than broad (Kores, 1989).

Kores (1989) has established that *D. minor* J.J. Smith is not conspecific with *D. micradenia* as previously stated by Lewis & Cribb (1989).

We have included *D. papuana* as a synonym on the basis of Schlechter's drawing, and *D. fimbriata* on Schlechter's description. The holotypes of both of these names have been destroyed and we have not been able to locate any isotypes.

23. EPIPOGIUM J.G. Gmelin

Saprophytic herbs lacking leaves. *Rhizomes* fleshy. *Stems* erect, fleshy. *Inflorescence* ephemeral, brittle and if broken giving off a fungus-like odour, terminal, few- to many-flowered. *Flowers* non-resupinate, with sepals and petals of more or less equal length, narrow; lip concave, entire, with minutely warty ridges, shortly spurred; column short; pollinia 2.

A genus of two species widely distributed from western Europe to Japan, tropical Africa, SE. Asia to the Pacific and Australia, a single species being recorded.

E. roseum *(D. Don) Lindley* in Journ. Linn. Soc., Bot. 1: 177 (1878). Type: Nepal, *Wallich* (holotype BM!; isotype K!).
Limodorum roseum D. Don, Prodr. Fl. Nepal: 30 (1825).

See Clements (1989) for full synonymy.

A saprophytic herb 5–60 cm tall. *Rhizomes* embedded only shallowly in the top layer of humus, ovoid, up to 6 cm long, 4 cm wide, but usually much less, lacking roots. *Stems* leafless, brittle, pale buff to dull yellow. *Inflorescence* 5–20 cm long; axis hollow, sheathed by bracts; rhachis usually nodding at first, few- to many-flowered. *Flowers* not opening widely, cream to yellow with rose-pink spots on the lip, apparently self-fertilising; sepals similar, narrowly oblong, acute, 7–12 mm long, 1–2 mm wide; petals slightly shorter and broader than sepals; lip triangular-obovate, concave, more or less enclosing the column, 1 cm long, 6–7 mm wide, the upper surface minutely warty in 2 rows which converge towards the apex, the margins erose, decurved at apex, spurred, the spur cylindrical, obtuse, 2–4 mm long; column 2–2.5 mm long. (See fig. 11 A–H).

DISTRIBUTION: Widely distributed from Africa, Asia and the Malay Peninsula to New Guinea, New Caledonia and Australia.

HABITAT: Variety of habitats from bush to rain forest and from the coast to hills.

COLLECTIONS: No specimens seen.

Although we have not seen any collections of *E. roseum* from the Solomon Islands and Bougainville we have included it because it is found in New Guinea,

---

Fig. 11. Epipogium roseum A-H. A, habit x 2/3; **B**, flower x 3; **C**, column and spur x 6; **D**, anther-cap x 8; **E**, lip x 4; **F**, dorsal sepal x 4; **G**, petal x 4; **H**, lateral sepal x 4. **Stereosandra javanica J-L. J**, habit x 2/3; **K**, flower x 3; **L**, lip x 6. **Didymoplexis micradenia M-O. M**, habit x 2/3; **N**, flower x 3; **O**, lip x 6. **A-H** drawn from *Bregulla* s.n.; **J-L** from *Mitchell* 35; **M-O** from *Cribb & Wheatley* 3. All drawn by Sue Wickison.

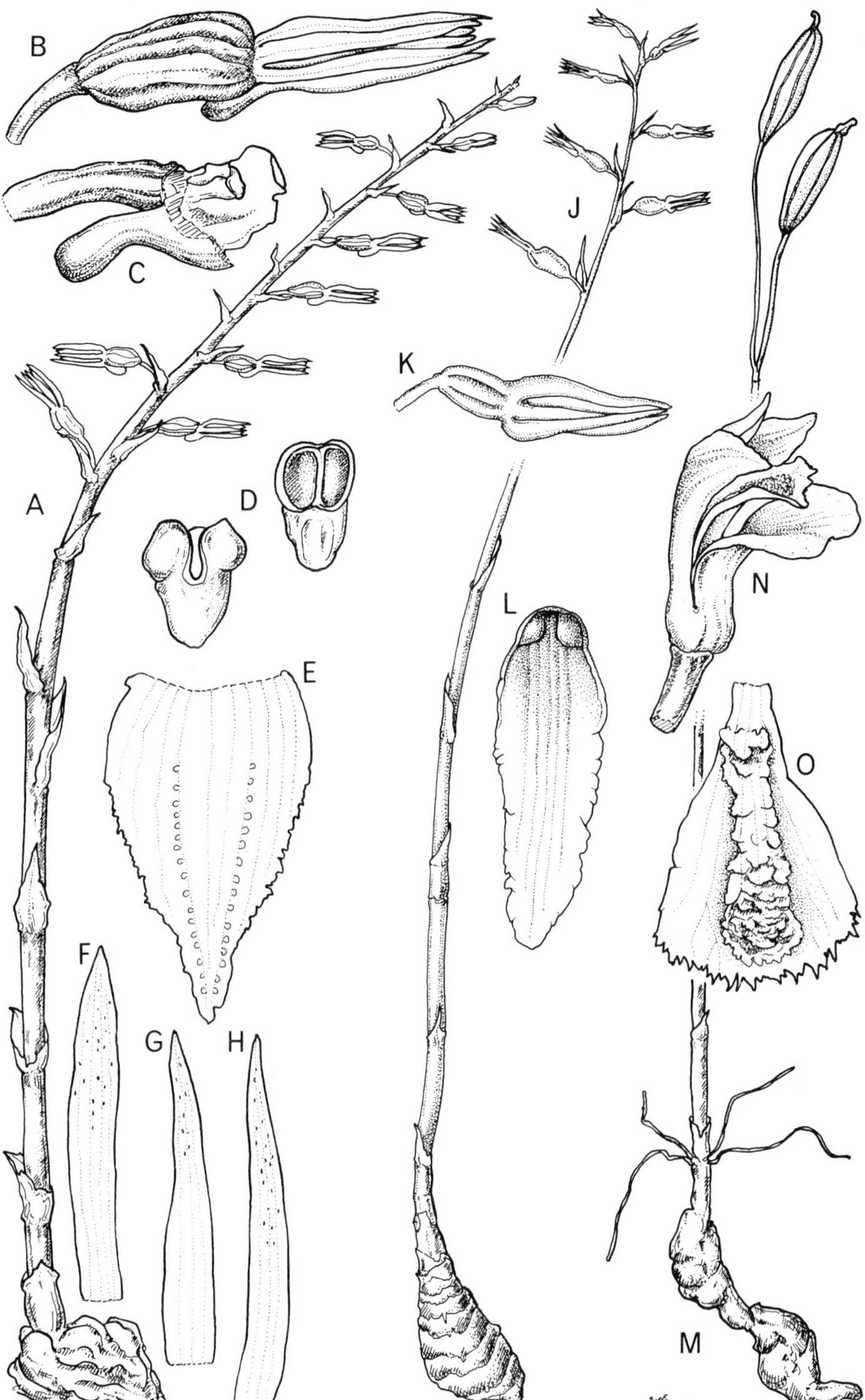
B
C
A
D
E
F
G
H
J
K
L
N
O
M

Australia, Vanuatu and New Caledonia and we anticipate that it will be found in the islands.

(For a discussion on the biology of *E. roseum* see Docters van Leeuwen, 1937.)

24. STEREOSANDRA Blume

Saprophytic herbs lacking leaves. *Rhizomes* tuberous. *Stems* erect. *Inflorescence* terminal, few- to many-flowered. *Flowers* with sepals and petals more or less equal, narrow; lip narrow, entire, with 2 glands at base; column short; anther erect on a broad filament rising from the back of the column; pollinia 2.

A small genus of about five species from SE. Asia and New Guinea, a single species being recorded.

S. javanica *Blume* in Mus. Bot. Lugd. Bat. 2: 176 (1856). Type: Java, *Blume* s.n. (holotype L!).

A saprophytic herb, 20–25 cm tall, arising from a fleshy, erect rhizome. *Rhizome* cylindrical to ovoid, 2–3.5 cm long, 0.4–1.3 cm wide, covered with scaly bracts. *Stems* erect, pale buff. *Inflorescence* very short-lived, terminal, laxly few-flowered; rhachis 3–5 cm long; peduncle slender, 14–16 cm long; bracts sheathing, lanceolate, acuminate, 3–6 mm long. *Flowers* white with tips of sepals and petals purple and with a white lip tinged with purple; pedicel and ovary purple-striped, 2–3 mm long; dorsal sepal lanceolate, acute, 8 mm long, 2 mm wide; lateral sepals lanceolate, acute, 8 mm long, 2 mm wide; petals lanceolate, acute, 7.5 mm long, 1.5 mm wide; lip cymbiform, obtuse, 8–9 mm long, 2.5 mm wide, with erose margins and 2 fleshy round calli at base; column 4 mm long. *Fruit* ellipsoidal, 7 mm long with the perianth persistant. (See fig. 11 J–L).

DISTRIBUTION: Choiseul. Also in the Malay Peninsula, Borneo, Java, Sumatra and New Guinea.

HABITAT: In deep leaf litter in dense shade under palms and ferns in clay soil, 600 m.

COLLECTION: *Mitchell* 35 (K).

See Mitchell (1989) for further details of this collection.

*25. ARUNDINA Blume

Terrestrial herbs. *Stems* erect, narrowly terete, leafy. *Leaves* distichous, narrow, lanceolate. *Inflorescence* terminal, few-flowered, the flowers borne on spreading branches of the lax panicle. *Flowers* opening 1 or 2 at a time, medium-sized, showy; sepals equal; petals broader than the sepals; lip trilobed, somewhat tubular; column long and straight, expanded in the apical part; pollinia 8.

A small genus of about five species native to Ceylon, India, China to Malaysia and Indonesia. Some of these species are also widely cultivated and a few have become naturalised elsewhere. A single species is recorded.

*__A. graminifolia__ *(D. Don) Hochr.* in Bull. New York Bot. Gard. 6: 270 (1910). Type: Nepal, *Buchanan- Hamilton* s.n. (holotype BM!).
Cymbidium bambusifolium Roxb., Hort. Beng. 63 (1814), nom. nud.; Fl. Ind. ed. 2, 3: 460 (1832), nom. illeg.
Bletia graminifolia D. Don., Prodr. Fl. Nepal.: 29 (1825).
Arundina speciosa Blume, Bijdr. Fl. Ned. Ind.: 401 (1825). Type: Java, *Blume* (holotype L!).

A terrestrial herb up to 1 m high. *Stems* erect, terete, leafy. *Leaves* distichous, narrowly oblong, acute, 15–30 cm long, 0.8–1.5 cm wide, sheathing at base. *Inflorescence* terminal, with few, large, showy flowers opening 1 or 2 at a time. *Flowers* rosy-purple with a deeper purple apex to the lip; sepals equal, oblong-lanceolate, 3.5–4 cm long, 1–1.2 cm wide; petals rhombic-elliptic, acute, 3.5–4 cm long, 2.3 cm wide; lip trilobed, 4.5 cm long, 2.5–3 cm wide, the lateral lobes inrolled, enclosing the column, the mid-lobe divided into two rounded lobules, the disc with 2–3 keels; column 2–2.2 cm long, straight with an expanded apical portion.

DISTRIBUTION: Bougainville (introduced). Distributed in Sri Lanka, northern India, southern China and Malaysia as far east as Sulawesi. The species has become widely naturalised in the Pacific.
HABITAT: Planted as an ornamental.
COLLECTION: *Paimuru* in BCL 377 (LAE).

This species has been planted as an ornamental at Arawa and Panguna in Bougainville.

26. ACANTHEPHIPPIUM Blume

Terrestrial herbs. *Pseudobulbs* few-noded, fleshy, subcylindrical, green, covered by sheaths when immature, 2- to 3-leaved at apex. *Leaves* suberect, large, plicate. *Inflorescence* lateral, shorter than the leaves, few-flowered. *Flowers* erect, large, urn-shaped; sepals fleshy, connate with free apices; lateral sepals forming a chin-like mentum with the column-foot; petals included within the sepals, much narrower; lip trilobed, articulated with the column-foot; column with a long foot; pollinia 8.
A genus of about fifteen species from tropical Asia to the Pacific Islands, a single species being recorded.

A. papuanum *Schltr.* in Fedde, Repert. Spec. Nov. Regni Veg., Beih. 1: 371 (1912). Type: New Guinea, *Schlechter* 16894 (holotype B†, isotypes AMES!, L, NSW).
Acanthephippium vitiense L.O. Williams in Amer. Orchid Soc. Bull. 10: 169 (1941). Type: Fiji, *Degener* 14605 (holotype AMES!).

See Kores (1989) for synonmy and a detailed explaination of the status of this species.

A terrestrial herb. *Pseudobulbs* few-noded, fleshy, subcylindric, up to 26.5 cm long, 2-leaved at apex. *Leaves* plicate, lanceolate, up to 65 cm long, 11 cm wide, long-petiolate. *Inflorescence* lateral, up to 25 cm long, erect, 3- to 6-flowered; bracts lanceolate, up to 3.5 cm long. *Flowers* large, urn-shaped, yellow lined and flushed with red, sweetly scented; sepals and petals connate with free apices; dorsal sepal ovate, 3.5 cm long, 1.5 cm wide; lateral sepals obliquely ovate, 3.5 cm long, 2 cm wide, forming a 1.4 cm long mentum; petals spathulate, 3.2 cm long, 1 cm wide; lip trilobed, articulated with the column-foot, 1.6 cm long, 1.6 cm wide, the lateral lobes erect, obliquely subquadrate, the mid-lobe obovate, obtuse, recurved at apex, the callus obscure, of 3 interrupted ridges; column 1.5 cm long, with an erect, 1.3 cm long column-foot.

DISTRIBUTION: Guadalcanal. Also in New Guinea, Vanuatu, New Caledonia, Fiji, Tonga and the Horne Islands.
HABITAT: Ridge-top forest, 500–1000 m.
COLLECTION: *Cribb & Morrison* 1868 (K).

Kores (1989) notes that this species is similar to *A. splendidum* J.J. Smith, from New Guinea, and *A. javanicum* Blume, a widespread species in other parts of Malesia, and it should be critically compared with them.

27. CALANTHE R. Brown

Terrestrial herbs with thick, long, hairy roots. *Stems* pseudobulbous, often clustered and hidden by leaf bases. *Leaves* persistent or rarely deciduous, distichous, lanceolate to broadly lanceolate, up to 1 m long, plicate, petiolate. *Inflorescences* arising from the base of the pseudobulb, erect, long or short, unbranched, few- to many-flowered; bracts persistent or deciduous. *Flowers* often showy, usually turning blue when damaged; sepals and petals similar, free; lip entire or 3- to 4-lobed, adnate to column at base, with or without a basal callus, spurred; column short, fleshy; pollinia 8.

A genus of about one hundred species in the tropics and warm temperate regions distributed from Africa to Asia and the Pacific Islands, with a single species in the tropical Americas. Five species are recorded.

1. Bracts longer than the flowers and almost hiding them; lip entire, orange-red; sepals and petals white .... **1. C. apostasioides**
   Bracts shorter than the flowers and not hiding them; lip three- or four-lobed 2
2. Lip four-lobed 3
   Lip three-lobed 4
3. Sepals and petals pale green, 6–7 mm long; lip with a hairy basal callus, the spur 6 mm long .... **3. C.** aff. **torricellensis**
   Sepals and petals white, more than 8 mm long; lip with a glabrous callus, the spur more than 1.5 cm long .... **4. C. triplicata**
4. Flowers yellow or orange-yellow; spur clavate, 5–7 mm long, appressed to ovary .... **5. C. ventilabrum**
   Flowers white; spur subclavate, c. 1 cm long, slightly S-shaped .... **2. C. hololeuca**

1. C. apostasioides *Schltr.* in Fedde, Repert. Spec. Nov. Regni Veg., Beih. 1: 377 (1912). Type: New Guinea, *Schlechter* 18731 (holotype B†).

A terrestrial herb up to 40 cm tall. *Stems* pseudobulbous, hidden by leaf bases. *Leaves* plicate, distichous, narrowly lanceolate, up to 28 cm long, less than 2 cm wide, narrowed at base to form a slender, 4 cm long petiole. *Inflorescence* lateral, arising from the base of the pseudobulb, short, 7–9 cm long, densely many-flowered; bracts longer than the flowers, c. 4 cm long, 7–8 mm wide, white veined with green. *Flowers* white or cream with an orange-red lip; dorsal sepal ovate, 1 cm long, 4 mm wide; lateral sepals obliquely ovate, 8–9 mm long, 3 mm wide; petals obliquely ovate, 8 mm long, 3 mm wide; lip entire, oblong, 5 mm long, 1.5–2 mm wide, triangular at recurved apex, the spur cylindrical, obtuse, 4.5–7 mm long, 2 mm wide; column 3 mm long. (See fig. 12 A–K).

DISTRIBUTION: Guadalcanal. Also in New Guinea.
HABITAT: Mossy forest, 2000–2300 m.
COLLECTION: *Hunt* in RSS 2955 (K).

2. C. hololeuca *Reichb. f.* in Seem., Fl. Vit.: 298 (1868). Type: Fiji, *Seeman* 607 (holotype W, isotype K!).
Calanthe neohibernica Schltr. in K. Schum. & Laut., Nachtr. Fl. Deutsch. Südsee: 142. (1905) Type: New Ireland, *Schlechter* 14707 (holotype B†).
Calanthe vaupeliana Kraenzl. in Notizbl. Bot. Gart. Berlin 5: 111 (1909). Type: Samoa, *Vaupel* 358 (holotype B†, isotypes AMES!, K!).
Calanthe neocaledonia Rendle in Journ. Linn. Soc., Bot. 45: 251 (1921). Types: New Caledonia, *Compton* 1409 & 1609 (syntypes BM!).

A terrestrial plant up to 70 cm tall. *Stems* pseudobulbous, hidden by leaf bases. *Leaves* plicate, distichous, lanceolate, up to 65 cm long, 7.5 cm wide, the petiole up to 29 cm long. *Inflorescence* lateral, arising from the base of the pseudobulb, up to 40 cm long, erect, densely or subdensely many-flowered, the basal flowers opening first; the rhachis 12 cm long; peduncle sheathed in bracts; bracts deciduous, white. *Flowers* white, sometimes with green lateral lobes to the lip; dorsal sepal ovate, 1.1–1.2 cm long, 5.5–8 mm wide; lateral sepals ovate, 11–12 mm long, 5–6 mm wide; petals ovate, 10–11 mm long, 5–6 mm wide; lip trilobed, 7 mm long, 4 mm wide, the lateral lobes auriculate, the mid-lobe oblong, obtuse, fleshy at base, the spur slightly S-shaped, 1 cm long, 2–3 mm wide; column 4 mm long. (See fig. 13 M–O and plate 4B).

DISTRIBUTION: Guadalcanal, New Georgia and the Santa Cruz Islands (Vanikoro). Also in New Ireland, Vanuatu, New Caledonia, the Horne Islands, Fiji, Tonga and Samoa.
VERNACULAR NAME: Qola ase (Kusage language, New Georgia).
HABITAT: Rain forest, 260–500 m.
COLLECTIONS: *Dickson* A3 (K); *Douglas* K15 (K); *Hadley* 25 (BM); *Hunt* in RSS 2046 (K); *Wickison* 45 & 124 (K).

3. C. aff. **torricellensis** *Schltr.* in K. Schum. & Laut., Nachtr. Fl. Deutsch. Südsee: 144 (1905). Type: New Guinea, *Schlechter* 14498 (holotype B†).

A terrestrial plant up to 60 cm tall. *Stems* pseudobulbous, hidden by leaf bases. *Leaves* plicate, distichous, oblanceolate, acuminate, up to 29 cm long, 5.5 cm wide, petiole 5–7 cm long. *Inflorescence* lateral, arising from the base of the pseudobulb, up to 60 cm long, laxly many-flowered; bracts persistent, lanceolate, up to 1.3 cm long, 3 mm wide. *Flowers* pale green with a cream-coloured lip and column, pubescent on outer surface; pedicel and ovary pubescent, 8–10 mm long; dorsal sepal ovate, 6–7 mm long, 4 mm wide; lateral sepals ovate, 6–7 mm long, 3–4 mm wide; petals oblong-ovate, 6–7 mm long, 3 mm wide; lip 4-lobed, 6–8 mm long, 1–1.4 cm wide, the lateral lobes spreading, rounded, the middle lobes sub-parallel, united for half their length, with a short pubescent basal callus, the spur straight, 6 mm long, 1–2 mm wide; column 2 mm long.

DISTRIBUTION: Guadalcanal. Also in New Guinea (see note).
HABITAT: Mossy forest, 2000–2200 m.
COLLECTION: *Hunt* in RSS 2956 (K).

This collection differs from the type illustration of *C. torricellensis* in having parallel middle-lobes to the lip and a shorter pubescent callus. More material is needed of this species to confirm the identification.

4. C. triplicata *(Willemet) Ames* in Philipp. Journ. Sci., Bot. 2: 326 (1907). Type: Amboina, *Rumphius* s.n. (holotype L-lost, lectotype illustration in Rumphius, Herb. Amb. 6: t.52, f.2!).
Flos triplicatus Rumph., Herb. Amb. 6: 115, t.52 f.2. (1750) nom. illeg. pre 1753.
Orchis triplicata Willemet, in Usteri, Ann. Bot. 18: 52 (1796).

For full synonymy see Clements (1989).

A terrestrial plant up to 1.3 m tall. *Stems* pseudobulbous, hidden by leaf bases. *Leaves* plicate, distichous, lanceolate, up to 75 cm long, 16 cm wide, dark green sometimes with a paler venation; petiole up to 27 cm long. *Inflorescence* lateral, arising from the base of the pseudobulb, pyramidal to cylindric, up to 1.3 m long, many-flowered; bracts persistent, lanceolate, 1.5 cm long, 5 mm wide. *Flowers* white with a yellow or red basal callus on the lip, rarely cleistogamous, the sepals and petals finely pubescent on outer surfaces; dorsal sepal ovate, 8–14 mm long, 5–6 mm wide; lateral sepals ovate, acuminate, 9.5–14 mm long, 5–6 mm wide; petals ovate-elliptic, 8.5–14 mm long, 4–5.5 mm wide; lip 4-lobed, 1–2 cm long, 1.3–2.3 cm wide, the lobes subequal, oblong, rounded, the basal callus of 3–5 tapering verrucose ridges, the spur decurved, cylindrical, 1.5–4 cm long, 1–3 mm wide; column 4–6 mm long. (See fig. 12 L–M and plate 4C).

Fig. 12. Calanthe apostasioides. A-K. A, habit x 2/3; **B**, flower x 4; **C**, side view of lip, spur and column x 4; **D**, pollinia x 8; **E**, dorsal and ventral view of anther-cap x 8; **F**, dorsal sepal x 4; **G**, petal x 4; **H**, lateral sepal x 4; **J**, column x 12; **K**, lip x 6. **C. triplicata L-M. L**, flower x 5/3; **M**, lip x 5/3. **A-K** drawn from *Hunt* 2955; **L-M** from *Wickison* 245. All drawn by Sue Wickison.

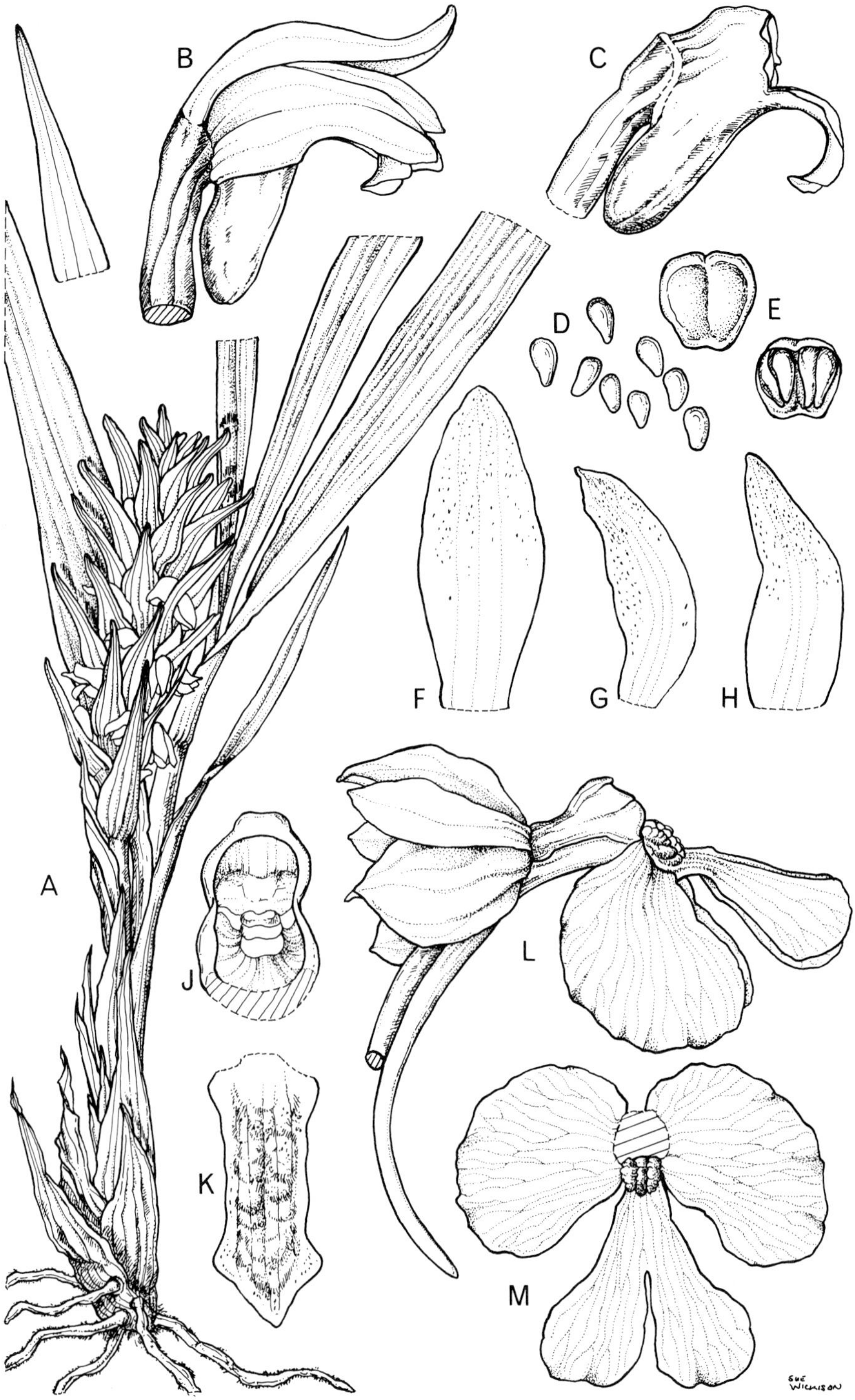
A
B
C
D
E
F
G
H
J
K
L
M
SUE WICKISON

DISTRIBUTION: Choiseul, Guadalcanal, Malaita and Santa Isabel. Widely distributed from Madagascar, Asia to the Malay Archipelago, New Guinea, Vanuatu, Guam, Fiji, New Caledonia, Lord Howe Island and Australia.

HABITAT: Rain forest, 300–800 m.

COLLECTIONS: *Brass* C (G); *Cribb, Campbell & Dennis* 5002 (K); *Corner* 3035 (K); *Dennis* s.n. & in RSS 2103 (K); *Mitchell* 36 (K); *Wickison* 245 & 246 (K).

Calanthe triplicata is a variable and widespread species which would benefit from a detailed systematic study.

5. C. ventilabrum *Reichb. f.* in Seem., Fl. Vit.: 298 (1868). Type: Fiji, *Seemann* 606 (holotype W-probably lost, isotype K!).

Calanthe langei F. Muell. in Wing's South Sci. Rec.: 1. (1885). Type: New Caledonia, cult. *Lange* s.n. (holotype MEL).

Calanthe chrysantha Schltr. in K. Schum. & Laut., Nachtr. Fl. Deutsch. Südsee: 141 (1905). Type: New Guinea, *Schlechter* 14494 (holotype B†).

Calanthe bigibba Schltr. in Fedde, Repert. Spec. Nov. Regni Veg. 9: 99 (1911). Type: Samoa, *Vaupel* 359 (holotype B†, isotype AMES!).

A terrestrial herb up to 1 m tall. *Stems* pseudobulbous, hidden by leaf bases. *Leaves* plicate, distichous, lanceolate, 40–100 cm long, 5–9.5 cm wide, the petiole up to 15 cm long. *Inflorescence* lateral, arising from the base of the pseudobulb, up to 55 cm long; rhachis 6–7 cm long, densely many-flowered; bracts deciduous. *Flowers* yellow to orange; dorsal sepal ovate, 1 cm long, 6 mm wide; lateral sepals ovate, 1 cm long, 5–6 mm wide; petals elliptic, 1–1.1 cm long, 5–6 mm wide; lip trilobed, 5.5–6.5 mm long, 4–5 mm wide, the lateral lobes small and auriculate, the mid-lobe rounded and obtuse, the basal callus of 2 short, low fleshy ridges, the spur clavate, parallel to ovary, 5–7 mm long, 2 mm wide; column 4 mm long. (See fig. 13 A–L and plate 4A).

DISTRIBUTION: Bougainville, Guadalcanal, Kolombangara, New Georgia and the Santa Cruz Islands (Vanikoro). Also in New Guinea, Vanuatu, New Caledonia, Fiji and Samoa.

HABITAT: Montane rain forest, 580–1350 m.

COLLECTIONS: *Chaplin* 845 (K); *Cribb* S51 (K); *Hadley* 24 (BM); *Hunt* in RSS 2035 (LAE) & 2480 (K); *Lavarack & Ridsdale* in NGF 31390 (K, LAE) & 31411 (LAE); *Schodde & Craven* 3717 (K); *Wickison* 36 (K).

We have included *C. chrysantha* as a synonym based on Schlechter's published illustration and description.

Fig. 13. Calanthe ventilabrum A-L. A, habit x 2/3; **B**, inflorescence x 2/3; **C**, flower x 3; **D**, lip, spur and column x 4; **E**, dorsal and ventral view of anther-cap x 8; **F**, column x 8; **G**, pollinia x 8; **H**, lip x 6; **J**, dorsal sepal x 4; **K**, petal x 4; **L**, lateral sepal x 4. **C. hololeuca M-O. M**, flower x 3; **N**, spur, lip and column x 3; **O**, lip x 4. **A-L** drawn from *Wickison* 36; **M-O** from *Wickison* 124. All drawn by Sue Wickison.

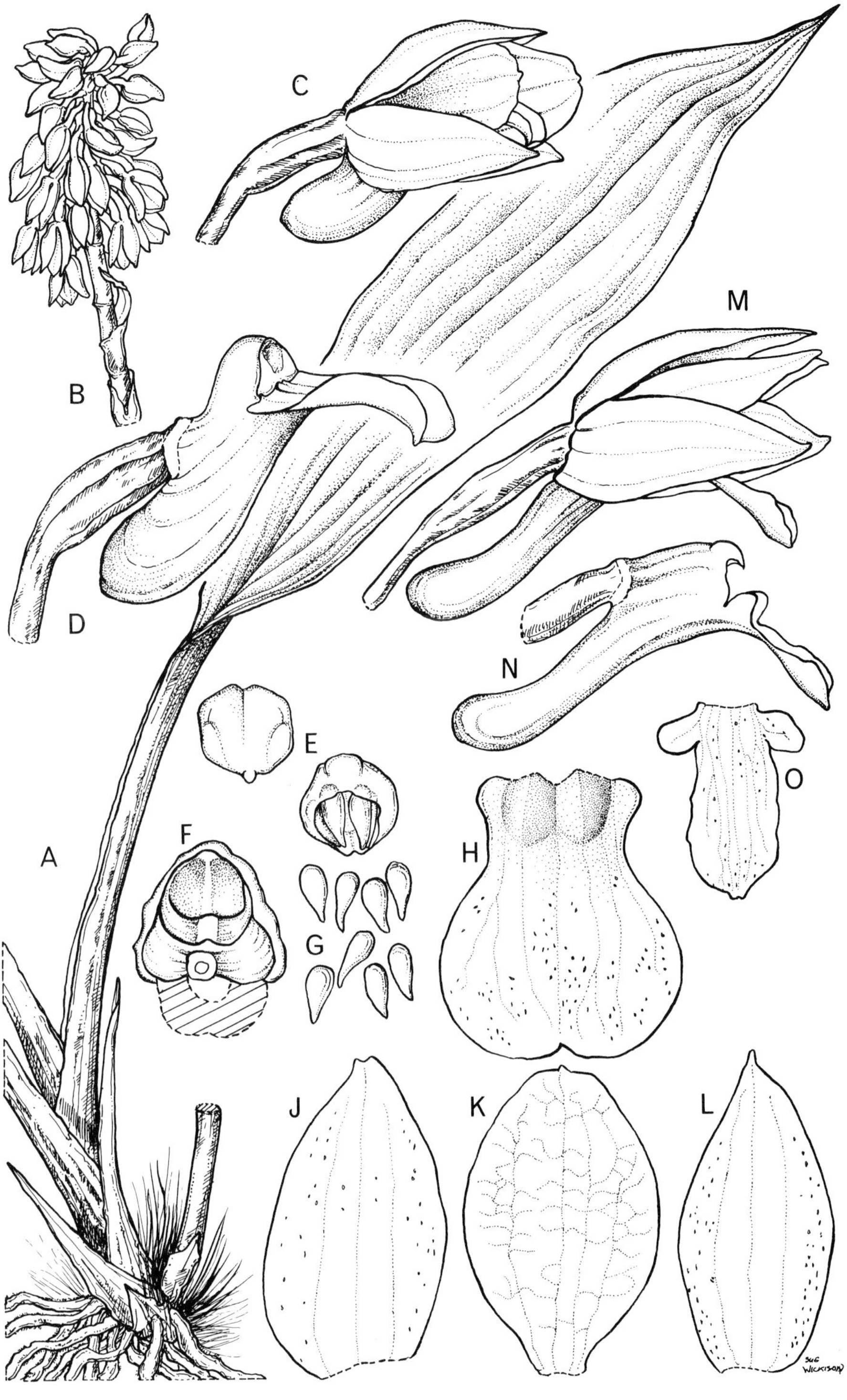

A
B
C
D
E
F
G
H
J
K
L
M
N
O

28. **PHAIUS** Loureiro

Fairly large terrestrial herbs. *Stems* clustered, short, elongate, cylindrical or pseudobulbous. *Leaves* plicate, distichous, several, narrowly obovate, acuminate, medium to large in size, petiolate. *Inflorescence* arising from the base of the pseudobulb or lateral from the side of the stem, erect, loosely racemose. *Flowers* rather large and showy, turning blue when damaged; sepals and petals similar, free; lip free, obscurely trilobed, with the lateral lobes inrolled to form a tube which enfolds the column, the base of the lip shortly spurred or ± saccate; column long, slender and curved; pollinia 8.

A genus of about forty species in the Old World Tropics, two species being recorded.

Stem elongate; peduncle produced from side of stem; flowers white, with a yellow lip which may be marked with red-purple
.... 1. **P. amboinensis** var. **papuanus**

Stem thickened into a short, many-jointed pseudobulb; peduncle arising from the base of the pseudobulb; flowers pale rose to brown-orange, with a white lip tinged with rose to purple...**2. P. tankervilleae**

1. P. amboinensis *Blume*, Mus. Bot. Lugd. Bat. 2: 180 (1856). Type: Ambon, *Blume* s.n. (holotype L-lost, lectotype illustration in Rumphius, Herb. Amb.!).

var. **papuanus** *(Schltr.) Schltr.* in Fedde, Repert. Spec. Nov. Regni Veg., Beih. 1: 375 (1912). Type: New Guinea, *Schlechter* 14595 (holotype B†).

Phaius papuanus Schltr. in K. Schum. & Laut., Nachtr. Fl. Deutsch. Südsee: 139 (1905).

Phaius graeffei Reichb. f. in Seem., Fl. Vit.: 299 (1868). Type: Samoa, *Graeffe* s.n. (holotype W).

A terrestrial herb. *Stems* clustered, elongate, terete, up to 1.3 m tall, 1–1.3 cm diameter at the base. *Leaves* plicate, 4–6, distichous, erect, lanceolate, acuminate, 20–60 cm long, 5–8 cm wide; petiole 15–22 cm long, sheathing at base. *Inflorescences* lateral, from side of stem, up to 1 m long, laxly few-flowered. *Flowers* creamy-white with a yellow lip which may be marked with red-purple; dorsal sepal oblong-ovate, 2.6–3.5 cm long, 8 mm wide; lateral sepals narrowly elliptic, 2.7–3.7 cm long, 1.2 mm wide; petals linear, 2.7–3.4 cm long, 5 mm wide; lip trilobed in apical half, 3–3.4 cm long, 2.5–3.4 cm wide, finely pubescent, the lateral lobes obliquely oblong, rounded, the mid-lobe small, rounded, with a central callus of 2 pubescent, thickened ridges in the centre of the lip, lacking a spur; column 2.3 cm long. (See fig. 14 A–K and plate 3F).

Fig. 14. Phaius amboinensis var. **papuanus A-K. A**, habit x 1/9; **B**, flower x 2/3; **C**, side view of lip x 4/3; **D**, lip x 4/3; **E**, dorsal sepal x 4/3; **F**, petal x 4/3; **G**, lateral sepal x 4/3; **H**, column x 4/3; **J**, pollinia x 3; **K**, dorsal and ventral view of anther-cap x 3. **Phaius tankervilleae. L**, flower reduced. **A & C-K** drawn from *Wickison* 92; **B** drawn from slide; **L** drawn from slide. All drawn by Sue Wickison.

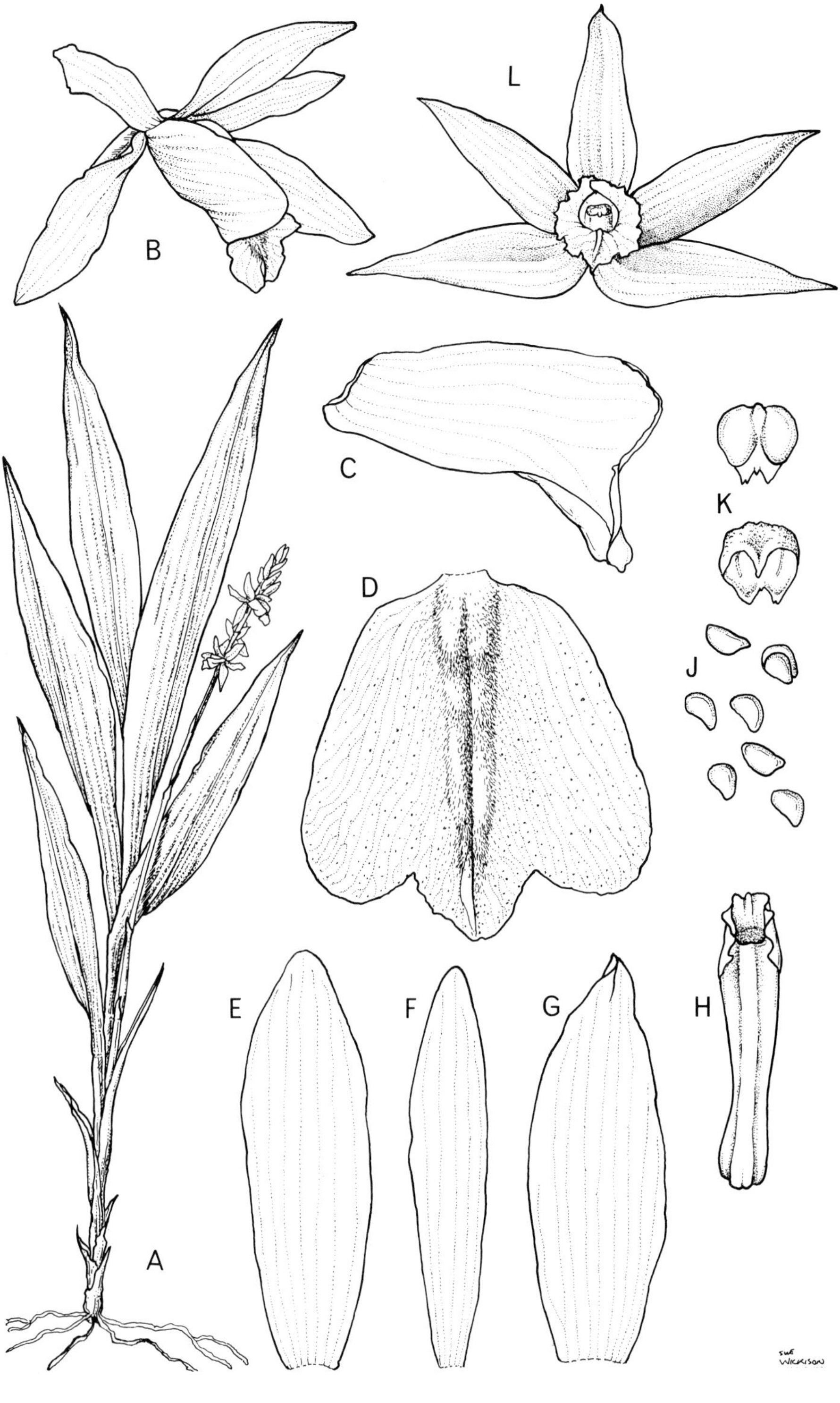
B
L
C
K
D
J
E
F
G
H
A

DISTRIBUTION: Bougainville (Buka Island), Choiseul, Guadalcanal, Kolombangara, New Georgia. Also in New Guinea, Vanuatu, Fiji and Samoa.
HABITAT: Rain forest, 80–600 m.
COLLECTIONS: *Eadley* 3 (K); *Mauriasi et al.* in BSIP 8806 (K); *Waterhouse* 48 (K); *Wickison* 92 (K).

Herbarium specimens from Samoa which agree well with the description of *P. graeffei* cannot be distinguished from *P. amboinensis* var. *papuanus*. *Phaius neocaledonicus* Rendle, from New Caledonia is similar and with further study may prove to be conspecific.

2. P. tankervilleae *(Banks ex L'Her.) Blume*, Mus. Bot. Lugd. Bat. 2: 177 (1856).
Type: ex China, cult. Kew, 1791, *herb. J.E. Smith* 1403.1 (lectotype LINN).
Limodorum tankervilleae Banks ex L'Her. in Sert. Angl.: 28 (1789) (as *L. tancarvilleae*).

For full synonymy and discussion see Clements (1989).

A terrestrial herb. *Stems* thickened into short, many-jointed pseudobulbs, 6–7 cm long, 1 cm wide. *Leaves* plicate, 3–5, suberect, lanceolate or oblanceolate, acuminate, 30–80 cm long, 4–12 cm wide. *Inflorescences* lateral, from base of pseudobulb, up to 1.5 m tall, laxly few-flowered. *Flowers* white on exterior of sepals and petals, tinged pale rose or yellow-brown inside and with a yellow-white lip tinged rose to purple, sometimes self-pollinating; dorsal sepal oblong-elliptic, obtuse, 4.2 cm long, 1 cm wide; lateral sepals oblong to narrowly elliptic, obtuse, 4.3 cm long, 1.5 cm wide; petals oblanceolate, obtuse, 4 cm long, 1.3 cm wide; lip trilobed, 4.3 cm long, 3.3 cm wide, the lateral lobes obtuse, oblong, rounded, with erose margins, the mid-lobe transversely oblong, shortly spurred at base; column 2 cm long. (See fig. 14 L).

DISTRIBUTION: Bougainville. Widely distributed from Asia and the Malay Archipelago to New Guinea, Vanuatu, New Caledonia, the Horn Islands, Fiji, Samoa and Australia.
HABITAT: Edge of cliff in rain forest, 900–1000 m.
COLLECTIONS: *Lavarack & Ridsdale* in NGF 30579 & 36567 (K).

29. SPATHOGLOTTIS Blume

Terrestrial herbs. *Pseudobulbs* clustered, small to large. *Leaves* plicate, lanceolate to broadly lanceolate, up to 1 m long, petiolate. *Inflorescences* lateral, arising from a basal leaf axil, tall and slender, with a few sheaths along its length; rhachis bearing a succession of many flowers. *Flowers* pink, purple or white; pedicel long and slender; sepals and petals free, about equal, spreading widely, often finely pubescent on exterior surface; lip trilobed, with a callus at the base of the mid-lobe, lacking a spur; column slender, curved, without a foot; pollinia 8.

A genus of about forty species from tropical Asia to Australia and the Pacific Islands, two species being recorded.

Lip mid-lobe as wide as and shorter than the lateral lobes; basal callus finely pubescent to strongly hairy .... **1. S. pacifica**
Lip mid-lobe narrower and longer than the lateral lobes; basal callus glabrous .... **2. S. plicata**

1. S. pacifica *Reichb.f.* in Seem., Fl. Vit.: 300 (1868). Types: Fiji, *Seemann* 585 & s.n. (syntypes W), & *Harvey* s.n. (syntype W, isosyntypes AMES, K!).

See Cribb & Tang (1982) for full synonymy.

A terrestrial plant. *Pseudobulbs* ovoid. *Leaves* plicate, lanceolate to broadly lanceolate, acuminate, 25–100 cm long, 1.5–5 cm wide, with 6–25 cm long petioles. *Inflorescences* lateral from a basal leaf axil, up to 1 m long, pubescent; raceme densely many-flowered. *Flowers* white to pink with a white or yellow callus on the lip; exterior of sepals and petals finely pubescent; pedicel and ovary c. 4 cm long, densely covered with brown hairs; dorsal sepal narrowly obovate, 1.6–2.2 cm long, 0.8–1 cm wide; lateral sepals ovate, 1.6–2.2 cm long, 9 mm wide; petals ovate, 1.6–1.8 cm long, 8–9 mm wide; lip T-shaped when flattened, c. 6 mm long, 4 mm wide, with the lateral lobes making a right-angle with the longitudinal axis of the lip and slightly curved forward in apical half, the mid-lobe c. 1 cm long, with the apical lamina transversely oblong with a crisped margin, 0.6–1 cm wide, the base with 2 long, attenuate, hairy, fleshy calli, extending from the base of the claw onto the mid-lobe; column slender, curved, 1.2 cm long. (See fig. 15 K–L).

DISTRIBUTION: Kolombangara, Krihi Island (off Santa Isabel), New Georgia and Rennell. Also in Vanuatu, the Horn Islands, Fiji, Samoa, Tahiti and Tonga.
HABITAT: Rain forest, sea level to 400 m.
COLLECTIONS: *Cribb* 6049 (K); *Cribb & Campbell* 5102 (K); *Dickson* MP1 (K); *Hunt* in RSS 2733 (K).

2. S. plicata *Blume*, Bijdr. Fl. Ned. Ind. 5: 401 (1825). Type: Java, *Blume* s.n. (holotype L, isotype P!).
Bletia angustifolia Gaud. in Freyc., Voy. Bot.: 421 (1829). Type: Moluccas, *Freycinet* s.n. (holotype P!).
Spathoglottis angustifolia (Gaud.) Benth. & Hook. f., Gen. Pl. 3: 512 (1883).
Spathoglottis daenikeri Kraenzl. in Viertelj. Nat. Ges. Zür. 74: 80 (1929). Type: New Caledonia, *Daeniker* 1622 (holotype Z!).
Spathoglottis vieillardii Reichb. f. in Linnaea 41: 85 (1877). Type: New Caledonia, *Vieillard* 1302 (holotype P!).
Spathoglottis deplanchei Reichb. f., l.c.: 86 (1877). Type: New Caledonia, *Deplanche* 1302 (holotype K!).

See Cribb & Tang (1982) for more information.

A terrestrial plant. *Pseudobulbs* ovoid or slender. *Leaves* plicate, lanceolate to broadly lanceolate, acuminate, 50–90 cm long, 2–6 cm wide, attenuate and tapered into petiole at the base. *Inflorescence* lateral from a basal leaf axil, 60–100 cm long, hairy, few- to many-flowered. *Flowers* white to pinkish-purple with a yellow callus on the lip; finely pubescent on outer surfaces of sepals and petals; pedicel and ovary 3–5 cm long; dorsal sepal narrowly ovate to ovate, 1.6–3 cm long, 1–1.3 cm wide; lateral sepals obliquely ovate, acute, 1.6–2.8 cm long, 1–1.2 cm wide; petals ovate, 2–3 cm long, 1.3–1.8 cm wide; lip T-shaped when flattened, the lateral lobes at right-angles to mid-lobe, oblong, 7–8 mm long, the mid-lobe much longer than lateral lobes, 1–1.3 cm long, with a long narrow claw, dilated at the apex, the basal callus fleshy, glabrous, bipartite, with each lobe triangular-ovoid, attenuated in front, with 2 usually hairy triangular lateral teeth in front of the calli; column slender, curved, 1.2–1.3 cm long. (See fig. 15 A–H and plate 3C).

DISTRIBUTION: Bougainville, Guadalcanal, Kolombangara, Malaita, New Georgia, Rennell, Santa Ana, Santa Cruz Islands (Nendo), Treasury Island. Widely distributed from Asia and the Malay Penisula to New Guinea, Vanuatu, New Caledonia, the Horn Islands, Samoa, Tonga and Niue.

VERNACULAR NAMES: Tongoruha (Bougainville), Molamola (New Georgia) and Laulau Ngwane (Kwara'ae name).

USES: Hancock and Henderson (1988) note that this species is used as a medicinal plant.

HABITAT: Occurring in a variety of habitats (see Dennis, 1976), sea level to 1500 m.

COLLECTIONS: *Comins* 230 (K); *Cribb* S25 (K); *Danish Noona Dan Expedition* 2877 (K); *Dennis* in RSS 2356 & 11, 14 & 15 (K); *Guppy* s.n. (K); *Hunt* in RSS 2094, 2410, 2411 & 2939 (K); *Lever* s.n. (K); *Officers of HMS Penguin* s.n. (K); *Ridsdale & Lavarack* in NGF 30551 & 30614 (K); *Smith & Newmarch* 4/005 (K); *Waterhouse* 91 & 162 (K).

Spathoglottis plicata has long been cultivated, and several varieties, including white ones, have been selected. It has also been introduced into the Hawaiian Islands and elsewhere in SE. Asia, Africa and the Americas.

G. Dennis (pers. comm.) has noted that some of the variation in *S. plicata* seems to warrant taxonomic recognition but from the herbarium material it has not proved possible to do this. More critical work needs to be done on this taxon.

Dennis (1976) also notes that *S. plicata* grows on the banks of Lake Waipiapia, Santa Ana, which is a remnant of the original coral lagoon. In this unusual habitat seedlings are produced on partly submerged tufts of sedges, and adult plants which are subjected to temporary or permanent submergence rarely develop the usual onion-shaped pseudobulbs of dry land forms but, instead, produce thick elongated stems. These can proliferate and produce new plants from the nodes of the stems which remain above the water level, the submerged section subsequently dying. In addition these wet-growing plants bear flowers with a faint, fragrant perfume.

Fig. 15. Spathoglottis plicata A-H. A, habit reduced; **B**, flower x 1; **C**, anther-cap x 6; **D**, lip x 2; **E**, column x 2; **F**, dorsal sepal x 2; **G**, lateral sepal x 2; **H**, petal x 2. **S. pacifica K-L. K**, flower x 2; **L**, lip x 3. **A-H** drawn from *Hunt* 2939; **K-L** drawn from *Simmonds* s.n. All drawn by Sue Wickison.

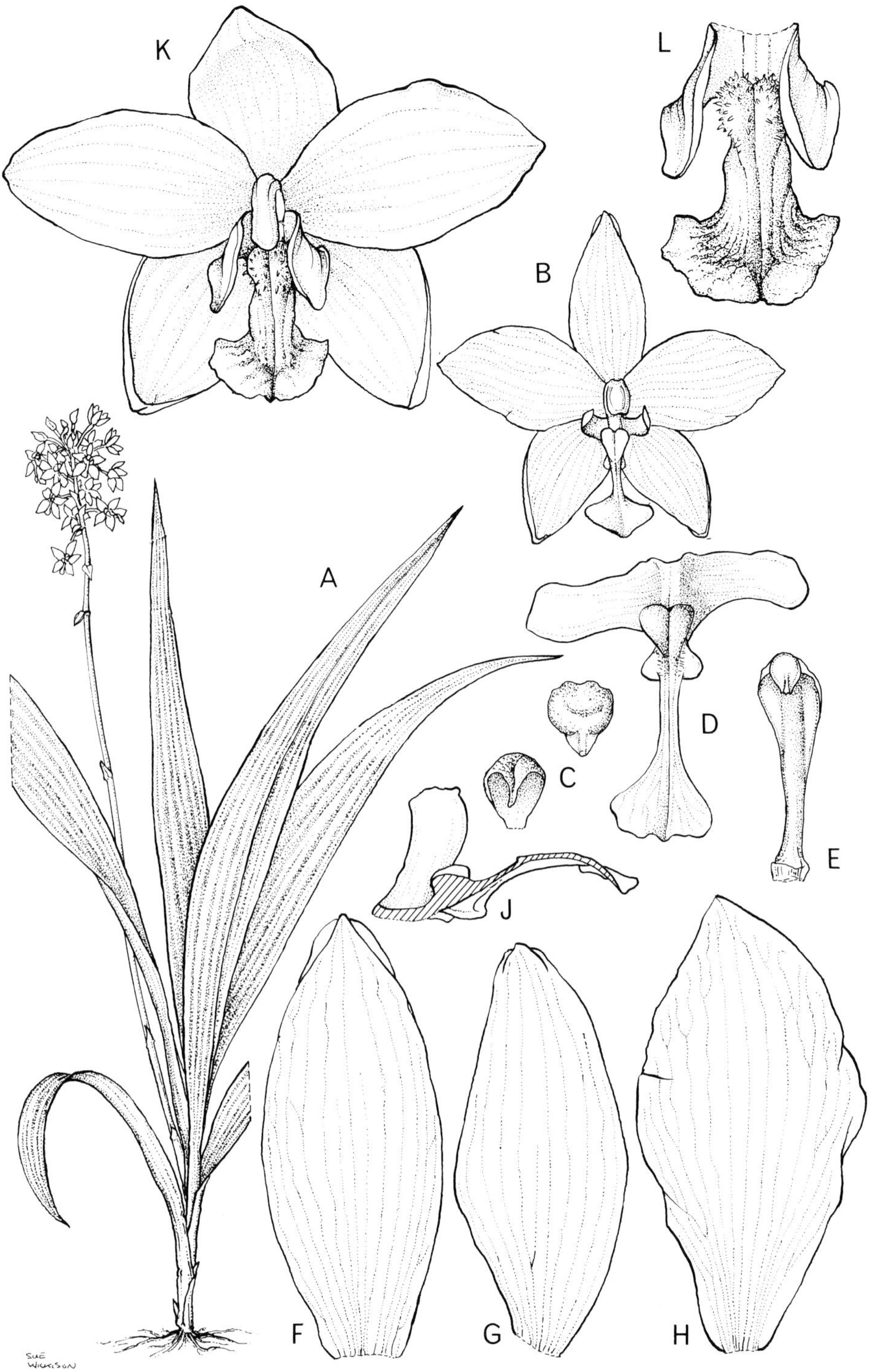
K
L
B
A
C
D
E
J
F
G
H
SUE WICKISON

30. PLOCOGLOTTIS Blume

Terrestrial or rarely epiphytic herbs. *Stems* slenderly pseudobulbous and unifoliate, or elongate and several-leaved (all species from the Solomon Islands and Bougainville have the latter habit). *Leaves* plicate, lanceolate, sheathing at base. *Inflorescences* arising from the base of the pseudobulb, erect, the rhachis bearing a succession of flowers with a few open at one time; peduncle long; bracts rather small to long, persistent. *Flowers* with sepals and petals similar but the sepals pubescent on outer surface; lateral sepals forming a very short mentum with the column-foot; lip short, fleshy, convex, usually joined to the sides and tip of the short column-foot thus forming a sac at the base, usually with an elastic joint that springs when the lip is touched; column rather short; pollinia 4.

A genus of about thirty-five species from the Malay Peninsula to the Solomon Islands, three species being recorded.

1. Column glabrous .... **1. P. latifrons**
   Column pubescent 2
2. Lip with two basal callus swellings .... **2. P. papuana**
   Lip smooth, lacking basal swellings .... **3. P. torana**

1. P. latifrons *J.J. Smith* in Fedde, Repert. Spec. Nov. Regni Veg. 11: 135 (1912). Type: New Guinea, *Gjellerup* 1006 (holotype BO!).

An epiphytic herb. *Stems* up to 1 m tall, leafy. *Leaves* plicate, distichous, lanceolate, acuminate, 15–24 cm long, 2–4 cm wide, deep green, sheathing at base. *Inflorescence* from the base of the pseudobulb, up to 1.2 m long, several-flowered; bracts lanceolate, up to 1 cm long, 2 mm wide. *Flowers* orange-yellow, spotted with red and with a white lip with a yellow apex; sepals pubescent on outer surface; dorsal sepal ovate, acute, concave, 1.1–1.5 cm long, 4–5 mm wide; lateral sepals ovate, acuminate, 1.2–1.5 cm long, 5–6 mm wide; petals lanceolate, acute, 1.3 cm long, 3 mm wide; lip fleshy, convex, oblong, widened near apex, 5–7 mm long, 5 mm wide, with the acute apex recurved; column glabrous, 4–5 mm long.

DISTRIBUTION: Makira. Also in New Guinea.
HABITAT: Rain forest, 130–170 m.
COLLECTIONS: *Hunt* in RSS 2252 & 2346 (K).

2. P. papuana *Schltr.* in Fedde, Repert. Spec. Nov. Regni Veg., Beih. 1: 407 (1912). Type: New Guinea, *Schlechter* 16425 (syntype B†, isosyntypes K!, L, NSW, S), & 17006 (syntype B†, isosyntypes AMES, K!, L, MO, NSW, S).

A terrestrial herb. *Stems* up to 80 cm tall, leafy. *Leaves* plicate, distichous, lanceolate, acute, up to 21 cm long, 6 cm wide, mid-green, sheathing at base. *Inflorescence* from the base of the pseudobulb, up to 60 cm long, several-flowered; bracts lanceolate, 1 cm long, 2 mm wide. *Flowers* yellow or orange, spotted with maroon, with a white lip with 2 purple spots, a white column with a maroon band and a pale yellow anther-cap; sepals pubescent on outer surface; dorsal sepal ovate, acute, 1.1–1.3 cm long, 5 mm wide; lateral sepals obliquely ovate, acuminate, 1.1–1.4 cm long, 5–6 mm wide; petals obliquely linear, acute, 1–1.2 cm long, 3 mm wide; lip appressed to column, broadly oblong, convex, 6–7 mm long, 5 mm wide, fleshy, with an acute, reflexed apex, the disc with 2 ovoid basal calli; column pubescent, 6 mm long.

DISTRIBUTION: Bougainville, Choiseul, Guadalcanal, Makira and Malaita. Also in New Guinea and New Britain.
HABITAT: Primary rain forest, 200–600 m.
COLLECTIONS: *Cribb* S22 (K); *Dennis* s.n. & in RSS 4650 (K); *Hunt* in RSS 2194 (K); *Wickison* 232 (K); *Ridsdale & Lavarack* in NGF 30576 (K).

3. P. torana *J.J. Smith* in Fedde, Repert. Spec. Nov. Regni Veg. 11: 135 (1912). Type: New Guinea, *Gjellerup* 769 (holotype BO!).

A terrestrial herb. *Stems* up to 1.3 m tall, leafy. *Leaves* plicate, distichous, lanceolate, acuminate, up to 35 cm long, 4 cm wide, sheathing at base. *Inflorescence* from the base of the pseudobulb, several-flowered; bracts lanceolate, up to 8 mm long, 3 mm wide. *Flowers* orange, heavily spotted with red, with a yellow lip, a white column with red in the middle, and a yellow anther-cap; sepals pubescent on outer surface; dorsal sepal ovate, 1.6 cm long, 4–5 mm wide; lateral sepals ovate, acuminate, 1.7 cm long, 4 mm wide; petals linear, acute, 1.5 cm long, 2–3 mm wide; lip appressed to column, oblong, 6–7 mm long, 5 mm wide, fleshy, with an acute, recurved apex; column pubescent, 4 mm long.

DISTRIBUTION: San Jorge (off Santa Isabel). Also in New Guinea.
HABITAT: Forest at the edge of garden land, sea level.
COLLECTION: *Hunt* in RSS 2720 (K).

31. MISCHOBULBUM Schlechter

Terrestrial plants. *Pseudobulbs* slender, unifoliate. *Leaf* apical, broadly ovate. *Inflorescence* arising from the base of a recently matured pseudobulb, with few to many, well-spaced, rather large flowers. *Flowers* with sepals and petals free, elliptic to lanceolate; lateral sepals joined to the column-foot to form a mentum; lip entire, with prominent keels on the upper surface; column with a foot to which the lip and lateral lobes are joined; pollinia 8.

A genus of about seven species from China and India to the Malay Peninsula and Archipelago, New Guinea and the Solomon Islands, a single species being recorded.

M. papuanum *(J.J. Smith) Schltr.* in Fedde, Repert. Spec. Nov. Regni Veg., Beih. 1: 98 (1911). Types: New Guinea, *Von Roemer* 944 & 966 (syntypes BO!).
Tainia papuana J.J. Smith in Bull. Dép. Agric. Indes Néerl. 39: 21 (1910) non Gagnep. (1932).

See Turner (in press) for a revision of this genus.

A terrestrial herb. *Stems* arising from a creeping rhizome, up to 15 cm tall. *Leaf* terminal, solitary, ovate, acuminate, 6.5–16 cm long, 3–9 cm wide, dark green with purple veins, lighter green below. *Inflorescence* arising from the base of the pseudobulb, up to 33 cm long, laxly 3- to 9-flowered. *Flowers* pale brown with a purple lip and column; dorsal sepal elliptic, acute, 1.5–2.2 cm long, 2.5–5 mm wide; lateral sepals lanceolate, acute, 1.6–2.4 cm long, 5–9 mm wide; petals elliptic, acute, 1.5–2 cm long, 4–7.5 mm wide; lip entire, ovate, acute to acuminate, 1.7–2.4 cm long, 9–13 mm wide, the distal margins undulating, the disc with 3 undulate longitudinal keels, the median keel smaller and raised only in the apical half; column 6.5–8.5 mm long, with a 6–7 mm long foot. (See fig. 16).

DISTRIBUTION: Santa Isabel. Also in New Guinea.
HABITAT: Montane forest, 600 m.
COLLECTIONS: *Brass* 3235 (AMES) & 3333 (AMES, L).

32. TAINIA Blume

Terrestrial plants. *Pseudobulbs* slender, fleshy, unifoliate, each pseudobulb often developing one to several fleshy roots from the base. *Leaf* apical, broad, smooth or plicate, petiolate. *Inflorescence* arising from the base of a recently matured pseudobulb, with few to many, well-spaced, rather large flowers. *Flowers* with sepals and petals triangular to obovate; lateral sepals joined to the column-foot to form a mentum; lip entire or trilobed, with prominent keels on the upper surface; column with a foot to which the lip and lateral lobes are joined; pollinia 8.

A genus of about twenty-five species from India and China to Malaysia, New Guinea, the Solomon Islands and Australia. A single species is recorded from the Solomon Islands and Bougainville.

T. trinervis *(Blume) Reichb. f.* in Bonplandia 5: 54 (1857). Type: Java, *Korthals* s.n. (holotype L).
Mitopetalum trinerve Blume, Mus. Bot. Lugd. Bat. 2:185 (1856).
Tainia parviflora Schltr. in Fedde Repert. Spec. Nov., Regni Veg., Beih. 1: 100 (1911). Type: New Guinea, *Schlechter* 20132 (holotype B†).

See Turner (in press) for a revision of this genus.

A terrestrial herb. *Pseudobulbs* clustered, slender, slightly swollen towards the base, up to 7 cm long, 6 mm wide, bearing a single terminal leaf. *Leaf* elliptic, acute, 9.5–26 cm long, 3–8 cm wide; petiole 3–13 cm long. *Inflorescence* arising

---

Fig. 16. Tainia trinervis A-G. A, flower x 2; **B**, column x 4; **C**, lip x 4; **D**, anther-cap x 10; **E**, dorsal sepal x 4; **F**, petal x 4; **G**, lateral sepal x 4. **Mischobulbum papuanum H-Q. H**, flower x 2; **J**, habit reduced; **K**, column x 4; **L**, lip x 2; **M**, pollinia x 8; **N**, anther-cap x 8; **O**, dorsal sepal x 2; **P**, petal x 2; **Q**, lateral sepal x 2. **A-G** drawn from LAE 64078; **H-Q** from *Brass* 3333. All drawn by Sue Wickison.

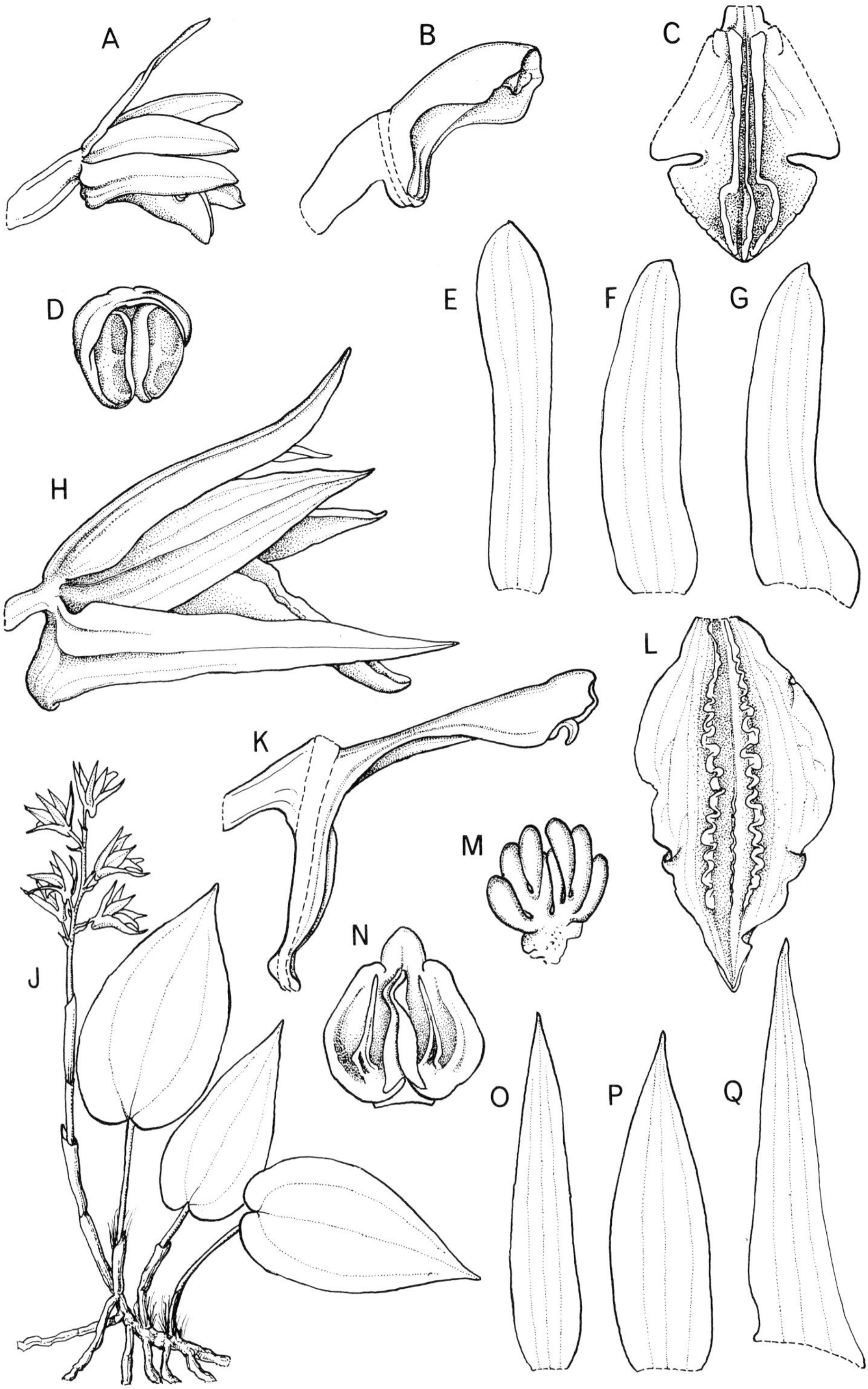
A
B
C
D
E
F
G
H
J
K
L
M
N
O
P
Q

from the base of the pseudobulb, up to 60 cm long, laxly few- to many-flowered. *Flowers* olive-yellow, the petals with a broad crimson streak, the lip yellow suffused with purple and the column purple; sepals and petals opening widely; dorsal sepal elliptic, 9–12.5 mm long, 1.5–2 mm wide; lateral sepals elliptic to triangular, 8–11 mm long, 1.5–3 mm wide; petals obliquely elliptic-triangular, 8.5–11 mm long, 1.5–2 mm wide; lip trilobed, 6.5–8 mm long, 5–6.5 mm wide, the lateral lobes erect, obliquely triangular, the mid-lobe triangular to semi-circular, the disc with 2 parallel keels which extend from the base to the apex of the mid-lobe, with the third ridge between them extending from the centre of the mid-lobe to its apex; column 3.5–5 mm long. (See fig. 16).

DISTRIBUTION: Bougainville. Also in the Moluccas, New Guinea and Australia.
HABITAT: Montane forest, 450–1200 m.
COLLECTIONS: *Cribb* S62 (K); *Lavarack & Ridsdale* in NGF 31592 (K).

33. COELOGYNE Lindley

Epiphytic or rarely lithophytic or terrestrial herbs. *Pseudobulbs* clustered or remote on rhizome, usually large, cylindrical-conical to ovoid, 1- or 2-leaved at apex. *Leaves* plicate, elliptic, petiolate. *Inflorescence* terminal, produced from old or new growths, erect or pendulous, racemose, with one to many large or fairly large flowers; bracts persistent or deciduous. *Flowers* produced one at a time or all together; sepals often concave, free; petals narrower than sepals; lip usually trilobed, lacking a spur, with erect lateral lobes, the mid-lobe of lip usually with longitudinal keels which are often papillose or verrucose; column long, clavate, winged around the apex; pollinia 4.

A large genus of about three hundred species from tropical Asia to the Pacific Islands. Four species being recorded.

1. Bracts deciduous, dropping as flower opens 2
   Bracts persistent 3
2. Callus of 3 well separated ridges; sepals and petals 2–3 cm long
   .... **2. C. carinata**
   Callus of 2 main and several subsidiary ridges; sepals and petals 4.5–5.5 cm long .... **3. C. susanae**
3. Inflorescence pendent; flowers white; lip obscurely trilobed, lacking a callus; column and ovary shortly pubescent.................................. **4. C. veitchii**
   Inflorescence arcuate; flowers creamy marked with orange on lip; lip trilobed, with a pustular red-orange callus; column and ovary very shortly pubescent .... **1. C. asperata**

Fig. 17. Coelogyne asperata A-C. A, side view of lip x 5/3; **B**, inflorescence x 1/3; **C**, lip x 5/3. **C. carinata D-N. D**, habit x 2/3; **E**, column x 2; **F**, dorsal and ventral view of anther-cap x 4; **G**, pollinia x 4; **H**, dorsal sepal x 5/3; **J**, petal x 5/3; **K**, lateral sepal x 5/3; **L**, cross-section of ovary x 4; **M**, lip x 5/3; **N**, side view of lip x 5/3. **A-C** drawn from *Wickison* 125; **D-N** from *Wickison* s.n. All drawn by Sue Wickison.

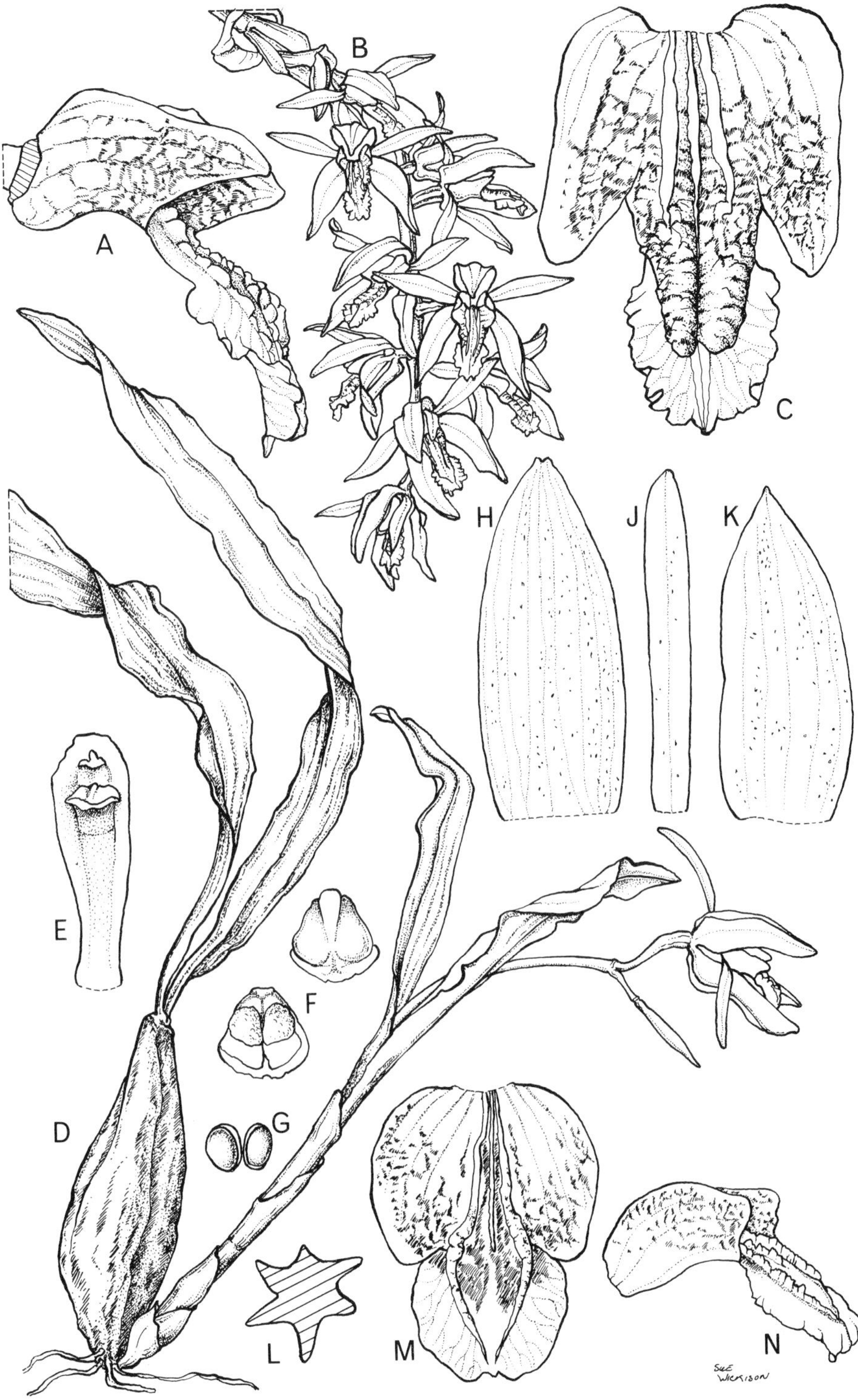
A
B
C
D
E
F
G
H
J
K
L
M
N
SUE WICKISON

1. C. asperata *Lindley* in Journ. Hort. Soc. 4: 221 (1849). Type: Malaya, cult. *Twisden Hodges* (holotype K!).

An epiphytic, lithophytic or terrestrial plant forming large clumps. *Pseudobulbs* elongate, tapering, to 15 cm long, 2-leaved at apex. *Leaves* elliptic, up to 1 m long, 4.5–20 cm wide. *Inflorescence* arcuate, from apex of new growth; bracts large, persistent, papyraceous. *Flowers* cream-coloured to yellowish with a white lip marked with red-orange, fragrant; ovary shortly pubescent; dorsal sepal ovate, 3 cm long, 1.6 cm wide; lateral sepals ovate, obscurely oblique, 3 cm long, 1.5 cm wide; petals linear, 3 cm long, 7 mm wide; lip trilobed, 3 cm long, 2.3 cm wide, the lateral lobes oblong, the callus of 3–5 coalescent pustulose ridges; column finely pubescent, 1.3 cm long. (See fig. 17 A–C and plate 5B).

DISTRIBUTION: Bougainville, Choiseul, Guadalcanal, New Georgia, Malaita and the Santa Cruz Islands. Also in the Malay Peninsula and Archipelago.
VERNACULAR NAME: Makapauna (Bougainville).
HABITAT: On open rock faces or epiphytic in riverine or submontane forest, 270–950 m.
COLLECTIONS: *Hadley* 106 (BM); *Hunt* in RSS 3056 (K); *Officers of HMS Penguin* s.n. (K); *Ridsdale & Lavarack* in NGF 30560 (K); *Waterhouse* 672 & 728 (K); *Wickison* 125 & 239 (K).

2. C. carinata *Rolfe* in Bull. Misc. Inf., Kew 1895: 191 (1895). Type: New Guinea, cult. *Sander* (holotype K!).

An epiphytic or rarely terrestrial plant. *Pseudobulbs* ovoid, 2-leaved at apex. *Leaves* elliptic, up to 40 cm long, 5 cm wide. *Inflorescence* terminal, produced with new leaves, 1– to 5-flowered; bracts deciduous. *Flowers* produced at the same time, pale green with a white lip marked with orange-brown; ovary 6-angled in cross-section; dorsal sepal ovate, 2–3 cm long, 1–1.5 cm wide; lateral sepals ovate, 2–3 cm long, 1–1.5 cm wide; petals linear, 2–3 cm long, 4–6 mm wide; lip trilobed, 2–2.2 cm long, 1.7–2 cm wide, the callus of three scarcely verrucose keels, the outer two extending almost to the lip apex, the central terminating at the base of the mid-lobe; column 1.6 cm long. (See fig. 17 D–N and plate 5C).

DISTRIBUTION: Bougainville, Guadalcanal, Makira and New Georgia. Also in New Guinea.
HABITAT: Rain forest, 60 m.
COLLECTIONS: *Chaplin* 793 & 794 (K); *Dennis* in RSS 4637 (K); *Dickson* 18 (K); *Hunt* in RSS 2202 (K); *Mason* 1650 (K); *Wickison* s.n. (K).

Coelogyne truncicola Schltr. from New Guinea is similar and may prove to be conspecific.

Fig. 18. Coelogyne susanae. A, habit reduced; **B**, habit x 2/3; **C**, part of inflorescence x 2/3; **D**, lip x 2; **E**, side view of lip x 2; **F**, dorsal and ventral view of anther-cap x 2; **G**, pollinia x 2; **H**, column x 2; **J**, dorsal sepal x 2; **K**, petal x 2; **L**, lateral sepal x 2. Drawn from *Wickison* 40 by Sue Wickison.

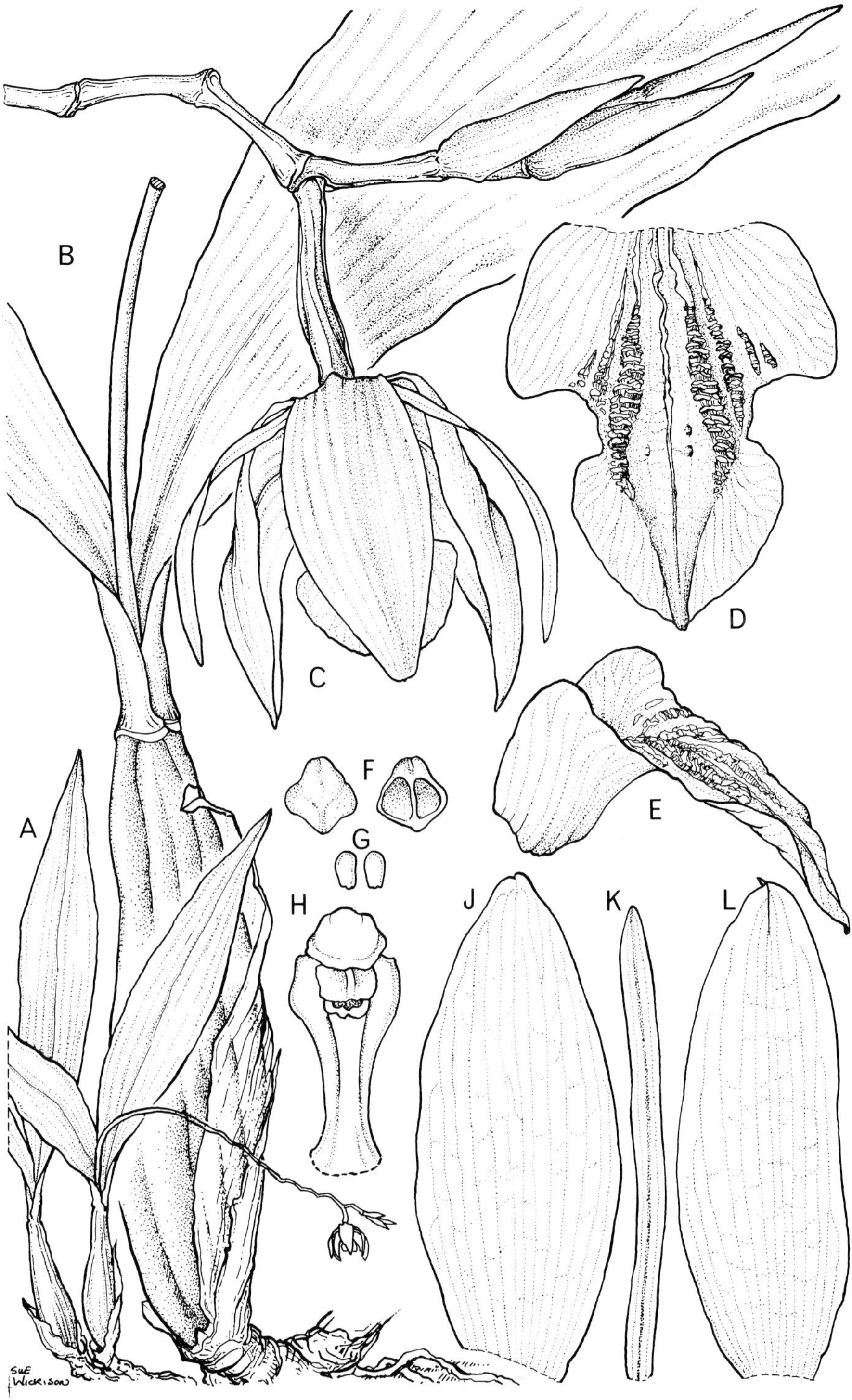
B
D
C
A
F
E
G
H
J
K
L
SUE WICKISON

3. C. susanae *Cribb & B. Lewis* in Kew Bull. 46(2) (in press). Type: Solomon Islands, New Georgia, *Wickison* 40 (holotype K!).

An epiphytic herb often forming large clumps. *Pseudobulbs* elongate, ovate, 7.5 cm long, 2.5 cm wide, 4- to 5-angled, 2-leaved, covered by sheaths when young. *Leaves* lanceolate, acute, up to 45 cm long, 4.8–5.2 cm wide, prominently 5-veined. *Inflorescence* produced with the new leaves before the pseudobulb swells, erect, up to 28 cm long, several-flowered, the flowers produced successively; rhachis zig-zag; bracts deciduous, lanceolate. *Flowers* pale yellow-green or creamy-green with a red-brown lip with a pale yellow or white apex, fragrant; pedicel and ovary 6-angled in cross-section, 2–4 cm long; dorsal sepal elliptic-lanceolate, 4.5–6.5 cm long, 1.7–2.5 cm wide, keeled at base on back; lateral sepals oblong-lanceolate, obtuse, keeled at back, 4.5–6.5 cm long, 1.5–2 cm wide; petals linear, 4.5–6.3 cm long, 3.5–5 mm wide; lip trilobed, 4–5 cm long, 2.8–3 cm wide, the lateral lobes erect, obliquely oblong, the mid-lobe subcircular, obtuse, the callus of 2 main and several subsidiary verrucose ridges; column whitish, brown on ventral surface, winged towards apex, up to 2.5 cm long. (See fig. 18 and plate 5D).

DISTRIBUTION: Bougainville, Guadalcanal, Kolombangara, Makira, New Georgia and the Shortland Islands.
HABITAT: Rain forest, 160–500 m.
COLLECTIONS: *Cribb* 1922 (K); *Dennis* s.n. (K); *Hunt* in RSS 2250 & 2510 (K); *Lever* 22.10.36 (K); *Wickison* 40, 104A & 134B (K).

This species is similar to *C. beccarii* Reichb. f., from New Guinea, but that differs in having 2 raised, smooth, convex callus ridges. *C. fragrans* Schltr., from New Guinea, is also similar, but that has unifoliate pseudobulbs and smaller, paler flowers with an obovate mid-lobe to the lip.

4. C. veitchii *Rolfe* in Bull. Misc. Inf., Kew 1895: 282 (1895). Type: New Guinea, cult. *Veitch* (holotype K, probably destroyed).

An epiphytic herb. *Pseudobulbs* 2-leaved, conical. *Leaves* elliptic, up to 40 cm long, 6 cm wide, longly petiolate. *Inflorescence* pendent, laxly many-flowered, produced from new growth; bracts large, persistent, papery. *Flowers* small, white, all opening at the same time; pedicel and ovary shortly pubescent; dorsal sepal ovate, up to 1.8 cm long, 8 mm wide, keeled at back; lateral sepals ovate, up to 1.8 cm long, 7 mm wide, keeled at back; petals ovate, acute, up to 1.8 cm long, 4 mm wide; lip obscurely trilobed, 1.8–2 cm long, 1.4–1.6 cm wide, ecallose, recurved at apex, the lateral lobes erect; column c. 1 cm long. (see fig. 19 and plate 5A).

DISTRIBUTION: Guadalcanal. Also in New Guinea.
HABITAT: Montane forest, 800–1100 m.
COLLECTIONS: *Cribb, Dennis & Campbell* 5042 (K); *Hunt* in RSS 2992 (K).

Fig. 19. Coelogyne veitchii. A, habit reduced; **B**, flower x 2; **C**, column x 3; **D**, dorsal and ventral view of anther-cap x 4; **E**, pollinia x 4; **F**, dorsal sepal x 2; **G**, petal x 2; **H**, lateral sepal x 2; **J**, side view of lip x 2; **K** lip x 2. **A** drawn from *Hunt* 2992; **B-K** from *Cribb, Dennis & Campbell* 5042. All drawn by Sue Wickison.

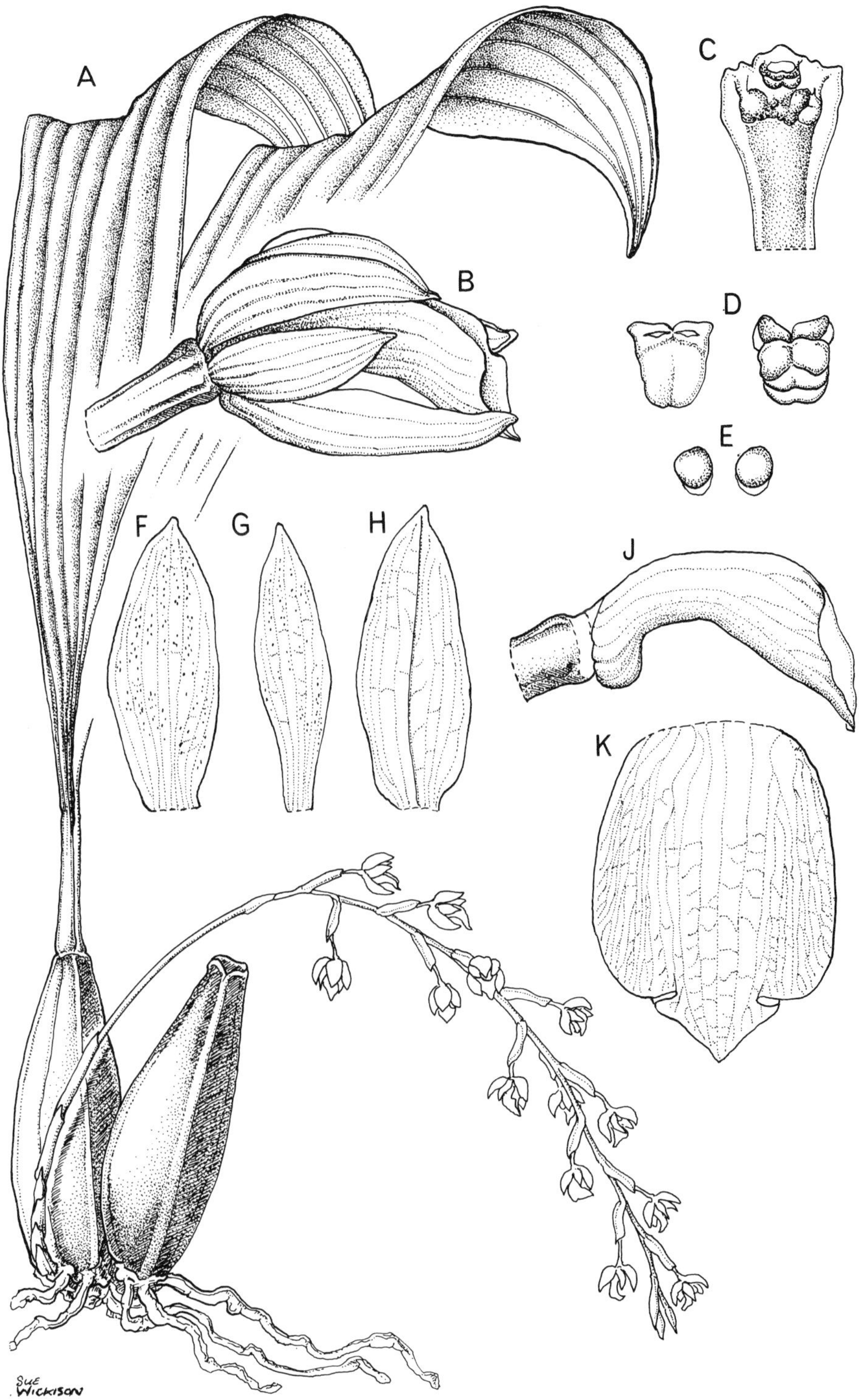
A
B
C
D
E
F
G
H
J
K
SUE WICKISON

34. PHOLIDOTA Lindley ex Hooker

Pendulous or erect epiphytic or lithophytic herbs. *Pseudobulbs* slender or swollen. *Leaf* solitary from apex of pseudobulb, lanceolate, ovate or oblong. *Inflorescence* terminal, emerging from apex of pseudobulb, pendent, slender, bearing many small flowers alternately in 2 ranks; the subtending bracts large and concave, persistent. *Flowers* small, fleshy; dorsal sepal concave or convex, broadly ovate to elliptic; lateral sepals concave or convex, ovate to ovate-oblong; petals ovate to linear; lip sessile, with a saccate base and a small, deflexed apical blade which may be bilobed or entire; column short with a wide hood around the anther; pollinia 2.

A genus of twenty-nine species from SE. Asia, Malesia, Australia and the Pacific, two species being recorded. The species are arranged in two sections according to de Vogel (1988).

Rhizome pendulous; inflorescence laxly flowered; rhachis zig-zag; sepals convex .... **1. P. gibbosa**
Rhizome creeping; inflorescence densely flowered; rhachis straight; sepals concave .... **2. P. imbricata**

Section CHELONANTHERA

1. P. gibbosa *(Blume) de Vriese*, Ill. Orch.: 5 (1854). Type: Java, *Blume* s.n. (lectotype L).
Chelonanthera gibbosa Blume, Bidjr. Fl. Ned. Ind.: 383 (1825).

For full synonymy see de Vogel (1988).

An epiphytic herb up to 50 cm high, pendulous. *Rhizome* pendulous. *Pseudobulbs* spirally arranged along rhizome, slender, 7 cm long, 1 cm wide. *Leaf* solitary from apex of pseudobulb, ovate-oblong, acuminate, 18–25 cm long, 2.5–5 cm wide; petiole up to 5 cm long. *Inflorescence* terminal; rhachis pendulous, about 19 cm long, laxly many-flowered; peduncle terete, up to 35 cm long; bracts elliptic-ovate, 10–11 mm long, 9 mm wide. *Flowers* non-resupinate, creamy-green; pedicel and ovary 8–9 mm long; dorsal sepal convex, ovate-elliptic, 6–8 mm long, 3–4 mm wide; lateral sepals convex, broadly ovate, 7–8 mm long, 3–4.5 mm wide; petals reflexed, linear, rounded at apex, 5–8 mm long, 0.8–1.3 mm wide; lip trilobed, saccate at base, 7–10 mm long, 5–9 mm wide, the lateral lobes triangular, erect, the mid-lobe bilobed, with lobes more or less semi-orbicular, reflexed at the apex; column broadly spathulate, 5 mm long, 5 mm wide. (See fig. 20 M–N).

DISTRIBUTION: Guadalcanal. Also in the Malay Peninsula and Archipelago.
HABITAT: Ridge-top forest, 980 m.

Fig. 20. Pholidota imbricata A-L. A, habit x 2/3; **B**, flower x 4; **C**, lip x 4; **D**, side view of lip x 4; **E**, column x 4; **F**, anther-cap x 4; **G**, pollinia x 4; **H**, dorsal sepal x 4; **J**, petal x 4; **K**, lateral sepal x 4; **L**, inflorescence x 4. **P. gibbosa M-N. M**, flower x 3; **N**, lip x 4. **A-L** drawn from *Wickison* 91; **M-N** from *Hunt* 2944. All drawn by Sue Wickison.

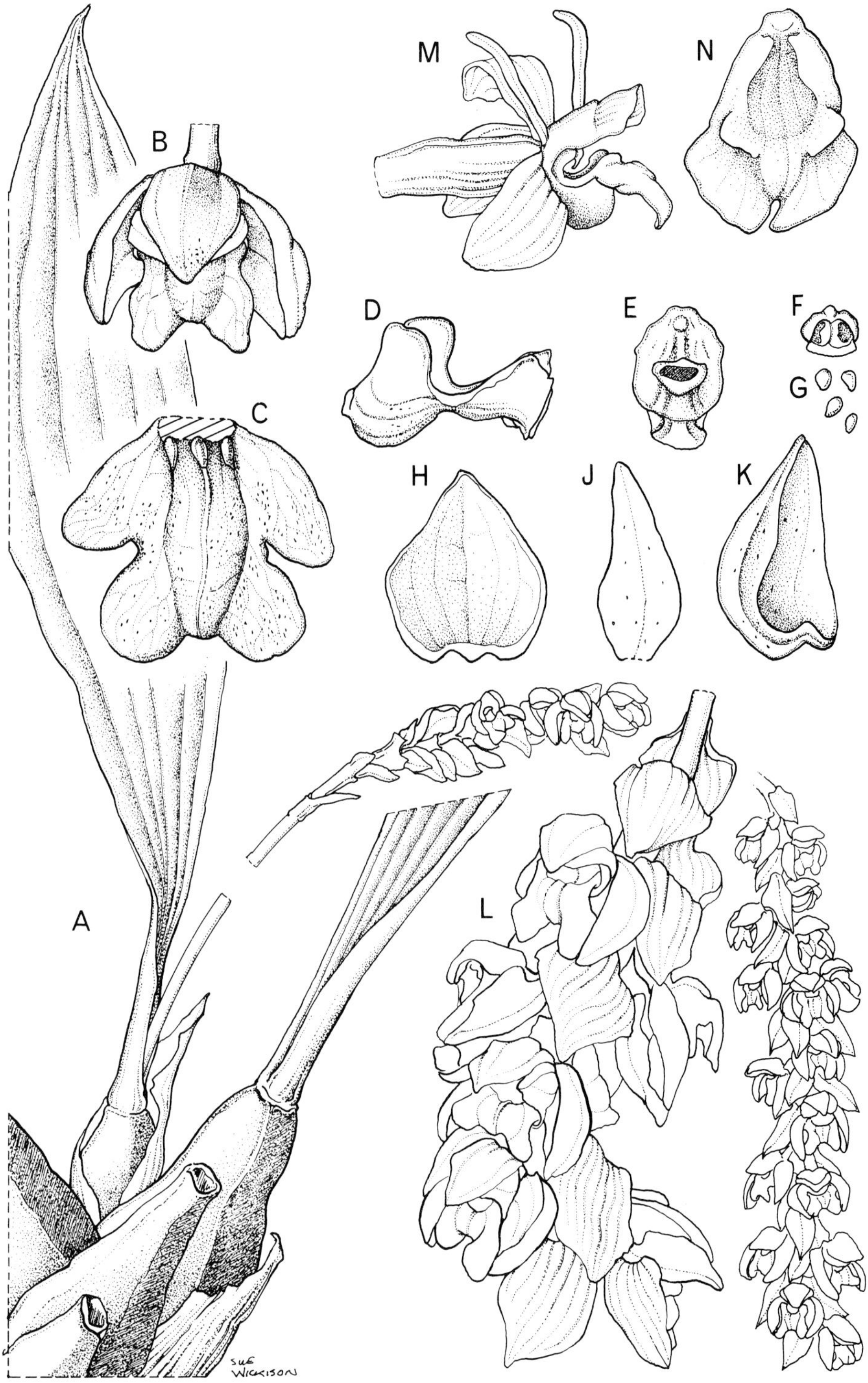
M
N
B
D
E
F
G
C
H
J
K
A
L
SUE WICKISON

COLLECTION: *Hunt* in RSS 2944 (K).

It is strange that *P. gibbosa* occurs in Guadalcanal, so remote from its main range, and has never been found in New Guinea.

Section PHOLIDOTA

2. P. imbricata *Hook.*, Exotic Fl. 2: 138 (1825). Type: Himalayas, *Wallich* s.n., cult. Kew 1824 (probably not preserved; lectotype: the drawing in Exotic Fl.: 138, 1825).
Pholidota grandis Kraenzl. in Bull. Soc. Bot. Fr. 76: 301 (1929), non Ridley 1908.. Type: Vanuatu, Efate, *Levat* s.n. (holotype B†).
Pholidota spectabilis Kraenzl. ex Guillaumin in Ann. Mus. Col. Marseille, ser. 2, 15 (1948). Type: as for *P. grandis.*

For full synonymy see de Vogel (1988).

An epiphytic herb up to 60 cm high, erect. *Rhizome* creeping. *Pseudobulbs* close together, sturdy, swollen, 4–7 cm long, 1.5–4 cm wide, mid-green to dull grey-green. *Leaf* solitary from apex of pseudobulb, thickly coriaceous, oblong or obovate-oblong to linear, 17–37 cm long, 3–5.5 cm wide, mid-green to dark green above, lighter green below; petiole 2–5 cm long. *Inflorescence* terminal; peduncle terete, up to 40 cm long; rhachis 12 cm long, usually pendulous, many-flowered; bracts imbricate, persistent, ovate, 7 mm long, 6.5 mm wide. *Flowers* non-resupinate, white to cream-coloured sometimes tinged yellowish at the base of the lip; pedicel and ovary 2–5 mm long; dorsal sepal and petals forming a hood, concave; dorsal sepal ovate, 5–6 mm long; lateral sepals obliquely ovate, 5–7 mm long, keeled; petals falcate, 4.5–6 mm long; lip trilobed, saccate at base, 6 mm long, 5 mm wide, the lateral lobes erect, triangular to almost semi-circular, the mid-lobe with a bilobulate apex, with semi-orbicular lobules, reflexed, the base of lip with 3 short yellow calli; column short, orbicular, 3mm long and wide. (See fig. 20 A–L and plate 7A).

DISTRIBUTION: Guadalcanal and Makira. Widely distributed from Asia and the Malay Penisula to New Guinea, Vanuatu, the Loyalty Islands, New Caledonia, Fiji and Australia.
HABITAT: Rain forest, 50–150 m.
COLLECTIONS: *Hunt* in RSS 2231 (K); *Wickison* 91 (K).

35. HIPPEOPHYLLUM Schlechter

Epiphytic herbs. *Rhizome* long, creeping and rigid. *Stems* short, lacking pseudobulbs, usually spaced 2–6 cm apart on the rhizome. *Leaves* distichous, bilaterally flattened, sheathing at base. *Inflorescence* terminal, densely many-flowered. *Flowers* small, non-resupinate; sepals and petals free, with the sepals broader than the petals; lip usually trilobed, sometimes with a bilobed callus at the base; column slender; pollinia 4.
A small genus of about five species from the Malay Peninsula to the Solomon Islands.

H. micranthum *Schltr.* in K. Schum. & Laut., Nachtr. Fl. Deutsch. Südsee: 108 (1905). Type: New Guinea, *Schlechter* 13809 (holotype B†).

An epiphytic herb. *Rhizomes* long, creeping and rigid. *Stems* short, up to 2 cm long, 2–4 cm apart on the rhizome. *Leaves* 2–5, distichous, bilaterally flattened, oblong-lanceolate, 2.5–7 cm long, 4–6 mm wide, bright green, sheathing at base. *Inflorescence* terminal, up to 16 cm long, densely many-flowered; bracts lanceolate, persistent, 2 mm long, 0.8 mm wide. *Flowers* white with a yellow-green lip; pedicel and ovary 3–4 mm long, pubescent; sepals and petals reflexed; dorsal sepal ovate, 1.8 mm long, 0.75 mm wide; lateral sepals ovate, 1.4 mm long, 0.7 mm wide; petals linear, obtuse, 1.5 mm long, 0.3–0.4 mm wide; lip trilobed, decurved, 1.5 mm long, 1.1 mm wide, the lateral lobes small, triangular, erect, the mid-lobe triangular with a bilobed callus at the base; column slender, 0.9 mm long.

DISTRIBUTION: New Georgia. Also in New Guinea.
HABITAT: Growing on trees around a football pitch, sea level.
COLLECTION: *Wickison* 3 (K).

36. LIPARIS L.C. Richard

Epiphytic, lithophytic or terrestrial plants. *Pseudobulbs* ovoid to cylindrical, fleshy, 1– to several-leaved at the apex. *Leaves* plicate or conduplicate, narrowly lanceolate or ovate. *Inflorescence* terminal, racemose, several- to many-flowered. *Flowers* resupinate, small- to medium-sized, sometimes self-fertilising, white, yellow, green, orange or purple; sepals and petals free, often reflexed; sepals ovate to lanceolate; petals usually narrow, linear; lip variable, entire or trilobed, often deflexed, ± with a fleshy basal callus; column rather elongate, curved or bent at apex, sometimes swollen at base, lacking a foot; pollinia 4.

A cosmopolitan genus of about two hundred and fifty species. Seven species being recorded.

1. Leaves broadly ovate, 4–8.5 cm wide; flowers maroon with a green column .... **5. L. layardii**
   Leaves narrowly lanceolate, less than 3.5 cm wide; flowers not maroon 2
2. Bracts distichous and imbricate; flowers opening one at a time .... **4. L. gibbosa**
   Bracts not distichous; several to many flowers open at a time 3
3. Pseudobulbs elongated, 7–24 cm long; lip with a ciliate basal margin .... **3. L. condylobulbon**
   Pseudobulbs compact, up to 6 cm high; margins of lip not ciliate 4
4. Lip subcircular and entire; pedicels more than 1.7 cm long 5
   Lip oblong or trilobed; pedicels less than 9 mm long 6
5. Lip 6 mm long, 6.75 mm wide; sepals 7.5 mm long .... **6. L. orbiculata**
   Lip 14 mm long, 13 mm wide; sepals 13 mm long .... **2. L. chlorantha**
6. Leaf 6–10 cm long, 3–5 mm wide; lip oblong........ **1. L. caespitosa**
   Leaf 15–19 cm long, 2.8–3.4 cm wide; lip trilobed .... **7. L. truncicola**

1. L. caespitosa *(Thouars) Lindley* in Bot. Reg. 11: 882 (1825). Type: Mauritius, *Thouars* s.n. (holotype P).
Malaxis caespitosa Thouars, Orch. Iles Austr. Afr.: t.90 (1822).

For synonymy see Seidenfaden (1976).

A small epiphytic or lithophytic herb. *Pseudobulbs* less than 1 cm high, 3–5 mm wide. *Leaf* solitary from apex of pseudobulb, lanceolate, 6–10 cm long, 0.7–1.1 cm wide, tapering at base to 1–2 mm wide, light green. *Inflorescence* terminal, 5–9 cm long, few- to many-flowered; bracts lanceolate, 5 mm long, 1 mm wide. *Flowers* usually self-pollinating, yellow to light green; pedicel and ovary 4 mm long, the ovary triangular in cross-section; dorsal sepal ovate-lanceolate, reflexed, 2.2 mm long, 1 mm wide; lateral sepals reflexed, ovate, 2.2 mm long, 1 mm wide; petals reflexed, linear, 2.2 mm long, 0.3 mm wide; lip decurved, fleshy, oblong, 2.5 mm long, 1.5 mm wide; column 1.2 mm long. (See fig. 21 K–S).

DISTRIBUTION: Bougainville and Guadalcanal. Widely distributed from Madagascar and E. Africa to Asia, the Malay Archipelago, New Guinea, Vanuatu, New Caledonia, Fiji and the Austral Islands.
HABITAT: Montane rain forest, 520–670 m.
COLLECTIONS: *Cribb, Dennis & Campbell* 5048 (K); *Ridsdale & Lavarack* in NGF 30674 (K, LAE); *Wickison* 98 & 138 (K).

Kores (1989) notes that this is a very variable as well as geographically widespread species and M. Clements (pers. comm.) has noted that another Pacific species, *L. revoluta* Hook. f. & Arn., originally described from Tahiti, appears very similar to *L. caespitosa* and its status should be re-examined.

2. L. chlorantha *Schltr.* in Fedde, Repert. Spec. Nov. Regni Veg., Beih. 1: 201 (1911). Type: New Guinea, *Schlechter* 17147 (holotype B†, isotypes E, K!, L, MO, NSW, S).

A terrestrial herb. *Pseudobulbs* narrowly conical, bilaterally compressed, up to 6 cm long, 0.9 cm wide, covered by leaf sheaths. *Leaf* solitary, from apex of pseudobulb, erect, lanceolate-elliptic, acute, up to 15 cm long, 2–3 cm wide, with 4–5 prominent longitudinal veins. *Inflorescence* terminal, 13–22 cm long, several- flowered. *Flowers* green with an orange-buff lip and a white column; pedicel and ovary 1.7–2 cm long; dorsal sepal reflexed, lanceolate, 1.3 cm long, 2–4 mm wide; lateral sepals reflexed, lanceolate, 1.3 cm long, 2–3 mm wide; petals reflexed, narrow, linear, 1.3 cm long, 2–3 mm wide; lip subcircular, 1.4 cm long, 1.3 cm wide, with a small fleshy basal callus; column 4.5 mm long.

Fig. 21. Liparis condylobulbon A-J. A, habit x 2/3; **B**, lip x 10; **C**, flower x 10; **D**, dorsal sepal x 10; **E**, lateral sepal x 10; **F**, petal x 10; **G**, column x 14; **H**, dorsal and ventral view of anther- cap x 14; **J**, pollinia x 14. **L. caespitosa K-S. K**, flower x 10; **L**, pollinia x 14; **M**, dorsal and ventral view of anther-cap x 14; **N**, lip x 10; **O**, column x 14; **P**, dorsal sepal x 10; **Q**, lateral sepal x 10; **R**, petal x 10; **S**, habit x 2/3. **A** drawn from *Hallé* 6394; **B-J** from *Wickison* 59; **K-R** from *Cribb, Dennis & Campbell* 5048; **S** from *Ridsdale & Lavarack* in NGF 30674. All drawn by Sue Wickison.

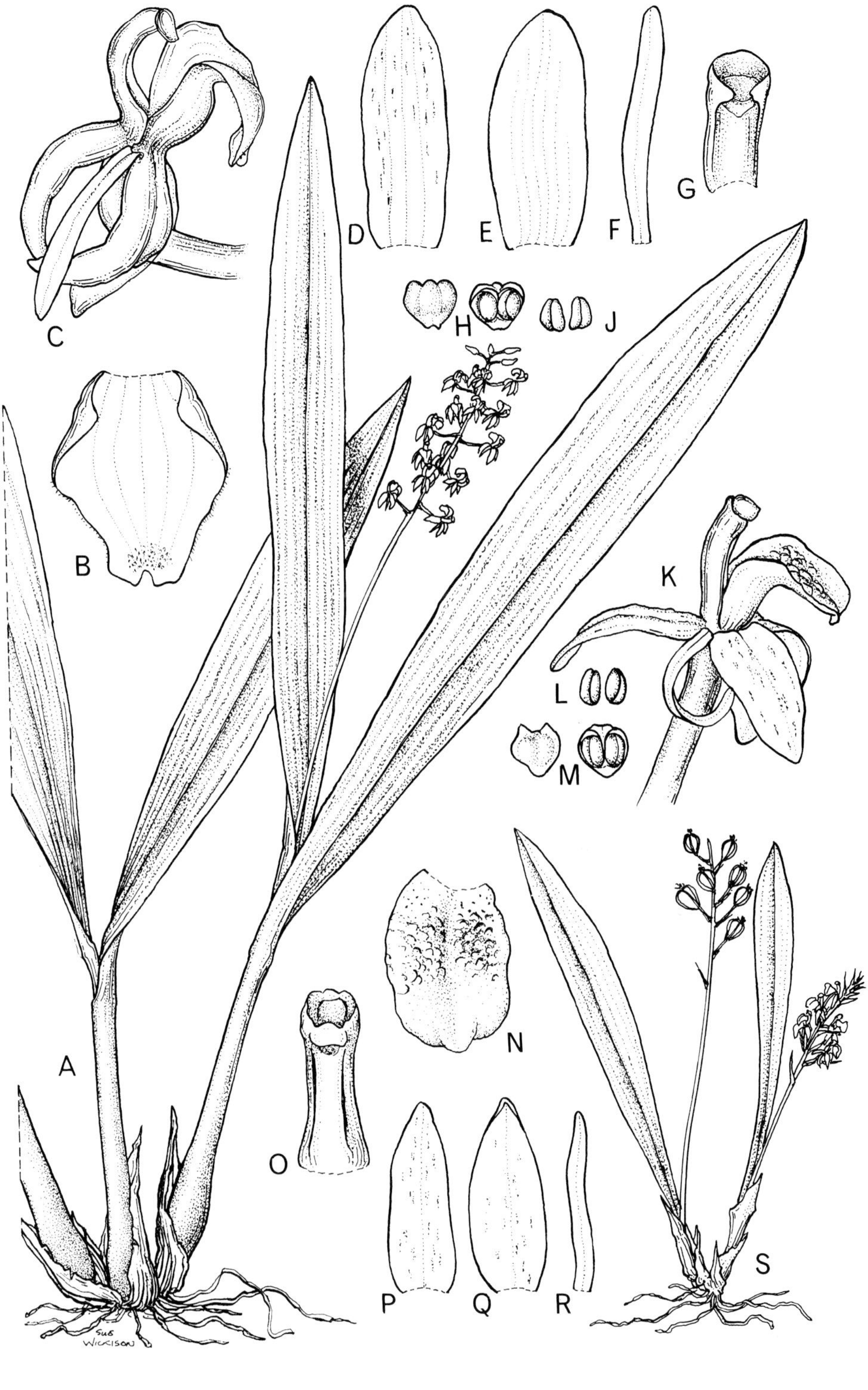
A
B
C
D
E
F
G
H
J
K
L
M
N
O
P
Q
R
S
SUE WICKISON

DISTRIBUTION: Guadalcanal. Also in New Guinea.
HABITAT: Montane ridge-top forest, 1000 m.
COLLECTION: *Cribb, Dennis & Campbell* 5096 (K).

3. L. condylobulbon *Reichb. f.* in Hamb. Gartenz. 18: 34 (1862). Type: cult. ex Batavia, *Schiller* s.n. (holotype W).
Liparis nesophila Reichb. f. in Otia Bot. Hamb. 1: 56 (1878). Type: Fiji, *Seemann* 614 (holotype W, isotype K!).
Liparis confusa J.J. Smith, Orch. Java: 275, fig. 211 (1908). Type: Java, *J.J. Smith* 903 (holotype BO!).
Liparis longipes auct. non Lindley, Gen. Sp. Orch. Pl.: 30 (1830).
Liparis elegans sensu Ames, in Journ. Arn. Arb. 14: 105 (1933) non Lindley.

An epiphytic herb forming large clumps. *Pseudobulbs* elongated, up to 24 cm long, swollen at base, 1.2–2 cm wide. *Leaves* usually 2 from apex of pseudobulb, erect, lanceolate, 12–20 cm long, 1.3–2.5 cm wide. *Inflorescence* terminal, 6–22 cm long, extending as the fruits ripen, many-flowered. *Flowers* usually self-pollinating, pale green, sometimes with an orange-pink lip, usually scented; pedicel and ovary 6–10 mm long; dorsal sepal reflexed, lanceolate-oblong, 2.5 mm long, 1.3 mm wide; lateral sepals reflexed, ovate-lanceolate, 3.1 mm long, 1.4 mm wide; petals reflexed, linear, 2.6 mm long, 0.5 mm wide; lip deflexed, oval, bifid at apex, 3 mm long, 2 mm wide, with a ciliate basal margin; column 1.5 mm long. (See fig. 21 A–J and plate 6A).

DISTRIBUTION: Bougainville, Choiseul, Guadalcanal, Kolombangara, Makira, New Georgia, Rennell, the Santa Cruz Islands and Santa Isabel. Widely distributed from Asia to the Malay Archipelago, New Guinea, Vanuatu, New Caledonia, the Horn Islands, Fiji and Samoa.
VERNACULAR NAMES: Hanaso (Mono Island & Treasury Island); Manapauna (NE. Bougainville); Susui & Fi'i Fari (Kwara'ae name).
HABITAT: A common epiphyte in rain forest, 20–600 m.
COLLECTIONS: *Beer* in BSIP 7297 (K); *Cribb* 6016 (K); *Dennis* in RSS 8173 (K); *Dickson* A4 (K); *Dockrill & Coode* in NGF 34187 (K); *Kere* in BSIP 5086 (K); *Maenu'u* in BSIP 5971 (K); *Mauriasi et al.* in BSIP 9762 (K) & 16651 (K); *Mitchell* 37 (K); *Sirute* in BSIP 9989 (K); *Waterhouse* 171A & 688 (K); *Wickison* 59, 93A & 93B (K); *Whitmore* 6262 (K).

4. L. gibbosa *Finet* in Bull. Soc. Bot. Fr. 55: 342, fig. 32–44 (1908). Type: Java, *Blume* s.n. (holotype P!).
Liparis disticha sensu Schltr. in Engler, Bot. Jahrb. 39: 60 (1906), non Lindley (1825).

An epiphytic or lithophytic plant growing in clumps. *Pseudobulbs* ovoid, up to 4 cm high, 1.2 cm wide, pale green, sometimes sheathed in cataphylls. *Leaf* solitary from apex of pseudobulb, erect, lanceolate, 11–25 cm long, 0.7–1.6 cm wide, pale green to rich green. *Inflorescence* terminal, 10–25 cm long; raceme 2–3 cm long; bracts distichous and imbricate, green to orange-brown. *Flowers* usually opening one at a time, bright orange, dark red or brown; pedicel and ovary 1 cm long; dorsal sepal erect or reflexed, ovate, 5 mm long, 2.3 mm wide; lateral sepals reflexed, ovate, 5 mm long, 2.6 mm wide; petals reflexed, linear, 5.2 mm long, 1.5 mm wide; lip reflexed, ovate, 4.5 mm long, 2 mm wide, with lateral margins erect; column swollen at base, 2 mm long, 2.5 mm wide, pale green. (See fig. 22 A–J).

DISTRIBUTION: Bougainville, Guadalcanal, Makira and Rendova. Widely distibuted from SE. Asia, the Malay Archipelago to New Guinea, Vanuatu, New Caledonia, Fiji and Samoa.
HABITAT: Rain forest, 150–900 m.
COLLECTIONS: *Chaplin* 763 (K); *Cribb, Dennis & Campbell* 5065 (K); *Hunt* in RSS 2213, 2950, 2990 & 3008 (K); *Lavarack & Ridsdale* in NGF 31100 (K); *Wickison* 105A (K).

5. L. layardii *F. Muell.* in Wing's South Sci. Rec.: 1 (1885). Type: New Caledonia, *Layard* s.n. (holotype MEL).
Liparis stricta Schltr. in Fedde, Repert. Spec. Nov. Regni Veg. 9: 95 (1910), non J.J. Smith (1907). Type: Samoa, *Vaupel* 134 (holotype B†, isotypes AMES, K!).
Liparis mataanensis J.J. Smith in Bull. Jard. Bot. Buitenz. 8: 56 (1912). Type: as for *L. stricta* Schltr. (1910).

A terrestrial herb. *Stems* up to 25 cm high, sheathed in cataphylls. *Leaves* 2–4 from apex of stem, ovate, acute, 7–13 cm long, 4–8.5 cm wide, glossy, mid-green, paler beneath. *Inflorescence* terminal, 18–30 cm long, purple, few- to several-flowered. *Flowers* purple with a green column; pedicel and ovary 2 cm long, purple; dorsal sepal reflexed, lanceolate, 12.5–14 mm long, 1 mm wide; lateral sepals lanceolate, 11 mm long, 1 mm wide; petals reflexed, linear, 12 mm long, 0.7 mm wide; lip ovate, 10 mm long, 8–9 mm wide; column slender, 5 mm long. (See fig. 22 K–S and plate 6B).

DISTRIBUTION: Guadalcanal, Kolombangara and Makira. Also in Vanuatu, New Caledonia, Fiji and Samoa.
HABITAT: Montane rain forest, 530–2200 m.
COLLECTIONS: *Cribb & Campbell* 5124 (K); *Cribb, Dennis & Campbell* 5016 & 5055 (K); *Dennis* in RSS 2037 & 2040 (K); *Hunt* in RSS 2360, 2957, 2987 & 2988 (K); *Mitchell* 5 (K); *Wickison* 228 (K).

6. L. orbiculata *L.O. Williams* in Amer. Orchid Soc. Bull. 10: 201 (1941). Type: Fiji, *Degener* 14810 (holotype AMES!, isotype K!).

An epiphytic herb 20 cm tall. *Pseudobulbs* 2 cm long, 5–8 mm wide, sheathed in bracts. *Leaves* 1 or 2 from apex of pseudobulb, linear-oblanceolate, acute, up to 16 cm long, 1.7 mm wide, with 3–5 prominent nerves. *Inflorescence* terminal, racemose, about equalling the leaves in length, loosely 5- to 10-flowered. *Flowers* white; pedicel and ovary 2.6 cm long; dorsal sepal broadly lanceolate, 8 mm long, 2 mm wide; lateral sepals broadly elliptic, slightly oblique, 7.5 mm long, 2.7 mm wide; petals linear, 7 mm long, 1 mm wide; lip broadly orbicular, 6 mm long, 6.75

mm wide, with an emarginate apex, the disc with a basal callus; column 3 mm long, swollen at the base.

DISTRIBUTION: Makira. Also in Vanuatu and Fiji (see note).
HABITAT: *Casuarina* forest, 550 m.
COLLECTION: *Dennis* in RSS 2354 (K).

This collection closely resembles the type of *L. orbiculata* (syn. *L. pullei* sensu B. Lewis & Cribb in 'Orchids of Vanuatu') but differs in being a smaller plant and in having smaller white flowers and a somewhat undulate margin to the lip. Further material may show that it warrants recognition as a distinct taxon.

7. L. truncicola *Schltr.* in Fedde, Repert. Spec. Nov. Regni Veg., Beih. 1: 199 (1911). Type: New Guinea, *Schlechter* 16674 (holotype B†, isotypes E, K!, L, MO, NSW, S).

An epiphytic herb. *Pseudobulbs* bilaterally flattened, 4 cm long, 8–10 mm wide, light green. *Leaf* solitary from apex of pseudobulb, erect, elliptic-lanceolate, 15–19 cm long, 2.8–3.4 cm wide, light green. *Inflorescence* terminal, 9–13.5 cm long, many-flowered; bracts lanceolate, 7 mm long, 2 mm wide, pale yellow-green. *Flowers* pale orange-green, becoming yellow-green as they mature; pedicel and ovary 6–9 mm long; dorsal sepal erect or reflexed, ovate, 5.6 mm long, 2.6 mm wide; lateral sepals widely spreading, obliquely ovate, 6 mm long, 2.8 mm wide; petals reflexed, linear, obtuse, 5.2 mm long, 0.9 mm wide; lip trilobed, 4.6 mm long, 4.2 mm wide, with a fleshy basal callus, the lateral lobes rounded and erect, the mid-lobe ovate-orbicular and reflexed; column slender, 4 mm long, 1.3 mm wide.

DISTRIBUTION: Guadalcanal. Also in New Guinea.
HABITAT: Rain forest, 700 m.
COLLECTION: *Wickison* 135 (K).

37. MALAXIS Solereder ex Swartz

Terrestrial or rarely epiphytic herbs. *Rhizome* creeping. *Stems* erect, leafy. *Leaves* plicate, thin-textured, broad, ovate, acute, often asymmetrical; the petioles sheathing at base. *Inflorescence* terminal, erect, racemose, with many small flowers. *Flowers* non-resupinate, occasionally self-pollinating; sepals and petals free; petals usually narrower than the sepals; lip entire or lobed, often

---

Fig. 22. Liparis gibbosa A-J. A, habit x 2/3; **B**, flower x 6; **C**, petal x 8; **D**, dorsal sepal x 8; **E**, lateral sepal x 8; **F**, lip x 8; **G**, column x 10; **H**, dorsal and ventral view of anther-cap x 10; **J**, pollinia x 10. **L. layardii K-S. K**, habit x 2/3; **L**, flower x 3; **M**, lip x 3; **N**, column x 3; **O**, lateral sepal x 3; **P**, dorsal sepal x 3; **Q**, petal x 3; **R**, dorsal and ventral view of anther-cap x 8; **S**, pollinia x 8. **A-J** drawn from *Wickison* 105A; **K** from *MacKee* 16275; **L-S** from *Mitchell* 5. All drawn by Sue Wickison.

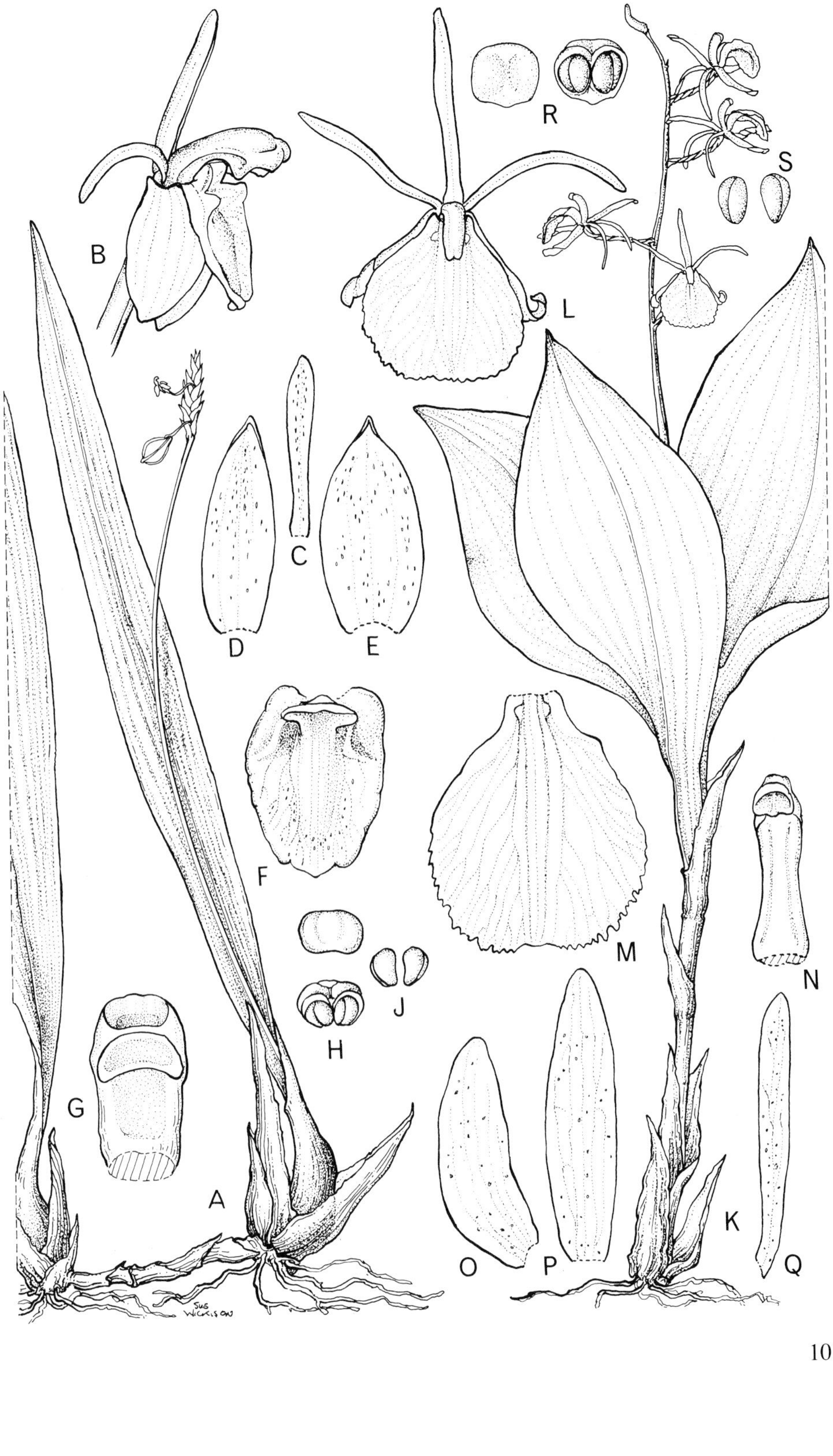
B
R
S
L
C
D
E
F
M
J
H
N
G
A
K
O
P
Q

with a hollow near the base, which may contain nectar, usually with 2 basal auricles close to the sides of the column, apex often toothed; column very short.

A cosmopolitan genus of about three hundred species, six species are recorded from the Solomon Islands and Bougainville.

The similar vegetative morphology, and small flowers which preserve poorly when pressed, make the identification of *Malaxis* species difficult and few of them have been adequately collected throughout the archipelago. Therefore the present treatment of *Malaxis* should be regarded as provisional.

The species are primarily distinguished on the basis of characters associated with the lip, although the shape of the petals is also important.

1. Lip entire; petals spathulate, more than 1.5 mm wide .. .. **2. M. latipetala**
   Lip with marginal teeth; petals linear or lanceolate, less than 1 mm wide 2
2. Pedicel and ovary 8–21 mm long .. .. **4. M. pectinata**
   Pedicel and ovary less than 6 mm long 3
3. Leaves less than 5 cm long 4
   Leaves more than 5 cm long 5
4. Leaves ovate-lanceolate, bronze-green .. .. **3. M.** aff. **lawleri**
   Leaves obliquely ovate, dark green, purple below .. .. **1. M. decumbens**
5. Teeth on lip shorter than the mid-lobe, less than 0.5 mm long .. .. **6. M. xanthochila**
   Teeth on lip longer than the mid-lobe, more than 1 mm long .. .. **5. M. resupinata**

1. M. decumbens *(Schltr.) P.F. Hunt* in Kew Bull. 24: 79 (1970). Type: New Guinea, *Schlechter* 17371 (holotype B†, isotype K!).

Microstylis decumbens Schltr. in Fedde, Repert. Spec. Nov. Regni Veg., Beih. 1: 137 (1911).

A terrestrial herb. *Rhizome* creeping. *Stems* erect, leafy, purple. *Leaves* obliquely ovate, acute, 3.5–5 cm long, 1.5–2.5 cm wide, dark green, purple below; petiole 1.5–2 cm long. *Inflorescence* terminal, slender, up to 10 cm long, lax, bearing several small flowers; bracts reflexed. *Flowers* non-resupinate, yellow with a purple disc on the lip and a purple column; pedicel and ovary 1 mm long; dorsal sepal oblong, 4 mm long, 1.7 mm wide; lateral sepals obliquely ovate, 4 mm long, 2 mm wide; petals lanceolate, 3 mm long, 1 mm wide; lip obscurely trilobed to horse-shoe shaped, 3.5 mm long, 4 mm wide, the basal auricles rounded, the apical margin with 4–6 short teeth, the mid-lobe elliptic; column 1 mm long.

DISTRIBUTION: Guadalcanal. Also in New Guinea.
HABITAT: Rain forest, 900 m.
COLLECTION: *Cribb, Dennis & Campbell* 5064 (K).

The type material examined is poorly preserved and therefore the identification of the Solomon Island collections is provisional.

2. M. latipetala *(J.J. Smith) P.F. Hunt* in Kew Bull. 24: 81 (1970). Type: New Guinea, *Djibda* 383 (holotype BO!).
Microstylis latipetala J.J. Smith in Bull. Dép. Agric. Indes Néerl. 39: 17 (1910).

A terrestrial herb. *Rhizome* creeping. *Stems* fleshy, erect, 4 cm high, leafy. *Leaves* thin, asymmetrical, ovate, acute, 4–8 cm long, 1.2–3 cm wide, with wavy edges; petiole up to 3.5 cm long, sheathing at base. *Inflorescence* terminal, 15–30 cm long, lax, bearing many small flowers; bracts reflexed. *Flowers* non-resupinate, sometimes cleistogamous, yellow to pale orange-green with paler petals, a yellowish-green to purple lip and a purple apex to the column, often scented; pedicel and ovary 5–8 mm long; dorsal sepal ovate, 3.5 mm long, 2 mm wide; lateral sepals obliquely ovate, 3.5 mm long, 1.5–2.5 mm wide; petals spathulate, 2.5–3 mm long, 1.5–2.5 mm wide; lip entire, 2–2.5 mm long, 1.5 mm wide, the disc with a hollow, basal auricles oblong, rounded; column short, 1 mm long, usually with a papillose base.

DISTRIBUTION: Bougainville, Choiseul, Guadalcanal, Kolombangara, New Georgia, San Cristobal and Santa Isabel. Also in New Guinea.
HABITAT: Lowland rain forest to moss forest, 30–1000 m.
COLLECTIONS: *Beer's collectors* in BSIP 7364 (K); *Cribb & Campbell* 5105 (K); *Dennis* in RSS 2051, 2357 & 8508 (K); *Hunt* in RSS 2220 & 2646 (K); *Mitchell* 50 (K); *Ridsdale & Lavarack* in NGF 30676 (LAE); *Robinson* K192 & K196 (K); *Susui* in BSIP 8263 (K); *Wickison* 61 (K).

This species is readily distinguished by its entire lip and spathulate petals. It is similar to *M. latisegmenta* C. Schweinf. from Fiji.

3. M. aff. **lawleri** *Lavarack & Gray* in Orchadian 8(1): 6 (1984). Type: Australia, *Lawler* in BG 3378 (holotype QRS).

A small terrestrial herb. *Stems* up to 4 cm high, leafy. *Leaves* 3–5, ovate-lanceolate, 3.5–6 cm long, 1–2 cm wide, bronze-green, with 3 prominent veins; petiole up to 3 cm long. *Inflorescence* terminal, up to 14 cm long, lax, bearing few small flowers; bracts reflexed. *Flowers* non-resupinate, pale brown or translucent light green with lavender tips; column dark green with a yellow anther-cap; pedicel and ovary up to 1.5 mm long; dorsal sepal oblong-linear, 3–4 mm long, 1–2 mm wide; lateral sepals obliquely ovate, 3–4 mm long, 2 mm wide; petals narrowly linear, 3–3.5 mm long, 0.8 mm wide; lip horse-shoe-shaped, 3.5–4.5 mm long, 3–3.5 mm wide, the basal auricles triangular, rounded, the apical margin with short, triangular teeth; column 1 mm long.

DISTRIBUTION: Bougainville, Ranongga and Santa Ana Island.
HABITAT: Damp moss overlying leaf litter, sea level up to 900 m.
COLLECTIONS: *Dennis* 7 (K); *Hadley* 77 (BM); *Mitchell* 72 (K); *Womersley* in NGF 48679 (K, LAE).

This species is readily distinguished from other species in the Solomon Islands and Bougainville by its small habit and shortly lobed lip. The material agrees well with the description of *M. lawleri* from Australia but this identification is provisional as we have not seen any Australian material of that species.

4. M. pectinata *(J.J. Smith) P.F. Hunt* in Kew Bull. 24: 83 (1970). Type: New Guinea, *Branderhorst* 149 (holotype BO!).
Microstylis pectinata J.J. Smith in Bull. Dép. Agric. Indes Néerl. 19: 29 (1908).

A terrestrial herb. *Rhizome* creeping. *Stems* erect, 3–13 cm high, leafy. *Leaves* obliquely ovate, acute, 4–11 cm long, 1.3–5 cm wide; petioles up to 4 cm long, sheathing at base. *Inflorescence* terminal, 9–22 cm long, lax, bearing many small flowers; bracts reflexed. *Flowers* deep purple (sometimes with a pale green disc on the lip); pedicel and ovary 0.8–2.1 cm long, elongating up to 3 cm long as the seed capsule matures; dorsal sepal oblong, obtuse, 3 mm long, 1 mm wide; lateral sepals obliquely ovate, 3 mm long, 1.3 mm wide; petals narrowly oblong-linear, 3 mm long, 0.5 mm wide; lip horse-shoe-shaped, 3.5 mm long, 3.5 mm wide, the basal auricles obliquely triangular; the disc concave, the apical margin fringed by 8–10 teeth; column 1 mm long.

DISTRIBUTION: Bougainville, Kolombangara and New Georgia. Also in New Guinea.
VERNACULAR NAME: Mengonirung (Bougainville).
HABITAT: Rain forest, 30–900 m.
COLLECTIONS: *Hunt* in RSS 2441 (K, LAE); *Waterhouse* 291 (K); *Wickison* 60 (K); *Ridsdale & Lavarack* in NGF 31008 (K).

This species is readily distinguished by its long pedicels. It is similar to *M. setipes* (Schltr.) C. Schweinf. from Palau in the Caroline Islands.

5. M. resupinata *(Forst.f.) Kuntze*, Rev. Gen. Pl. 2: 673 (1891). Type: Tahiti, *J.R. & G. Forster* s.n. (lectotype BM!).
Epidendrum resupinatum Forst. f., Fl. Ins. Austr. Prodr.: 61 (1786).
Microstylis resupinata (Forst. f.) Drake, Ill. Fl. Ins. Mar. Pac.: 305 (1892).

A terrestrial herb. *Rhizome* creeping. *Stems* erect, fleshy, leafy. *Leaves* obliquely ovate, acute, 11–16 cm long, 3.5–5 cm wide, pale green with wavy margins; petiole up to 5.5 cm long, sheathing at base. *Inflorescence* terminal, erect, up to 41 cm long, purple, lax to dense, bearing many small flowers; bracts reflexed. *Flowers* non-resupinate, greenish-yellow or purple with a yellow anther-cap; pedicel and ovary 4–6 mm long; dorsal sepal ovate, 2.5–3 mm long, 1.7–2 mm wide; lateral sepals oblong-elliptic, 2.5–3 mm long, 2 mm wide; petals narrowly oblong, 2.5–3 mm long, 0.7–1 mm wide; lip obscurely trilobed to horse-shoe-shaped, 3–4 mm long, 3–4 mm wide, the basal auricles obliquely triangular, the apical margin fringed with 6–8, 1–2 mm long, teeth, the mid-lobe elliptic, shortly emarginate; column 1 mm long.

DISTRIBUTION: Guadalcanal, Makira, Ranongga and Santa Isabel. Also in Vanuatu, Fiji, Tonga, Samoa, Niue, the Cook Islands, the Tubai Islands and the Society Islands.
VERNACULAR NAME: Ongi'ongi (Kwara'ae name, Guadalcanal).
HABITAT: Rain forest, sea level to 500 m.
COLLECTIONS: *Baea* in BSIP 8102 (K); *Cribb, Dennis & Campbell* 5003 & 5172 (K); *Hunt* in RSS 2223, 2362, 2687 & 3023 (K); *Mitchell* 75 (K); *Nakisi* in BSIP 8061 (K); *Nakisi & Mauriasi* in BSIP 8068 (K).

This species is similar to *M. stenostachys* (Schltr.) P.F. Hunt from New Guinea.

6. M. xanthochila *(Schltr.) Ames & C. Schweinf.* in Ames, Orch. 5: 67 (1915). Type: New Guinea, *Schlechter* 13678 (holotype B†).
Microstylis xanthochila Schltr. in K. Schum. & Laut., Nachtr. Fl. Deutsch. Südsee: 102 (1905).
Microstylis sordida J.J. Smith in Bull. Dép. Agric. Indes Néerl. 19: 30 (1908). Type: New Guinea, *Versteeg* 1333 (holotype L).

A terrestrial herb. *Stems* fleshy, 20–30 cm high, leafy. *Leaves* obliquely ovate, acuminate, 5.5–10 cm long, 2–3.7 cm wide; petiole 2.5–5 cm, sheathing at base. *Inflorescence* terminal, 11–15 cm long, lax, bearing many small flowers; bracts reflexed. *Flowers* non-resupinate, yellow, yellowish-orange, greenish buff, or purple; pedicel and ovary 4–6 mm long; dorsal sepal elliptic-oblong, obtuse, 3–3.8 mm long, 1–1.5 mm wide; lateral sepals obliquely elliptic-ovate, obtuse, 3–3.5 mm long, 2 mm wide; petals linear, obtuse, 3–4 mm long, 1 mm wide; lip obscurely trilobed to horse-shoe-shaped, 4–4.5 mm long, 2.5–3.5 mm wide, the basal auricles rounded, shortly 6–8-toothed on front margin, the teeth c. 0.5 mm long, the mid-lobe small, semi-elliptic, shortly emarginate; column 1 mm long. (See plate 6D).

DISTRIBUTION: Kolombangara, New Georgia, San Cristobal, the Shortland Islands and Treasury Island. Also in New Guinea, Vanuatu and Australia.
HABITAT: Rain forest, 55–800 m.
COLLECTIONS: *Guppy* 82 (K); *Hunt* in RSS 2261, 2262, 2324 & 2444 (K); *Mitchell* 5 (K); *Runikera* in BSIP 13188 (K); *Wickison* 57 (K).

38. OBERONIA Lindley

Epiphytic herbs. *Stems* short or long, lacking pseudobulbs. *Leaves* bilaterally flattened, either arranged distichously on an elongated stem or grouped in a fan on a short stem, sheathing at base. *Inflorescence* terminal, densely many-flowered, with the flowers in whorls and the basal usually opening last. *Flowers* small, non-resupinate, white, pale green, yellow, orange, red or brown, apparently sessile; sepals and petals free with the sepals broader than the petals; lip usually 3-lobed with 2 basal auricles that may or may not enfold the base of the column; column short, terete, lacking a foot; pollinia 4.

A genus of about three hundred and thirty species in the Old World tropics, seven species being recorded.

1. Leaves grouped at the base of the very short stem (less than 6 cm long) .... **3. O. heliophila**
   Leaves distichous on an elongate stem (usually more than 6 cm long) 2
2. Leaves less than 2 cm long, appressed to the stem .... **4. O. imbricata**
   Leaves more than 2 cm long, not appressed to the stem 3
3. Stem less than 15 cm long 4
   Stem more than 20 cm long 5

4. Leaves up to 10 cm long, 3–5 mm wide .. .. **1. O. bifida**
   Leaves 2–3 cm long, 3–4 mm wide .. .. **7. O. sp. B**
5. Inflorescence pendent, up to 30 cm long, laxly many-flowered .. .. .. **6. O. sp. A**
   Inflorescence erect, less than 20 cm long, densely many-flowered 6
6. Leaves up to 30 cm long; pedicel, ovary and bracts glabrous; lip obcordate, with a bilobed apex with semicircular lobes .. .. **2. O. djamuensis**
   Leaves less than 12 cm long; pedicel, ovary and bracts finely pubescent; lip oblong, slightly constricted in the middle, with a minutely bilobed apex with obtuse, rounded lobes .. .. **5. O. longicaulis**

1. O. bifida *Schltr.*, in K. Schum. & Laut., Nachtr. Fl. Deutsch. Südsee: 109 (1905). Type: New Guinea, *Schlechter* 14081 (holotype B†).

An epiphytic herb. *Stems* up to 4 cm long. *Leaves* distichous, up to 6, linear-lanceolate, acute, up to 10 cm long, 3–5 mm wide, light green, with sheathing bases. *Inflorescence* terminal, up to 10 cm long, densely many-flowered; bracts lanceolate with serrate margins, 2 mm long, 0.4 mm wide. *Flowers* in whorls of up to 8, non-resupinate, pale pink-brown; dorsal sepal ovate, 0.8 mm long, 0.6 mm wide; lateral sepals ovate, 0.7 mm long, 0.5 mm wide; petals linear-ovate, 0.7 mm long, 0.3 mm wide; lip oblong, 1 mm long, 0.5 mm wide, bifid at apex, with obtuse lobes; column short, terete. (See fig. 23 C–D).

DISTRIBUTION: Bougainville and Guadalcanal. Also in New Guinea and Vanuatu.
HABITAT: Rain forest and exposed ridges, 730–950 m.
COLLECTION: *Ridsdale & Lavarack* in NGF 31608 (K); *Wickison* 109 (K).

Material of this species, (*Lewis & McDonagh* 149), has recently been collected on Efate, Vanuatu which extends its known distribution southeastwards.

2. O. djamuensis *Schltr.* in Fedde, Repert. Spec. Nov. Regni Veg., Beih. 1: 174 (1911). Type: New Guinea, *Schlechter* 16633 (holotype B†).

An epiphytic herb. *Stems* 30 cm long, slightly zig-zag. *Leaves* distichous, linear, acute, up to 30 cm long, 1.1 cm wide, mid-green, with sheathing bases. *Inflorescence* terminal, up to 17 cm long, densely many-flowered. *Flowers* in whorls, non-resupinate, brown; dorsal sepal ovate, 1.8 mm long, 1 mm wide; lateral sepals ovate, 1.8 mm long, 1 mm wide; petals obliquely ovate-lanceolate, subacute, 1.7 mm long, 0.8 mm wide; lip obcordate, 2 mm long, 2 mm wide, fleshy at base, the apex bilobed with semi-circular lobes; column short, terete.

Fig. 23. Oberonia heliophila A-B, A, habit x 2/3; **B**, flower x 14. **O. bifida C-D. C**, habit x 2/3; **D**, flower x 14. **O. imbricata E-F. E**, habit x 2/3; **F**, flower x 14. **A-B** drawn from *Cribb & Morrison* 1770; **C-D**, from *Wickison* 109; **E** from *Hallé* 16370; **F** from *Robinson* K133. All drawn by Sue Wickison.

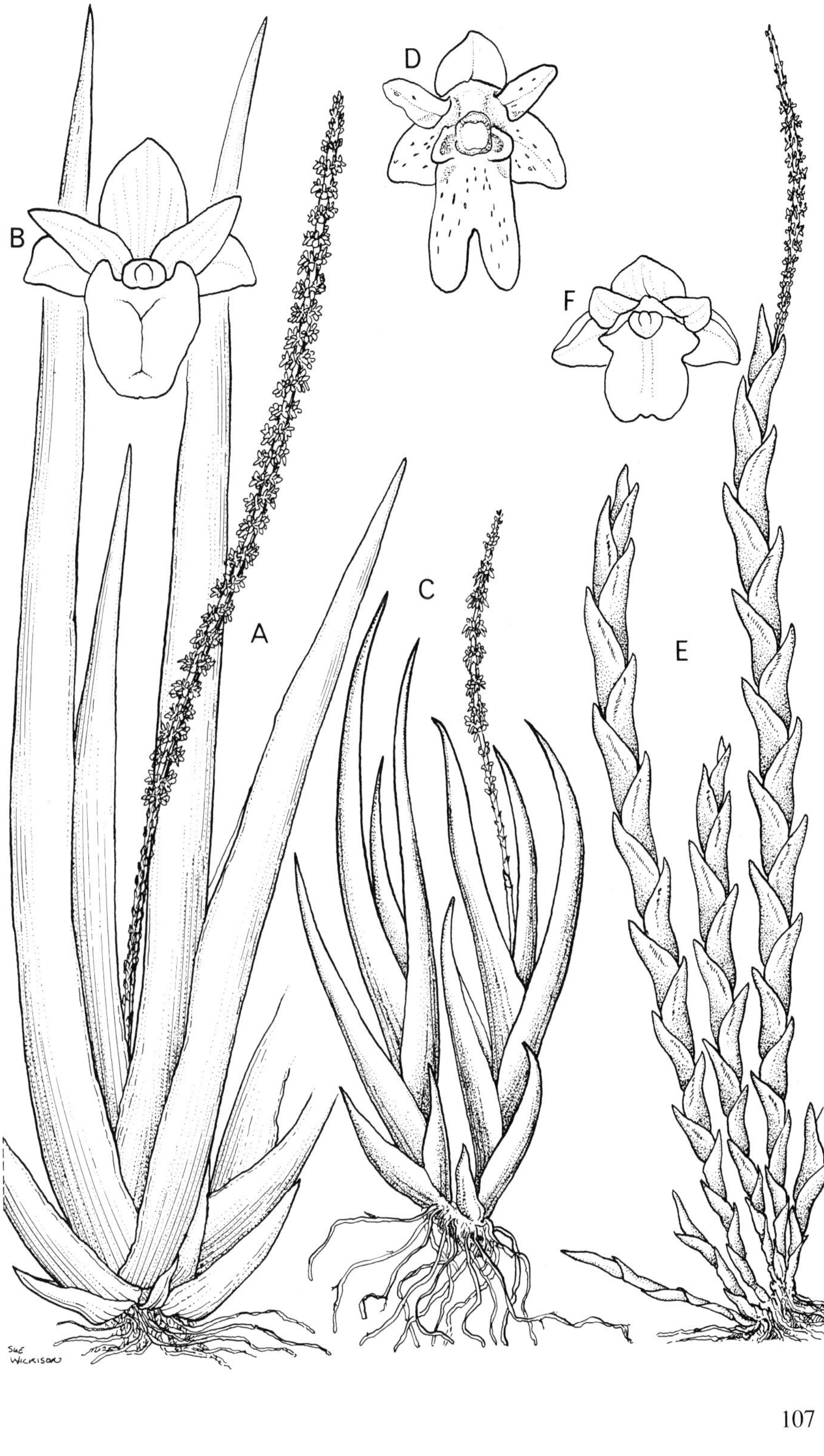
A
B
C
D
E
F

DISTRIBUTION: Bougainville. Also in New Guinea.
HABITAT: Rain forest, 1100–1670 m.
COLLECTIONS: *Lavarack & Ridsdale* in NGF 31328 & 31588 (K).

3. O. heliophila *Reichb. f.* in Otia Bot. Hamb.: 56 (1878). Type: Samoa, *U.S. Expl. Exped.* (lectotype W, isolectotypes AMES, US).
Malaxis heliophila Reichb. f. in Otia Bot. Hamb.: 56 (1878) nom. alt.
Oberonia betchei Schltr. in Bull. Herb. Boiss., ser 2, 6: 303 (1906). Type: Samoa, *Betche* 38 (holotype B†, isotype MEL).

For further information see Kores (1989).

An epiphytic herb. *Stems* up to 6 cm long. *Leaves* grouped at the base of the stem, linear, acuminate, 10–35 cm long, 1–1.5 cm wide, yellow-ochre to dark green. *Inflorescence* terminal, 20–30 cm long, densely many-flowered. *Flowers* in whorls, non-resupinate, white to pale brown with a yellow or bright green lip; dorsal sepal ovate, 1 mm long, 0.75 mm wide; lateral sepals ovate, 1 mm long, 0.8–0.9 mm wide; petals lanceolate, 1 mm long, 0.5 mm wide; lip entire or obscurely trilobed, obovate, obtuse, 1.5 mm long, 1.2 mm wide, with 2 rounded basal auricles which enclose the base of the column; column short, terete. (See fig. 23 A–B).

DISTRIBUTION: Guadalcanal and Makira. Also in Vanuatu, Fiji and Samoa.
HABITAT: Rain forest, 50–630 m.
COLLECTIONS: *Hunt* in RSS 2226 (K); *Wickison* 87 (K).

4. O. imbricata *(Blume) Lindley*, Gen. Sp. Orch. Pl.: 17 (1830). Type: Java, *Blume* s.n. (holotype L, isotype P!).
Malaxis imbricata Blume, Bijdr. Fl. Ned. Ind.: 395 (1825).

An epiphytic herb. *Stems* erect or pendent, up to 40 cm long. *Leaves* distichous, appressed to stem, falcate-lanceolate, 1–2 cm long, 4–7 mm wide, with imbricate sheathing bases, mid-green becoming paler at margins and apex. *Inflorescence* terminal, 7–15 cm long, densely many-flowered. *Flowers* in whorls, non-resupinate, green or brown to yellow-orange; dorsal sepal ovate, 1 mm long, 0.6–0.7 mm wide; lateral sepals ovate, 1 mm long, 0.6–0.7 mm wide; petals ovate-oblong, 0.8 mm long, 0.4 mm wide; lip oblong, 1 mm long, 0.8 mm wide, with a rounded, minutely apiculate apex, and erect basal auricles which enclose the base of the lip; column short, terete. (See fig. 23 E–F).

DISTRIBUTION: Bougainville, Guadalcanal, Kolombangara and Makira. Also in the Malay Archipelago, New Guinea and Vanuatu.
HABITAT: Rain forest, sea level to 1100 m.
COLLECTIONS: *Chaplin* 838 (K); *Dennis* in RSS 4437 (K); *Hunt* in RSS 2233, 2369 & 2930 (K); *Ridsdale & Lavarack* in NGF 31590 & 31607 (K); *Wickison* 219 (K).

5. O. longicaulis *Schltr.* in K. Schum. & Laut., Nachtr. Fl. Deutsch. Südsee: 112 (1905). Type: New Ireland, *Schlechter* 14666 (holotype B†).

An epiphytic herb. *Stems* pendent, slightly zig-zag, up to 50 cm long. *Leaves* distichous, 3–8, linear, acute, incurved at apex, up to 12 cm long, 1.1 cm wide, pale yellow-green to mid-green, with sheathing bases. *Inflorescence* terminal, up to 14 cm long, with many, dense flowers; bracts lanceolate, with serrate margins, finely pubescent on outer surface. *Flowers* in whorls, non-resupinate, yellow to brown; pedicel and ovary 2.5 mm long, pubescent; dorsal sepal oblong-ovate, obtuse, 2 mm long, 0.75 mm wide; lateral sepals obliquely oblong-ovate, 1.5 mm long, 0.75 mm wide; petals linear, obtuse, 2 mm long, 0.5 mm wide; lip oblong, 1.5 mm long, 1 mm wide, slightly constricted in the middle, with a minutely bilobed apex with obtuse, rounded lobes; column short, terete. *Fruits* yellow, cylindrical, ribbed, 4 mm long, 2 mm wide.

DISTRIBUTION: Bougainville, Choiseul, Guadalcanal, Makira and New Georgia. Also in New Guinea and New Ireland.
HABITAT: Rain forest, 50–600 m.
COLLECTIONS: *Hunt* in RSS 2192 & 2230; *Kajewski* 2154 (G); *Mitchell* 40 (K); *Peach* 12 (K); *Waterhouse* 171B; *Wickison* 90 (K).

6. O. sp. A.

An epiphytic herb. *Stem* up to 40 cm long, slightly zig-zag. *Leaves* 7–9, distichous, linear, acute, up to 17 cm long, 6 mm wide, with sheathing bases. *Inflorescence* terminal, pendent, up to 30 cm long, laxly many-flowered. *Flowers* not observed. *Fruits* cylindrical, ribbed, 4 mm long, 2 mm wide.

DISTRIBUTION: Guadalcanal.
HABITAT: Rain forest, 545 m.
COLLECTION: *Wickison* 133 (K).

This collection is similar in habit to *Oberonia longicaulis* Schltr. and *O. djamuensis* Schltr. but it differs from them in having long, narrow leaves and a long, pendent, laxly-flowered inflorescence. Flowering material of this species is needed for identification.

7. O. sp. B.

An epiphytic herb. *Stem* up to 12 cm long, slightly zig-zag. *Leaves* 6–12, distichous, linear-lanceolate, acute, 2–3 cm long, 3–4 mm wide, pale green, sheathing at base. *Inflorescence* terminal, up to 5 cm long, densely many-flowered. *Flowers* not observed. *Fruits* cylindrical, ribbed, 1.5 mm long, 1.5 mm wide.

DISTRIBUTION: Guadalcanal and Kolombangara. Also in Vanuatu.
HABITAT: Rain forest, 660–730 m.
COLLECTIONS: *Hunt* in RSS 2546 (K); *Wickison* 109B (K).

The collections we have seen of this species are in fruit but the small growth habit with the well-spaced, small leaves is distinctive.

39. ERIA Lindley

Epiphytic plants. *Pseudobulbs* short or long, slender or thick and fleshy, 1– to several-leaved, basal part always covered by thin-textured sheaths. *Leaves* distichous or near apex only. *Inflorescences* lateral or apparently terminal, rarely 1-flowered but usually many-flowered. *Flowers* hairy or glabrous on the outside, usually rather small and not very showy; sepals and petals not greatly elongated; dorsal sepal and petals free; lateral sepals usually fused to the column-foot to form a saccate to spur-like mentum; lip entire or trilobed, the disc with or without keels; column short or moderately long, with a column-foot; pollinia 8.

A large genus of about three hundred and seventy-five species mostly in tropical Asia extending to New Guinea, the Pacific and Australia; seven species being recorded.

1. Flowers, pedicel and rhachis glabrous or very sparsely pubescent 2
   Flowers, pedicel and rhachis pubescent 4
2. Sepals and petals 1.5 cm long or more; lip with 3 low raised calli towards the apex .. .. **3. E. kaniensis**
   Sepals and petals 1 cm long or less; lip with 2 or 3 raised callus ridges in basal half 3
3. Lip entire, ovate-lanceolate, acute, with 2 callus ridges .. .. **7. E. rostriflora**
   Lip trilobed, with 3 callus ridges .. .. **5. E. oligotricha**
4. Mentum 4–5 mm long, cylindrical, longer than the sepals; ovary and sepals covered in stellate hairs; lip oblong to obscurely trilobed and ecallose .. .. **6. E. robusta**
   Mentum shorter than the sepals; sepals and ovary covered in simple hairs; lip not as above 5
5. Pseudobulbs orangish-brown when dry; lip entire; callus of 3 longitudinal low ridges .. .. **2. E. fitzalanii**
   Pseudobulbs green; lip trilobed 6
6. Pseudobulbs cylindrical and stem-like; leaves 2(-3), slenderly petiolate .. .. **1. E. eriaeoides**
   Pseudobulbs fleshy, fusiform-cylindrical; leaves 3(-4), tapering towards the base .. .. **4. E. kingii**

1. E. eriaeoides *(Bailey) Rolfe* in Orchid Review 17: 95 (1909). Type: Australia, *Bancroft* s.n. (holotype BRI, isotype K!).
Dendrobium eriaeoides Bailey, Syn. Qd. Fl. Suppl. 2: 56 (1888).
Eria diphylla Schltr. in Fedde, Repert. Spec. Nov. Regni Veg., Beih. 1: 663 (1912). Type: New Guinea, *Schlechter* 16582 (holotype B†, isotypes AMES!, E!, K!, L!, NSW, S).

An epiphytic plant. *Pseudobulbs* cylindrical, stem-like, 4–6 cm long, 2–3.5 mm wide, clustered, covered by loose, short, obtuse to acute sheaths. *Leaves* 2(-3) near apex of pseudobulb, erect, lanceolate or oblanceolate, acute, slenderly petiolate, 15–25 cm long, 1.5–2 cm wide. *Inflorescence* lateral from upper nodes, much shorter than the leaves, erect, laxly few-flowered, 5–8 cm long; bracts elliptic, 2–3 mm long, erose. *Flowers* creamy white to pale pink, with a maroon mentum and lateral lobes of the lip, pubescent; dorsal sepal ovate, obtuse, 3 mm long, 1 mm wide; lateral sepals obliquely triangular-ovate, acute, 3–4 mm long, 3–4 mm wide; mentum oblong-conical, 2.5 mm long; petals elliptic, obtuse, 3 mm long, 1 mm wide; lip trilobed and recurved in middle, 3.5 mm long, 2.5 mm wide, the lateral lobes rounded, the mid-lobe ovate, obtuse, the disc with a callus of 2 fleshy ridges in the basal half; column 1 mm long, the foot 2.5 mm long. (See fig. 24 C–D).

DISTRIBUTION: Guadalcanal, Makira and New Georgia. Also in New Guinea and Australia.
HABITAT: Rain forest, 70–500 m.
COLLECTIONS: *Dennis* in RSS 2266 (K); *Hunt* in RSS 2935 (K); *Wickison* 95 & s.n. (K).

2. E. fitzalanii *F. Muell.* in Wing's South. Sci. Rec. 2: 252 (1882). Type: Australia, *Fitzalan* (holotype MEL but not found), *Clarkson* 2604 (neotype BRI).
Eria solomonensis Rolfe in Bull. Misc. Inf., Kew 1909: 63 (1909). Type: Solomon Islands, Malaita, *Woodford* 11, cult. Kew (holotype K!).

An epiphytic herb. *Pseudobulbs* elliptic to cylindrical, several-noded, 6–13 cm long, 1–2 cm wide, orange-brown when dried, 3- to 5-leaved. *Leaves* near apex of pseudobulb, suberect, plicate, lanceolate, acuminate, 16–30 cm long, 1–2.5 cm wide. *Inflorescence* lateral from upper nodes, 13–30 cm long, often drooping, laxly to densely many-flowered; bracts 5–6 mm long. *Flowers* lemon-yellow, covered in brown curly hairs; pedicel and ovary 7–8 mm long, hairy; dorsal sepal oblong, obtuse, 7 mm long, 2 mm wide; lateral sepals oblong-triangular, obtuse, 6 mm long, 4 mm wide; mentum oblong-conical, 3.5 mm long; petals oblong, rounded or obtuse, 6 mm long, 2 mm wide, erose at apex; lip variable, entire to obscurely trilobed near apex, 6 mm long, 3 mm wide, papillose, the disc with a callus of 3 fleshy ridges, extending from the base to the apex, reflexed at apex; column 2 mm long. (See fig. 24 E–F and plate 5E).

DISTRIBUTION: Guadalcanal, Malaita and New Georgia. Also in Australia.
HABITAT: Sea level to lower montane forest.
COLLECTIONS: *Chaplin* 813, 819 & 821A (K); *Cribb & Morrison* 1832 & 1873 (K); *Im Thurn* 354 (K); *Woodford* 11 (K).

3. E. kaniensis *Schltr.* in Fedde, Repert. Spec. Nov. Regni Veg., Beih. 1: 651 (1912). Type: New Guinea, *Schlechter* 17642 (holotype B†).

An epiphytic herb. *Pseudobulbs* 4-angled, ovoid-ellipsoid, 3–4 cm long, 1.3–2 cm wide. *Leaves* 2–3 near apex of pseudobulb, erect or suberect, oblanceolate-elliptic, acute or shortly acuminate, 18–25 cm long, 1.7–2.3 cm wide, petiolate.

Inflorescence lateral from upper nodes, arcuate to pendent, laxly many-flowered, up to 20 cm long; peduncle bilaterally flattened; bracts lanceolate, acuminate, 9–10 mm long. *Flowers* translucent pale yellow-green with a white column and anther; pedicel and ovary triangular in cross section; dorsal sepal lanceolate, acute, 16–19 mm long, 4.5 mm wide; lateral sepals obliquely lanceolate, acute, 18–20 mm long, 4.5 mm wide; mentum rounded, parallel to ovary, 2.5–4 mm long; petals lanceolate, acute, 16–17 mm long, 4 mm wide; lip elliptic, acute, erose, 11–15 mm long, 5.5 mm wide, recurved at apex, with a callus of 3 raised erose ridges; column 3.5 mm long, the foot incurved, 4 mm long. (See fig. 24 G–H).

DISTRIBUTION: Bougainville. Also in New Guinea.
HABITAT: No information from Bougainville; the type collection is from the forests of the Kani Range, c. 800 m.
COLLECTION: *Mason* RHS 182 (K).

4. E. kingii *F. Muell.* in Wing's South Sci. Rec. 2(4): 71 (1882). Type: Solomon Islands, *Goldfinch* s.n. (holotype MEL).

For full synonymy see Clements (1989).

An epiphytic herb. *Pseudobulbs* fleshy, fusiform-cylindrical, several-noded, 5–18 cm long, 1.2–2 cm wide. *Leaves* 3(-4) near apex of pseudobulb, arcuate to suberect, oblanceolate to oblong-lanceolate, acute, 11–42 cm long, 1.2–4 cm wide, plicate. *Inflorescences* 1–3 from upper nodes, erect or suberect, densely many-flowered, 13–25 cm long; bracts reflexed, lanceolate, subacute, 5–7 mm long, pubescent; peduncle pubescent, up to 25 cm long. *Flowers* buff yellow with pink margins to petals, a bright yellow lip and a purple anther-cap, covered in red-brown hairs; pedicel and ovary 7–16 mm long, pubescent; dorsal sepal elliptic, obtuse, 4.5 mm long, 3 mm wide; lateral sepals oblong-triangular, 5 mm long, 4 mm wide; mentum roundly conical, 3–3.5 mm long; petals elliptic, obtuse, 4 mm long, 2 mm wide; lip trilobed, 5 mm long, 4 mm wide, saccate at base, the disc with a central callus and 2 obscure short calli at the base, the lateral lobes oblong, erect, the mid-lobe semicircular to transverse elliptic, recurved at apex; column 1.5 mm long. (See fig. 24 J–K and plate 5F).

DISTRIBUTION: Guadalcanal, Malaita and Santa Isabel. Also in New Guinea and Australia.
HABITAT: Rain forest, sea level to 530 m.
COLLECTIONS: *Cribb & Morrison* 1845 (K); *Dennis* in RSS 4617 (K); *Hunt* in RSS 2812 & 3081 (K).

Fig. 24. Eria rostriflora A-B. A, flower x 4; B, lip x 6. **E. eriaeoides C-D. C,** flower x 4; D, lip x 6. **E. fitzalanii E-F. E**, flower x 4; **F**, lip x 6. **E. kaniensis G-H. G**, flower x 2; **H**, lip x 4. **E. kingii J-K. J**, flower x 4; **K**, lip x 6. **E. robusta L-M. L,** flower x 4; **M**, lip x 6. **A-B** drawn from *Smith & Newmarch* 1/006; **C- D** from *Wickison* 95; E-**F** from *Cribb & Morrison* 1832; **G-H** from *Mason* RHS 182; **J-K** from *Cribb & Morrison* 1845; **L-M** from *Wickison* 156. All drawn by Sue Wickison.

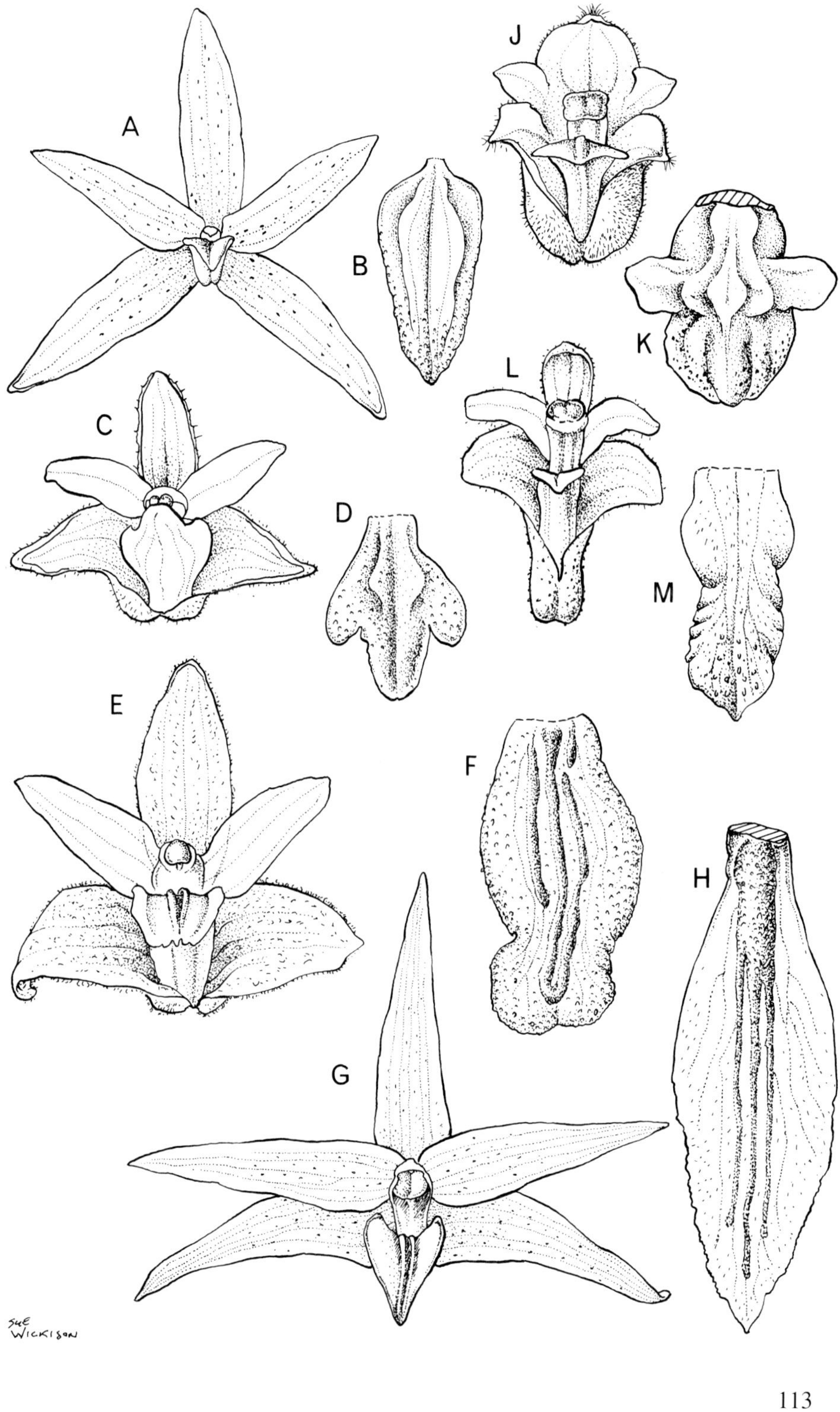
A
B
C
D
E
F
G
H
J
K
L
M
SUE WICKISON

5. E. oligotricha *Schltr.* in K. Schum. & Laut., Nachtr. Fl. Deutsch. Südsee: 181 (1905). Type: New Guinea, *Schlechter* 14353 (holotype B†).

An epiphytic herb. *Pseudobulbs* clustered on short rhizome, 7–24 cm long, 7–25 mm wide, swollen, several-noded. *Leaves* 4–6 near apex of pseudobulb, suberect, lanceolate, acute, 22–35 cm long, 3–4.3 cm wide. *Inflorescence* from upper nodes, erect to arcuate, densely many-flowered, c. 16 cm long; peduncle and rhachis sparsely pubescent; bracts 8.5–10 mm long, 5.5–6 mm wide, yellow-ochre. *Flowers* buff to pale yellow with a darker yellow lip and anther-cap; pedicel and ovary 7–10 mm long, sparsely pubescent; dorsal sepal ovate, obtuse, concave, 7.5–9 mm long, 4 mm wide; lateral sepals oblong-triangular to ovate, obtuse, 7.5–8 mm long, 5 mm wide; mentum 3 mm long; petals falcate, elliptic, obtuse, 6.5–8.5 mm long, 3 mm wide; lip trilobed, 8 mm long, 5 mm wide, the lateral lobes porrect, oblong, the mid-lobe transversely oblong, the disc with a callus of 3 parallel fleshy ridges extending onto the base of the mid-lobe, fleshy at apex; column 3 mm long, column-foot 4 mm long. (See fig. 25).

DISTRIBUTION: Guadalcanal, Malaita and New Georgia. Also in New Guinea.
HABITAT: Lower montane to montane forest, c. 560 m.
COLLECTIONS: *Lavis* 1 (K); *Wickison* 154 & s.n. (K).

6. E. robusta *(Blume) Lindley*, Gen. Sp. Orch. Pl.: 69 (1830). Type: Java, *Blume* s.n. (holotype L).
Dendrolirium robustum Blume, Bijdr. Fl. Ned. Ind.: 347 (1825).
Eria aeridostachya Reichb. f. ex Lindley in Journ. Linn. Soc. Bot. 3: 48 (1859). Type: Java, *Loddiges* 229 (holotype K!).

For full synonymy see Seidenfaden (1982).

An epiphytic herb. *Pseudobulbs* fusiform-cylindric, 4–6 cm long, 1.5 cm wide, covered by 4–8 acute distichous reduced leaves, 2- to 3-leaved at apex. *Leaves* erect, coriaceous, oblanceolate to narrowly oblong, acute or obtuse, 16–37.5 cm long, 2.3–6 cm wide, tapering into a stout or slender petiole at base. *Inflorescence* lateral, emerging from the axil of a sheath, erect, 16–30 cm long, densely many-flowered on a cylindrical rhachis. *Flowers* cream, covered with stellate hairs; pedicel and ovary 6 mm long, covered with stellate hairs; dorsal sepal ovate, subacute, concave, 3 mm long, 2 mm wide; lateral sepals oblong-elliptic, oblique, obtuse, 6.5 mm long, 2 mm wide; mentum cylindric-rounded, 3–4 mm long; petals linear, obtuse, 2.5 mm long, 1 mm wide; lip concave, oblong, very obscurely trilobed, 4.5 mm long, 1.6 mm wide, reflexed at apex, ecallose; column 1 mm long, the column-foot 3–4 mm long. (See fig. 24 L–M).

DISTRIBUTION: Guadalcanal, Makira and New Georgia. Also in India, SE. Asia, the Malay Archipelago, New Guinea, New Caledonia, Fiji and Samoa.

Fig. 25. Eria oligotricha. A, habit x 2/3; **B**, flower x 3; **C**, dorsal and ventral view of anther-cap x 6; **D**, column x 6; **E**, pollinia x 6; **F**, lip x 6; **G**, dorsal sepal x 4; **H**, petal x 4; **J**, lateral sepal x 4. Drawn from *Wickison* 154 by Sue Wickison.

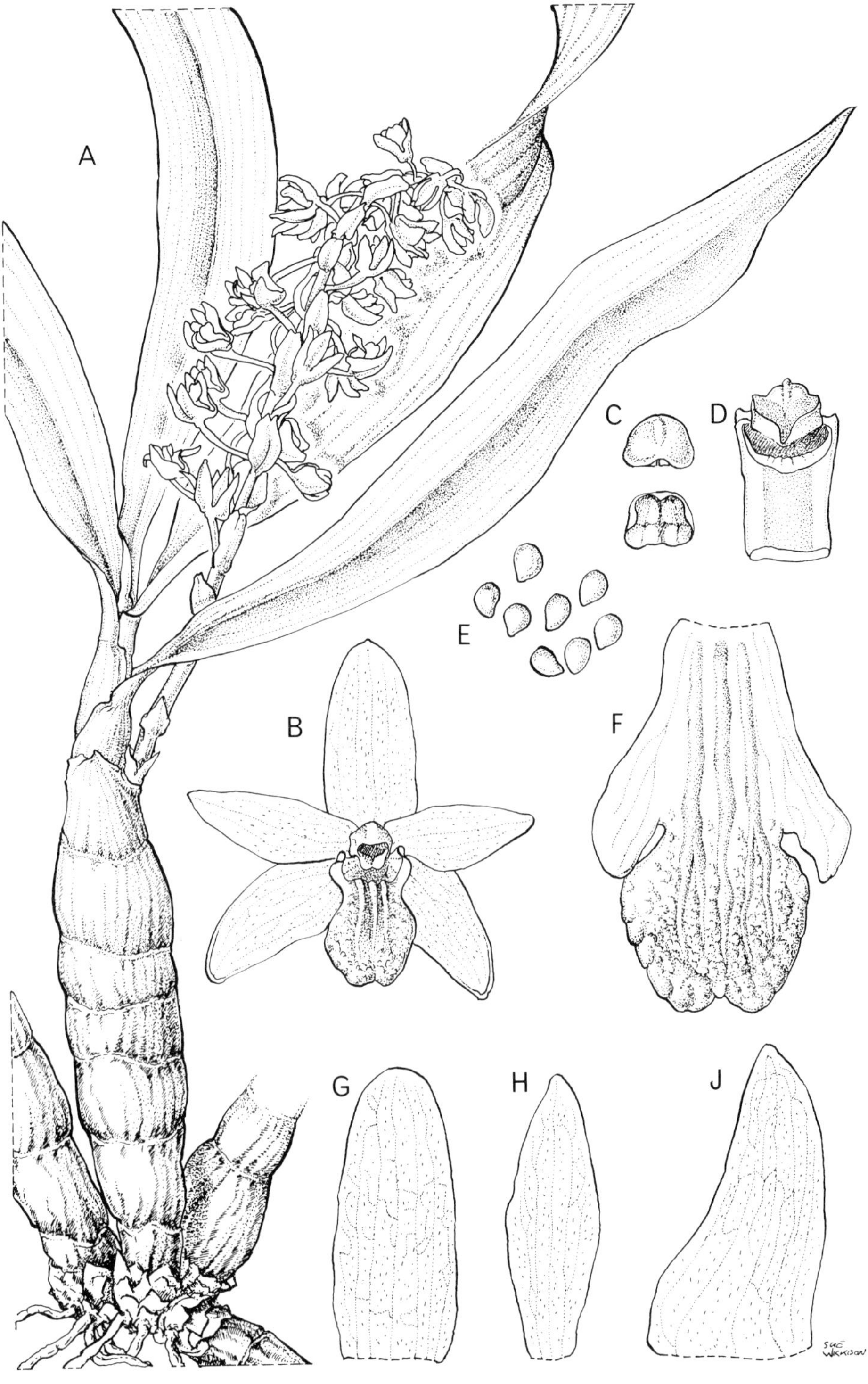
A
B
C
D
E
F
G
H
J
SUE WICKISON

HABITAT: Rain forest, 300–440 m.
COLLECTIONS: *Chaplin* 818 (K); *Dennis* in RSS 2274 (K); *Wickison* 156 (K).

7. E. rostriflora *Reichb. f.* in Seem., Fl. Vit. 301 (1868). Type: Fiji, *Seeman* 615 (holotype W, isotypes K!, isotypes AMES!, P!).
Eria vieillardii Reichb. f. in Linnaea 41: 86 (1877). Type: New Caledonia, *Vieillard* 1335 (holotype P!).
Eria drakeana Kraenzl. in Viertelj. Nat. Ges. Zür. 74: 92 (1929). Type: New Caledonia, *Daeniker* 929 (holotype Z!).
Eria kajewskii Ames in Journ. Arn. Arb. 13: 135 (1932). Type: Vanuatu, Anatom, *Kajewski* 820 (holotype AMES, isotype K!).

For full synonymy see Kores (1989).

An epiphytic herb. *Pseudobulbs* 8–25 cm long, cylindrical to subfusiform, ± covered in papery sheaths. *Leaves* (3-)5–8, near apex of pseudobulb, plicate, suberect, spreading, 6.5–20 cm long, 9–24 mm wide. *Inflorescence* from upper nodes, shorter than the leaves, 5–13 cm long, suberect, arcuate, densely many-flowered, ± glabrous; bracts lanceolate, acute, 4–8 mm long. *Flowers* cream to lemon-yellow with a pale yellow or purple lip; pedicel and ovary slightly pubescent, up to 1 cm long; dorsal sepal lanceolate, acute or subacute, 5–9 mm long, 2 mm wide; lateral sepals oblong-lanceolate, subacute to obtuse, 5–10 mm long, 2.5 mm wide; mentum not prominent, 1–1.5 mm long; petals lanceolate, acute, 4–8 mm long, 2 mm wide; lip entire, lanceolate or oblong-lanceolate, acute or subacute, 4–4.5 mm long, 1.5 mm wide, the sides upcurved, the disc with a callus of 2 ridges in the basal half of the lip, recurved at apex; column 1 mm long, with a 1–1.5 mm long column-foot. (See fig. 24 A–B).

DISTRIBUTION: Guadalcanal, New Georgia and the Santa Cruz Islands. Also in the Mariana Islands (Guam), Vanuatu, New Caledonia, Fiji, Samoa and Tahiti.
HABITAT: Rain forest, 370–500 m.
COLLECTIONS: *Hunt* in RSS 3015 (K); *Kajewski* 820 (K, AMES); *Smith & Newmarck* 1/006 (K); *Wickison* s.n. (K).

We have seen two herbarium specimens named as *E. kajewskii* with the collection number *Kajewski* 820. One is from Anatom, the southernmost island in Vanuatu and the other is from Vanikoro in the Santa Cruz Islands. Ames lists only the former collection, from Anatom, in his type description of *E. kajewski* and therefore this specimen must represent the holotype.

M. Clements (pers. comm.) has noted that specimens of *E. rostriflora* from Fiji are self-pollinating whilst Vanuatu and Solomon Island specimens are not. Further study of living material of this group is needed but at the present time we have chosen to follow Kores (1989) interpretation of the use of this name.

40. TRICHOTOSIA Blume

Epiphytic plants covered in red-brown hairs. *Stems* long or short, leafy except at base. *Leaves* distichous, red-hairy or glabrous. *Inflorescences* lateral, short, few-flowered; bracts at right-angles to the rhachis, large and concave. *Flowers* not opening widely, covered externally in red-brown hairs; sepals and petals free; lateral sepals forming a prominent mentum; lip trilobed or entire, with or without callus ridges; column slender with a column-foot; pollinia 8.

A genus of about fifty species from tropical Asia to the Pacific Islands, three species being recorded.

1. Leaves not markedly hairy; lip lacking a callus .... **3. T. sp. A**
   Leaves markedly red-hairy especially on margins; lip with a fleshy apical callus 2
2. Disc of lip with 4 nerves, ecarinate .... **2. T. vanikorensis**
   Disc of lip with 2 callus ridges .... **1. T. collina**

1. T. collina *(Schltr.) P.F. Hunt* in Kew Bull. 24: 92 (1970). Type: New Guinea, *Schlechter* 19294 (holotype B†, isotype K!).
Eria collina Schltr. in Fedde, Repert. Spec. Nov. Regni Veg., Beih. 1: 678 (1912).

An epiphytic herb covered in red-brown hairs. *Stems* pendulous, up to 1 m tall, leafy, sheathed by leaf bases. *Leaves* distichous, lanceolate, up to 12 cm long, 2.5 cm wide. *Inflorescences* lateral on the upper part of the stem opposite the leaves, up to 5 cm long; bracts lanceolate, up to 9 mm long, 3 mm wide. *Flowers* covered in red hairs on outer surface, creamy-coloured with pink-purple blotches, the underside of lip and column pink-purple, the centre of the lip with white hairs; dorsal sepal ovate, acute, 9 mm long, 4 mm wide; lateral sepals obliquely triangular, 1.2 cm long, 8 mm wide, forming a prominent, obtuse mentum; petals linear-oblong, 8 mm long, 2 mm wide; lip trilobed, 1.3 cm long, 8 mm wide, the lateral lobes erect, crescent-shaped, the mid-lobe constricted 2–3 mm from the apex, reniform, notched, the disc with a short fleshy, pubescent apical callus and 2 callus ridges on either side in the middle of the lip; column 7 mm long. (See fig. 26 and plate 4F).

DISTRIBUTION: Bougainville, Kolombangara and New Georgia. Also in New Guinea.
HABITAT: Montane rain forest, 320–850 m.
COLLECTIONS: *Craven & Schodde* 179 (K); *Hunt* in RSS 2560 (K); *Wickison* 37 (K).

2. T. vanikorensis *(Ames) Cribb & B. Lewis* **comb. nov**. Type: Santa Cruz Islands, Vanikoro, *Kajewski* 512 (holotype AMES!).
Eria vanikorensis Ames in Journ. Arn. Arb. 13: 135 (1932).

An epiphytic herb covered in red-brown hairs. *Stems* pendulous, up to 1 m long, leafy, sheathed by leaf bases. *Leaves* distichous, oblong-lanceolate, up to 12.5 cm long, 1.7–2.5 cm wide. *Inflorescences* lateral on the upper part of the stem opposite the leaves, up to 3.5 cm long; bracts lanceolate, up to 1 cm long, 3 mm wide. *Flowers* covered in red hairs on outer surface; dorsal sepal lanceolate, acute, 1 cm long, 4 mm wide; lateral sepals triangular-lanceolate, acute, 1 cm long, 5 mm wide, forming a prominent, obtuse mentum; petals linear-oblong, rounded at apex, 8 mm long, 1.5 mm wide; lip obscurely trilobed, 1 cm long, 3 mm wide, the disc oblong, contracted towards the base, 4-nerved, ecarinate, the mid-lobe constricted 2–3 mm from the apex, forming a transversely elliptic lobe with a short fleshy, apical callus in the middle; column 6 mm long.

DISTRIBUTION: Santa Cruz Islands (Vanikoro). Also in Vanuatu.
HABITAT: Rain forest, 50 m.
COLLECTION: *Kajewski* 512 (AMES).

Kajewski (28/9/1928) notes that this species is common on Vanikoro.

We mistakenly considered *T. vanikorensis* to be conspecific with *T. vulpina* (Reichb. f.) Kraenzl. (see Lewis & Cribb, 1989). However, we now consider the two species to be distinct. *T. vulpina* differs in having a longer inflorescence (10–20 cm long) and a different shaped lip which is unguiculate and dilated into a somewhat quadrate, obscurely trilobed blade.

3. T. sp. A

An epiphytic herb covered in red-brown hairs on the stems and inflorescences. *Stems* pendulous, leafy in apical half, sheathed by leaf bases. *Leaves* distichous, lanceolate, glabrous or finely pubescent, up to 8 cm long, 2 cm wide. *Inflorescences* lateral on the upper part of the stem opposite the leaves, up to 8 cm long; bracts triangular, up to 4 mm long, 3 mm wide. *Flowers* covered in red hairs on outer surface, pale-green; dorsal sepal ovate, acute, 4–5 mm long, 3 mm wide; lateral sepals obliquely ovate, 5 mm long, 3–4 mm wide, forming an obtuse mentum; petals linear-oblong, 4 mm long, 1.5 mm wide; lip entire, 5–6 mm long, 2–3 mm wide, lacking a callus; column 3 mm long.

DISTRIBUTION: Kolombangara.
HABITAT: Montane moss forest, 1500 m.
COLLECTION: *Cribb & Campbell* 5120 (K).

This collection may be a peloric form and more material is needed to ascertain its identity.

Fig. 26. Trichotosia collina. A, habit x 2/3; **B**, flower x 2; **C**, habit reduced; **D**, lip x 4; **E**, dorsal sepal x 4; **F**, lateral sepal x 4; **G**, ventral and dorsal view of anther-cap x 6; **H**, pollinia x 6; **J**, petal x 4; **K** column x 6. Drawn from *Wickison* 37 by Sue Wickison.

A
B
C
D
E
F
G
H
J
K
SUE WICKISON

41. MEDIOCALCAR J.J. Smith

Small epiphytic or rarely terrestrial herbs growing in montane or cloud forest. *Rhizomes* creeping, concealed by brownish bracts when young. *Pseudobulbs* conical or obliquely conical, close together or well spaced along the rhizome, 1– or 2-leaved at the apex, enveloped at base by several bracts. *Leaves* apical, slightly succulent, mid-green. *Inflorescences* terminal, 1–2, 1-flowered; peduncle and pedicel elongating after fertilization. *Flowers* urceolate, orange or red with green, white or yellow tips; pedicel and ovary often red; sepals ovate-lanceolate, connate in basal half; lateral sepals saccate at base; petals linear-lanceolate, free; lip entire, ovate, acute, clawed, saccate to spurred; pollinia 8.

A genus of about thirty-five species from New Guinea and the SW. Pacific. Three species have been recorded and are often found growing together. The species are arranged in two sections according to J.J. Smith (1929). See Kores (1989) for a discussion of the sections.

1. Base of the lip spurred, the spur as long as the claw; pseudobulbs unifoliate .... **1. M. robustum**
   Base of the lip saccate, much shorter than the claw; pseudobulbs 1– or 2-leaved 2
2. Leaves oblong to oblong-oblanceolate, acute or obtuse, up to 2.5 cm wide .... **3. M. paradoxum**
   Leaves linear-lanceolate, acute, up to 1.2 cm wide .... **2. M. alpinum**

Section MEDIOCALCAR

1. M. robustum *Schltr.* in Fedde, Repert. Spec. Nov. Regni Veg., Beih. 1: 225 (1911). Type: New Guinea, *Schlechter* 20278 (holotype B†).

An epiphytic herb, forming mats. *Pseudobulbs* obliquely oblong-ovoid, 1–2 cm long, 3–5 mm wide, 2–6 cm apart on the rhizome, unifoliate. *Leaves* apical, oblong-elliptic, 7–16 cm long, 1.5–2.7 cm wide. *Inflorescences* 1–2, apical, 1-flowered; pedicel and ovary up to 8 cm long. *Flowers* red with yellow tips and pale yellow petals, lip and column; sepals ovate-lanceolate, connate in basal half, 1–1.4 cm long; petals linear, lanceolate, free, 1–1.3 cm long, 2–3 mm wide; lip ovate, acute, 6 mm long, 5 mm wide, clawed, spurred at base, the spur conical, 6 mm long; column 6–7 mm long.

DISTRIBUTION: Guadalcanal and Kolombangara. Also in New Guinea.
HABITAT: Montane forest, 800–1000 m.
COLLECTIONS: *Cribb & Campbell* 5119 (K); *Hunt* in RSS 2053 (K).

Section BREVICALCAR

2. M. alpinum *J.J. Smith.* in Bull. Jard. Bot. Buitenz. ser. 2, 13: 62 (1914). Types: New Guinea, *Kock* 68 & 143 (syntypes BO!).
Mediocalcar bifolium var. *validum* J.J. Smith in Nova Guinea 12: 30 (1913). Type: as for *M. alpinum*.

An epiphytic herb forming mats on tree trunks and branches. *Pseudobulbs* obliquely oblong-ovoid, 0.5–1.5 cm long, 3–4 mm wide, 3–5 cm apart on the rhizome, 1- or 2-leaved. *Leaves* apical, linear-lanceolate, acute, 7–9 cm long, 6–12 mm wide. *Inflorescences* apical, 1-flowered; pedicel and ovary 1–2 cm long. *Flowers* red to orange with yellow tips and a yellow lip; sepals ovate-lanceolate, connate in basal half, 10 mm long; petals free, linear, lanceolate, 9–10 mm long, 1–2 mm wide; lip ovate, acute, 7 mm long, 4 mm wide, clawed, saccate at base; column 5–6 mm long. (See fig. 27 H–P).

DISTRIBUTION: Bougainville, Guadalcanal and Kolombangara. Also in New Guinea and Vanuatu.
HABITAT: Montane and ridge top forest, 600–2240 m.
COLLECTIONS: *Cribb* s.n. (K); *Cribb, Dennis & Campbell* 5052A & 5135 (K); *Hunt* in RSS 2034 & 2959 (K); *Mitchell* 10 (K); *Schodde & Craven* 3763 (K); *Waterhouse* 619B (K); *Wickison* 79 (K).

3. M. paradoxum *(Kraenzl.) Schltr.* in Fedde, Repert. Spec. Nov. Regni Veg. 9: 96 (1910). Type: Samoa, *Reinecke* 300 (holotype B†, isotype G).
Eria paradoxa Kraenzl. in Engler, Bot. Jahrb. 25: 606 (1898).
Mediocalcar vanikorense Ames in Journ. Arn. Arb. 13: 136 (1932). Type: Solomon Islands, Santa Cruz Islands (Vanikoro), *Kajewski* 641 (holotype AMES!).

An epiphytic herb forming mats. *Pseudobulbs* obliquely oblong-conical, 1–1.5 cm long, up to 6 mm wide, 1–4 cm apart on the rhizome, 1- or 2-leaved. *Leaves* apical, variable, oblong or oblong-oblanceolate, acute to obtuse, 3.7–9 cm long, 0.8–2.5 cm wide, dark green. *Inflorescence* apical, 1-flowered; pedicel and ovary 2–3 cm long. *Flowers* red or orange with yellow tips; sepals ovate-lanceolate, connate in basal half, 10 mm long; petals linear-lanceolate, free, 3–4 mm long, 1–1.5 mm wide; lip ovate, acuminate or acute, 7 mm long, 3 mm wide, clawed, saccate at base; column 5–6 mm long. (See fig. 27 A–G and plate 7B).

DISTRIBUTION: Bougainville, Guadalcanal, Kolombangara, New Georgia, Ranongga and the Santa Cruz Islands (Vanikoro). Also in Vanuatu, Fiji and Samoa.
HABITAT: Montane forest, 240–1700 m.
COLLECTIONS: *Cribb & Campbell* 5130 (K); *Cribb, Dennis & Campbell* 5052 (K); *Hunt* in RSS 2472 (K); *Mitchell* 9 & 70 (K); *Waterhouse* 619A (K); *Whitmore* in BSIP 2063 (K); *Wickison* 22B (K).

42. CERATOSTYLIS Blume

Small epiphytes. *Stems* clustered, slender, with thin brown sheaths at the base. *Leaf* solitary, apical, fleshy, narrow, flat or almost terete. *Inflorescences* terminal, short, few, each of one flower, usually with a few open together. *Flowers* small, often finely pubescent on outer surface; dorsal sepal and petals about equal; lateral sepals forming a mentum enclosing the base of the lip; lip narrow at base, the blade thickened, entire, with longitudinal calli; column short,

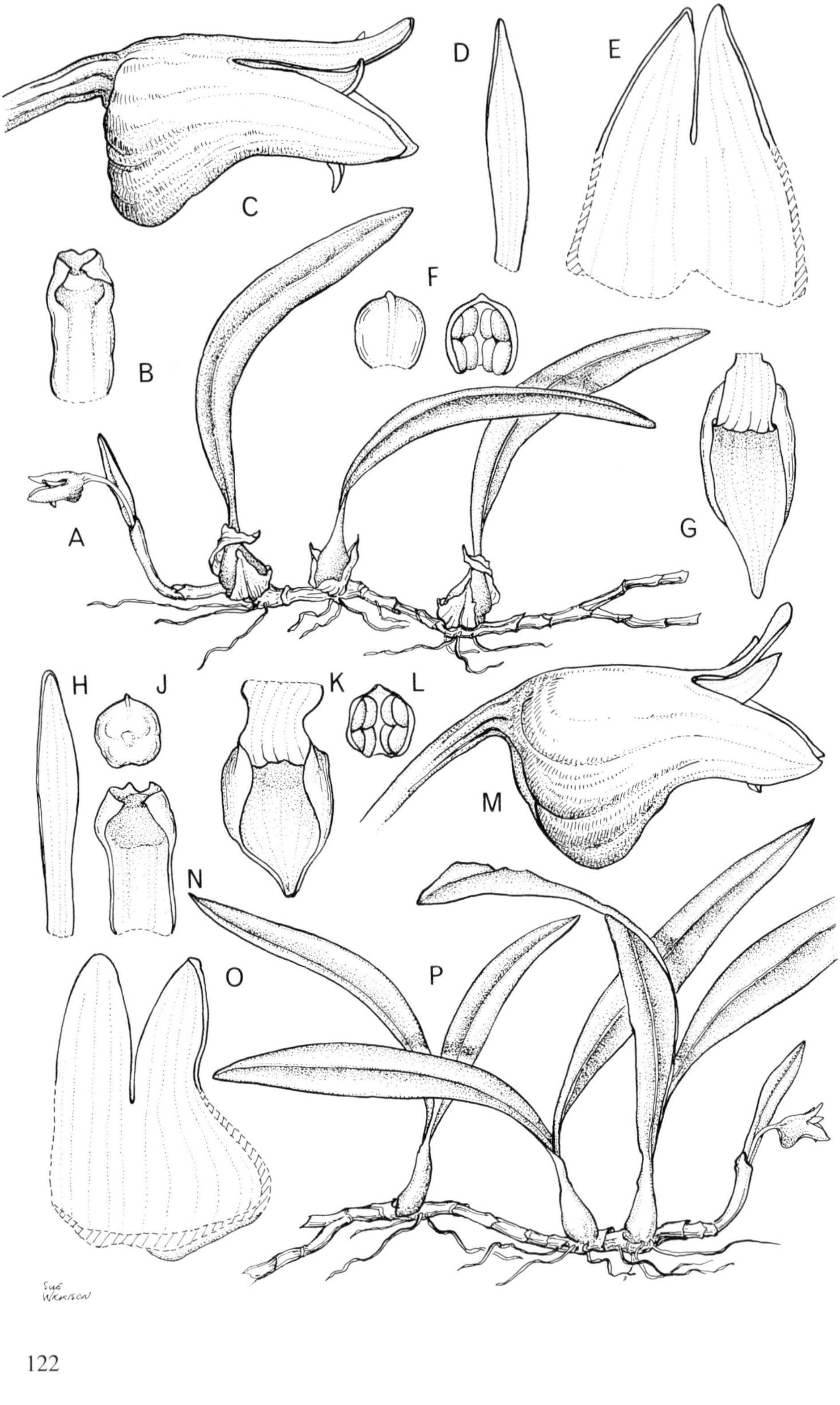
A
B
C
D
E
F
G
H
J
K
L
M
N
O
P
SUE WICKISON

divided at the apex into 2 erect arms which carry the stigma lobes on their inner surfaces; pollinia 8.

Young plants may have additional short green leaves at the base of the stem in place of some of the usual brown sheaths.

A genus of about one hundred species from tropical Asia to the Pacific Islands, two species being recorded.

Stems up to 4 mm in diameter; bracts 5–10 mm long; lip 6 mm long; the spur cylindrical, 4 mm long .... **1. C. longipes**
Stems 0.3–1.5 mm in diameter; bracts 3–4 mm long; lip 3.5 mm long; the spur globose, 1 mm long .... **2. C. subulata**

1. C. longipes *Schltr.* in Fedde, Repert. Spec. Nov. Regni Veg., Beih. 1: 250 (1912). Type: New Guinea, *Schlechter* 18469 (holotype B†).

An epiphytic herb. *Stems* closely tufted, 8–30 cm long, 1.5–4 mm in diameter. *Leaf* solitary, almost terete, broadly grooved on one side, 4–7 cm long, 2–6 mm wide. *Inflorescences* terminal, 3–5; bracts papery, lanceolate, 5–10 mm long. *Flowers* off-white with a bright yellow lip, the column tinged with purple, pubescent on outer surface of sepals; pedicel and ovary pubescent, 3 mm long; dorsal sepal ovate, 3 mm long, 1 mm wide, reflexed at tip; lateral sepals lanceolate, 6 mm long, 1 mm wide, recurved at tips and fused at base to form a 4 mm long, conical mentum; petals obliquely lanceolate, acute; lip narrow at the base, ovate-oblong, 6 mm long, 1 mm wide at fleshy apex, the disc with 2 obscure calli, the spur cylindrical, 4 mm long, pubescent; column 1 mm long, with a 2 mm long column-foot. (See fig. 28 A–M).

DISTRIBUTION: Bougainville, Guadalcanal, Kolombangara, Makira, New Georgia and Rendova. Also in New Guinea.

HABITAT: Rain forest, sea level to 1200 m.

COLLECTIONS: *Chaplin* 759 (K); *Craven & Schodde* in NGF 335 (K, G, LAE); *Cribb & Campbell* 5153 (K); *Cribb, Dennis & Campbell* 5046 (K); *Dennis* in RSS 2052 (K); *Hunt* in RSS 2475 & 2975 (K); *Lavarack & Ridsdale* in NGF 31289 (K); *Maenu'u* in BSIP 6044 (K, LAE); *Wickison* 16a, 16e, 16f, 16g & 223 (K); *Whitmore* in RSS 6263 (K, LAE).

2. C. subulata *Blume*, Bijdr. Fl. Ned. Ind.: 306 (1825). Type: Java, *Blume* s.n. (holotype L!).

Ceratostylis cepula Reichb. f. in Bonplandia 5: 53 (1857). Type: Java, *Zollinger* 3185 (holotype W!).

Fig. 27. Mediocalcar paradoxum A-G. A, habit x 2/3; **B**, column x 4; **C**, flower x 4; **D**, petal x 4; **E**, dorsal and lateral sepal x 4; **F**, dorsal and ventral view of anther-cap x 6; **G**, lip x 4. **M. alpinum H-P. H**, petal x 4; **J**, dorsal view of anther-cap x 6; **K**, lip x 4; **L**, ventral view of anther-cap x 6; **M**, flower x 4; **N**, column x 4; **O**, dorsal and lateral sepal x 4; **P**, habit x 2/3. **A** drawn from *Wickison* 22B; **B-G** from *Mitchell* 9; **H-L** & **N-P** from *Hunt* 2034; **M** from *Robinson* K197. All drawn by Sue Wickison.

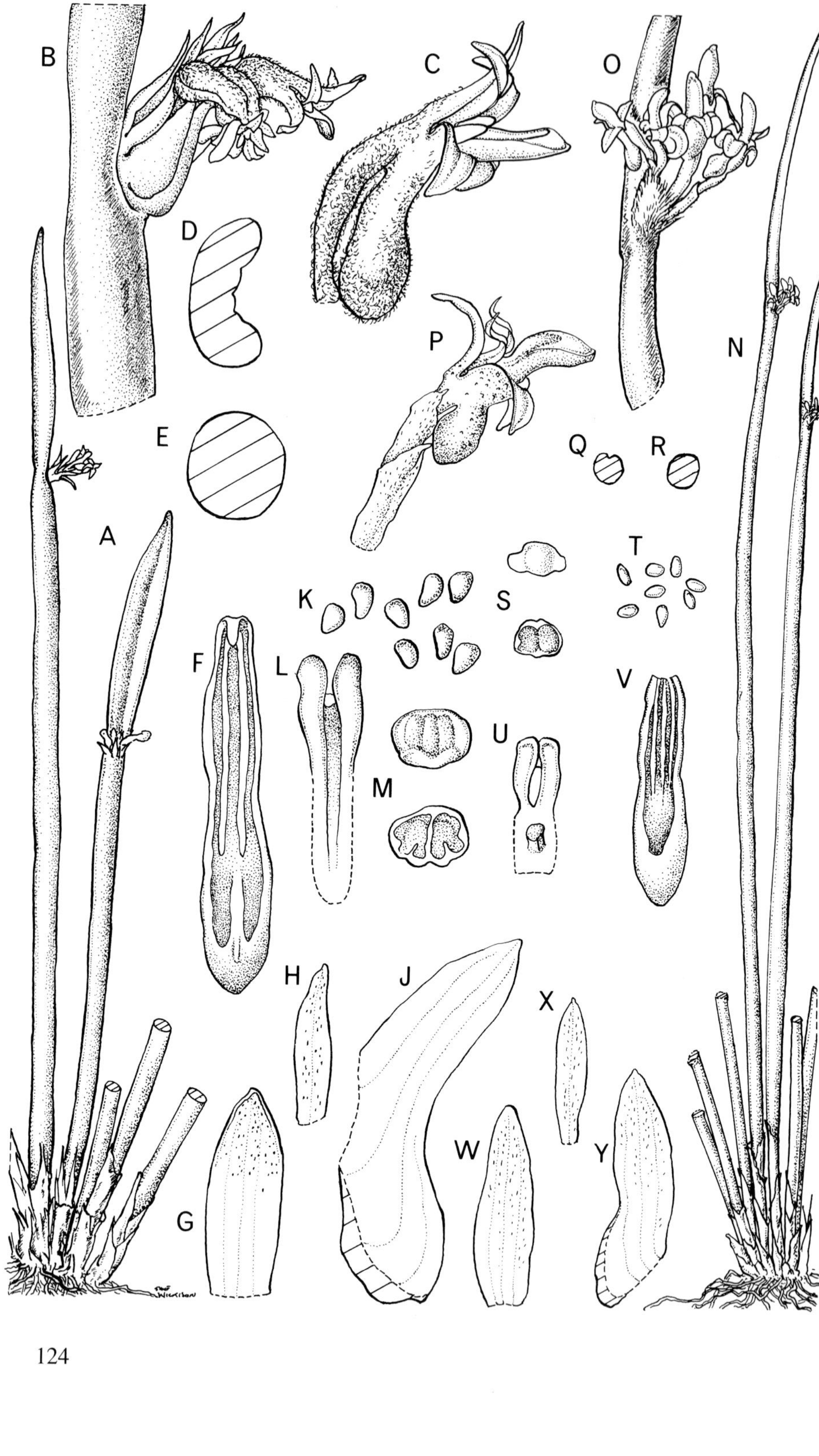
B
C
O
D
E
A
P
N
Q
R
T
K
S
F
L
V
U
M
H
J
X
W
Y
G

Ceratostylis malaccensis Hook. f., Fl. Brit. Ind. 5: 823 (1890). Type: Malay Peninsula, *Scortechini* (holotype K!).

An epiphytic herb. *Stems* closely tufted, 7–15 cm long, variable in thickness from wiry to slender, 0.3–1.5 mm in diameter, sometimes finely pubescent, dark green. *Leaf* solitary, almost terete, broadly grooved on one side, 3–8 cm long, 1–2.5 mm wide, sometimes finely pubescent. *Inflorescences* terminal, 1–3; bracts papery, lanceolate, 3–4 mm long. *Flowers* dull red-purple to pale purple with a pinkish white mentum and a yellow lip, finely pubescent; pedicel and ovary pubescent, 3 mm long; dorsal sepal ovate, 3 mm long, 1 mm wide, recurved at tip; lateral sepals ovate, 3.5 mm long, 1 mm wide, recurved at tips and fused at base to form a finely hairy, 1 mm long, globose mentum; petals unguiculate, acute, 2 mm long, 0.75 mm wide; lip narrow at the base, narrowly ovate, blunt, 3.5 mm long, 1.5 mm wide at fleshy apex, the disc with 3 obscure calli at base, the spur globose, 1 mm long; column 1 mm long, with a short column-foot, 0.75 mm long. (See fig. 28 N–Y and plate 3D).

DISTRIBUTION: Bougainville, Choiseul, Guadalcanal, New Georgia, Makira, Malaita, Rennell and the Santa Cruz Islands. Also in the Malay Archipelago, New Guinea and Vanuatu.

HABITAT: Rain forest, sea level to 800 m.

COLLECTIONS: *Cribb* 1875 (K); *Cribb, Dennis & Campbell* 5035 (K); *Dickson* MD29 (K); *Dissing* 2794; *Hunt* in RSS 2185 & 3075 (K); *Lavarack & Ridsdale* in NGF 31362 (K, LAE); *Mauriasi et al.* in BSIP 16686 (K, LAE); *Mitchell* 41 (K); *Peach* 7 (K); *Wickison* 48 & 110 (K).

Ceratostylis micrantha Schltr., from New Caledonia, and *C. kaniensis* Schltr., from New Guinea, are similar and may prove to be conspecific.

43. EPIBLASTUS Schlechter

Large, sturdy, epiphytic herbs. *Pseudobulbs* superposed, enclosed by increasingly larger imbricate bracts, unifoliate at apex. *Leaves* linear, flat, but at the base folded along the midrib. *Inflorescences* several, of 1 flower each, terminal, enveloped by base of leaf. *Flowers* waxy, reddish-orange, subglobose to campanulate; pedicel and ovary elongate; dorsal sepal elliptic; lateral sepals obliquely triangular, forming a blunt open mentum; petals free; lip clawed, trilobed to ovate, the lateral lobes curving upwards, the mid-lobe usually with a fleshy ridge, the apex recurved; column short and stout; pollinia 8.

A genus of about fifteen species in New Guinea and the SW. Pacific, a single species being recorded.

Fig. 28. Ceratostylis longipes A-M. A, habit x 2/3; **B**, inflorescence x 3; **C**, flower x 6; **D**, cross section of leaf blade x 3; **E**, cross section of stem x 3; **F**, lip x 8; **G**, dorsal sepal x 8; **H**, petal x 8; **J**, lateral sepal x 8; **K**, pollinia x 14; **L**, column x 8; **M**, dorsal and ventral view of anther-cap x 14. **C. subulata N-Y. N**, habit x 2/3; **O**, inflorescence x 3; **P**, flower x 6; **Q**, cross section of leaf blade x 3; **R**, cross section of stem x 3; **S**, dorsal and ventral view of anther-cap x 14; **T**, pollinia x 14; **U**, column x 8; **V**, lip x 8; **W**, dorsal sepal x 8; **X**, petal x 8; **Y**, lateral sepal x 8. **A-M** drawn from *Wickison* 16a; **N-Y** from *Wickison* 110. All drawn by Sue Wickison.

E. sciadanthus *(F. Muell.) Schltr.* in K. Schum. & Laut., Nachtr. Fl. Deutsch. Südsee: 137 (1905). Type: Samoa, *Betche* s.n. (holotype MEL).
Bulbophyllum sciadanthum F. Muell. in Wing's South. Sci. Rec. 2: 95 (1882).
Eria ornithoides Kraenzl. in Engler, Bot. Jahrb. 25: 606 (1898), non *Epiblastus ornithidioides* Schltr. (1905).

An epiphytic herb growing in large clumps up to 1 m long. *Pseudobulbs* ascending or pendulous, superposed. *Leaves* solitary, arising from the apex of the pseudobulb, strap-like, linear-oblong, acute, up to 35 cm long, 3.2 cm wide, mid- to yellow-green. *Inflorescences* terminal, up to 8, of 1 flower each. *Flowers* red at base fading to pink at tips; pedicel and ovary 1.5–2 cm long; dorsal sepal ovate, 5–8 mm long, 4 mm wide; lateral sepals obliquely triangular, 6–9 mm long, 4–4.5 mm wide; petals elliptic, 7–8 mm long, 2.6–3 mm wide; lip obscurely trilobed, obovate, 7.8–8 mm long, 4–5 mm wide when flattened, the basal part narrowly attached to the column-foot, the lateral lobes erect, broadly rounded with the margins slightly incurved, the mid-lobe raised and fleshy with a large, transverse, plate-like callus; column 2.8–3 mm long.

DISTRIBUTION: Guadalcanal and Malaita. Also in Vanuatu, Fiji and Samoa.
HABITAT: Rain forest, 520–800 m.
COLLECTIONS: *Cribb, Dennis & Campbell* 5052B (K); *Hunt* in RSS 2943 (K); *Mauriasi et al.* in BSIP 12078 (K); *Wickison* 96 & 237 (K).

This species is similar to *E. ornithidioides* Schltr., from New Guinea and New Ireland but that species apparently differs in having a lip with a naked disc.

44. AGROSTOPHYLLUM Blume

Epiphytic herbs. *Stems* clustered, erect or pendent, bilaterally flattened, leafy in the apical parts. *Leaves* in 2 ranks, narrow, rather thin, usually twisted at the base to lie in one plane, with overlapping sheaths. *Inflorescence* terminal, with the flowers in a globose head, in clusters on an elongate axis, or in a panicle, surrounded by bracts. *Flowers* small, often self-pollinating; sepals and petals similar, the petals narrower; lateral sepals forming a mentum which contains the saccate base of the lip; lip entire or trilobed; column short or relatively long, the foot scarcely developed; pollinia 8.

A genus of forty to fifty species from tropical Asia and the Malay Archipelago to the Pacific Islands. Seven species are recorded from the Solomon Islands and Bougainville, arranged in 4 sections according to Schlechter (1912).

Agrostophyllum is a taxonomically difficult genus as many of the herbarium specimens from this region are sterile or in fruit and the present treatment should be regarded as provisional.

1. Inflorescence elongated 2
   Inflorescence a globose head 3
2. Inflorescence with long branches arranged in whorls .... **6. A. paniculatum**
   Inflorescence unbranched, with the flowers in sessile clusters
   .... **5. A. megalurum**

3. Leaves at right-angles to the stem, 1.8–2.2 cm long; stems branched. .... **4. A. elongatum**
   Leaves erect or spreading but at an acute angle to the stem, more than 6 cm long; stems not branched 4
4. Stems robust and sturdy, up to 1 m long, 1.5 cm wide; leaves strap-like .... **3. A. majus**
   Stems short or long, slender or fleshy; leaves linear-lanceolate or elliptic 5
5. Leaves elliptic-oblong .... **1. A. curvilabre**
   Leaves linear-lanceolate 6
6. Rhizome elongated and stem-like, covered in involucral bracts; stems arising from the axils of the bracts, 2.5–4 cm long .... **7. A. superpositum**
   Rhizome not as above; stems clustered, up to 60 cm long .. **2. A. graminifolium**

Section AGROSTOPHYLLUM

1. A. curvilabre *J.J. Smith* in Fedde, Repert. Spec. Nov. Regni Veg., Beih. 11: 274 (1912). Type: New Guinea, *Gjellerup* 512 (holotype BO).

An epiphytic herb. *Stems* clustered, up to 15 cm tall, 1.3 cm wide, leafy, sheathed by leaf bases. *Leaves* distichous, elliptic-oblong, acute, 9–14 cm long, 2.2–3 cm wide, articulated to sheathing bases 4–6 cm long. *Inflorescence* a terminal, globose head, 2.5 cm in diameter, many-flowered. *Flowers* white, surrounded by chaffy bracts; pedicel and ovary pubescent; dorsal sepal ovate, acute, 6 mm long, 5 mm wide; lateral sepals obliquely ovate, acuminate, 1 cm long, 5 mm wide, forming an obtuse mentum; petals linear, acute, 5 mm long, 1.5 mm wide, with an erose apical margin; lip recurved, broadly ovate, apiculate, 7 mm long, 8 mm wide, saccate at the base; column 2 mm long.

DISTRIBUTION: Bougainville. Also in New Guinea.
HABITAT: In mossy *Pandanus* forest, 1730 m.
COLLECTION: *Lavarack & Ridsdale* in NGF 31323 (LAE).

Unfortunately this specimen is in fruit and the description of the flowers has been taken from New Guinea material.

2. A. graminifolium *Schltr.* in Fedde, Repert. Spec. Nov. Regni Veg., Beih. 1: 266 (1912). Type: New Guinea, *Schlechter* 18594 (holotype B†, isotypes AMES, K!).

A tall, slender epiphyte. *Stems* clustered, up to 60 cm tall, 0.5 cm wide, slender, leafy in apical half, sheathed in tubular leaf bases. *Leaves* distichous, linear, acute, 12–18 cm long, 6–9 mm wide, dark green; leaf bases 3–6 cm long. *Inflorescence* terminal, globose, c. 3 cm in diameter. *Flowers* white; dorsal sepal ovate, acute, 3 mm long, 1.2 mm wide; lateral sepals obliquely lanceolate, keeled, 3 mm long, 1.2 mm wide, forming a mentum; petals elliptic, 3 mm long, 1.2 mm wide; lip obscurely trilobed in the apical half, saccate at base, 3 mm long, 1.5 mm wide, with erect lateral lobes; column 3 mm long.

DISTRIBUTION: Bougainville and Guadalcanal. Also in New Guinea and Vanuatu.
HABITAT: Primary rain forest, 580–900 m.
COLLECTIONS: *Kajewski* 2140 (G); *Wickison* 102 (K).

3. A. majus *Hook. f.*, Fl. Brit. Ind. 5: 824 (1890). Type: India, *King* s.n. (holotype K!).

A robust, pendent epiphyte. *Stems* up to 1 m long, 1.5 cm wide, thick and robust, leafy in apical half. *Leaves* distichous, strap-like, obtuse, 10–14 cm long, 1.3–2 cm wide, mid-green; leaf bases up to 8 cm long, sheathing and imbricate. *Inflorescence* terminal, a globose head of many flowers, 3–4 cm in diameter. *Flowers* white turning creamy-yellow as they mature; sepals similar, ovate, acute, 2–2.5 mm long, 1.5 mm wide; lateral sepals forming a mentum; petals narrowly ovate, 2 mm long, 1 mm wide; lip saccate at base, the apical lamina obovate, 2.5–3 mm long, 1.5 mm wide, with the lateral margins inrolled; column 2 mm long.

DISTRIBUTION: Bougainville, Kolombangara, New Georgia and Santa Isabel. Widespread from India across to SE. Asia and Vanuatu.
VERNACULAR NAME: Fifari (Kwara'ae name).
HABITAT: Wide variety of habitats from dense shade to almost full sun in lowland forests, 100–520 m.
COLLECTIONS: *Beer's collectors* in BSIP 7302 (K); *Cowmeadow's collectors* in BSIP 3270 (K, LAE); *Hunt* in RSS 2538 (K); *Lavarack & Ridsdale* in NGF 30566 (LAE) & 31263 (K, LAE); *Maenu'u* in BSIP 5986 (K); *Wickison* 18B, 18D, & 211 (K).

Agrostophyllum parviflorum J.J. Smith from New Guinea is similar and may be conspecific.

Section APPENDICULOPSIS

4. A. elongatum *(Ridley) Schuiteman* in Blumea 35: 165 (1990). Type: Malay Peninsula, *Ridley* s.n. (holotype BM).
Appendicula elongata Ridley, Trans. Linn. Soc. London, Bot. II, 3: 375 (1893).
Agrostophyllum costatum J.J. Smith in Bull. Dép. Agric. Indes Néerl. 19: 1 (1908). Type: New Guinea, *Lorentz* 205 (holotype BO!).

For full synonymy see Schuiteman (1990).

An epiphytic herb. *Rhizome* creeping and elongated. *Stems* erect or pendulous, densely leaved, branched, up to 60 cm long, sheathed in imbricate leaf bases which make a characteristic zig-zag pattern on the stem. *Leaves* distichous, at right-angles to the stem, twisted at base to lie in one plane, thin-textured, oblong with a rounded minutely bilobed apex, 1.8–2.2 cm long, 6–10 mm wide, mid-green. *Inflorescences* terminal on lateral branches of the stem, globose; bracts 10 mm long. *Flowers* white with a hint of green with a white lip and a dark maroon apical margin to the column; dorsal sepal ovate, 3.5 mm long, 2 mm wide; lateral sepals obliquely triangular, 2.8 mm long, 3 mm wide, forming a mentum; petals linear, 2.6 mm long, 8 mm wide; lip trilobed, 3.5 mm long, 3.5 mm wide, the lateral lobes triangular to rounded, erect, saccate at base; column c. 2.5 mm long.

DISTRIBUTION: Bougainville, Kolombangara, Makira, Malaita, New Georgia and Rendova. Also in Malaya, Philippines, New Guinea, Vanuatu and the Caroline Islands (Palau).

HABITAT: Lowland forest in fairly open habitats, often epiphytic on trees overhanging a river, up to 500 m.
COLLECTIONS: *Chaplin* 755 (K); *Dockrill & Coode* in NGF 34189 (LAE); *Hunt* in RSS 2380 (K); *Waterhouse* 626 (K); *Whitmore* 6265 (K, LAE); *Wickison* 35 & 215 (K).

Section DOLICHODESME

5. A. megalurum *Reichb. f.* in Seem., Fl. Vit.: 296 (1868). Type: Samoa, *E. Graeffe* s.n. (holotype W!).

An epiphytic herb. *Stems* up to 30 cm tall, leafy except for base, sheathed by imbricate leaf bases. *Leaves* distichous, lanceolate, 4.5–11 cm long, 5–10 mm wide; leaf bases tubular and sheathing. *Inflorescence* terminal, elongate, 8–17 cm long, with the flowers in sessile clusters of 2–4 from axils of bracts; bracts lanceolate, 1–1.5 cm long, imbricate. *Flowers* white with a pink anther-cap; dorsal sepal ovate, 3.5 mm long, 2 mm wide; lateral sepals ovate, acute, 3.5 mm long, 2 mm wide, forming a mentum; petals subovate, 3 mm long, 1 mm wide; lip entire, ovate, 4 mm long, 2 mm wide; column 2.5 mm long.

DISTRIBUTION: Bougainville and Guadalcanal. Also in the Caroline Islands, Fiji and Samoa.
HABITAT: Rain forest, 40–730 m.
COLLECTIONS: *Dockrill* in NGF 34182 (LAE); *Wickison* 111 (K).

This species is similar to *A. mucronatum* J.J. Smith from New Guinea.

6. A. paniculatum *J.J. Smith* in Bull. Dép. Agr. Indes Néerl. 19: 2 (1908). Type: New Guinea, *G.M. Versteeg* 1128 (holotype BO!).

A robust epiphyte. *Stems* up to 1 m tall, sturdy, leafy in apical half. *Leaves* distichous, oblong, unequally bilobed at the apex, 12–23 cm long, 1.5–2 cm wide; leaf bases sheathing the stem, imbricate, 5–7 cm long. *Inflorescence* terminal, paniculate, 18–30 cm long; branches in whorls of up to 5, each bearing up to 4 flowers, whorls 0.7–2 cm apart; bracts small and appressed to the branch, 3–4 mm long. *Flowers* pale green to white, the disc often with a hint of pale mauve; dorsal sepal ovate, 4 mm long, 1.2 mm wide; lateral sepals lanceolate, 4.5 mm long, 1.5 mm wide, forming a mentum; petals subovate, 4 mm long, 1 mm wide; lip obscurely trilobed to broadly ovate, 4.5 mm long, 4 mm wide, margins inrolled, with a saccate base; column 3 mm long.

DISTRIBUTION: Guadalcanal and Makira. Also in New Guinea.
HABITAT: Rain forest, 270 m.
COLLECTIONS: *Hunt* in RSS 2248 & 2978 (K); *Wickison* 123 (K).

Section OLIGANTHE

7. A. superpositum *Schltr.* in Fedde, Repert. Spec. Nov. Regni Veg., Beih. 1: 278 (1912). Type: New Guinea, *Schlechter* 19534 (holotype B†, isotypes AMES, K!).

A pendulous epiphyte. *Rhizome* elongated and stem-like, covered in involucral bracts which are lanceolate, acute, 2.8–3.9 cm long, 4–5 mm wide. *Stems* arising from the axils of the bracts, 2.5–4 cm long, leafy towards the apex. *Leaves* distichous, erect or slightly spreading, lanceolate, 6–9.5 cm long, 9–11 mm wide; leaf bases sheathing and imbricate. *Inflorescence* terminal, capitate, few-flowered, surrounded by chaffy bracts, 2–2.5 cm long. *Flowers* cream, apparently not opening; dorsal sepal oblong-ovate, 6 mm long, 2 mm wide; lateral sepals obliquely oblong, 8 mm long, 2 mm wide, fused at the base to form a short globose mentum; petals linear, 6 mm long, 1 mm wide; lip trilobed, 8.5 mm long, 6.5 mm wide, the lateral lobes rounded, the mid-lobe semi-circular, saccate at base; column 1.3 mm long.

DISTRIBUTION: Guadalcanal. Also in New Guinea.
HABITAT: Montane moss forest, up to 1500 m.
COLLECTION: *Hunt* in RSS 2985 (K).

This collection differs from the type in having slightly broader leaves.

45. APPENDICULA Blume

Epiphytic, lithophytic or terrestrial herbs. *Stems* not swollen at base, closely tufted, simple or branched, slender, leafy, sheathed in leaf bases. *Leaves* distichous, twisted at base to lie in one plane. *Inflorescences* lateral or terminal, few- to many-flowered. *Flowers* small, green, white or yellow, sometimes flushed with maroon; dorsal sepal free; lateral sepals fused at base to form a mentum; petals smaller than sepals; lip obscurely trilobed, saccate at base; column short; pollinia 6.

A genus of about sixty species from tropical Asia to the Pacific Islands, seven species being recorded.

1. Inflorescences lateral, less than 1.5 cm long .... **7. A. reflexa**
   Inflorescences terminal or lateral, more than 2.5 cm long 2
2. Bracts longer than the flowers, up to 15 mm long .... **2. A. disticha**
   Bracts shorter than the flowers, less than 5.5 mm long 3
3. Lip narrow at base, wedge-shaped with a fleshy basal callus; inflorescence simple .... **4. A. neohibernica**
   Lip trilobed to oblong with 2 fleshy callus ridges which may converge at base; inflorescence simple or branched 4
4. Bracts ovate, reflexed, 3.5–5.5 mm long, 1–2.5 mm wide .... **1. A. bracteosa**
   Bracts linear-ovate or lanceolate, sometimes reflexed, up to 2 mm long, 0.5 mm wide 5
5. Inflorescence branched basally, c. 12 cm long .... **6. A. polystachya**
   Inflorescence simple, up to 7 cm long 6

6. Inflorescence terminal; flowers pale to bright yellow, sometimes marked with purple, with a yellow lip with or without red spots on either side of the mid-lobe; lip oblong, constricted in middle .. .. **3. A. lutea**
Inflorescence terminal or lateral; flowers creamy-white to pale yellow-green, with a red spot on either side of the lip mid-lobe; lip oblong, narrowed at base, slightly expanded at the apex .. .. **5. A. pendula**

1. A. bracteosa *Reichb. f.* in Seem., Fl. Vit.: 299 (1868). Type: Fiji, *Seemann* 592 (holotype W, isotypes AMES, K!).
Lobogyne bracteosa (Reichb. f.) Schltr. in Mem. Herb. Boiss. 21: 65 (1900).

An epiphytic herb. *Stems* up to 40 cm long, concealed by tubular bases of leaves, light green. *Leaves* distichous, oblong-lanceolate, acute, 3–4 cm long, 0.8–1.2 cm wide, dark green. *Inflorescence* terminal, branched basally or not, 7–11 cm long, many-flowered; bracts ovate, reflexed, 3.5–5.5 mm long, 1–2.5 mm wide. *Flowers* pale yellow-green with a yellow lip with a reddish base and a reddish-purple column; dorsal sepal ovate, 4 mm long, 1.5 mm wide; lateral sepals obliquely triangular, 5 mm long, 2–3 mm wide, joined at base to form a blunt mentum; petals oblong-lanceolate, 4 mm long, 1 mm wide; lip obscurely trilobed, 5.5 mm long, 3 mm wide, the lateral lobes erect, semicircular, the mid-lobe oblong, with a basal horse-shoe-shaped callus.

DISTRIBUTION: Bougainville, Makira, New Georgia and the Shortland Islands. Also in Vanuatu, Fiji and Samoa.
VERNACULAR NAME: Togarupia (Bougainville).
HABITAT: Lowland rain forest, c. 80 m.
COLLECTION: *Dickson* 8 (K); *Guppy* s.n., Nov. 1893 (K); *Hunt* in RSS 2214 (K, LAE); *Waterhouse* 562 (K).

2. A. disticha *Ridley* in Journ. Bot., Lond. 24: 354 (1886). Type: New Guinea, *Forbes* s.n. (holotype BM!).

An epiphytic herb forming small clumps. *Stems* up to 30 cm long, slender, leafy. *Leaves* thin, distichous, oblong, 1.7–2.7 cm long, 4.5–7 mm wide, minutely unequally bilobed at the apex, sheathing at the base. *Inflorescence* terminal, short, 2.5–3 cm long, several-flowered; bracts distichous, lying in one-plane, longer than the flower, lanceolate, up to 1.5 cm long, 3 mm wide. *Flowers* white; dorsal sepal ovate, 3.5 mm long, 2.5 mm wide; lateral sepals obliquely ovate, 4 mm long, 2.5 mm wide, forming an obtuse mentum; petals lanceolate, 3 mm long, 1 mm wide, with a serrated margin; lip obscurely trilobed, 4 mm long, 2 mm wide, the apex reflexed, orbicular, apiculate, with a serrate margin, the lateral lobes erect, oblong, saccate at base; column 0.75 mm long.

DISTRIBUTION: Guadalcanal. Also in New Guinea and New Ireland.
HABITAT: Primary rain forest, 535 m.
COLLECTION: *Wickison* 99A (K).

3. A. lutea *Schltr.* in Fedde, Repert. Spec. Nov. Regni Veg., Beih. 1: 350 (1912). Type: New Guinea, *Schlechter* 18577 (holotype B†).

An epiphytic or terrestrial herb, forming pendulous or erect clumps. *Stems* up to 30 cm long, slender, leafy. *Leaves* thin-textured, distichous, oblong-lanceolate, 1.8–2.5 cm long, 4–8 mm wide, sheathing at base. *Inflorescence* terminal, solitary, up to 7 cm long, several- to many-flowered. *Flowers* pale to bright yellow, sometimes marked with purple with a yellow lip with or without red spots on either side of the mid-lobe; dorsal sepal ovate, 3 mm long, 2 mm wide; lateral sepals obliquely triangular-ovate, keeled, 4 mm long, 2 mm wide, forming an obtuse mentum; petals oblong, obtuse, 2.5 mm long, 1.5 mm wide; lip oblong, constricted in the middle, 4 mm long, 2 mm wide, with 2 raised callus ridges in the basal half converging at base, the lateral lobes erect, the apex deflexed, obtuse; column 2 mm long.

DISTRIBUTION: Bougainville, Guadalcanal and Makira. Also in New Guinea.
HABITAT: Rain forest, 50–1200 m.
COLLECTIONS: *Cribb, Dennis & Campbell* 5071 (K); *Hunt* in RSS 2260 (K, LAE) & 2970 (K); *Wickison* 224 (K).

4. A. neohibernica *Schltr.* in Fedde, Repert. Spec. Nov. Regni Veg., Beih 1: 352 (1912). Type: New Ireland, *Schlechter* 14706 (holotype B†).

An epiphytic herb forming pendulous clumps. *Stems* up to 40 cm long, slender, leafy. *Leaves* thin, distichous, lanceolate, minutely apiculate, 2–2.3 cm long, 4–5 mm wide, sheathing at base. *Inflorescence* terminal, solitary, short, up to 5 cm long, several-flowered. *Flowers* pale yellow; dorsal sepal ovate, 2.5 mm long, 2 mm wide; lateral sepals obliquely triangular, 3 mm long, 2–2.5 mm wide, forming a long, obtuse mentum; petals linear, obtuse, 2.2 mm long, 1 mm wide; lip narrow at base, wedge-shaped, 4 mm long, 3 mm wide, with a fleshy basal callus, the apex reflexed, triangular; column 3–3.5 mm long.

DISTRIBUTION: Choiseul and Guadalcanal. Also in New Ireland.
HABITAT: In disturbed village sites, 600 m.
COLLECTIONS: *Mitchell* 38 (K); *Wickison* 106 (K).

5. A. pendula *Blume*, Bijdr. Fl. Ned. Ind.: 298 (1825). Type: Java, *Blume* s.n. (holotype L).

A pendent, epiphytic herb. *Stems* slender, up to 40 cm long, leafy. *Leaves* distichous, lanceolate, acute, minutely apiculate, 2–2.5 cm long, 4–6 mm wide, thin-textured, sheathing at base. *Inflorescence* terminal or lateral, short, up to 7 cm long, several-flowered. *Flowers* creamy-white to pale yellow-green with a red spot on either side of the mid-lobe; dorsal sepal ovate, 4 mm long, 2 mm wide; lateral sepals obliquely triangular, 4 mm long, 3 mm wide; petals ovate, obtuse, 3 mm long, 1.5 mm wide; lip oblong, narrowed at base, 5 mm long, 2–2.5 mm wide, the apex slightly expanded, reflexed, the disc with 2 longitudinal callus ridges converging at base; column 3 mm long.

DISTRIBUTION: Bougainville and Guadalcanal. Also in Thailand, the Malay Archipelago, New Guinea and Fiji.

HABITAT: Rain forest, 300–1600 m.

COLLECTIONS: *Craven & Schodde* 329 (LAE); *Cribb, Dennis & Campbell* 5017 (K); *Hunt* in RSS 2031 & 2131 (K); *Ridsdale & Lavarack* in NGF 31423 (LAE) & 31586 (K, LAE); *Schodde & Craven* 3752 (LAE).

6. A. polystachya *(Schltr.) Schltr.* in Fedde, Repert. Spec. Nov. Regni Veg., Beih. 1: 356 (1912). Type: New Ireland, *Schlechter* 14664 (holotype B†).
Podochilus polystachyus Schltr. in K. Schum. & Laut., Nachtr. Fl. Deutsch. Südsee: 121 (1905).

An epiphytic herb. *Stems* up to 1 m long, erect or pendulous, concealed by tubular bases of leaves. *Leaves* thin, distichous, oblong-lanceolate, acute, 4–4.5 cm long, 1–1.2 cm wide, light green to dark green. *Inflorescence* terminal, branched basally, c. 12 cm long, many-flowered; bracts linear-ovate, reflexed, 1–1.5 mm long, 0.5 mm wide. *Flowers* greenish-white, usually flushed with maroon, with a green lip with a white apex; dorsal sepal ovate, 2.5–3 mm long, 2 mm wide; lateral sepals ovate, acute, 3.5–4 mm long, 2 mm wide, forming a blunt mentum; petals ovate, 2–2.5 mm long, 1.5 mm wide; lip oblong, broadest at apex, 3–3.5 mm long, 2–2.5 mm wide, the disc with 2 broad, fleshy callus ridges converging at base; column 1.5 mm long. (See fig. 29 A–H).

DISTRIBUTION: Bougainville, Guadalcanal, Kolombangara, Makira and New Georgia. Also in New Ireland and Vanuatu.

HABITAT: Rain forest, 20–1500 m.

COLLECTIONS: *Chaplin* 837 (K); *Cribb & Campbell* 5125 (K); *Dickson* A1 (K); *Hunt* in RSS 2211 (K, LAE), 2238, 2436, 2498, & 2945 (K); *Iromea* 2540 (K); *Schodde & Craven* 3941 (K); *Wickison* 46A, 89, 103 & 210 (K).

7. A. reflexa *Blume*, Bijdr. Fl. Ned. Ind. 1(7): 301 (1825). Type: Java, *Blume* s.n. (holotype L, isotype P!).
Appendicula vanikorensis Ames in Journ. Arn. Arb. 13: 138 (1932). Type: Santa Cruz Islands, Vanikoro, *Kajewski* 594 (holotype AMES!).

For synonymy see Kores (1989).

An epiphytic herb. *Stems* erect, up to 60 cm long, concealed by sheathing tubular bases of the leaves. *Leaves* distichous, 2-ranked, twisted at the base so that the blades lie in one plane, oblong-lanceolate, 3.5–5.5 cm long, 8–15 mm wide, c. 1 cm apart, bluntly bilobed at apex. *Inflorescences* lateral, opposite leaves, reduced, c. 1 cm long, 2– to 4-flowered. *Flowers* white, maturing yellowish; dorsal sepal concave, narrowly elliptic, obtuse, 2.5 mm long, 1.5 mm wide; lateral sepals concave, broadly triangular, 3 mm long, 2–2.5 mm wide, forming a blunt mentum; petals oblong-lanceolate, 2 mm long, 1–1.5 mm wide; lip simple, elliptic, obtuse, 3 mm long, 2 mm wide, reflexed at apex, the spur globose, 1 mm long, 1 mm wide, with a large transversely elliptic callus in the centre; column 2 mm long. (See fig. 29 J–T).

B
C
D
E
F
G
H
J
K
A
L
M
N
O
P
Q
R
S
T

DISTRIBUTION: Bougainville, Guadalcanal, Kolombangara, Malaita, New Georgia, Rendova, the Santa Cruz Islands and Santa Isabel. Widely distributed from the Malay Penisula and Archipelago to New Guinea, New Ireland, the Caroline Islands (Palau), the Horn Islands, Vanuatu, New Caledonia and Fiji.

HABITAT: Rain forest, sea level to 850 m.

COLLECTIONS: *Beer et al.* in BSIP 6671 (K, LAE) & 7359 (K); *Chaplin* 806 (K); *Comins* 296 (K); *Cribb & Campbell* 5148 (K); *Hunt* in RSS 2404, 2931 & 3073 (K); *Im Thurn* 346 (K); *Kajewski* 594 (AMES) & 1550 (G); *Nakisi* in BSIP 8086 (K); *Ridsdale & Lavarack* in NGF 31125 (K, LAE); *Schodde & Craven* 3952 (LAE); *Wickison* 44 & 46b (K).

Ames (1932) distinguished *A. vanikorensis* from *A. reflexa* chiefly on the shape of the petals. Having studied the type material of *A. vanikorensis* we find that these differences cannot be upheld and we consider the two to be conspecific.

46. CHILOPOGON Schlechter

Small epiphytic herbs. *Stems* short, slender, leafy. *Leaves* distichous, sheathing at base. *Inflorescence* terminal, several- to many-flowered; bracts distichous, usually longer than the flower and sometimes attractively coloured. *Flowers* small; dorsal sepal ovate to ovate-lanceolate; lateral sepals obliquely ovate, forming a mentum; petals linear, obtuse; lip trilobed or obscurely trilobed, shortly spurred; column short; pollinia 6.

A small genus of about five species centred on New Guinea and adjacent islands, a single species being recorded.

C. oxysepalum *(Schltr.) Schltr.* in Fedde, Repert. Spec. Nov. Regni Veg., Beih. 1: 333 (1912). Type: New Britain, *Schlechter* 13696 (holotype B†).

Podochilus oxysepalus Schltr. in K. Schum. & Laut., Nachtr. Fl. Deutsch. Südsee: 120 (1905).

A small epiphytic herb forming clumps on branches. *Stems* up to 19 cm long, leafy. *Leaves* distichous, lanceolate, minutely acuminate at tip, 1.6–2 cm long, 4–7 mm wide, thin, light green, tinged purple near the top of the stem, sheathing at base. *Inflorescence* terminal, up to 4 cm long, several- to many-flowered; bracts distichous, twisted at base to lie in one plane, longer than the flower, ovate, acuminate, with serrated margins, up to 7 mm long, 3 mm wide, purple with white margins. *Flowers* white or pale yellow with a purple spot on the anther-cap; dorsal sepal ovate, apiculate, 4 mm long, 2 mm wide; lateral sepals obliquely ovate, apiculate, 5 mm long, 2.5 mm wide; petals linear, 3 mm long, 1 mm wide; lip obscurely trilobed, 4 mm long, 3 mm wide, the lateral lobes rounded, the mid-

Fig. 29. Appendicula polystachya A-H. A, habit x 2/3; **B**, dorsal sepal x 10; **C**, lateral sepal x 10; **D**, column x 10; **E**, flower x 8; **F**, lip x 10; **G**, petal x 10; **H**, column and lip x 10. **A. reflexa J- T. J**, habit x 2/3; **K**, lip and column x 10; **L**, column x 10; **M**, lip x 10; **N**, ventral view of anther-cap x 14; **O**, dorsal view of anther-cap x 14; **P**, pollinia x 14; **Q**, flower x 9; **R**, dorsal sepal x 10; **S**, petal x 10; **T**, lateral sepal x 10. **A-H** drawn from *Cribb & Wheatley* 23; **J-T** from *Wickison* 44. All drawn by Sue Wickison.

lobe fleshy, the apex fleshy and rounded, the spur globose, 1.4 mm long, 1.2 mm wide; column short, 2 mm long. (See fig. 30 L–T and plate 4E).

DISTRIBUTION: Bougainville, Guadalcanal, Kolombangara, Makira, New Georgia and Rendova. Also in New Guinea and New Britain.
HABITAT: Rain forest, 250–820 m.
COLLECTIONS: *Chaplin* 753 & 754 (K); *Dickson* 27 (K); *Hunt* in RSS 2228, 2359 & 2465 (K); *Mason* 1687 (K); *Wickison* 121 (K).

47. PODOCHILUS Blume

Small or rather small epiphytes, often forming dense mats on trees. *Stems* slender and leafy. *Leaves* usually small, distichous, lying in one plane by the twisting of the leaf bases, sheathing at base. *Inflorescences* terminal and/or lateral, short, few- to many-flowered. *Flowers* small, usually white or green, often with purple markings; sepals and petals either joined at the base or free; lateral sepals forming a mentum; petals slightly smaller than sepals; lip obscurely trilobed, joined to the column-foot, sometimes with basal appendages; column-foot usually rather long and often curved upwards; pollinia 4.
A genus of about sixty species from India, China, Malaysia and New Guinea to the Pacific, a single species being recorded.

P. bimaculatus *Schltr.* in Fedde, Repert. Spec. Nov. Regni Veg., Beih. 1: 328 (1912). Types: New Guinea, *Schlechter* 16497 (syntype B†, isosyntypes AMES, E, K!, S), 18177 (syntype B, isosyntype K!) & 19193 (syntype B†, isosyntype AMES).

Epiphytic, forming creeping mats on tree trunks, moss-like. *Stems* slender and leafy, branching, up to 20 cm long, 0.5–1 mm wide. *Leaves* distichous, lanceolate, c. 5 mm long, 1 mm wide, thin, light green, lying in one plane. *Inflorescences* terminal, short, few-flowered. *Flowers* 2–3, white, not opening widely; dorsal sepal ovate, 1 mm long, 0.6 mm wide; lateral sepals ovate, fused in basal half, 2 mm long, 1 mm wide; petals elliptic, 0.9 mm long, 0.5 mm wide; lip obovate, 1.7 mm long, 0.8 mm wide, saccate at base, the spur globose, minutely bilobed at apex, 1.3 mm long and wide; column 1 mm long, the column-foot 1 mm long, curved upwards at apex.

DISTRIBUTION: Bougainville, Guadalcanal and Makira. Also in New Guinea.
VERNACULAR NAME: Tukuritukuri (Bouganville).
HABITAT: Montane forest, 530–730 m.
COLLECTIONS: *Dennis* in RSS 2349 (K); *Waterhouse* 604 (K); *Wickison* 108 (K).

Fig. 30. Octarrhena angraecoides A-K. A, habit x 2/3; **B**, flower x 14; **C**, lateral sepal x 18; **D**, dorsal sepal x 18; **E**, petal x 18; **F**, lip x 18; **G**, side view of lip x 18; **H**, ventral and dorsal view of anther-cap x 18; **J**, pollinia x 18; **K**, column x 18. **Chilopogon oxysepalum L-T. L**, habit x 2/3; **M**, flower x 6; **N**, side view of column and lip x 10; **O**, lip x 10; **P**, dorsal and ventral view of anther-cap x 10; **Q**, pollinia x 10; **R**, dorsal sepal x 10; **S**, petal x 10; **T**, lateral sepal x 10. **A-K** drawn from *Wickison* 15; **L-T** from *Wickison* 121. All drawn by Sue Wickison.

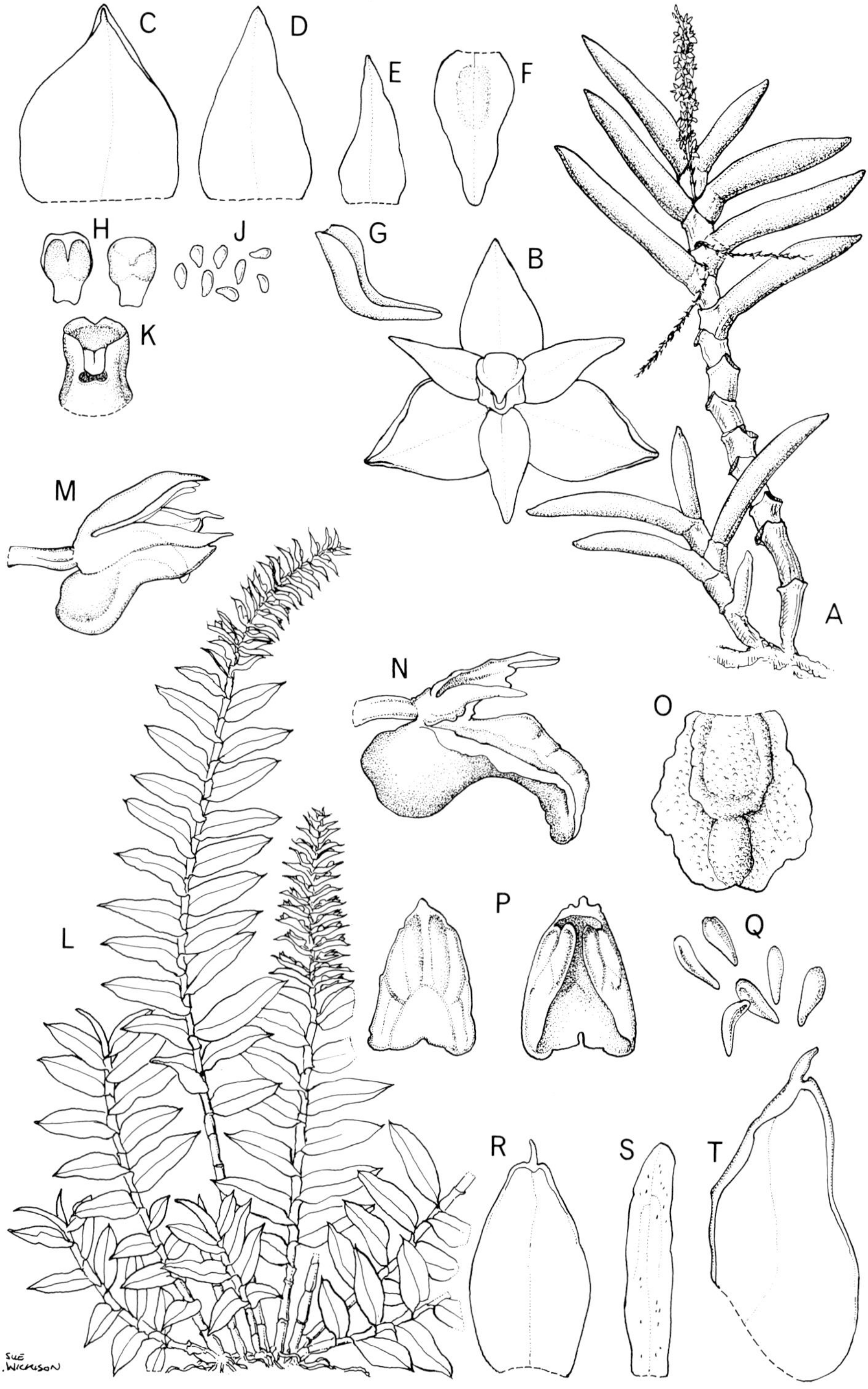
C
D
E
F
H
J
G
B
K
M
A
N
O
L
P
Q
R
S
T
SUE WICKISON

48. OCTARRHENA Thwaites

Small epiphytic or rarely terrestrial herbs. *Stems* elongate, leafy, branching from the root base. *Leaves* distichous, terete or bilaterally flattened. *Inflorescences* lateral, simple, with many small, clustered flowers. *Flowers* with sepals and petals free, spreading; lip entire, about equal in length to the petals; column lacking a foot; pollinia 8.

A genus of about twenty species from tropical Asia to the Pacific Islands, a single species being recorded.

O. angraecoides *(Schltr.) Schltr.* in Fedde, Repert. Spec. Nov. Regni Veg., Beih. 1: 217 (1911). Type: New Guinea, *Schlechter* 14502 (holotype B†).
Phreatia angraecoides Schltr. in K. Schum. & Laut., Nachtr. Fl. Deutsch. Südsee: 185 (1905).

A small epiphytic herb. *Stems* erect, up to 20 cm tall, sometimes appearing zig-zag, leafy. *Leaves* distichous, fleshy, bilaterally flattened, 3–5 cm long, 1.5–3 mm wide, dark green, sheathing at base. *Inflorescences* up to 5, lateral, 4–6 cm long, pale green, with many small dense flowers. *Flowers* yellow, rarely white; dorsal sepal ovate, 0.8 mm long, 0.2–0.3 mm wide; lateral sepals broadly ovate, 0.8 mm long, 0.7 mm wide; petals elliptic, acute, 0.6 mm long, 0.2 mm wide; lip narrowly obovate, acute, 0.6 mm long, 0.2 mm wide, uncinate; column 0.5 mm long. (See fig. 30 A–K and plate 6C).

DISTRIBUTION: Bougainville, Guadalcanal, Kolombangara and New Georgia. Also New Guinea, New Ireland and Vanuatu.
HABITAT: Montane and moss forest, 100–1900 m.
COLLECTIONS: *Cribb, Dennis & Campbell* 5050 (K); *Dickson* MD28 (K); *Hunt* in RSS 2548 (K); *Lavarack & Ridsdale* in NGF 31445 (K); *Wickison* 15 & 227 (K).

This species is similar to *O. condensata* (Ridley) Holttum, from the Malay Peninsula.

49. PHREATIA Lindley

Epiphytic herbs, pseudobulbous or caulescent, the former usually with 1–3 leaves, the latter with as many as 12 leaves. *Stems* very short or elongated. *Leaves* linear, erect to spreading, terminal, distichous or arranged in a fan. *Inflorescences* lateral or arising from the base of the pseudobulb, with many tiny flowers opening together. *Flowers* very small, often self-pollinating, white or green; sepals and petals similar; lateral sepals usually forming a distinct mentum; lip entire or obscurely trilobed, spurred or not; column short, usually with a well developed column-foot; pollinia 8.

A genus of about one hundred and fifty species from SE. Asia to Australia and the Pacific Islands, nine species being recorded.

1. Plants with distinct pseudobulbs; leaves 1 or 2 from apex of pseudobulb; inflorescence from base of pseudobulb 2
   Plants lacking pseudobulbs; leaves 4–12, distichous or arranged in a fan; inflorescences lateral 5
2. Lip spurred 3
   Lip lacking a spur 4
3. Lip oblong, obtuse, 1.5 mm long, 1 mm wide, shortly hairy .... **1. P. cylindrostachya**
   Lip obovate, acute or shortly apiculate, 2–3 mm long, 2–3 mm wide, papillate at base .... **7. P. paleata**
4. Leaves 1 or 2, linear to oblanceolate (2.5-)7–11 cm long, 4–6 mm wide .... **3. P. hypsorhynchos**
   Leaf solitary, oblanceolate-elliptic, 4–6.5 cm long, 10–15 mm wide .... **9. P. virescens**
5. Plant robust, fan-shaped; leaves ribbon-like, 7–35 cm long, 1.5–2.5 cm wide .... **5. P. micrantha**
   Plant delicate or elongate and not fan-shaped; leaves distichous, linear to oblanceolate, less than 13 cm long, 1.5 cm wide. 6
6. Plant up to 70 cm tall; stem elongate, fleshy; lip with a long claw more than 2 mm long .... **2. P. elongata**
   Plant less than 15 cm tall; stem reduced, sheathed in leaf bases; lip shortly clawed or not, about 1 mm long 7
7. Plant minute, up to 2.5 cm long .... **6. P. minima**
   Plant larger, 5–14 cm long 8
8. Leaves 3–7 cm long, 0.3–0.6 cm wide; inflorescence equal to or longer than the leaves; lip shortly clawed, broadly ovate, obtuse at apex .... **8. P. stenostachya**
   Leaves 2.5–8.5 cm long, 0.1–0.3 cm wide; inflorescence equal to or shorter than the leaves; lip square to oblong usually with a mucronate apex .... **4. P. matthewsii**

1. P. cylindrostachya *Schltr.* in Fedde, Repert. Spec. Nov. Regni Veg., Beih. 1: 944 (1913). Type: New Guinea, *Schlechter* 20140 (holotype B†).

An epiphytic herb. *Pseudobulbs* clustered, ovoid, up to 2 cm long, 1 cm wide. *Leaf* terminal, solitary, oblong to oblanceolate, 10–19 cm long, 1.5–2.8 cm wide, narrowed at base to form a petiole up to 6 cm long. *Inflorescence* arising from the base of the pseudobulb, up to 35 cm long, densely many-flowered. *Flowers* white; dorsal sepal ovate, 1.5 mm long, 0.8 mm wide; lateral sepals broadly ovate, 1.5 mm long, 1.2 mm wide, forming a rounded mentum; petals linear-ovate, 1.2 mm long, 0.8 mm wide; lip oblong, obtuse, 1.5 mm long, 1 mm wide, hairy, with a short, cylindrical spur appressed to the lip; column 0.5 mm long.

DISTRIBUTION: Kolombangara. Also in New Guinea.
HABITAT: Montane forest, 530–1500 m.
COLLECTIONS: *Cribb & Campbell* 5137 (K); *Whitmore* in BSIP 2080 (K); *Wickison* 77 (K).

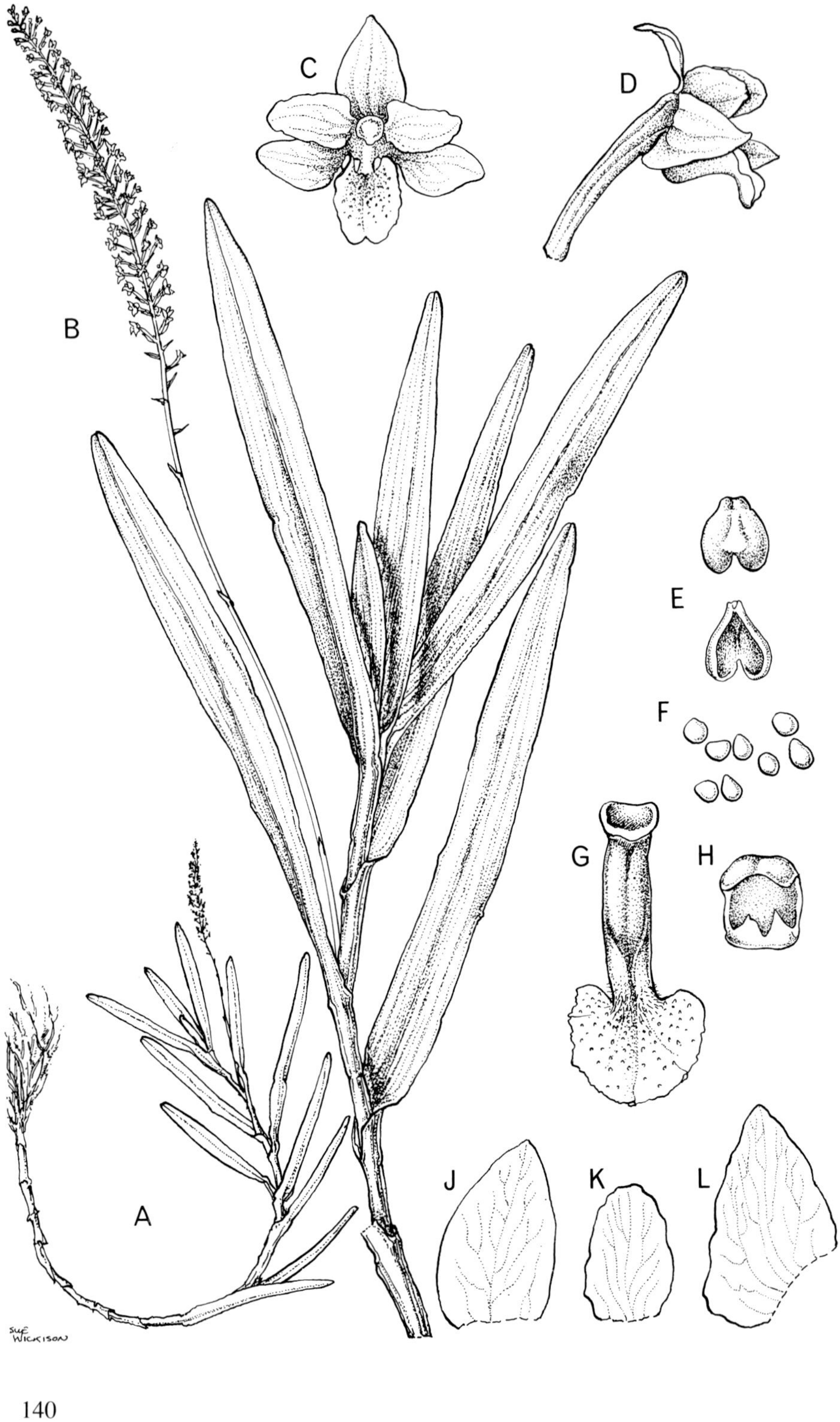
A
B
C
D
E
F
G
H
J
K
L
SUE WICKISON

2. P. elongata *Schltr.* in Fedde, Repert. Spec. Nov. Regni Veg., Beih. 1: 933 (1913). Type: New Guinea, *Schlechter* 16987 & 18243 (syntypes B†).

An epiphytic herb, up to 70 cm long. *Stems* elongate, fleshy, leafy in apical half. *Leaves* distichous, oblong-lanceolate, 11–12 cm long, 1.2–1.5 cm wide, sheathing at base. *Inflorescence* lateral, arising from the axil of a leaf near the apex of the stem, up to 22 cm long, densely many-flowered. *Flowers* white; dorsal sepal ovate, 2–2.5 mm long, 1.5 mm wide; lateral sepals obliquely ovate, 3 mm long, 1.5 mm wide, forming a rounded obtuse, mentum; petals oblong, obtuse, 1.5 mm long, 1.2 mm wide; lip clawed at base, 4 mm long, 2 mm wide, the apex expanded, rounded, finely papillose-pubescent, with an undulate margin; column short, 1 mm long. (See fig. 31).

DISTRIBUTION: Guadalcanal. Also in New Guinea and New Ireland.
HABITAT: Mossy forest, 1900 m.
COLLECTION: *Hunt* in RSS 2962 (K).

3. P. hypsorhynchos *Schltr.* in Engler, Bot. Jahrb. 39: 77 (1906). Type: New Caledonia, *Schlechter* 15427 (holotype B†, isotypes K!, P!, Z).
Phreatia comptonii Rendle in Journ. Linn. Soc. Bot. 45: 249, t. 13, fig. 10–12 (1921). Type: New Caledonia, *Compton* 580 (holotype BM!).

An epiphytic herb, forming large clumps. *Pseudobulbs* proximate, ovoid, 4–7 mm high, 3–6 mm wide. *Leaves* apical, 1 (or 2), linear to oblanceolate, erect, (2.5-) 7–11 cm long, 4–6 mm wide. *Inflorescence* arising from base of pseudobulb, 10–22 cm long, longer than the leaves, many-flowered. *Flowers* self-pollinating, white or greenish; dorsal sepal elliptic-ovate, 1–1.5 mm long, 1 mm wide; lateral sepals obliquely triangular, 1.5 mm long, 1.5 mm wide, the mentum short, blunt and obscure; petals elliptic, obtuse, 1 mm long, 0.5 mm wide; lip spade-shaped, 1.2 mm long, 1.5 mm wide, shortly clawed; column 1 mm long. (See fig. 32 A–J).

DISTRIBUTION: Guadalcanal, Kolombangara and Rendova. Also in Vanuatu and New Caledonia.
HABITAT: Montane and mossy cloud forest, 800–2300 m.
COLLECTIONS: *Cribb & Campbell* 5074, 5138 & 5140 (K); *Dennis* in RSS 2039 (K); *Dickson* 40 (K); *Hunt* in RSS 2479 & 2958 (K); *Mitchell* 12 (K); *Wickison* 75 & 107 (K).

This species is similar to *P. bismarckiensis* Schltr. from New Guinea.

4. P. matthewsii *Reichb. f.* in Otia Bot. Hamb.: 55 (1878). Type: Society Islands, *Matthews* s.n (holotype W!).

Fig. 31. Phreatia elongata. A, habit reduced; **B**, habit x 2/3; **C**, flower x 6; **D**, side view of flower x 6; **E**, dorsal and ventral view of anther-cap x 14; **F**, pollinia x 14; **G**, lip x 10; **H**, column x 14; **J**, dorsal sepal x 10; **K**, petal x 10; **L**, lateral sepal x 10. Drawn from *Hunt* 2962 by Sue Wickison.

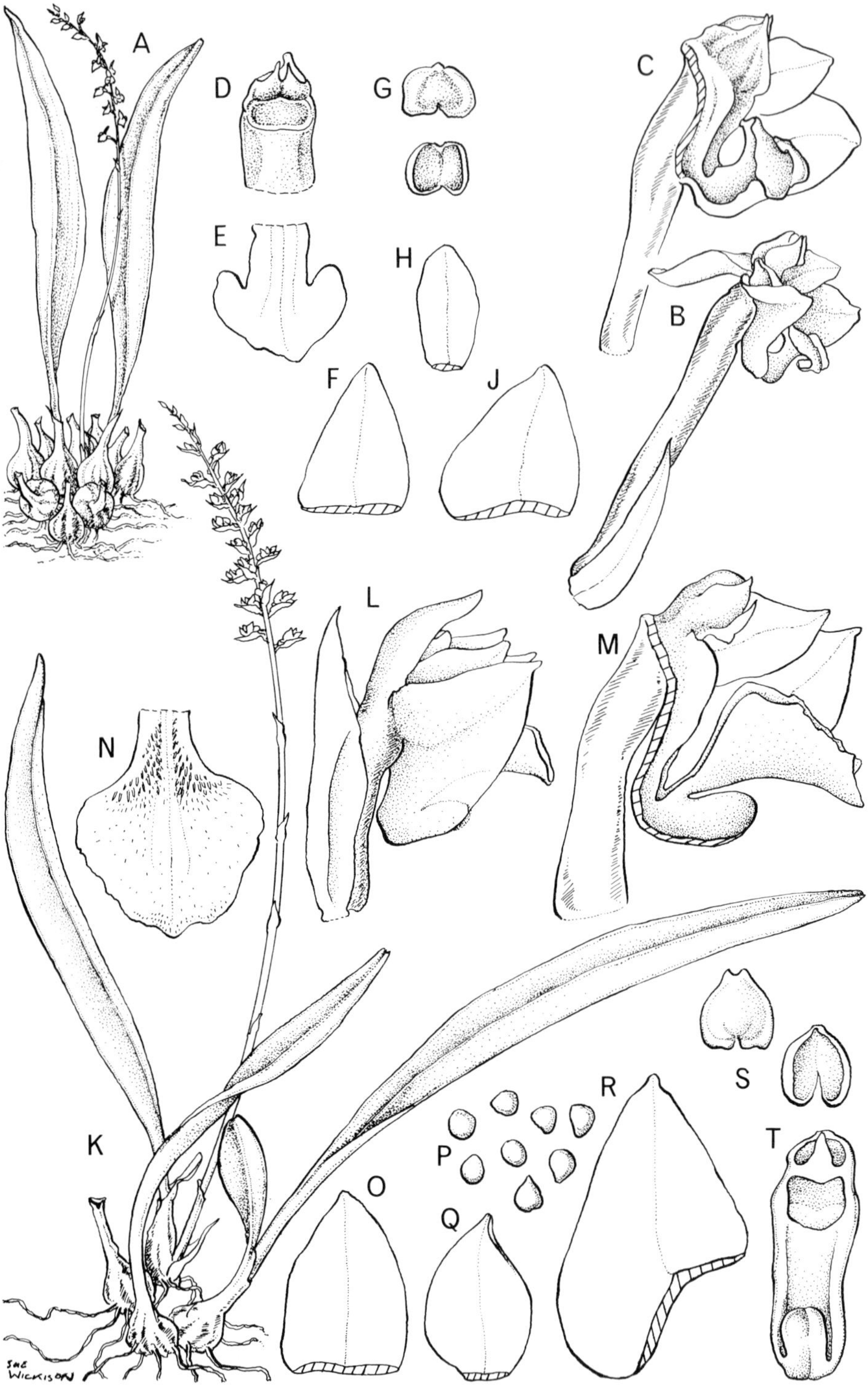
A
B
C
D
E
F
G
H
J
K
L
M
N
O
P
Q
R
S
T
SUE WICKISON

Phreatia inversa Schltr. in K. Schum. & Laut., Nachtr. Fl. Deutsch Südsee: 187 (1905). Type: New Ireland, *Schlechter* 14644 (holotype B†, isotype K!).
Phreatia neocaledonica Schltr. in Engler, Bot. Jahrb. 39: 78 (1906). Types: New Caledonia, *Schlechter* 14755 (syntype B†, isosyntype BM!), *Schlechter* 15228 (syntype B†, isosyntypes BM!, K!, L, NSW, P, Z).
Phreatia reineckei Schltr. in Fedde, Repert. Spec. Nov. Regni Veg. 9: 109 (1911). Type: Samoa, *Vaupel* 658 (holotype B†, isotype K!).
Phreatia minutiflora sensu Kraenzl., in Engler, Bot. Jahrb. 25: 607 (1898), non Lindley (1859).

A small epiphytic herb. *Stems* short, clustered, up to 1.5 cm long. *Leaves* arranged in a fan, linear, obtuse, slightly fleshy, 2.5–8.5 cm long, 0.1–0.3 cm wide, sheathing at base. *Inflorescence* lateral, equalling or shorter than leaves, 2.5–8 cm long, many-flowered; peduncle 1–3 cm long. *Flowers* white or green, self-pollinating; dorsal sepal ovate, 1–1.3 mm long, 0.6–0.7 mm wide; lateral sepals obliquely ovate, acute, 1–1.3 mm long, 0.6–0.7 mm wide, forming a short, obtuse mentum; petals linear, 0.8 mm long, 0.3 mm wide; lip entire, square or oblong, 1 mm long, 0.5–0.7 mm wide, with a fleshy, mucronate apex; column 0.5 mm long. (See fig. 33 A–K and plate 6F).

DISTRIBUTION: Bougainville, Guadalcanal, Kolombangara, Makira and New Georgia. Also in New Ireland, Vanuatu, New Caledonia, Fiji, the Society Islands and Samoa.
HABITAT: Rain forest, 250–1100 m.
COLLECTIONS: *Balodis* 2 (K); *Cribb & Campbell* 5104 (K); *Cribb, Dennis & Campbell* 5070 & 5076 (K); *Dennis* in RSS 4438 (K); *Hunt* in RSS 2188 (K); *Lavarack & Ridsdale* in NGF 31143 (K); *Officers of HMS Penguin* s.n. (K); *Waterhouse* 522 (K); *Whitmore* in BSIP 2052 & 6264 (K); *Wickison* 128 & 132 (K).

5. P. micrantha *(A. Rich.) Schltr.* in Fedde, Repert. Spec. Nov. Regni Veg., Beih. 1: 919 (1913). Type: Santa Cruz Islands, Vanikoro, *Lesson* s.n. (holotype P).
Oberonia micrantha A. Rich. in Sert. Astrol.: t. 3 (1833).
Phreatia macrophylla Schltr. in Engler, Bot. Jahrb. 39: 78 (1906). Type: New Caledonia, *Schlechter* 15465 (holotype B†, isotypes BM, K!, P!, W, Z).
Phreatia sarcothece Schltr. in Fedde, Repert. Spec. Nov. Regni Veg. 9: 438 (1911). Type: Vanuatu, Anatom, *Morrison* s.n. (holotype B†).
Phreatia collina Schltr. in Fedde, Repert. Spec. Nov. Regni Veg., Beih. 1: 919 (1913), Non J.J. Smith (1911). Type: New Guinea, *Schlechter* 16438 (holotype B).
Phreatia robusta R.S. Rogers in Trans. Roy. Soc. Austr. 54: 39 (1930). Type: cult. Brisbane, Jan. 1930, *C.T. White*, ex Queensland, *A. Beck* s.n. (holotype AD).

Fig. 32. Phreatia hypsorhynchos A-J. A, habit x 2/3; **B**, flower, ovary, pedicel and bract x 10; **C**, flower with dorsal sepal, petal and lateral sepal removed x 14; **D**, column x 14; **E**, lip x 14; **F**, dorsal sepal x 14; **G**, dorsal and ventral view of anther-cap x 14; **H**, petal x 14; **J**, lateral sepal x 14. **P. paleata K-T. K**, habit x 2/3; **L**, flower, ovary, pedicel and bract x 8; **M**, side view of flower with dorsal sepal, right petal and lateral sepal removed x 10; **N**, lip x 10; **O**, dorsal sepal x 10; **P**, pollinia x 14; **Q**, petal x 10; **R**, lateral sepal x 10; **S**, dorsal and ventral view of anther-cap x 14; **T**, column x 10. **A-J** drawn from *Wickison* 107; **K-T** from *Wickison* 17. All drawn by Sue Wickison.

Rhynchophreatia micrantha (A. Rich.) N. Hallé in Fl. Nouv. Caled. 8: 341 (1977). For full synonymy see Kores (1989) and Clements (1989).

An epiphytic herb forming a single fan-like growth, up to 40 cm tall. *Stem* vestigial, continuing to grow throughout the life of the plant, covered by persistent, imbricate leaf bases. *Leaves* alternate, 4–10, arranged in a fan, linear, blunt, 7–35 cm long, 1.5–2.5 cm wide, with an unequally bilobed apex, overlapping and persisting at base. *Inflorescences* 1-several, from axils of the upper leaves, exceeding the leaves in length, erect or decurved; rhachis usually longer than the peduncle, many-flowered. *Flowers* minute, white; dorsal sepal ovate, cucullate, 1.5 mm long, 1 mm wide; lateral sepals obliquely ovate, 1.5 mm long, 1.5 mm wide, forming a short rounded mentum with the column-foot; petals ovate, 1 mm long, 0.7 mm wide; lip broadly clawed, subrhomboidal, 1.2 mm long, 1.2 mm wide, with a shallow depression containing viscid fluid near the base; column 1.5 mm long.

DISTRIBUTION: Bougainville, Guadalcanal, Kolombangara, Makira, New Georgia and the Santa Cruz Islands (Vanikoro). Also in New Guinea, New Ireland, Vanuatu, New Caledonia, Fiji, Samoa, the Horn Islands and Australia.
VERNACULAR NAME: Kikilapa Ngirisi (New Georgia).
HABITAT: A common rain forest epiphyte, sea level to 520 m.
COLLECTIONS: *Chaplin* 773, 802, 803 & 830 (K); *Dickson* B2 (K); *Hanson* 6 (K); *Hunt* in RSS 2336, 2363, 2416 & 2983 (K); *Im Thurn* 353 (K); *Lavarack & Ridsdale* in NGF 31028 & 31353 (K); *Mauriasi et al.* in BSIP 16681 (K); *Waterhouse* 157 (K); *Wickison* 97 (K).

We consider *P. sarcothece* to be synonymous with *P. micrantha* based on Schlechter's description and examination of living plants seen on Anatom (B.L.).

6. P. minima *Schltr.* in Fedde, Repert. Spec. Nov. Regni Veg., Beih. 1: 922 (1913). Type: New Guinea, *Schlechter* 19734 (holotype B†).

A minute epiphytic herb, up to 2.5 cm long. *Stems* up to 4 mm long, sheathed in leaf bases. *Leaves* distichous on stem, 3–8, oblong, 1–1.5 cm long, 1–2 mm wide. *Inflorescence* lateral, up to 5 mm long, several-flowered. *Flowers* minute, pale green, probably cleistogamous; sepals and petals less than 1 mm long, 0.5 mm wide; dorsal sepal ovate; lateral sepals obliquely triangular, forming an obtuse mentum; petals linear; lip spade-shaped, 0.7 mm long, 0.5 mm wide; column very short, 0.5 mm long.

DISTRIBUTION: Guadalcanal. Also in New Guinea.
HABITAT: Montane ridge-top forest, 900 m.
COLLECTION: *Cribb, Dennis & Campbell* 5021 (K).

Fig. 33. Phreatia matthewsii A-K. A, habit x 2/3; **B**, flower, ovary, pedicel and bract x 20; **C**, flower with the dorsal sepal, petal and lateral sepal removed x 20; **D**, dorsal sepal x 20; **E**, petal x 20; **F**, lateral sepal x 20; **G**, lip x 20; **H**, column x 20; **J**, dorsal and ventral view of anther-cap x 20; **K**, pollinia x 20. **P. stenostachya L-U. L**, habit x 2/3; **M**, flower, ovary, pedicel and bract x 20; **N**, flower with the dorsal sepal, petal and lateral sepal removed x 20; **O**, dorsal sepal x 20; **P**, petal x 20; **Q**, lateral sepal x 20; **R**, lip x 20; **S**, column x 20; **T**, dorsal and ventral view of anther-cap x 20; **U**, pollinia x 20. **A-K** drawn from *Wickison* 128; **L-U** from *Wickison* 153. All drawn by Sue Wickison.

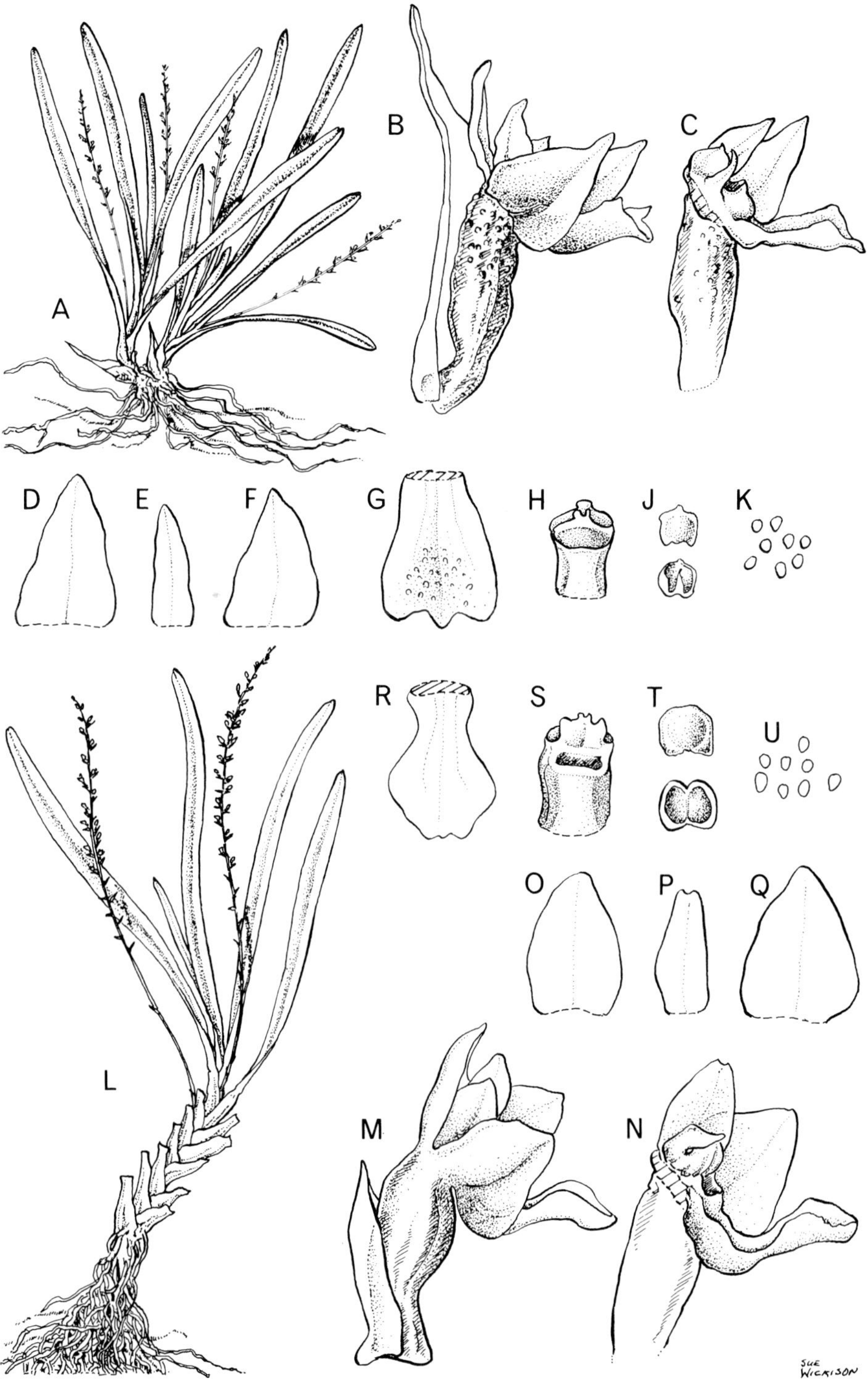
A
B
C
D
E
F
G
H
J
K
R
S
T
U
O
P
Q
L
M
N
SUE WICKISON

7. P. paleata *(Reichb. f.) Reichb. f.* in Linnaea 41: 653 (1877). Type: New Caledonia, *Vieillard* 1331 (holotype W!, isotype P!).
Eria paleata Reichb. f. in Linnaea 41: 87 (1877).
Phreatia pholidotoides Kraenzl. in Not. Syst. 4: 140 (1928). Type: New Caledonia, *Cribs* 1217 (holotype P!).

An epiphytic herb. *Pseudobulbs* clustered, ovoid, obtuse, 3–6 mm long, 3–5 mm wide. *Leaves* apical, 1–2, oblanceolate, acute, erect, 12–28 cm long, 1.7–2.2 cm wide, petiolate. *Inflorescence* from base of pseudobulb, erect, 20–24 cm long, longer than the leaves, densely many-flowered. *Flowers* white; dorsal sepal ovate, acute, 2–3 mm long, 1.5–2 mm wide; lateral sepals obliquely ovate, 4 mm long, 2–2.5 mm wide, forming a cylindrical, blunt mentum 1.5 mm long; petals ovate, acute, 2–2.5 mm long, 1–1.5 mm wide; lip obovate, acute or shortly apiculate, 2–3 mm long, 2–3 mm wide, shortly clawed, papillate at base, the spur shortly cylindrical, blunt, 1 mm long, appressed to lip; column 2.5 mm long (See fig. 32 K–T).

DISTRIBUTION: Kolombangara and New Georgia. Also in New Guinea, New Ireland, Vanuatu, New Caledonia and Norfolk Island.
HABITAT: Rain forest, 260–1700 m.
COLLECTIONS: *Mitchell* 11 (K); *Wickison* 17 (K).

Phreatia saccifera Schltr., from New Ireland, is similar and may be conspecific.

8. P. stenostachya *(Reichb. f.) Kraenzl.* in Engler, Pflanzenr. Orch.-Mon.-Thelasin.: 29 (1911). Type: Fiji, *Seemann* 589 (holotype W, isotypes AMES, BM, P!, K!,).
Eria stenostachya Reichb. f. in Seem., Fl. Vit.: 301 (1868).
Phreatia outbatchensis Schltr. in Engler, Bot. Jahrb. 39: 79 (1906). Type: New Caledonia, *Schlechter* 15393 (holotype B†, isotypes K!, P!).

For full synonymy see Kores (1989).

A small epiphytic herb. *Stems* clustered, covered by leaf bases, up to 6 cm long. *Leaves* in a fan, 4–8, linear to linear-oblanceolate, obtuse, 3–7 cm long, 0.3–0.6 cm wide. *Inflorescence* lateral, equal to or longer than the leaves, up to 12 cm long; peduncle short, 1–4 cm long. *Flowers* white, self-pollinating; dorsal sepal ovate, subacute, 0.8–1 mm long, 0.6–0.8 mm wide; lateral sepals obliquely ovate, 1 mm long, 0.7 mm wide, forming an obtuse, rounded mentum; petals elliptic, 0.6 mm long, 0.3 mm wide; lip shortly clawed, 1 mm long, 0.6 mm wide, the apex broadly ovate, obtuse; column 0.6 mm long. (See fig. 33 L–U and plate 6E).

DISTRIBUTION: Guadalcanal. Also in Vanuatu, New Caledonia, Fiji and Samoa.
HABITAT: Montane forest, 400 m.
COLLECTIONS: *Wickison* 120 & 153 (K); *Woodford* 8 (K).

9. P. virescens *Schltr.* in Fedde, Repert. Spec. Nov. Regni Veg., Beih. 1: 941 (1913). Type: New Guinea, *Schlechter* 19611 (holotype B†).

An epiphytic herb. *Pseudobulbs* spaced 1.5–3 cm apart on the rhizome, ovoid, 1 cm long, 5 mm wide, sheathed in bracts. *Leaf* apical, solitary, erect, oblanceolate-elliptic, 4–6.5 cm long, 1–1.5 cm wide; petiole up to 3 cm long. *Inflorescence* arising from the base of the pseudobulb, extending beyond the leaf, up to 30 cm long, bearing many small flowers. *Flowers* green to white; dorsal sepal ovate, obtuse, 1.5–2 mm long, 0.8 mm wide; lateral sepals obliquely ovate, acute, 2–2.5 mm long, 1.5 mm wide, forming an obtuse, rounded mentum; petals ovate, 1–1.5 mm long, 0.8 mm wide; lip spade-shaped, erect, 1.5–2 mm long, 1–1.5 mm wide, joined to the column-foot; column 2 mm long.

DISTRIBUTION: Bougainville and Guadalcanal. Also in New Guinea.
HABITAT: Moss forest, 1400–1840 m.
COLLECTIONS: *Cribb, Dennis & Campbell* 5075 (K); *Hunt* in RSS 2986 (K); *Lavarack & Ridsdale* in NGF 31331 & 31398 (K).

This species has flowers which are similar to those of *P. hypsorhynchos* but it is distinguished by its shorter, broader leaves.

50. THELASIS Blume

Small epiphytic herbs. *Stems* with or without visible pseudobulbs, bearing terminal leaves. *Leaves* 1 or 2, narrow and thin-textured. *Inflorescence* lateral with a long slender scape; peduncle always longer than the rhachis; rhachis bearing many small flowers. *Flowers* with sepals and petals about equal, not spreading except at the tips; lateral sepals joined at the base; lip entire, widest and hollow at the base, narrowing towards the tip; column without a column-foot; pollinia 8.

A genus of about nineteen species from India to the Solomon Islands, a single species being recorded.

T. pygmaea (Griffith) Lindley in Journ. Linn. Soc. Bot. 3: 63 (1855). Type: Nepal, *Lawrence* (holotype CAL).
Euproboscis pygmaea Griffith in Calc. Journ. Nat. Hist. 5: 372, t. 26 (1845).

For full synonymy see Comber (1990).

An epiphytic herb forming dense clusters. *Pseudobulbs* visible, clustered, 5–10 mm long, 5–10 mm wide, bearing 2 unequal leaves at the apex. *Leaves* oblong, gradually tapering at base, the larger 3–4 cm long, 5 mm wide, the other 1–2 cm long, 3–5 mm wide, with a prominent central mid-vein, glossy mid-green. *Inflorescences* arising from the base of the pseudobulb, 8–13 cm long; rhachis 2.5–4.5 cm long, many-flowered; bracts lanceolate, acute, 1–2 mm long, at right-angles to the rhachis, persistent. *Flowers* opening a few at a time, small, pale green at base becoming white towards the apex; ovary triangular in cross section; dorsal sepal oblong, 2.5 mm long, 1 mm wide; lateral sepals obliquely oblong, keeled, 3 mm long, 1.5 mm wide; petals oblong, 2–2.5 mm long, 0.8 mm wide; lip oblong, widest at the base, blunt, 2.3 mm long, 1 mm wide; column 1.5 mm long.

DISTRIBUTION: Florida Islands (Nggela), Gizo and Kolombangara. Widely distributed from India and throughout most of SE. Asia.
HABITAT: Strand forest, sea level.
COLLECTIONS: *Cribb & Campbell* 5116 (K); *Mitchell* 17 (K); *Wickison* 85 & 85B (K).

51. AGLOSSORHYNCHA Schlechter

Epiphytic or lithophytic herbs. *Stems* erect or pendent, branched. *Leaves* distichous, sheathing at base, the leaf sheaths usually not ciliate at apex. *Inflorescence* terminal, short, 1– or 2-flowered. *Flowers* with dorsal sepal ovate to oblong; lateral sepals free or fused; petals free; lip concave, entire, with margins inrolled embracing the column, lacking a spur; column relatively long and slender; pollinia 4.
A small genus of about fourteen species from eastern Malesia to the SW. Pacific, two species being recorded.

Leaves oblong-lanceolate, unequally bilobed at apex; flowers usually solitary, bright lemon-yellow to transparent yellow, deepening to orange with age .... **1. A. aurea**
Leaves oblong, equally bilobed at apex; inflorescences 2-flowered; flowers white to greenish .... **2. A. biflora**

1. A. aurea *Schltr.* in K. Schum. & Laut., Nachtr. Fl. Deutsch. Südsee: 133 (1905). Type: New Ireland, *Schlechter* 14651 (holotype B†, isotypes BR, K!, P).

An epiphytic herb, growing in clumps with the roots forming loose mats around the branches. *Stems* up to 40 cm tall, branching. *Leaves* distichous, oblong-lanceolate, 3–5.5 cm long, 4–7 mm wide, unequally bilobed at the apex, sheathing at base. *Inflorescence* terminal, usually 1-flowered. *Flowers* bright lemon-yellow or translucent yellow deepening to orange with age; dorsal sepal ovate, 10.5 mm long, 4.5 mm wide; lateral sepals obliquely ovate, 10 mm long, 4.5 mm wide; petals ovate, 9 mm long, 4 mm wide; lip obovate, 8 mm long, 6 mm wide, with inrolled margins; column 6 mm long, with a serrate apex.

DISTRIBUTION: Bougainville and New Georgia. Also in New Guinea and New Ireland.
HABITAT: Rain forest, 380–840 m.
COLLECTION: *Craven & Schodde* 150 (K); *Wickison* 41a (K).

The Solomon Island and Bougainville collections differs from the type of *A. aurea* in having free sepals. More material of this species is needed to confirm its identification.

2. A. biflora *J.J. Smith* in Bull. Dép. Agric. Indes Néerl. 39: 1 (1910). Type: New Guinea, *Von Römer* 1290 (holotype BO!).

An epiphytic or lithophytic herb. *Stems* sturdy, branched, pendent or spreading, yellow to olive-green, up to 1 m long, sheathed in leaf bases. *Leaves* many, distichous, oblong, 5–7 cm long, c. 8 mm wide, equally bilobed at the apex, mid-green on upper surface, light green on lower surface. *Inflorescence* terminal, 2-flowered. *Flowers* white to greenish, with a green lip; pedicel and ovary c. 6 mm long, finely pubescent; dorsal sepal ovate, 10 mm long, 3 mm wide; lateral sepals ovate, 10 mm long, 3 mm wide; petals ovate-lanceolate, 10 mm long, 2 mm wide; lip concave, obovate, acute, 10 mm long, 6 mm wide, with inrolled margins embracing the column; column 5 mm long, with a serrate apex.

DISTRIBUTION: Bougainville. Also in New Guinea, Vanuatu and Fiji.
HABITAT: Montane rain forest, 2000 m.
COLLECTION: *Lavarack & Ridsdale* in NGF 31405 (K, LAE).

52. GLOMERA Blume

Epiphytic herbs. *Stems* erect, sheathed by leaf bases. *Leaves* distichous, lanceolate. *Inflorescence* terminal, globose, with flowers in a dense cluster, each flower subtended by a small, non–sheathing bract. *Flowers* usually white or yellow with a rose apex to the lip; dorsal sepal ovate; lateral sepals connate at base, forming a short conical-saccate mentum, enveloping the spur; petals free; lip with a reflexed apex and a deeply saccate, semi-cylindrical spur; column short; pollinia 4.

A genus of about fifty species centred on New Guinea but extending from Java to the SW. Pacific, four species being recorded.

1. Leaf sheaths smooth 2
   Leaf sheaths warty 3
2. Stems slender; leaves 9.5–15 cm long, 0.9–1.5 cm wide; inflorescence 1–3 cm in diameter .... **1. G. montana**
   Stems stout; leaves 10–18 cm long, 1.6–2 cm wide; inflorescence 4 cm in diameter .... **4. G. pendulata**
3. Sepals 6–7 mm long; leaves 3–5.4 cm long, 5–10 mm wide .... **2. G. neohibernica**
   Sepals 9–10 mm long; leaves 1.7–4.5 cm long, 3–8 mm wide .... **3. G. papuana**

1. G. montana *Reichb.f.* in Linnaea 41: 77 (1877). Type: Fiji, *Milne* 214 (holotype W, isotypes AMES, K!).

Agrostophyllum reineckeanum Kraenzl. in Engler, Bot. Jahrb. 25: 602 (1898). Type: Samoa, *Reinecke* 297 (holotype B†, isotypes E!, G!).

Glomera samoensis Rolfe in Bull. Misc. Inf., Kew 1908: 414 (1908). Type: Samoa, *Funk* 11 (holotype K!).

Glomera gibbsiae Rolfe in Journ. Linn. Soc. Bot. 39: 176 (1909). Type: Fiji, *Gibbs* 807 (holotype BM!, isotype K!).

Glomera reineckiana (Kraenzl.) Schltr. in Fedde, Repert. Spec. Nov. Regni Veg. 9: 96 (1910).
Glomera rugulosa Schltr. in Fedde, Repert. Spec. Nov. Regni Veg., Beih. 1: 287 (1912). Type: New Guinea, *Schlechter 18200* (holotype B†).

An epiphytic herb. *Stems* up to 1 m high, branching. *Leaves* distichous, lanceolate, 9.5–15 cm long, 0.9–1.5 cm wide, dark green; leaf sheaths smooth. *Inflorescence* terminal, globose, 1–3 cm in diameter, many-flowered. *Flowers* white with a rose-red apex to the lip, the anther-cap black; dorsal sepal ovate, 9–10 mm long, 3.4 mm wide; lateral sepals fused, 8 mm long, 8 mm wide; petals oblong-ovate, 9–10 mm long, 3.4–3.6 mm wide; lip oblong, 2.5 mm long, 2.5 mm wide, rounded at apex, the spur globose, 3 mm long, 1.5 mm wide; column 2–3 mm long. (See fig. 34 A–K and plate 7C).

DISTRIBUTION: Bougainville, Guadalcanal, Makira, Malaita and New Georgia. Also in New Guinea, Vanuatu, Fiji and Samoa.
VERNACULAR NAME: Tupaida (New Georgia).
HABITAT: Rain forest, 100–720 m.
COLLECTIONS: *Dennis* in RSS 2030 (K); *Hunt* in RSS 2201 (K, LAE); *Kotali et al.* in BSIP 11347 (K); *Waterhouse* 130 (K); *Wickison* 14 & 235 (K).

2. G. neohibernica *Schltr.* in K. Schum. & Laut., Nachtr. Fl. Deutsch. Südsee: 135 (1905). Type: New Ireland, *Schlechter* 14661 (holotype B†, isotypes K!, P).

An epiphytic herb forming small clumps. *Stems* up to 30 cm tall, branching. *Leaves* distichous, lanceolate, 3–5 cm long, 5–10 mm wide; leaf sheaths warty. *Inflorescence* terminal, globose, 1.3–1.8 cm in diameter, many-flowered. *Flowers* white to pale yellow, apex of lip pink, the anther-cap brown; dorsal sepal oblong, acute, 6 mm long, 2.9 mm wide; lateral sepals fused, 6.8 mm long, 4.5 mm wide; petals ovate, 5.8 mm long, 2.7 mm wide; lip oblong, 2.3 mm long, 2 mm wide, the spur cylindrical, rounded at apex, 2 mm long, 2 mm wide; column 1.5 mm long.

DISTRIBUTION: Malaita. Also in New Ireland.
HABITAT: Primary forest, on exposed ridges, 580 m.
COLLECTION: *Wickison* 234 (K).

3. G. papuana *Rolfe* in Bull. Misc. Inf., Kew 1899: 111 (1899). Type: New Guinea, *A. Giulianetti* s.n. (holotype K!).

An epiphytic herb. *Stems* 10–30 cm high, 1.5–2 mm wide, branching towards the base and covered by verrucose sheathing leaf bases. *Leaves* coriaceous,

Fig. 34. Glomera montana A-K. A, habit x 2/3; **B**, flower x 3; **C**, dorsal sepal x 4; **D**, petal x 4; **E**, lateral sepal x 4; **F**, column x 6; **G**, dorsal and ventral view of anther-cap x 8; **H**, pollinia x 8; **J**, lip x 6; **K**, lip and column x 6. **Glomera papuana L-U. L**, habit x 2/3; **M**, flower x 3; **N**, lip and column x 6; **O**, column x 8; **P**, pollinia x 8; **Q**, dorsal and ventral view of anther-cap x 8; **R**, dorsal sepal x 4; **S**, petal x 4; **T**, lateral sepal x 4; **U**, lip x 6. **A-K** drawn from *Cribb & Wheatley* 12; **L-U** from *Hunt* 2961. All drawn by Sue Wickison.

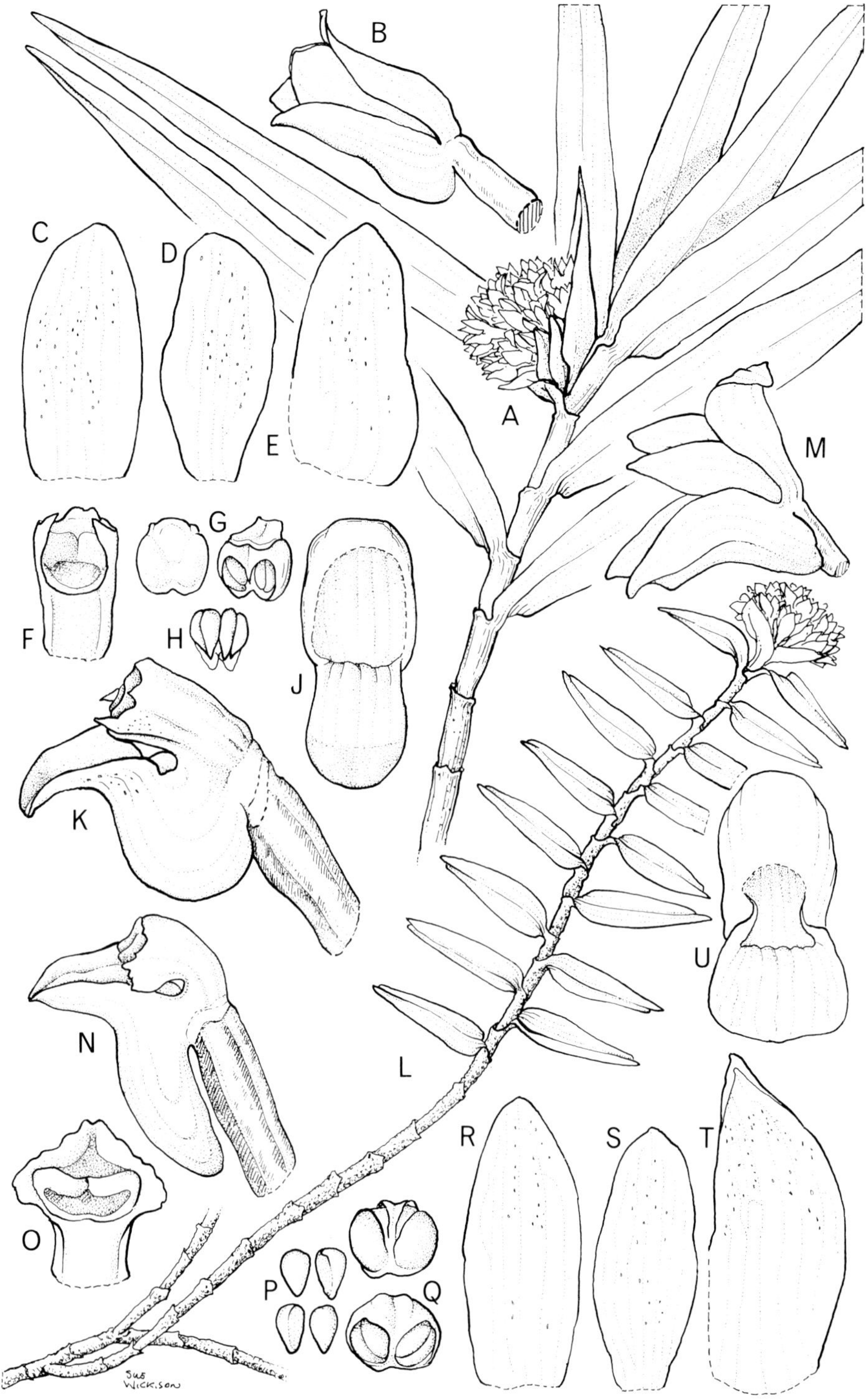
B
C
D
E
A
M
F
G
H
J
K
U
N
L
R
S
T
O
P
Q
SUE WICKISON

distichous, spreading, twisted at the base to lie in one plane, lanceolate, unequally bilobed at the subacute apex, 1.7–4.5 cm long, 3–8 mm wide. *Inflorescence* terminal, globose, 1–2 cm in diameter, many-flowered. *Flowers* white with a red apex to the lip, the anther-cap maroon; dorsal sepal oblong, obtuse 9–10 mm long, 4 mm wide; lateral sepals obliquely oblong-ovate, acute, 10 mm long, 4.5 mm wide; petals narrowly elliptic, obtuse, 8 mm long, 3.5 mm wide; lip oblong to subquadrate, obtuse, 2–2.5 mm long, 3–3.5 mm wide, the spur deeply saccate, semi-cylindrical, 2.5–3 mm long, 1.5–2 mm wide; column incurved, 2–2.5 mm long, with a short denticulate wing on each side at the apex. (See fig. 34 L–U).

DISTRIBUTION: Bougainville, Guadalcanal and Kolombangara. Also found in New Guinea and Vanuatu.
HABITAT: Montane and cloud forest, 1000–2340 m.
COLLECTIONS: *Hunt* in RSS 2482 & 2961 (K); *Iras* 2501 (K); *Schodde & Craven* 3749 (K, LAE); *Wickison* s.n. (K).

Hunt (no. 2961) notes that this species is 'probably the commonest orchid above 6,000 ft.' on Guadalcanal.

4. G. pendulata *(van Royen) J.J. Wood* in Kew Bull. 41(4): 819 (1986). Type: New Guinea, *van Balgooy* 539 (holotype CANB, isotype L!).
Agrostophyllum pendulatum van Royen, Alp. Fl. New Guinea 2: 686 (1979).

An epiphytic herb up to 70 cm long. *Stems* stout, 9–10 mm wide. *Leaves* distichous, lanceolate, 10–19 cm long, 1.6–2.3 cm wide, unequally bilobed at the apex, with smooth sheathing leaf bases. *Inflorescence* terminal, globose, 4–6 cm in diameter, many-flowered. *Flowers* cream with a pink-red lip; dorsal sepal ovate-elliptic, 9–11 mm long, 3–4 mm wide; lateral sepals obliquely elliptic-ovate, 9–11 mm long, 3–4 mm wide, fused in basal half; petals oblong-elliptic, 9–11 mm long, 3–3.5 mm wide; lip apex oblong, apiculate, 3.5 mm long, 3.5 mm wide, the spur rounded, obtuse, 3 mm long, 3 mm wide; column short, 3–4 mm long.

DISTRIBUTION: Bougainville. Also in New Guinea and New Ireland.
HABITAT: Riverine montane forest, 670 m.
COLLECTION: *Schodde & Craven* 3834 (K, LAE).

53. GLOSSORHYNCHA Ridley

Epiphytic or terrestrial herbs often forming dense spreading masses. *Stems* erect or pendulous, usually slender and much branched, stems in older parts covered with leaf sheaths only. *Leaves* distichous, slender, oblong-lanceolate to linear, with sheathing leaf bases which may be ciliate at the apex. *Inflorescence* terminal, solitary, each flower subtended by two opposite bracts. *Flowers* with sepals and petals similar, usually free; lip with a short or long spur; column short and stout; pollinia 4.

A genus of about sixty species in New Guinea, adjacent islands and the SW. Pacific, a single species being recorded.

G. macdonaldii *Schltr.* in Fedde, Repert. Spec. Nov. Regni Veg. 3: 19 (1906). Type: Vanuatu, Anatom, *Macdonald* s.n. (holotype B†).
Dendrobium sp. aff. *D. calcaratum* Lindley, sensu Reichb. f. in Seem., Fl. Vit.: 303 (1868).
Glomera macdonaldii (Schltr.) J.J. Smith in Nova Guinea 8: 44 (1909).
Dendrobium montismovi Kraenzl. in Viertelj. Nat. Ges. Zür. 74: 87 (1929). Types: New Caledonia, *Daeniker* 544a, 544b & 544c (syntypes Z!).
Dendrobium mouanum Guillaumin in Acta Horti Gotoburg. 18: 261, t. 26 (1950). Type: New Caledonia, *Skottsberg* 197 (holotype S).

An epiphytic herb forming large spreading masses. *Stems* erect, 12–50 cm tall, woody, branching; leaf sheaths warty, with ciliate margins. *Leaves* distichous, oblong-lanceolate, 1.5–3 cm long, 2–4 mm wide, with an unequally bilobed apex, sheathing at base. *Inflorescence* terminal, 1-flowered. *Flower* white with yellowish tips to sepals and petals, sweetly fragrant; dorsal sepal lanceolate, up to 12 mm long, 3–4 mm wide; lateral sepals lanceolate, up to 12 mm long, 3–4 mm wide; petals lanceolate, up to 11 mm long, 2.8–3.8 mm wide; lip triangular, acute, 6 mm long, 5.8 mm wide, with a cylindrical spur, 5 mm long, 2 mm wide; column 2.5 mm long. (See plate 7D).

DISTRIBUTION: Bougainville, Guadalcanal and Kolombangara. Also in Vanuatu, New Caledonia and Fiji.
HABITAT: Montane forest, 830–1400 m.
COLLECTIONS: *Craven & Schodde* 204 (K, LAE) & 3751 (K); *Cribb, Dennis & Campbell* 5080 (K); *Hunt* in RSS 2500 (K); *Wickison* 259 (K).

54. CADETIA Gaudichaud

Dwarf epiphytic plants. *Stems* either spaced along a rhizome or closely tufted, not very long, slender, often swollen at the base, circular, triangular or quadrangular in cross-section. *Leaf* solitary, terminal, usually flat and somewhat fleshy. *Inflorescences* terminal, 1–2, flowers borne from within a bract at the apex of the stem in the fold at the base of the leaf. *Flowers* white, on a slender pedicel; ovary smooth or echinate; dorsal sepal and petals free, the sepals rather broader; lateral sepals forming a mentum with the column-foot; petals narrower; lip trilobed, the base of the lip being joined to the apex of the column-foot and to at least part of its sides to form a distinct closed spur, the lateral lobes rather small, erect, the mid-lobe variable but usually decurved, often hairy; column usually hairy in front, apex with several erect teeth; pollinia 4.

A genus of about fifty species centred on New Guinea but extending from tropical Asia to Australia and the Pacific Islands, a single species being recorded.

C. hispida *(A. Rich.) Schltr.* in Fedde, Repert. Spec. Nov. Regni Veg., Beih. 1: 424 (1912). Type: Santa Cruz Islands, Vanikoro, *Lesson* s.n. (holotype P).
Dendrobium hispidum A. Rich., Sert. Astrol. 2: 13 (1834).
Cadetia adenantha Schltr. in Fedde, Repert. Spec. Nov. Regni Veg. 11: 46 (1912). Type: Solomon Islands, Shortland Islands, *Guppy* 152 (holotype K!).

Dendrobium guadalcanalense Guillaumin in Bull. Mus. Nat. Hist. ser. 2, 37(1): 199 (1965). Type: Solomon Islands, Guadalcanal, *Legand* 29 (holotype P!).

A small epiphytic herb up to 20 cm tall. *Stems* clustered, cylindrical, up to 10 cm tall, 1.5 mm wide. *Leaf* solitary, erect, oblong, up to 11 cm long, 7 mm wide. *Inflorescences* terminal, 1–2, borne from within a bract at the apex of the stem in the fold at the base of the leaf. *Flowers* white with a white or pale yellow lip, the column and anther-cap usually tinged with red-mauve, scented; pedicel and ovary 8 mm long, warty in apical half; dorsal sepal ovate, 5.5 mm long, 3.5 mm wide; lateral sepals obliquely ovate, 6 mm long, 4 mm wide; petals linear, 5.5 mm long, 1 mm wide; lip trilobed, 6 mm long, 3.5 mm wide, the lateral lobes small, triangular, the mid-lobe broadly semicircular, recurved at apex, the disc hairy; column 7 mm long, 2 mm wide, hairy in front. *Friuts* ovoid, covered in thick, recurved hairs, c. 1 cm long, 3–4 mm wide. (See plate 11D).

DISTRIBUTION: Bougainville, Choiseul, Gizo, Guadalcanal, Kolombangara, Makira, Malaita, Ranongga, Reef Islands (Ovau), Rennell, the Santa Cruz Islands (Vanikoro), Santa Isabel, the Shortland Islands, Tetepari (off New Georgia) and Tulaghi (near Guadalcanal). Also in Fiji.

VERNACULAR NAME: Fi'i Adi (Kwara'ae name).

HABITAT: A common epiphyte in primary and secondary forest, sea level to 550 m.

COLLECTIONS: *Beer's collectors* in BSIP 6756 (K); *Carment* s.n. (10.9.24) (K); *Cribb & Campbell* 5112 (K); *Chaplin* 789 (K); *de Koster* 32 (K); *Dennis* in RSS 2326 (K); *Dissing* 2812 (K); *Dockrill* in NGF 34184 (K, LAE); *Foreman* in NGF 45709 (K, LAE); *Gafui* in BSIP 14774 (K, LAE); *Guppy* 152 (K); *Hunt* in RSS 2168, 2417 & 2642 (K); *Im Thurn* 343 (K); *Mauriasi et al.* in BSIP 13378, 14418, 14519, 15815 & 15954 (K); *Millar* in NGF 38412 (LAE); *Milne* 540 (K); *Mitchell* 2 & 69 (K); *Nicholls* 4 (K); *Susui* in BSIP 8360 (K); *Wharton* s.n. (1894) (K); *Whitmore* in BSIP 2177 (K, LAE); *Wickison* 233 (K).

Kores (1989) notes that *C. hispida* and *C. adenantha* are closely related but he distinguishes them by their size and minor floral differences. Having examined a number of collections from the Solomon Islands and Bougainville we feel that these differences cannot be upheld as the variation in height and flower size is continuous.

Hancock and Henderson (1988) record the same Kwara'ae name for *Cadetia hispida*, *Diplocaulobium meckynosepalum* and *Diplocaulobium solomonense*.

55. DENDROBIUM Swartz

Epiphytic, lithophytic or occasionally terrestrial plants of extremely diverse form. *Pseudobulbs* present or absent. *Leaves* few to many, alternately arranged on stem or crowded near its apex, flat, bilaterally flattened or terete. *Inflorescences* lateral or terminal, 1- to many-flowered. *Flowers* with dorsal sepal and petals free; sepals and petals short and broad to filiform; lateral sepals forming a more-or-less conical mentum with the column-foot; lip variable, entire or trilobed, attached to the apex of the column-foot and hinged or fixed, the disc with or without raised callus ridges; column with a pronounced column-foot; pollinia 4.

A large genus of more than one thousand four hundred species from India across to Japan and south to Malaysia, New Guinea, the Pacific, Australia and New Zealand. *Dendrobium* is the largest genus in Bougainville and the Solomon Islands and fifty-five species have been recorded (but four are introduced). The species are arranged alphabetically in the following sections: Amblyanthus (1), Calyptrochilus (4), Conostalix (1), *Dichopus (1), Distichophyllum (1), Grastidium (9), Latouria (9), Monanthus (1), Oxyglossum (7), Oxystophyllum (1), Pedilonum (5), Platycaulon (1), Pleianthe (1), Rholoplanthe (3), Spatulata (9), Trachyrhizum (1). For a discussion on the sections see Schlechter (1912), Cribb (1983 & 1986), Seidenfaden (1985), Reeve & Woods (1989) and Comber (1990).

Some of these sections have been elevated to generic level (Brieger, 1981 and Rauschert, 1983) but until more research has been done on the generic limitation of *Dendrobium*, and we have a broader knowledge of all the species and their significant characters, we cannot justify segregating genera, although some of the sections may be found to warrant the status of genus on further critical research.

KEY TO SECTIONS OF DENDROBIUM*

* This key applies to *Dendrobium* in the Solomon Islands and Bougainville only.

1. Pseudobulbs large, 12–30 cm long, 1.5–4 cm wide, strongly bilaterally compressed; flowers not opening widely .... **section PLATYCAULON** (41. *D. platygastrium*)
   Pseudobulbs not bilaterally compressed; flowers closed or opening 2
2. Leaves bilaterally flattened 3
   Leaves dorsiventrally flattened 4
3. Leaves imbricate; stem terete .... **section OXYSTOPHYLLUM** (35. *D. tumoriferum*)
   Leaves not imbricate; stem swollen near the base, terete above .... **section RHOPALANTHE** (except **D. crumenatum*)
4. Petals longer and narrower than sepals, sometimes spirally twisted .... **section SPATULATA**
   Petals equal to or shorter than sepals, not twisted 5
5. Inflorescences terminal or apparently terminal 6
   Inflorescences lateral 8
6. Plants with a tufted, erect habit; pseudobulbs small; inflorescences usually 2-flowered but occasionally multi-flowered; ovary 3- to 10-winged .... **section OXYGLOSSUM**
   Plants erect or pendent; pseudobulbs long and cane-like, swollen near the base or stout; inflorescences multi-flowered; ovary not distinctly winged 7
7. Flowers cream; lip with 5 raised longitudinal callus ridges .... **section RHOPALANTHE** (*D. crumenantum* only)
   Flowers not as above .... **section LATOURIA** (in part)
8. Inflorescences 1-flowered or flowers opening one at a time 9
   Inflorescences 2- to many-flowered 11

9. Flowers cream, pale yellow or yellowish green with a dark purple-marked lip; lip without a central callus ridge
 .. .. **section MONANTHUS** (27. *D. bilobum*)
 Flowers white, yellow or orange with tinges and stripes of darker orange or pinkish white with yellow veins; lip with central callus ridges 10
10. Plant terrestrial, often growing in shallow water or marshy ground; lip with 3 callus ridges, the outer two warty, and with a warty apex
 .. .. **section CONOSTALIX** (6. *D. conostalix*)
 Plant epiphytic, growing in rain forest; lip with a 3-ridged fleshy callus
 .. .. **section DISTICHOPHYLLUM** (8. *D. austrocaledonicum*)
11. Lip apex strongly cucullate, erose, thin-textured
 .. .. **section CALYPTROCHILUS**
 Lip apex acute to rounded but if rounded then fleshy and only slightly cucullate 12
12. Lip entire, glabrous, with a saccate basal portion, apex usually acute
 .. .. **section PEDILONUM**
 Lip obscurely trilobed to trilobed, glabrous or hairy, sometimes with a saccate basal portion, apex acute or obtuse 13
13. Flowers covered in brown scales
 .. .. **section AMBLYANTHUS** (1. *D. melanostictum*)
 Flowers not covered in brown scales 14
14. Leaves markedly decreasing in size towards apex of stem; column with a motile appendage at base .. .. ***section DICHOPUS** (7. **D. insigne*)
 Leaves not markedly decreasing in size towards apex of stem; column not as above 15
15. Leaves crowded near apex of stem .. .. **section LATOURIA** (in part)
 Leaves distichous on stem 16
16. Inflorescences arising from the base of a leaf sheath; peduncle up to 10 cm long; lip with a 2-tailed mid-lobe
 .. ..**section TRACHYRHIZUM** (55. *D. prostheciglossum*)
 Inflorescences arising from laterally compressed protuberances on the side of the leaf sheath; peduncle reduced; mid-lobe of lip entire 17
17. Flowers in pairs, held facing in opposite directions
 .. .. **section GRASTIDIUM**
 Inflorescences 4- to 10-flowered, flowers facing in all directions
 .. .. **section PLEIANTHE** (42. *D. pleianthum*)

Section AMBLYANTHUS

1. D. melanostictum *Schltr.* in K. Schum. & Laut., Nachtr. Fl. Deutsch. Südsee: 167 (1905). Type: New Guinea, *Schlechter* 14441 (holotype B†).
Amblyanthus melanostictus (Schltr.) F.G. Brieger, in Schltr., Die Orchideen ed. 3, 1(11–12): 686 (1981).
Amblyanthe melanosticta (Schltr.) S. Rauschert in Fedde, Repert. Spec. Nov. Regni Veg. 94(7–8): 437 (1983).

An epiphytic herb. *Stems* erect, cane-like, fleshy, up to 70 cm long, leafy in apical half. *Leaves* distichous, lanceolate, up to 8 cm long, 1.5 cm wide, articulated to sheathing bases up to 1.5 cm long. *Inflorescences* lateral, arising on older stems, breaking through the leaf sheaths, densely 2- to 6-flowered; peduncle short, c. 5 mm long. *Flowers* not opening widely, white with a yellow lip with an orange-yellow centre, the sepals covered with thin brown scales; dorsal sepal ovate, 8.5 mm long, 5 mm wide; lateral sepals obliquely ovate, 1.2 cm long, 6–7 mm wide, forming a conical, obtuse, 5 mm long mentum; petals elliptic, 9 mm long, 4 mm wide; lip obscurely trilobed, narrowed at base, 1.4 cm long, 6.5 mm wide, the lateral lobes erect, small, triangular, the disc with a broad central fleshy callus, the mid-lobe obovate, densely hairy and fleshy at apex; column 6 mm long. (See fig. 42 J–P).

DISTRIBUTION: The Santa Cruz Islands. Also in Irian Jaya and New Guinea.
HABITAT: No information from the Solomon Islands but in Irian Jaya and New Guinea it grows in montane forest.
COLLECTION: *Dennis* 5 (K).

Section CALYPTROCHILUS

These species are bird-pollinated (see Slade, 1980, for an account of the pollination of *Dendrobium mohlianum*). See also Warren (1990d) for a discussion of this section.

1. Flowers opening widely; spur conical, tapering to apex 2
   Flowers not opening widely; spur cylindrical, obtuse 3
2. Flowers orange to bright red, often with a purple lip; mentum up to 8 mm long .. ..**4. D. mohlianum**
   Flowers white or lilac with a white lip; mentum up to 11 mm long .. ..**3. D. lawesii**
3. Flowers pink with white tips; leaves 5–10 cm long, 0.8–1.5 cm wide .. .. **2. D. erosum**
   Flowers dark red; leaves 8–10 cm long, 2–3 cm wide .. ..***5. D. rutriferum**

2. D. erosum *(Blume) Lindley*, Gen. Sp. Orch. Pl.: 86 (1830). Type: Java, *Blume* s.n. (holotype L).
Pedilonum erosum Blume, Bijdr. Fl. Ned. Ind.: 323 (1825).
Dendrobium aegle Ridley in Journ. Linn. Soc. 32: 260 (1896). Type: Malaya, *Ridley* s.n. (holotype BM!).
Dendrobium aemulans Schltr. in K. Schum. & Laut., Nachtr. Fl. Deutsch. Südsee: 176 (1905). Type: New Guinea, *Schlechter* 14329 (holotype B†).
Dendrobium inopinatum J.J. Smith in Bull. Jard. Bot. Buitenz. ser. 3, 5: 88 (1922). Type: Sumatra, *Bunnemeijer* 1132 (holotype BO!).

See Seidenfaden (1985) and Seidenfaden & Wood (in press) for synonymy and discussion.

An epiphytic herb. *Stems* clustered, pendent, ribbed, up to 40 cm long, yellow-green turning purple with age. *Leaves* alternate, lanceolate, acute, 5–10 cm long, 0.8–1.5 cm wide, sheathing at base. *Inflorescences* lateral, pendent, arising from nodes on leafless stems. *Flowers* clustered, 5–10, not opening widely, pink with white tips; pedicel and ovary c. 2 cm long; dorsal sepal broadly ovate, 5 mm long, 4.5 mm wide; lateral sepals obliquely ovate, 1.2 cm long, 6 mm wide, forming a conical, obtuse mentum 6 mm long; petals obovate, 4 mm long, 3 mm wide; lip spathulate, 1.1 cm long, 4 mm wide, the margins incurved, with the apical margin lacerate, the spur cylindrical, obtuse, 5 mm long; column 3 mm long, with a 9 mm long column-foot. (See fig. 44 A–K).

DISTRIBUTION: Kolombangara and Malaita. Also in Thailand, Malaya, Java, Sumatra, New Guinea and Vanuatu.
HABITAT: Montane forest, 620 m.
COLLECTIONS: *Hunt* in RSS 2550 (K); *Mitchell* 6 (K).

3. D. lawesii *F. Muell.* in Melb. Chemist, June (1884). Type: New Guinea, *Lawes* s.n. (holotype MEL).
Pedilonum lawesii (F. Muell.) S. Rauschert in Fedde, Repert. Spec. Nov. Regni Veg. 94(7–8): 461 (1983).

For full synonymy see Schlechter (1912).

An epiphytic herb. *Stems* pendent, clustered, ribbed, up to 50 cm long. *Leaves* distichous, lanceolate, acute, up to 6 cm long, 1 cm wide, sheathing at base. *Inflorescences* lateral, arising from a leafless stem, the flowers in clusters of 1–4. *Flowers* lilac-coloured with a white lip, white or creamy-white (purple in New Guinea); pedicel and ovary up to 2.5 cm long; dorsal sepal ovate, obtuse, up to 8.5 mm long, 5 mm wide; lateral sepals obliquely triangular, up to 2 cm long, 9 mm long, forming a tapering, obtuse mentum up to 1.1 cm long; petals elliptic, 8.5 mm long, 3.5 mm wide; lip spathulate, up to 1.8 cm long, 9 mm wide, cucullate at apex, the spur conical, obtuse, 9 mm long; column 3 mm long, column-foot 1.4 cm long. (See fig. 35 M–O).

DISTRIBUTION: Bougainville, Guadalcanal, Kolombangara and Rendova. Also in New Guinea.
HABITAT: Montane forest, 900–2000 m.
COLLECTIONS: *Cribb* 5019 (K); *Dickson* MD35 (K); *Lavarack & Ridsdale* in NGF 31386 (K) & 31420 (K, LAE); *Wickison* 72 & 222 (K).

4. D. mohlianum *Reichb. f.* in Bot. Zeit. 20: 214 (1862). Type: Fiji, *Seemann* 578 (holotype W!, isotypes BM, K!, P).
Dendrobium neoebudanum Schltr. in Bull. Herb. Boiss., ser. 2, 6:456 (1906). Types: Vanuatu, Anatom & Erromango, *Morrison* s.n. (syntypes B†, isosyntypes AMES, K!).
Dendrobium vitellinum Kraenzl. in Engler, Pflanzenr. Orch.-Mon.-Dendr. 45: 124, fig. 7 (1910). Type: Vanuatu, Anatom, *MacDonald* s.n. (holotype not located).
Pedilonum mohlianum (Reichb. f.) F.G. Brieger in Schltr., Die Orchideen ed. 3, 1(11–12) : 682 (1981).

Fig. 35. Dendrobium mohlianum A-L. A, habit x 2/3; **B**, inflorescence x 2/3; **C**, flower x 2; **D**, dorsal sepal x 2; **E**, petal x 2; **F**, lateral sepal x 2; **G**, lip and column x 3; **H**, lip x 3; **J**, pollinia x 4; **K**, anther-cap x 4; **L**, column x 3. **D. lawesii M-O. M**, flower x 2; **N**, lip x 3; **O**, lip and column x 3. **A-L** drawn from *Wickison* 70; **M-O** from *Wickison* 72. All drawn by Sue Wickison.

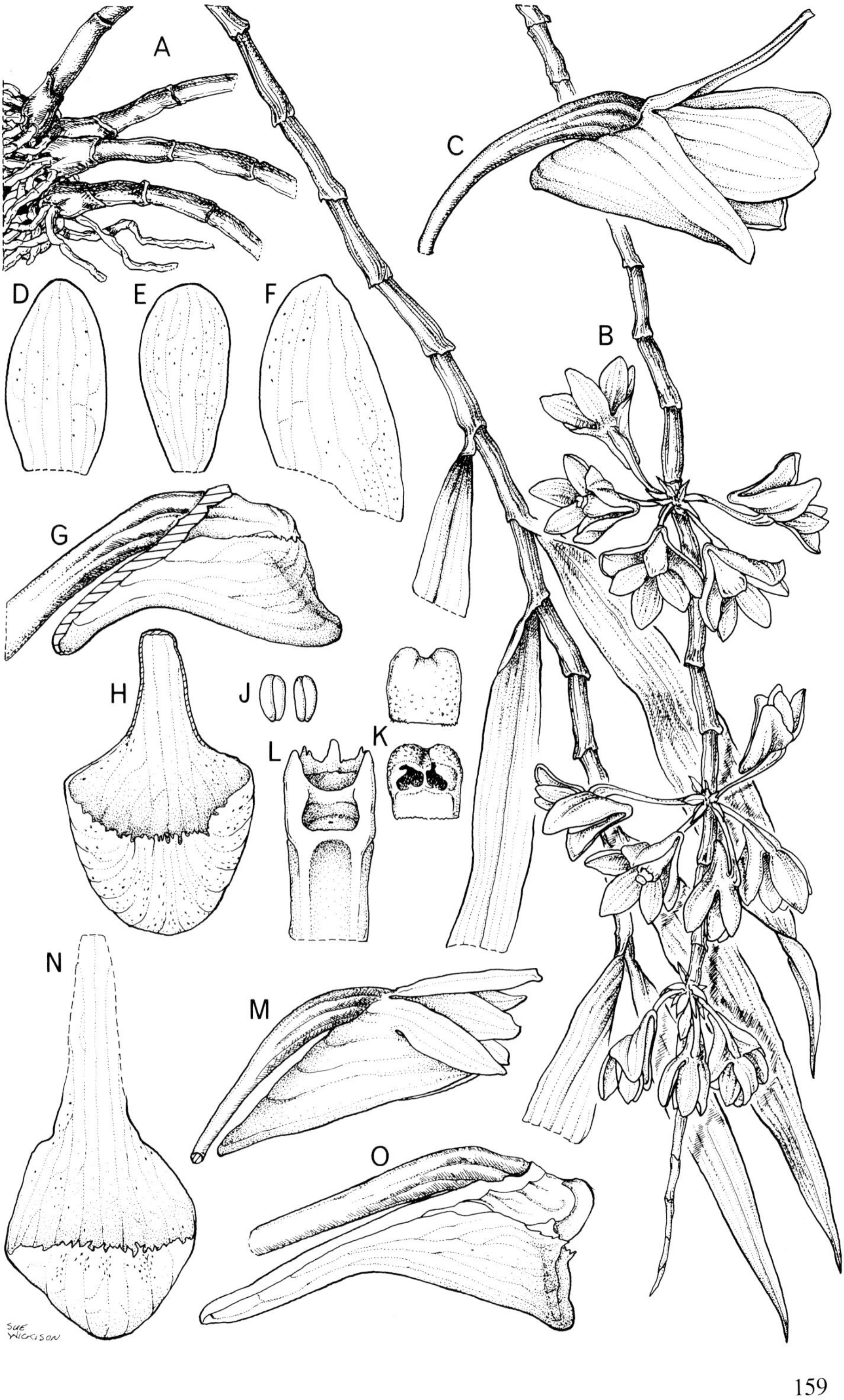
A
B
C
D
E
F
G
H
J
K
L
M
N
O
SUE WICKISON

Pedilonum neoebudanum (Schltr.) S. Rauschert in Fedde, Repert. Spec. Nov. Regni Veg. 94(7–8): 461 (1983).
Pedilonum vitellinum (Kraenzl.) S. Rauschert l.c.: 465.

An epiphytic herb. *Stems* spreading or pendent, clustered, ribbed, up to 50 cm long. *Leaves* distichous, lanceolate, 6–13 cm long, 1–2.5 cm wide, sheathing at base. *Inflorescences* lateral, arising from a leafless stem, the flowers in clusters of 4–6. *Flowers* orange to bright red, often with a mauve lip; pedicel and ovary 1.5–2 cm long; dorsal sepal ovate, obtuse, up to 1.2 cm long, 7 mm wide; lateral sepals obliquely oblong-ovate, up to 1.5 cm long, 8.5 mm long, forming a tapering, obtuse mentum up to 8 mm long; petals obovate-spathulate, up to 1.3 cm long, 7 mm wide; lip spathulate, up to 1.6 cm long, 1 cm wide, cucullate at apex, the spur conical, obtuse, 5.5 mm long; column 3 mm long, column-foot 1.3 cm long. (See fig. 35 A–L and plate 11A).

DISTRIBUTION: Guadalcanal, Kolombangara, Malaita and New Georgia. Also in Vanuatu, Fiji and Samoa.
HABITAT: Montane and cloud forest, 800–2150 m.
COLLECTIONS: *Chaplin* 825 (K); *Cribb & Campbell* 5132 (K); *Dennis* in RSS 4629 (K); *Hunt* in RSS 2033, 2474, 2478 & 2965 (K); *Im Thurn* s.n. (K); *Mitchell* 13 (K); *Whitmore* in BSIP 2081 (K); *Wickison* 70 (K).
We consider *D. vitellinum* to be conspecific with *D. mohlianum* based on Kraenzlin's description and living plants seen of this species on Anatom (B.L.).
Two interesting colour variations of this species have been observed in the field. One has completely orange to bright red flowers, the other has red sepals and petals with a purple lip.

***5. D. rutriferum** *Reichb. f.* in Gard. Chron., ser. 3, 2: 746 (1887). Type: New Guinea, ex *Messrs Linden* (holotype W!).
Dendrobium puniceum Rolfe in Bull. Misc. Inf., Kew 1901: 146 (1901), non Ridley (1886). Type: New Guinea, ex *Messrs Sanders* (holotype K!).
Dendrobium moirianum A.D. Hawkes in Lloydia 20: 121 (1957). Type: as for *D. puniceum* Rolfe.
Pedilonum rutriferum (Reichb. f.) S. Rauschert in Fedde, Repert. Spec. Nov. Regni Veg. 94(7–8): 463 (1983).
Pedilonum moirianum (A.D. Hawkes) S. Rauschert l.c.: 461.

An epiphytic herb. *Stems* pendent, long, ribbed. *Leaves* distichous, lanceolate, acute, 8–10 cm long, 2–3 cm wide, dark green, sheathing at base. *Inflorescences* lateral, arising from nodes, up to 3.5 cm long, many-flowered. *Flowers* dark red; dorsal sepal oblong, obtuse, 4 mm long, 2.8 mm wide; lateral sepals obliquely oblong, obtuse, 10.5 mm long, 4 mm wide, forming an obtuse mentum 6 mm long; petals oblong, obtuse, 4.5 mm long, 2.2 mm wide; lip spathulate, 9.5 mm long, 4 mm wide, thin-textured and cucullate at apex, the spur conical, obtuse, 5 mm long; column 2 mm long, column-foot 6.5 mm long.

DISTRIBUTION: Bougainville. Also in New Guinea.
HABITAT: Cultivated on *Leucaena leucocephala*, sea level.
COLLECTION: *Millar* in NGF 38407 (LAE).

This species is not native to Bougainville.

Section CONOSTALIX

See Seidenfaden (1985) for a discussion of this section.

6. D. lobbii *Teijsm. & Binnend.* in Natuurk. Tijdschr. Ned. Ind. 5: 491 (1853). Type: India, *Lobb* 482 (holotype K!, isotype W). Non Lindley (1859).

For full synonymy see Clements (1989).

A terrestrial herb forming sparse straggly clumps. *Stems* erect, slender, up to 80 cm tall, leafy at apex. *Leaves* distichous, held obliquely erect, lanceolate, 2–4 cm long, 2–4 mm wide, unequally bilobed at apex, sheathing at base, sheaths sparsely hairy. *Inflorescences* lateral, flowers opening one at a time. *Flowers* not opening widely, non-resupinate, the dorsal sepal and petals pinkish-white with yellow veins, the lateral sepals pale yellow, with yellow-orange veins, the lip white with a creamy-yellow base; dorsal sepal ovate, 5 mm long, 2.5 mm wide; lateral sepals oblong, incurved, up to 10 mm long, 3.5 mm wide; petals narrowly lanceolate, spreading, 4 mm long, 1 mm wide; lip trilobed, 11 mm long, 4 mm wide, the lateral lobes narrow, erect, the disc with 3 ridges the outer two warty, the mid-lobe ovate, the apex warty, notched; column 3 mm long, pink, with a 4 mm long column-foot; anther-cap yellow.

DISTRIBUTION: Santa Ana Island. Also in Indonesia, Malaysia, New Guinea and Australia.

HABITAT: Growing on the outer perimeter of tufts of wet sphagnum moss, sedges and ferns in peaty material at the edge of Waipiapia freshwater lagoon.

COLLECTION: *Dennis* 10 (K).

*Section DICHOPUS

***7. D. insigne** *(Blume) Miq.*, Fl. Ind. Bat. 3: 640 (1855). Type: New Guinea, *Zippelius* s.n. (holotype L!).

Dichopus insignis Blume, Mus. Bot. Lugd. Bat. 2: 176 (1856).

Dendrobium pentactis Kraenzl. in Engler, Pflanzenr. Orch.-Mon.-Dendr. 1: 200 (1910). Type: New Guinea, *Beccari* s.n. (holotype FI).

An epiphyte forming large straggly clumps. *Stems* swollen at base, erect, up to 80 cm long, leafy. *Leaves* distichous, ovate, thick, 1.5–9 cm long, 0.7–4 cm wide, the largest in the central part of the stem, becoming smaller towards the top, unequally bilobed at apex; leaf bases sheathing. *Inflorescences* lateral, arising from the nodes, borne in pairs, flowers short-lived. *Flowers* opening widely, yellow heavily blotched red-brown with a white to pale yellow lip, fragrant; dorsal sepal oblong, acute, erect, 2–2.5 cm long, 5 mm wide; lateral sepals obliquely triangular, 2–2.5 cm long, 8 mm wide, divergent, forming an obtuse mentum; petals linear, curved forwards, 1.2–2 cm long, 4 mm wide; lip trilobed, 1.8 cm long, 8 mm wide, the lateral lobes small, fimbriate, the mid-lobe lanceolate, recurved, the disc with two hairy ridges; column 7 mm long, with a long, narrow appendage projecting below the stigma.

DISTRIBUTION: Bougainville. Also in Indonesia, New Guinea, New Britain, New Ireland and Torres Strait Islands (Saibai Island).
HABITAT: Cultivated on *Leucaena leucocephala*, 10 m. Probably not native.
COLLECTIONS: *Millar* in NGF 38420 (LAE); *Millar & Vandenberg* in NGF 48511 (K, LAE).

We have followed Clements (1989) for the synonymy.

Section DISTICHOPHYLLUM

See Seidenfaden (1985) for a discussion of this section.

8. D. austrocaledonicum *Schltr.* in Fedde, Repert. Spec. Nov. Regni Veg. 3: 80 (1906). Type: New Caledonia, *Le Rat* 187 (holotype B†).
Dendrobium inaequale Finet in Bull. Soc. Bot. Fr. 50: 375 (1903), non Rolfe (1901). Type: New Caledonia, *Balansa* 761 (holotype P!).
Dendrobium cerinum Schltr. in Engler, Bot. Jahrb. 39: 72 (1906), non Reichb. f. (1879). Type: as for *D. inaequale*.
Dendrobium critaerubrae Guillaumin in Bull. Soc. Bot. Fr. 103: 281 (1956). Type: Vanuatu, Anatom, *L.E. Cheesman* s.n. (holotype BM!).
Dendrobium garayanum A.D. Hawkes & Heller in Lloydia 20: 19 (1957). Type: as for *D. inaequale*.
Dendrobium mendoncanum A.D. Hawkes in Orquidea 25: 19 (1963). Type as for *D. inaequale*.

An epiphytic herb. *Stems* pendent, clustered, ribbed, 25–60 cm long, slightly swollen at base. *Leaves* distichous, oblong, 2–4.5 cm long, 5–9 mm wide, light yellow-green, unequally bilobed at the apex, sheathing stem at base. *Inflorescences* lateral. *Flowers* solitary, white maturing yellow to pale orange with tinges and stripes of darker orange; sepals and petals reflexed; dorsal sepal ovate, acute, 8 mm long, 4.5 mm wide; lateral sepals obliquely ovate, 1.2 cm long, 6 mm wide, forming a short, conical, obtuse, 4 mm long mentum; petals ovate, 8 mm long, 4 mm wide; lip trilobed, 1.3 cm long, 1.2 cm wide, the lateral lobes erect, oblong, unequally bifid at apex, the mid-lobe semi-circular, slightly notched, the disc with a 3-ridged fleshy callus; column 4 mm long. (See fig. 36 A–K).

DISTRIBUTION: Bougainville, Guadalcanal and New Georgia. Also in Vanuatu and New Caledonia.
HABITAT: Lowland and montane rain forest, 240–900 m.
COLLECTIONS: *Cribb, Dennis & Campbell* 5067 (K); *Dickson* MCA2 (K); *Ridsdale & Lavarack* in NGF 30564 (K); *Wickison* 12 (K).

Fig. 36. Dendrobium austrocaledonicum A-K. A, habit x 2/3; **B**, flower x 3; **C**, dorsal sepal x 3; **D**, lateral sepal x 3; **E**, petal x 3; **F**, side view of flower with the lip and dorsal sepal removed x 3; **G**, pollinia x 6; **H**, dorsal and ventral view of anther-cap x 6; **J**, lip x 3; **K**, column x 4. **A** drawn from *Wickison* 12; **B-K** from *Sprunger* 129. All drawn by Sue Wickison.

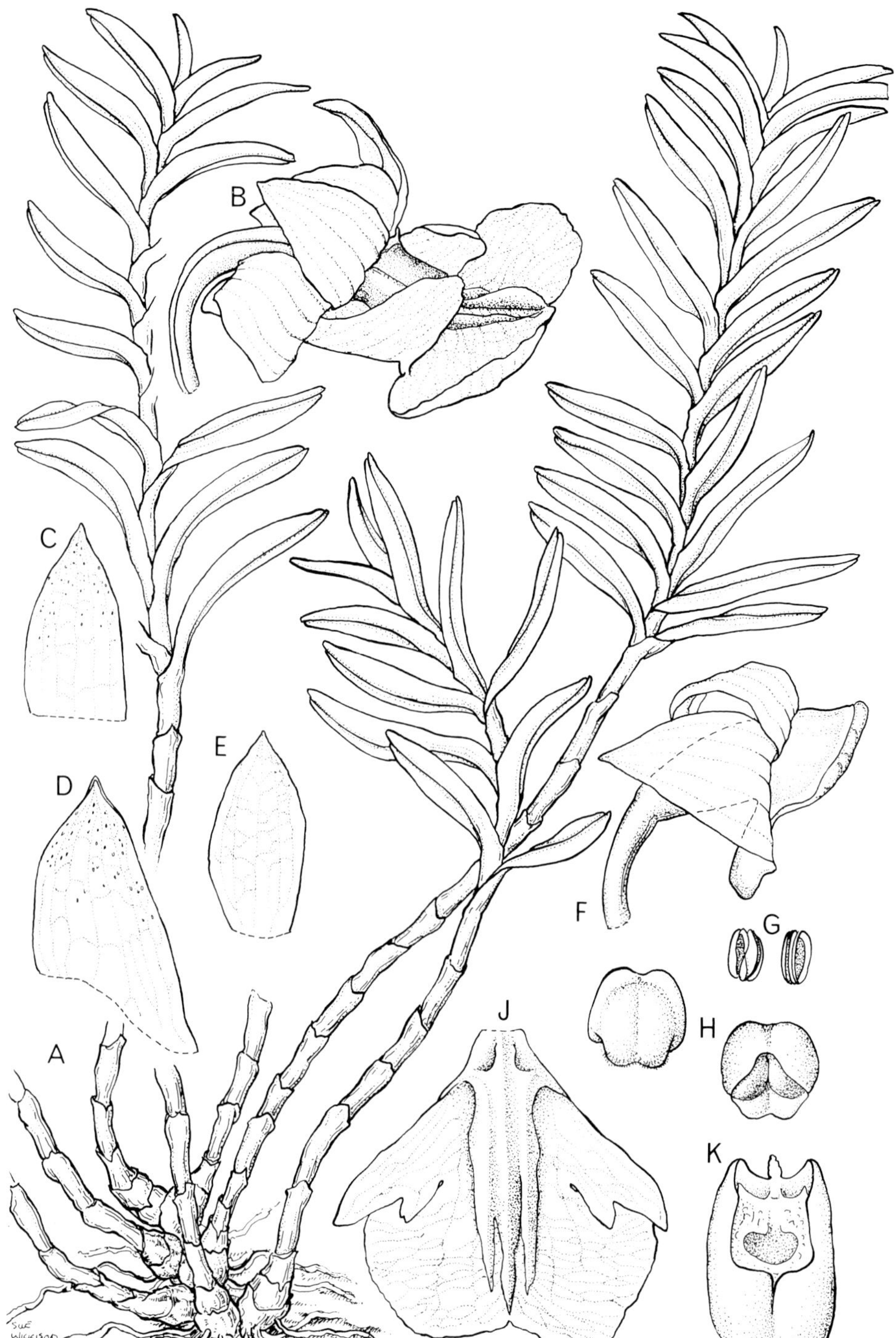
A
B
C
D
E
F
G
H
J
K
SUE WICKISON

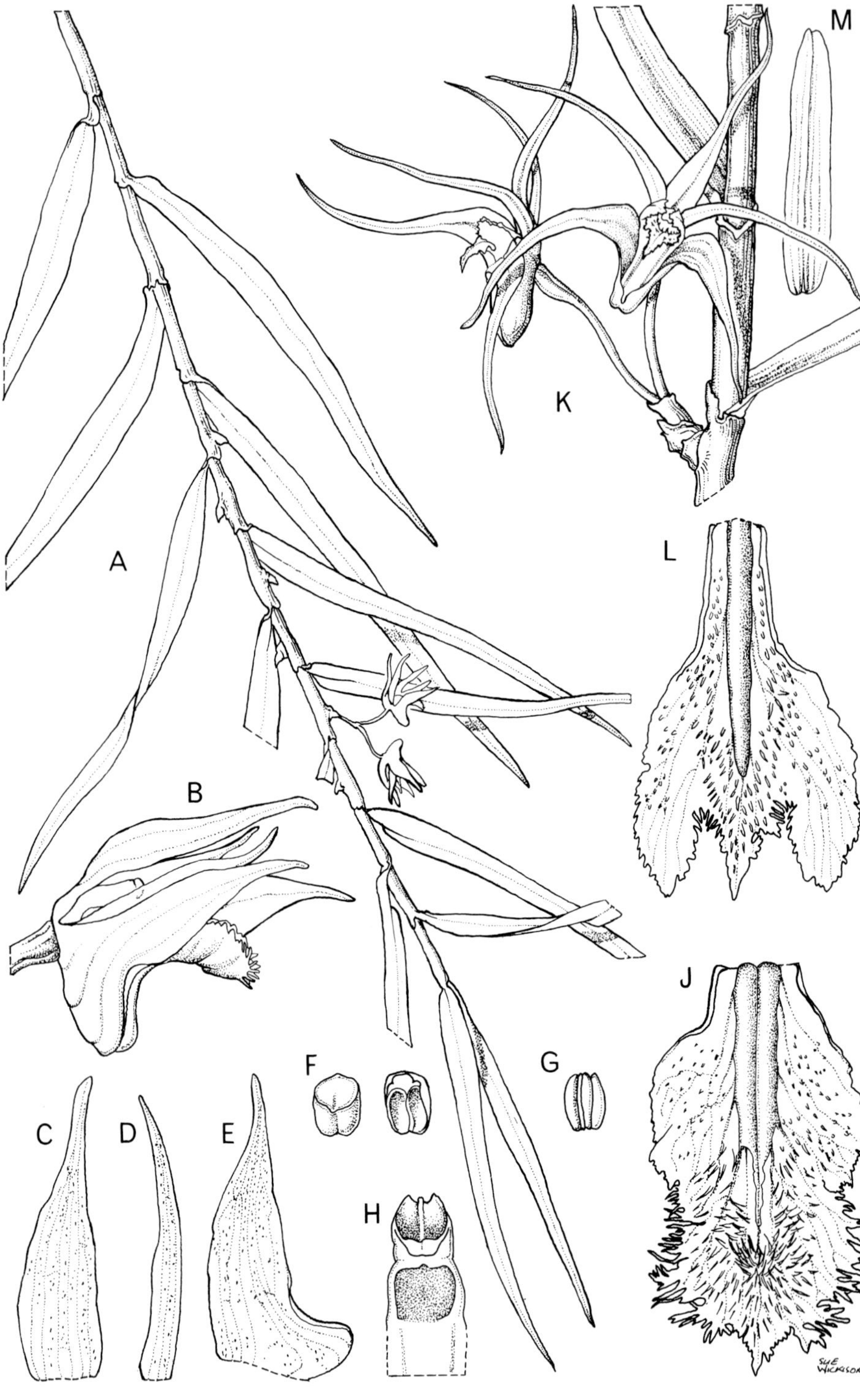
A
B
C
D
E
F
G
H
J
K
L
M
SUE WICKISON

Section GRASTIDIUM

See Seidenfaden (1985) for a discussion of this section.

1. Leaves less than 1 cm wide 2
   Leaves more than 1 cm wide 3
2. Flowers yellow-green densely spotted with purple; sepals and petals filiform, becoming tangled as the flowers age, 12–13 mm long .... **9. D.** aff. **baileyi**
   Flowers white; sepals and petals acuminate, not becoming tangled as the flowers age, 8–9.5 mm long .... **16. D. vanikorense**
3. Lip not papillose on the mid-lobe .... **12. D. kietaense**
   Lip papillose on the mid-lobe 4
4. Flowers maroon with an orange lip and a pale yellow column; sepals and petals filiform, 4–5 cm long .... **17. D. vernicosum**
   Flower colour not as above; sepals and petals not filiform, less than 2 cm long 5
5. Apex of sepals and petals cucullate .... **10. D. campbellii**
   Apex of sepals and petals not cucullate 6
6. Leaves elliptic, 14–18 cm long, 3–4.5 cm wide; peduncle arising in the fold at the base of the leaf .... **15. D. vandoides**
   Leaves lanceolate or ovate, less than 15 cm long and 3 cm wide; peduncle not concealed by the base of the leaf 7
7. Leaves lanceolate, up to 15 cm long, 2 cm wide .... **14. D. salomonense**
   Leaves ovate, 3–8 cm long, 1.4–3 cm wide 8
8. Lateral lobes of the lip more than half the length of the mid-lobe .... **13. D. rechingerorum**
   Lateral lobes of the lip less than a third of the length of the mid-lobe .... **11. D. cyrtosepalum**

9. D. aff. **baileyi** *F. Muell.*, Fragm. 8: 173 (1874). Type: Australia, *Bailey* s.n. (holotype MEL).
Callista baileyi (F. Muell.) Kuntze, Rev. Gen. Pl. 2: 654 (1891).

An epiphytic herb forming clumps. *Stems* pendent, slender and wiry, about 20 cm long, leafy. *Leaves* distichous, narrowly lanceolate, up to 8 cm long, 8 mm wide, unequally bilobed at the apex, pale green, sheathing at base. *Inflorescences* lateral, arising from bilaterally flattened protuberances, the flowers in pairs. *Flowers* short-lasting, yellow-green, densely spotted with purple, the base of the lip white; sepals and petals narrow, long, spreading widely but becoming tangled as the flowers age; dorsal sepal narrowly lanceolate, filiform above, 1.2 cm long, 2 mm wide; lateral sepals obliquely triangular, filiform above, 1.2 cm long, 5.5 mm wide, forming an obtuse 3 mm long mentum; petals linear, filiform above, 1.3 cm long, 1 mm wide; lip trilobed, 7.5 mm long, 4.5 mm wide, the lateral lobes erect, falcate, with serrate margins, the callus a central, sparsely hairy ridge

---

Fig. 37. Dendrobium vanikorense A-J. A, habit x 2/3; **B**, flower x 4; **C**, dorsal sepal x 4; **D**, petal x 4; **E**, lateral sepal x 4; **F**, dorsal and ventral view of anther-cap x 8; **G**, pollinia x 8; **H**, column x 8; **J**, lip x 6. **D.** aff. **baileyi K-M. K**, inflorescence x 3; **L**, lip x 6; **M**, leaf x 2/3. **A-J** drawn from *Wickison* 26; **K-M** from *Wickison* 208. All drawn by Sue Wickison.

extending from the base to the mid-lobe, the mid-lobe narrowly triangular, acuminate, hairy, recurved; column 2 mm long, the column-foot 4 mm long. (See fig. 37 K–M).

DISTRIBUTION: Choiseul, the Florida Islands and Malaita.
HABITAT: Rain forest, sea level to 350 m.
COLLECTIONS: *Mitchell* 52 (K); *Wickison* 208 (K); *Woodford* 10 (K).

This species is similar to *D. baileyi*, from Australia, but differs in having shorter sepals and petals and a shorter lip mid-lobe. We have not been able to match the specimens with any described material from New Guinea or neighbouring areas.

10. D. campbellii *Cribb & B. Lewis* sp. nov., *D. triste* Schltr. sepalis petalisque apicibus succulentis cucullatis affinis, sed floribus flavo-viridis, mento albo, sepalis glabris, labelli lobis lateralibus longioribus atque lobo medio minore facile distinguenda. Typus: Solomon Islands, Rendova, *Cribb & Campbell* 5145 (holotypus K!).

An epiphytic herb. *Stems* pendent, up to 2 m long, leafy. *Leaves* distichous, elliptic, up to 11. 5 cm long, 3 cm wide, sheathing at base. *Inflorescences* lateral, arising from bilaterally compressed protuberances. *Flowers* in pairs, yellow-green with a white mentum; dorsal sepal oblong, with a fleshy, cucullate apex, 1.1 cm long, 5 mm wide; lateral sepals fleshy, falcate, cucullate at the apex, 1 cm long, 6 mm wide, forming an obtuse mentum 3 mm long; petals spathulate, with fleshy, incurved lateral margins in the apical half, 8 mm long, 3.5 mm wide; lip trilobed, 8 mm long, 6 mm wide, the lateral lobes erect, rounded, 2 mm long, 2 mm wide, the disc with a central callus extending to the base of the mid-lobe, with transverse rugulose ridges radiating from it on to the lateral lobes, the mid-lobe triangular, acute, 2 mm long, 2 mm wide at base, papillose, recurved; column 3.5 mm long, column-foot slightly upcurved, 5 mm long. (See fig. 38).

DISTRIBUTION: Rendova and Makira.
HABITAT: Lowland rain forest, 50 m.
COLLECTION: *Cribb & Campbell* 5145 (K); *Sore* in BSIP 2291 (K).

This species is readily distinguished by the fleshy, cucullate apices to the sepals and petals. It is similar to *D. triste* Schltr. from New Guinea but differs in the colour of the flowers and in having a smooth exterior to the sepals. The shape of the lip is also distinct in having relatively larger lateral lobes and a smaller mid-lobe.

We have named this species for John Campbell of Pebbleton, New Zealand.

Fig. 38. Dendrobium campbellii. A, habit x 2/3; **B**, flower x 2; **C**, lip x 6; **D**, side view of lip x 6; **E**, dorsal sepal x 4; **F**, petal x 4; **G**, lateral sepal x 4; **H**, column x 6; **J**, pollinia x 6; **K**, dorsal and ventral view of anther-cap x 6. Drawn from *Cribb & Campbell* 5145 by Sue Wickison.

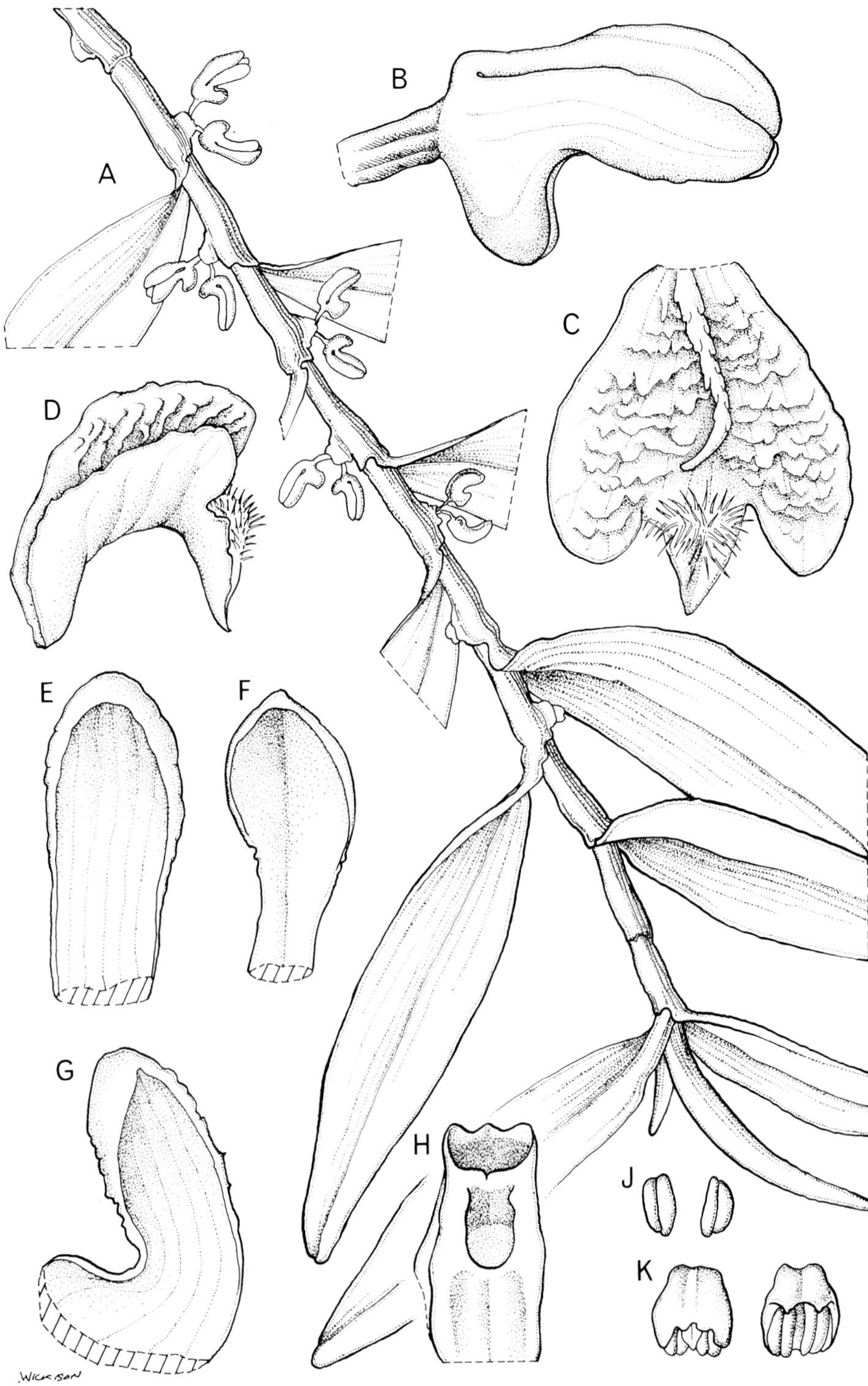
A
B
C
D
E
F
G
H
J
K
WICKISON

11. D. cyrtosepalum Schltr. in K. Schum. & Laut., Nachtr. Fl. Deutsch. Südsee: 169 (1905). Type: New Ireland, *Schlechter* 14621 (holotype B†).
Grastidium cyrtosepalum (Schltr.) S. Rauschert in Fedde, Repert. Spec. Nov. Regni Veg. 94(7–8): 448 (1983).

An epiphytic herb. *Stems* arching to pendent, up to 50 cm long, 5 mm wide. *Leaves* distichous, ovate, 3–5 cm long, 1.4–2.9 cm wide, unequally bilobed at the apex, sheathing at base. *Inflorescences* lateral, arising from bilaterally compressed protuberances. *Flowers* in pairs, white or yellow, the lip cream with a pink base, an orange callus and yellow papillae, fragrant; dorsal sepal lanceolate, 1.5 cm long, 3 mm wide; lateral sepals falcate, 1.5 cm long, 3 mm wide, forming an obtuse, 2.5 mm long mentum; petals linear-lanceolate, 1.3 cm long, 2 mm wide; lip trilobed, 0.9–1.1 cm long, 4–5 mm wide, the lateral lobes short, triangular, less than a third the length of the mid-lobe, the disc with a central erose callus extending to the base of the mid-lobe, becoming papillose on the mid-lobe, with transverse rugulose ridges, the mid-lobe narrowly triangular, acute, recurved; column 4 mm long, column-foot 3.5 mm long. (See fig. 39 K–M).

DISTRIBUTION: Bougainville, Florida Group (Small Nggela Island), Malaita, New Georgia, Rennell and Santa Ana. Also in New Ireland.
VERNACULAR NAME: Marapauna (Bougainville).
HABITAT: Rain forest, sea level to 900 m.
COLLECTIONS: *Cribb & Campbell* 5162 & 5168 (K); *Dennis* in *Cribb* 5186(K); *Mauriasi et al.* in BSIP 13731 (K); *Noona Dan Expedition* 2914 (K); *Ridsdale & Lavarack* in NGF 30594 & 30618 (K); *Waterhouse* 719 (K).

The specimens from the Solomon Islands and Bougainville differ slightly from Schlechter's illustration of the type collection which has a longer, acuminate apex to the lip.

12. D. kietaense *Schltr.* in Fedde, Repert. Spec. Nov. Regni Veg., Beih. 1: 614 (1912). Type: Bougainville, *Rechinger* 4824 (holotype B†).
Dendrobium malaitense Rolfe in Bull. Misc. Inf., Kew 1921: 55 (1921). Type: Solomon Islands, Malaita, *Im Thurn* 363 (holotype K!).
Grastidium kietaense (Schltr.) S. Rauschert in Fedde, Repert. Spec. Nov. Regni Veg. 94(7–8): 450 (1983).

An epiphytic herb. *Stems* clustered, pendulous, up to 1.5 m long. *Leaves* distichous, lanceolate, 9–17 cm long, 0.8–1.4 cm wide, obliquely unequally bilobed at apex, light green, sheathing at base. *Inflorescences* lateral, arising from bilaterally flattened protuberances. *Flowers* in pairs, fleshy, not opening widely, greenish-yellow to white; dorsal sepal oblong, 1.2–1.5 cm long, 5–6 mm wide; lateral sepals falcate, 1.2–1.5 cm long, 5–6 mm wide, forming an obtuse, 2 mm long mentum; petals elliptic, 1.2–1.5 cm long, 4–7 mm wide; lip trilobed,

---

Fig. 39. Dendrobium rechingerorum A-J. A, habit x 2/3; **B**, inflorescence x 1; **C**, dorsal and ventral view of anther-cap x 6; **D**, pollinia x 6; **E**, lip x 6; **F**, dorsal sepal x 3; **G**, petal x 3; **H**, lateral sepal x 3; **J**, column x 6. **D. cyrtosepalum K-M. K**, flower x 2; **L**, leaf x 2/3; **M**, lip x 6. **A-J** drawn from *Mitchell* 65; **K-M** from *Dennis* in *Cribb* 5186. All drawn by Sue Wickison.

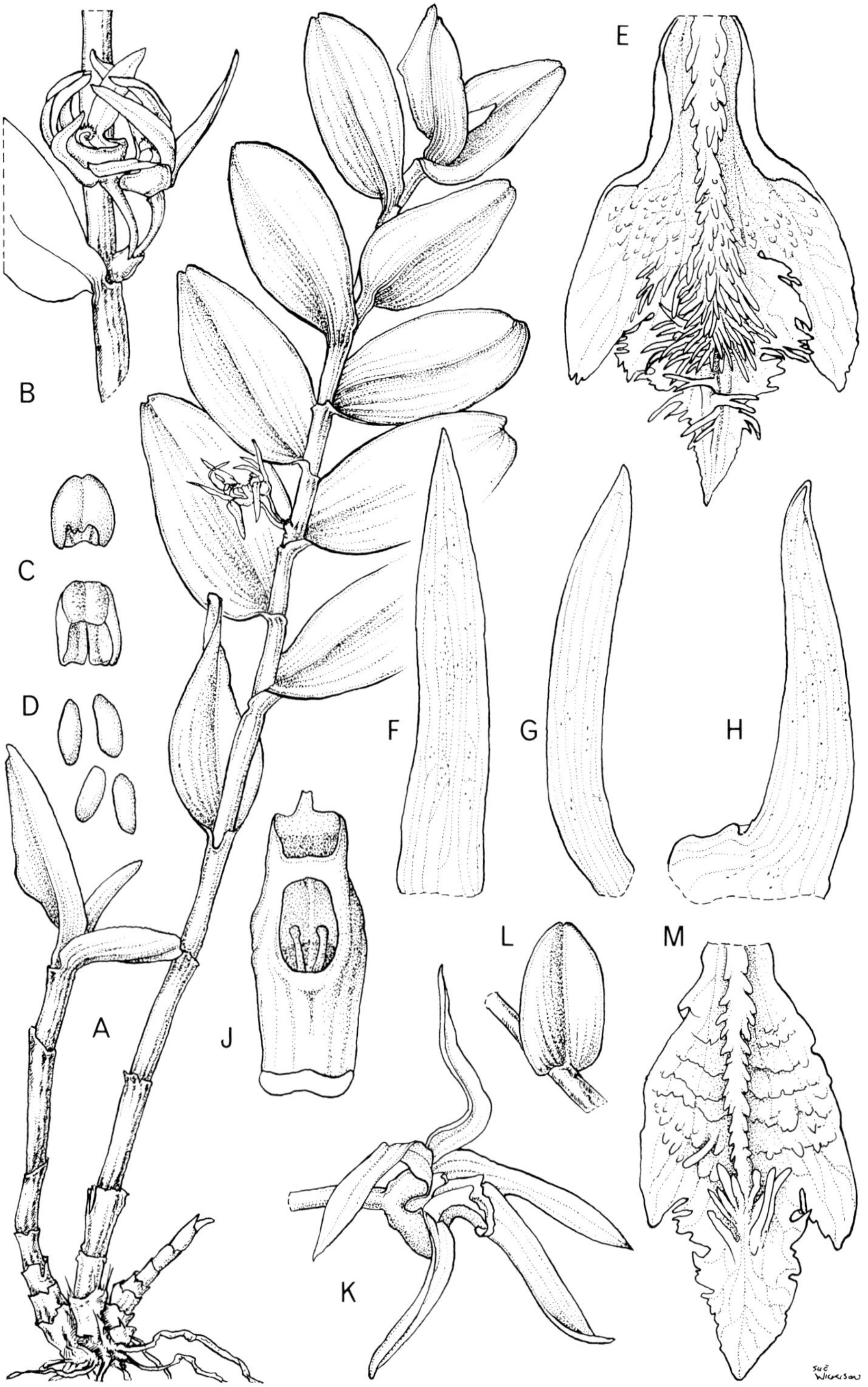
A
B
C
D
E
F
G
H
J
K
L
M

pandurate, 0.7–1 cm long, 4 mm wide, the lateral lobes erect, small, crescent-shaped to semi-circular, the disc with a central callus extending to halfway along the mid-lobe, the mid-lobe recurved, elliptic, apiculate; column 3.5 mm long, with a 2 mm long foot. (See fig. 40 L–M).

DISTRIBUTION: Bougainville, Guadalcanal, Jagi Island (off Santa Isabel), Kolombangara, Malaita and New Georgia. Also in Vanuatu.
VERNACULAR NAME: Hare (Kwara'ae name, Guadalcanal).
HABITAT: Rain forest, sea level to 300 m.
COLLECTIONS: *Chaplin* 822 (K); *Cribb & Campbell* 5113 (K); *Hunt* in RSS 2462, 2463, 2825 & 2946 (K); *Im Thurn* 363 (K); *Nakisi & Babala* in BSIP 8225 (K); *Waterhouse* 340 (K); *Wickison* 27 & 204 (K).

13. D. rechingerorum *Schltr.* in Fedde, Repert. Spec. Nov. Regni Veg., Beih. 1: 597 (1912). Type: Bougainville, *Rechinger* 4990 (holotype B†).
Grastidium rechingerorum (Schltr.) S. Rauschert in Fedde, Repert. Spec. Nov. Regni Veg. 94(7–8): 452 (1983).

An epiphytic herb. *Stems* up to 50 cm long. *Leaves* distichous, ovate, 4–8 cm long, 2–3 cm wide, unequally bilobed at the apex, sheathing at the base. *Inflorescences* lateral, arising from bilaterally flattened protuberances. *Flowers* in pairs, yellow striped with mauve, the petals lighter yellow; dorsal sepal lanceolate, acute, 2.1 cm long, 2.5 mm wide; lateral sepals falcate, acute, 1.7 cm long, wide, 3 mm wide; petals linear, acute, 1.8 cm long, 2 mm wide; lip trilobed, 6 mm long, 4 mm wide, the lateral lobes triangular, more than half the length of the triangular mid-lobe, the disc with a central erose callus extending to the base of the mid-lobe, becoming very papillose on the mid-lobe; column 3 mm long, column-foot 5 mm long. (See fig. 39 A–J).

DISTRIBUTION: Bougainville and Rannogga.
HABITAT: Coastal forest, sea level.
COLLECTIONS: *Mitchell* 65 (K); *Waterhouse* 8 (K).

This species is similar to *D. cyrtosepalum* but is distinguished by the lip which has longer lateral lobes and a more papillose mid-lobe. Schlechter's illustration of the type collection differs slightly in having an erose central callus.

14. D. salomonense *Schltr.* in Fedde, Repert. Spec. Nov. Regni Veg., Beih. 1: 588 (1912). Type: Bougainville, *Rechinger* 2563 (holotype B†).
Grastidium salomonense (Schltr.) S. Rauschert in Fedde, Repert. Spec. Nov. Regni Veg. 94(7–8): 452 (1983).

Fig. 40. Dendrobium salomonense A-K. A, habit x 2/3; **B**, flower x 2; **C**, dorsal sepal x 3; **D**, petal x 3; **E**, lateral sepal x 3; **F**, lip x 6; **G**, side view of lip x 6; **H**, dorsal and ventral view of anther-cap x 6; **J**, pollinia x 6; **K**, column x 6. **D. kietaense L-M. L**, flower x 2; **M**, lip x 6. **A-K** drawn from *Wickison* 24a; **L-M** from *Wickison* 27. All drawn by Sue Wickison.

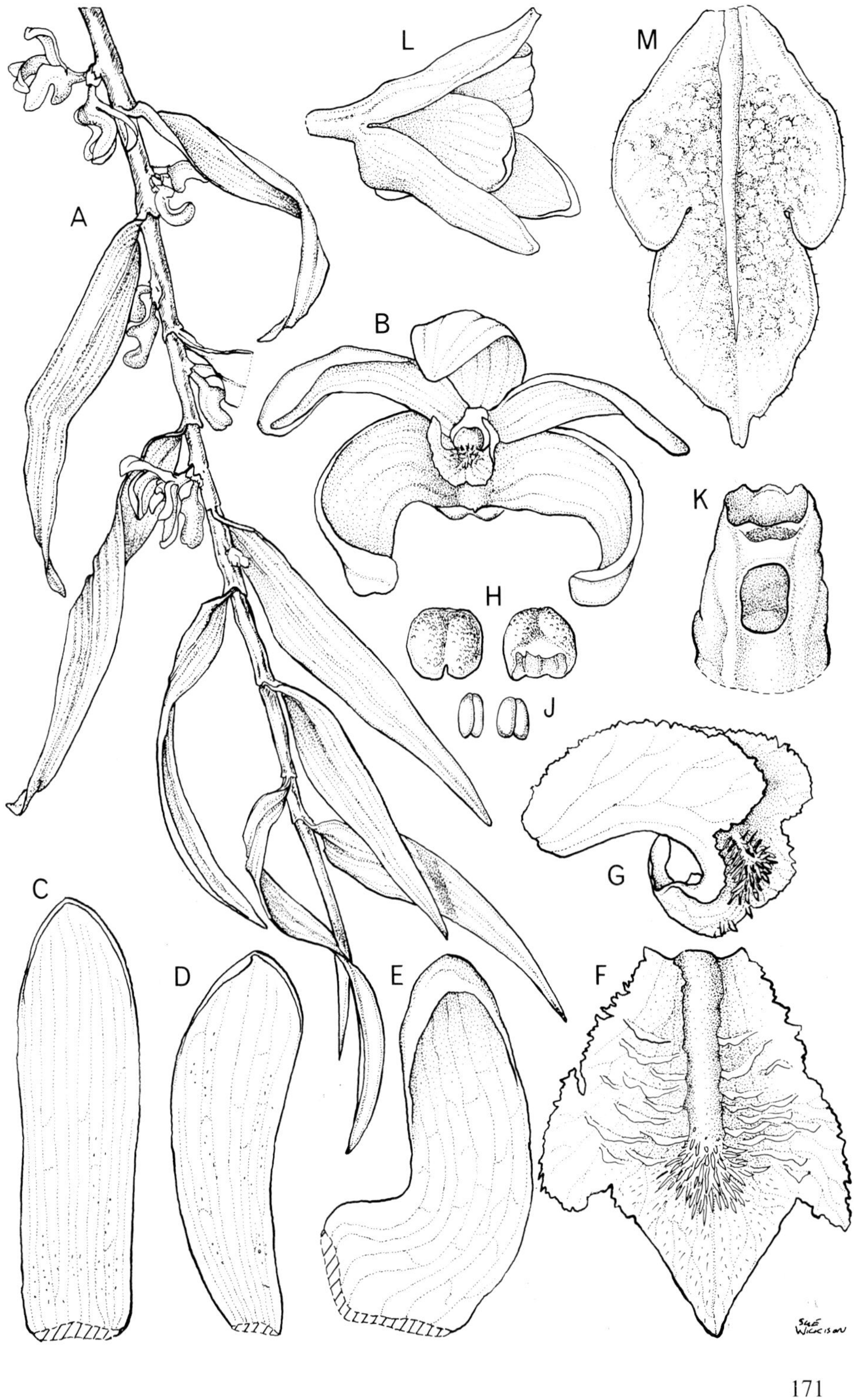
A
B
C
D
E
F
G
H
J
K
L
M

An epiphytic herb. *Stem* pendulous, up to 1.5 m long, leafy. *Leaves* distichous, lanceolate, up to 15 cm long, 2 cm wide, minutely unequally bilobed at apex, sheathing at base. *Inflorescences* lateral, arising from bilaterally flattened protuberances on the side of the leaf sheath. *Flowers* in pairs, white or yellow with a white or yellow lip veined with purple with a central orange callus; dorsal sepal oblong, obtuse, 1.3–1.4 cm long, 4–5 mm wide; lateral sepals falcate, 1–1.1 cm long, 5–7 mm wide, concave at apex, forming an obtuse, 3 mm long mentum; petals linear-falcate, obtuse, 1.2–1.4 cm long, 2.5–4 mm wide, concave at apex; lip trilobed, 8 mm long, 5–6 mm wide, the lateral lobes erect, small, triangular to crescent-shaped, the disc with a central callus extending to the base of the mid-lobe ending in a densely hairy patch, with transverse rugulose ridges, the mid-lobe triangular, recurved at apex; column 6 mm long, with a 4–6 mm long foot. (See fig. 40 A–K).

DISTRIBUTION: Bougainville, Choiseul, Gizo, Guadalcanal, Kolombangara, Makira and New Georgia.
HABITAT: Rain forest, sea level to 1150 m.
COLLECTIONS: *Comins* 155 (K); *Dennis* in RSS 2328 (K); *Dickson* 31 (K); *Hunt* in RSS 2334, 2345, 2368, 2435 & 2991 (K); *Mitchell* 47 (K); *Officers of HMS Penguin* s.n. (K); *Wickison* 24A, 24B & 25 (K).

***15. D. vandoides** *Schltr.* in Fedde, Repert. Spec. Nov. Regni Veg., Beih. 1: 603 (1912). Type: New Guinea, *Schlechter* 16694 (holotype B†).
Grastidium vandoides (Schltr.) F.G. Brieger in Schltr., Die Orchideen, ed. 3, 1(11–12) : 653 (1981).

A large, robust epiphytic herb. *Stems* pendulous, up to 70 cm long, 1–1.5 cm wide, stout, leafy. *Leaves* distichous, elliptic, 14–18 cm long, 3–4.5 cm wide, unequally bilobed at apex, dark green, sheathing at the base. *Inflorescences* lateral, arising in the fold at the base of the leaf, the flowers in pairs; peduncle 3 cm long. *Flowers* cream heavily spotted and mottled with dark red and with a white lip; pedicel and ovary 1.5–4 cm long, elongating after fertilization; dorsal sepal ovate, 1.6 cm long, 5 mm wide; lateral sepals obliquely ovate, 1.7 cm long, 9 mm wide, forming an obtuse mentum 5 mm long; petals lanceolate, 1.5 cm long, 5 mm wide; lip trilobed, 1–1.2 cm long, 4–5 mm wide, the lateral lobes narrowly triangular, the mid-lobe oblong, obtuse, hairy and fringed with hairs, the disc with a short basal callus; column 5 mm long, column-foot 5 mm long.

DISTRIBUTION: Bougainville. Also in New Guinea
HABITAT: Cultivated on *Leucaena leucocephala*, sea level.
COLLECTION: *Millar & Vandenberg* in NGF 48512 (K, LAE)

This species is probably not native to Bougainville.

16. D. vanikorense *Ames* in Journ. Arn. Arb. 13: 134 (1932). Type: Santa Cruz Islands, Vanikoro, *Kajewski* 663 (holotype AMES!).
Grastidium vanikorense (Ames) S. Rauschert in Fedde, Repert. Spec. Nov. Regni Veg. 94(7–8) : 453 (1983).

An epiphytic herb. *Stems* clustered, pendulous, up to 1 m long, slender, leafy. *Leaves* distichous, narrowly linear-lanceolate, tapering gradually to an acute apex, 7.5–11 cm long, 3–5 mm wide, sheathing at the base. *Inflorescences* lateral, arising from bilaterally flattened protuberances. *Flowers* in pairs, white or creamy yellow tinged with purple; dorsal sepal lanceolate, acuminate, 8–9 mm long, 2.5 mm wide; lateral sepals obliquely triangular, acuminate, 9–9.5 mm long, 3–4.5 mm wide, forming a conical, obtuse, 3–3.5 mm long mentum; petals lanceolate, acuminate, 8–9 mm long, 1 mm wide; lip obscurely trilobed to elliptic, 8 mm long, 4.5 mm wide, the lateral lobes small, erect, crescent-shaped, the disc with a longitudinal callus, extending midway along the lip, the mid-lobe oblong, papillose, with an erose margin, recurved; column 3 mm long, column-foot 4 mm long. (See fig. 37 A–J).

DISTRIBUTION: Bougainville, Makira, New Georgia, Rendova, Rennell and the Santa Cruz Islands (Vanikoro).
HABITAT: Lowland rain forest, 15–480 m.
COLLECTIONS: *Brass* 3067 (G); *Chaplin* 797 (K); *Cribb* 6036 (K); *Cribb & Campbell* 5155 (K); *Kajewski* 663 (AMES) & 1945 (G); *Wickison* 18 & 26 (K).

We originally considered *D. vanikorense* to be conspecific with *D. biflorum* Sw., from Vanuatu, Fiji, Samoa and Tahiti (see Cribb & Lewis, 1989) but *D. biflorum* differs in having longer, filiform sepals and petals.

17. D. vernicosum *Schltr.* in Fedde, Repert. Spec. Nov. Regni Veg., Beih. 1: 579 (1912). Type: New Guinea, *Schlechter* 17197 (holotype B†).
Grastidium vernicosum (Schltr.) S. Rauschert in Fedde, Repert. Spec. Nov. Regni Veg. 94(7–8): 453 (1983).

An epiphytic herb. *Stems* clustered, pendent, up to 1.4 m. *Leaves* distichous, lanceolate, up to 12 cm long, 2 cm wide, minutely unequally bilobed at apex, sheathing at the base. *Inflorescences* lateral, arising from bilaterally flattened protuberances. *Flowers* paired, maroon with an orange lip and a pale yellow column; dorsal sepal ovate, filiform, 4.3 cm long, 2.5 mm wide; lateral sepals falcate, filiform, 4.8–5 cm long, 0.9–1.2 cm wide, forming an obtuse mentum 3 mm long; petals linear, filiform, 4.2 cm long, 2.5 mm wide; lip trilobed, 1.8 cm long, 8 mm wide, the lateral lobes erect, small, rounded, the disc sparsely hairy, with a central callus extending to the base of the mid-lobe, the mid-lobe triangular, hairy with an erose margin, recurved at apex; column 4 mm long, column-foot 4–5 mm long.

DISTRIBUTION: Guadalcanal, Kolombangara and New Georgia. Also in New Guinea.
HABITAT: Mossy rain forest, 800–1500 m.
COLLECTIONS: *Dennis* in RSS 31785 (K); *Wickison* 31 & 71 (K).

Section LATOURIA

See Cribb (1983b) for a revision of this section. Also see Warren (1990a) for a discussion of this section.

1. Flowers with setose ovaries and sepals 2
   Flowers with glabrous ovaries and sepals 3
2. Pseudobulbs usually 2-leaved; lip side-lobes tapering to a round apex, always longer than broad; petals often reflexed .... **22. D. polysema**
   Pseudobulbs usually 3-leaved but up to 7-leaved; lip side-lobes more or less subquadrate, about as long as broad; petals spreading .... **20. D. macrophyllum**
3. Flowers white with purple markings; petals rhombic-spathulate to obovate 4
   Flowers not as above; petals tapering or more or less oblong 6
4. Petals narrowly clawed, rhombic above, acuminate, 4.2–5 cm long; lip mid-lobe lanceolate .... **19. D. johnsoniae**
   Petals spathulate to obovate, acuminate, less than 3.5 cm long; lip mid-lobe oblong to transversely oblong 5
5. Lip broader than long when flattened, more than 2.5 cm broad, spotted with purple on the margins of the side-lobes; the mid-lobe apiculate; petals more than 3 cm long; pseudobulbs circular in cross-section .... **24. D. rhodostictum**
   Lip longer than broad when flattened, less than 1.5 cm broad, usually unspotted on margins of side-lobes; the mid-lobe subacute; petals less than 2.5 cm long; pseudobulbs 4-angled .... **25. D. ruginosum**
6. Flowers somewhat contorted; tepals and mid-lobe with undulate margins .... **26. D. spectabile**
   Flowers not contorted; tepals and lip not contorted 7
7. Pseudobulb unifoliate; leaf lanceolate; inflorescence lateral .... **21. D. pachystele**
   Pseudobulb with 2 or more leaves; leaves elliptic; inflorescence sub-apical 8
8. Plant erect; peduncle stout, more than three times as long as the rhachis; flowers yellow or greenish with purple markings .... **18. D. bifalce**
   Plant usually pendent; peduncle equal to or shorter than the rhachis; flowers green .... **23. D. punamense**

18. D. bifalce *Lindley* in Hook., Journ. Bot., Lond. 2: 237 (1843). Type: New Guinea, *Hinds* s.n. (holotype K!).

Doritis bifalcis (Lindley) Reichb. f., Xen. Orch. 2: 7 (1862).

Dendrobium breviracemosum Bailey in Queensl. Agric. Journ. 3: 158 (1898). Type: New Guinea, *Bailey* s.n. (holotype BRI!).

Dendrobium chloropterum var. *striatum* J.J. Smith in Nova Guinea 8: 64 (1909). Types: New Guinea, *Branderhorst* 30 & 118, & *Koch* s.n. (syntypes BO!).

Sayeria bifalcis (Lindley) S. Rauschert in Fedde, Repert. Spec. Nov. Regni Veg. 94(7–8): 466 (1983).

For full synonymy see Cribb (1983b).

An epiphytic herb. *Pseudobulbs* erect, stout, 11–30 cm long, 0.8–1.2 cm wide, yellow, 5- to 7-noded below leaves. *Leaves* 2–4 at apex, coriaceous, suberect, elliptic-oblong, obtuse, up to 16 cm long, 3.5–6 cm wide. *Inflorescence* erect, lateral, emerging from just beneath leaf bases, 25–40 cm long, few- to many-flowered; peduncle terete, stout, very much longer than the rhachis. *Flowers* closely spaced, greenish brown to dull yellow with dark red stripes and with a white callus on the lip; dorsal sepal oblong-ovate, obtuse, 7 mm long, 4 mm wide; lateral sepals spreading, obliquely triangular, obtuse, 9 mm long, 9.5 mm wide, forming a conical mentum 7 mm long; petals oblong, obtuse, 7 mm long, 3 mm wide; lip reflexed, clawed at base, trilobed, 1–1.2 cm long, 1–1.2 cm wide, the lateral lobes erect, somewhat falcate, acute, the mid-lobe transversely oblong or reniform, the callus of 3 irregular ridges; column 2 mm long with a 7 mm long column-foot.

DISTRIBUTION: Bougainville. Also in Timor, New Guinea and Australia.
HABITAT: Coastal forest, sea level.
COLLECTION: *Crocker* 4 (K).

The specimen from Bougainville is incomplete and therefore the description of the habit is taken from Cribb (1983b).

19. D. johnsoniae *F. Muell.* in Wing's South Sci. Rec. 2: 95 (1882). Type: New Guinea, *J. Chalmers* s.n. (holotype MEL!, isotype K!).
Dendrobium macfarlanei Reichb. f. in Gard. Chron. n.s. 18: 520 (1882), non F. Muell (1875). Type: New Guinea, *McFarlane* ex *Veitch* (holotype W!).
Sayeria johnsoniae (F. Muell.) S. Rauschert in Fedde, Repert. Spec. Nov. Regni Veg. 94(7–8): 467 (1983).

For full synonymy see Cribb (1983b)

An erect epiphytic herb. *Pseudobulbs* stout, subclavate, 4- to 6-noded below the leaves, 20–30 cm long, 0.8–1.1 cm wide, yellow. *Leaves* 2–5 at apex, coriaceous, elliptic or oblong-elliptic, obtuse, 8–11 cm long, 3.5–4.5 cm wide. *Inflorescence* terminal or from below leaf bases, erect or suberect, 17–27 cm long, few- to many-flowered. *Flowers* large, white with purple lines on the lateral lobes of the lip and base of column; dorsal sepal lanceolate, acuminate, 2.6–2.8 cm long, 7–8 mm wide; lateral sepals obliquely lanceolate, acuminate, 2.6–3.4 cm long, 1.1 cm wide, forming an obliquely conical, 8–9 mm long mentum; petals narrowly clawed, rhombic above, acuminate, 4.2–5 cm long, 1.8–3 cm wide; lip trilobed, 3.5–4.6 cm long, 2–3 cm wide, the lateral lobes erect, subquadrate, the mid-lobe much longer than the side-lobes, ovate-lanceolate, acuminate, the callus fleshy, obscurely 2-ridged at base; column short, 3 mm long with 7–9 mm long foot. (See plate 9F).

DISTRIBUTION: Guadalcanal. Also in New Guinea.
HABITAT: Montane forest, epiphytic on *Calophyllum*, 500–600 m.
COLLECTIONS: Dennis (see Dennis, 1976); Cribb (sight records).

Dennis (1976) notes that *D. johnsoniae* occurs quite commonly on the higher slopes and summit (530 m) of Mt. Austen, Guadalcanal.

20. D. macrophyllum *A. Rich.* in Sert. Astrol.: 22 (1834). Type: New Guinea, *A. Lesson* (holotype P!).

Dendrobium ferox Hassk. in Retzia 1:1 (1855).Type: Java, *Hasskarl* s.n. (holotype BO!).
Dendrobium macrophyllum var. *veitchianum* (Lindley) Hook. f. in Curtis's Bot. Mag. 93: t.5649 (1867).
Dendrobium gordonii S. Moore ex Baker in Journ. Linn. Soc., Bot. 20: 372 (1883). Type: Fiji, *J. Horne* 942 (holotype K!).
Dendrobium brachythecum F. Muell. & Kraenzl. in Oest. Bot. Zeit. 44: 161 (1894). Type: New Guinea, *Anderson* s.n. (holotype MEL!).
Dendrobium ternatense J.J. Smith in Bull. Dép. Agr. Indes Néerl. 22: 26 (1909). Type: Moluccas, Ternate, *J.J. Smith* s.n. (holotype BO!).
Dendrobium psyche Kraenzl., in Engler, Pflanzenr. Orch.-Mon.-Dendr. 1: 246 (1910). Type: Vanuatu, *Braithwaite* s.n. (holotype B†).
Dendrobium tomohonense Kraenzl. in Engler, Pflanzenr. Orch.-Mon.-Dendr. 1: 244 (1910). Type: Celebes, *Sarrasin* 655 & 799 (syntypes B†).
Dendrobium musciferum Schltr. in Fedde, Repert. Spec. Nov. Regni Veg., Beih. 1: 494 (1914). Types: New Guinea, *Schlechter* 16920 & 19455 (syntypes B†); *Schlechter* 18958 (syntype B†, isosyntype K!).
Latouria macrophylla (A. Rich.) F.G. Brieger in Schltr., Die Orchideen ed. 3, 1(11–12) : 727 (1981).
Sayeria macrophylla (A. Rich.) S. Rauschert in Fedde, Repert. Spec. Nov. Regni Veg. 94(7–8) : 467 (1983).

An epiphytic herb. *Pseudobulbs* clustered, clavate or subclavate, rarely fusiform, up to 50 cm tall, drying yellow, 4- to 6-noded below leaves. *Leaves* usually 3 but occasionally up to 7, suberect, coriaceous, oblong or oblong-elliptic, obtuse, 15–31 cm long, 3.3–9 cm wide. *Inflorescence* apparently terminal, erect, up to 40 cm long, often densely many-flowered. *Flowers* rather variable in size, yellow or greenish with purple spots or stripes on the lateral lobes and mid-lobe of the lip; pedicel and ovary setose; sepals setose on outer surface; dorsal sepal oblong-ovate or oblong-lanceolate, acute, 2.1–2.6 cm long, 0.7–1 cm wide; lateral sepals obliquely triangular, acute or acuminate, 2.3–2.6 cm long, 1.2–1.4 cm wide, forming an obliquely conical, 1 cm long mentum; petals oblanceolate or oblong-oblanceolate, acute, 1.8–2.2 cm long, 0.8–1.1 cm wide, with somewhat undulate margins; lip trilobed, strongly recurved, 1–2 cm long, 1.6–2.7 cm wide, the lateral lobes erect, subquadrate-flabellate, dilated above, truncate, the mid-lobe transversely oblong, apiculate, conduplicate, the callus obscurely 3-ridged, white; column 3 mm long, with a 1 cm long foot. (See fig. 41 L–N and plate 10C).

DISTRIBUTION: Bougainville, Choiseul, Guadalcanal, Kolombangara, New Georgia and Rennell. Widespread from Java to New Guinea, Vanuatu, Fiji and Samoa.

HABITAT: Rain forest, sea level to 1070 m.

COLLECTIONS: *Chaplin* 770, 772 & 849 (K); *Craven & Schodde* 19 (K); *Dennis* in RSS 2026 & 2043 (K); *Gafui et al.* in BSIP 14761 (K); *Millar* in NGF 38393 (K); *Mitchell* 46 (K); *Ridsdale & Lavarack* in NGF 30609 (K); *Wickison* 38B (K).

---

* **Fig. 41. Dendrobium polysema A-K. A**, habit x 2/3; **B**, inside view of lateral sepal x 3; **C**, outside view of lateral sepal x 3; **D**, side view of lip x 3; **E**, lip x 3; **F**, column x 3; **G**, petal x 3; **H**, dorsal and ventral view of anther-cap x 3; **J**, pollinia x 3; **K**, dorsal sepal x 3. **D. macrophyllum L-N. L**, flower x 1; **M**, side view of lip x 2; **N**, lip x 2. **A-K** drawn from *Wickison* 38; **L-N** from *Wickison* 38B. All drawn by Sue Wickison.

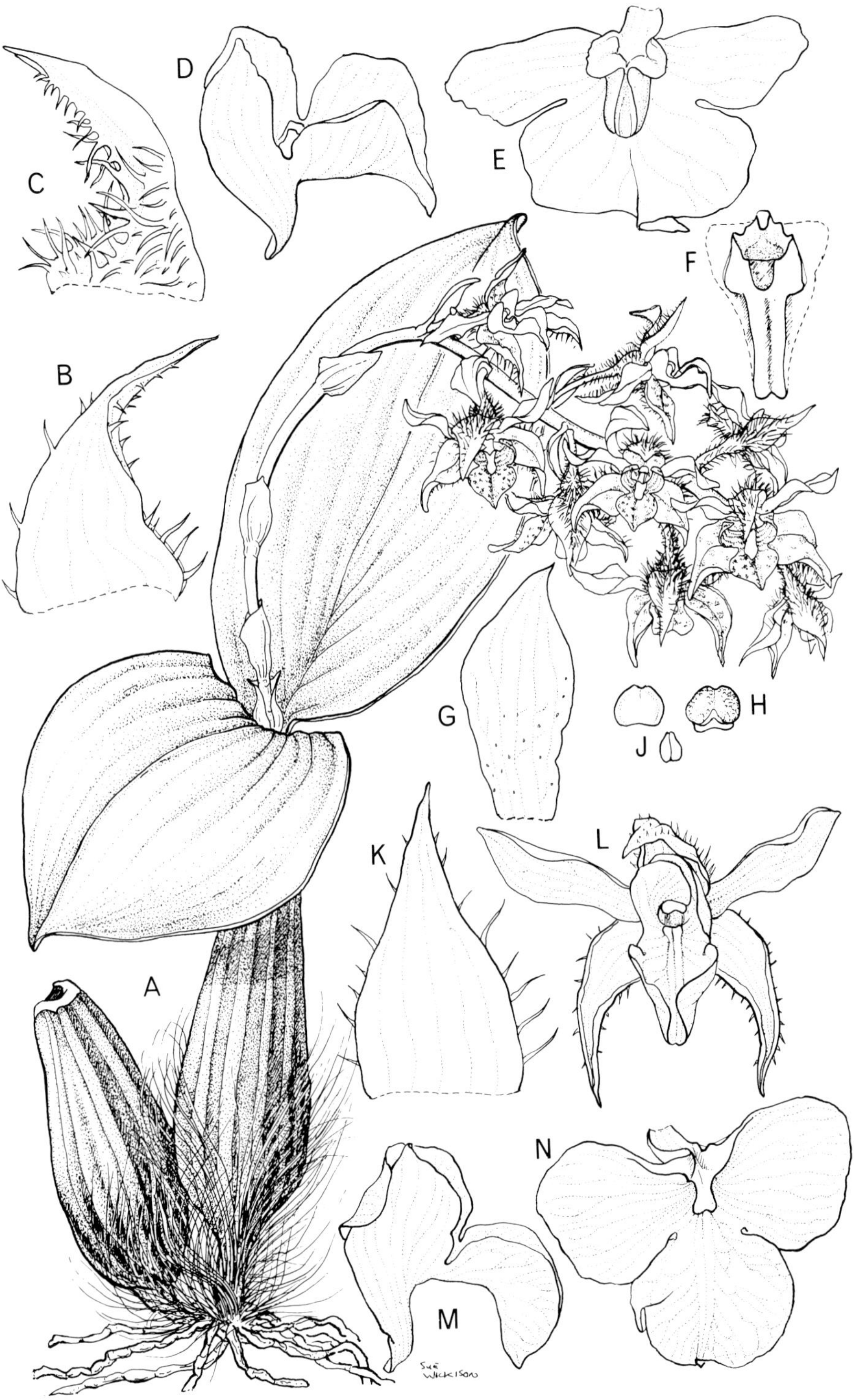
A
B
C
D
E
F
G
H
J
K
L
M
N
SUE WICKISON

21. D. pachystele *Schltr.* in Fedde, Repert. Spec. Nov. Regni Veg., Beih. 1: 485 (1912). Type: New Guinea, *Schlechter* 19182 (holotype B†).
Sayeria pachystele (Schltr.) S. Rauschert in Fedde, Repert. Spec. Nov. Regni Veg. 94(7–8) : 467 (1983).

An epiphytic herb. *Pseudobulbs* slender, 10–25 cm long, 7 mm wide, unifoliate at apex. *Leaf* coriaceous, lanceolate or narrowly elliptic-lanceolate, acute or acuminate, 18–30 cm long, 3.7–4.3 cm wide. *Inflorescence* lateral, short, c. 5 cm long, laxly 3- to 6-flowered. *Flowers* fleshy, whitish or creamy white; dorsal sepal ovate, obtuse, 1.2 cm long, 5–7 mm wide; lateral sepals obliquely triangular, obtuse, 1.2 cm long, 1 cm wide, forming a incurved, conical, 9 mm long mentum; petals oblong-elliptic, acute, 1 cm long, 4 mm wide; lip recurved, obscurely trilobed or entire, elliptic-obovate, 1 cm long, 1.2 cm wide, the callus 3-ridged, extending half to three-quarters of the way along the lip; column short and fleshy with a 9 mm long foot.

DISTRIBUTION: Kolombangara, Makira and New Georgia. Also in New Guinea and possibly the Caroline Islands (Palau).
HABITAT: Lowland rain forest, 25–350 m.
COLLECTIONS: *Dickson* A5 (K); *Hunt* in RSS 2298 & 2401 (K).

A sterile specimen (*Takamatsu* 1567), from Garasumao, Palau in the Caroline Islands, which has the same habit, may be attributed to *D. pachystele*.

22. D. polysema *Schltr.* in K. Schum. & Laut., Nachtr. Fl. Deutsch Südsee: 163 (1905). Type: New Guinea, *Schlechter* 14066 (holotype B†).
Dendrobium macrophyllum var. *stenopterum* Reichb. f. in Gard. Chron. ser. 3, 3: 393 (1888). Type: New Guinea, Hort. *Bull* (holotype W!).
Dendrobium pulchrum Schltr. in Fedde, Repert. Spec. Nov. Regni Veg., Beih. 1: 496 (1912). Type: New Guinea, *Schlechter* 19605 (holotype B†).
Dendrobium polysema var. *pallidum* Chadim in Austr. Orch. Review 29(1): 32 (1964). Type: New Guinea, *Chadim* D7 (holotype K!).
Sayeria polysema (Schltr.) S. Rauschert in Fedde, Repert. Spec. Nov. Regni Veg. 94(7–8): 468 (1983).

An epiphytic herb. *Pseudobulbs* clavate, 20–50 cm tall, drying yellow or orange, 3- to 5-noded below leaves. *Leaves* 2 at apex, spreading, coriaceous, elliptic or oblong-elliptic, obtuse, up to 22 cm long, 6–13 cm wide, shortly petiolate. *Inflorescence* apparently terminal, erect to arcuate, up to 45 cm long, few- to many-flowered. *Flowers* yellow or greenish heavily spotted maroon, the lateral lobes of lip striped maroon with a creamy-white callus; pedicel and ovary densely setose; sepals setose on outer surface; dorsal sepal ovate-lanceolate, acuminate, 2–2.4 cm long, 1.2–1.4 cm wide; lateral sepals obliquely falcate-lanceolate, acuminate, 2.3–3 cm long, 1.5 cm wide, forming an incurved, conical, 1.2 cm long mentum; petals often reflexed, oblanceolate or obovate, apiculate, 1.8–2.6 cm long, 0.7–1.6 cm wide, with ± strongly undulate margins; lip trilobed, recurved, 1.8–2.3 cm long, 2–2.6 cm wide, the lateral lobes smaller than the mid-lobe, erect or incurved, tapering to a rounded apex, the mid-lobe flat or with incurved margins, transversely oblong-ovate, subapiculate, the callus 3-ridged;

column 3-3.5 mm long, with an incurved, 1.2 cm long foot. (See fig. 41 and plate 10D).

DISTRIBUTION: Bougainville, Choiseul, Guadalcanal, Kolombangara, New Georgia, the Santa Cruz Islands and Santa Isabel. Also in New Guinea and Vanuatu.
HABITAT: Rain forest, sea level to 1000 m.
COLLECTIONS: *Cribb* S28 (K); *Cribb & Campbell* 5110 (K); *Hunt* in RSS 2647 (K); *Mauriasi et al.* in BSIP 16679 (K, LAE); *Mitchell* 26 (K); *Wickison* 38 (K); *Womersley* in NGF 24871 (LAE).

23. D. punamense *Schltr.* in K. Schum. & Laut., Nachtr. Fl. Deutsch. Südsee: 163 (1905). Type: New Ireland, *Schlechter* 14690 (holotype B†).
Dendrobium waterhousei Carr in Bull. Misc. Inf., Kew 1934: 379 (1934). Type: Bougainville, *Waterhouse* 333 (holotype K!).
Sayeria punamensis (Schltr.) S. Rauschert in Fedde, Repert. Spec. Nov. Regni Veg. 94(7–8) : 468 (1983).

See Wood (1978) for more information on this species.

An erect or pendulous epiphytic herb. *Pseudobulbs* clustered, cylindrical, 3- to 6-noded below the leaves, 15–30 cm long, 4–7 mm wide, green drying yellow. *Leaves* 2 at apex, subcoriaceous, twisted at base to lie in one plane, oblong-lanceoloate or lanceolate, acuminate, 15–30 cm long, 3–4.3 cm wide. *Inflorescence* subterminal, 5–10 cm long, few-flowered. *Flowers* fleshy, white to pale green; dorsal sepal ovate, acuminate, 1–1.3 cm long, 5.5–6 mm wide; lateral sepals falcate, obliquely broadly triangular, acuminate, up to 1.4 cm long, 1.5 cm wide, forming a large, incurved, conical, 1–1.4 cm long mentum; petals elliptic, acuminate, 1–1.3 cm long, 4 mm wide; lip trilobed towards apex, narrowly clawed below, 1.3–1.5 cm long, 1.4–1.6 cm wide, the lateral lobes as long as the mid-lobe, narrowly oblong, rounded at apex, the mid-lobe transversely oblong, apiculate, the callus 3-ridged; column 3 mm long with an incurved foot up to 1.4 cm long.

DISTRIBUTION: Bougainville and Guadalcanal. Also in Manus Island, New Britain and New Ireland.
VERNACULAR NAME: Tugorohia or Togoruha (Bougainville).
HABITAT: Rain forest, 220–1100 m.
COLLECTIONS: *Hunt* in RSS 3004 (K); *Ridsdale & Lavarack* in NGF 31611 (LAE); *Waterhouse* 333 & 590 (K); *Wickison* 101A & B (K).

A specimen from Makira (*Hunt* in RSS 2244) differs from *D. punamense* in having a shorter mentum and lateral lobes, and a subquadrate mid-lobe. It agrees with Samoan material at Kew (*Whistler* W3867) but we have refrained from describing it until we have seen better preserved material.

24. D. rhodostictum *F. Muell. & Kraenzl.* in Oest. Bot. Zeit. 44: 300 (1894). Type: New Guinea, *Sayer* s.n. (holotype MEL!, isotype B†).

Dendrobium madonnae Rolfe in Bull. Misc. Inf., Kew 1906: 32 (1906). Type: New Guinea, *Sander* s.n. (holotype K!).
Sayeria rhodosticta (F. Muell. & Kraenzl.) S. Rauschert. in Fedde, Repert. Spec. Nov. Regni Veg. 94(7–8) : 468 (1983).

An erect epiphytic or terrestrial herb. *Pseudobulbs* clavate, 3- to 5-noded below the leaves, 9–25 cm long, 0.3–1.2 cm wide, yellow or reddish. *Leaves* 2–4 at apex, suberect or spreading, coriaceous, ovate-lanceolate to linear-lanceolate, minutely emarginate at acute to obtuse apex, 5–11 cm long, 1–2.5 cm wide. *Inflorescences* apparently terminal or subterminal, erect to arcuate, few-flowered. *Flowers* white with purple stripes or spots on the lateral lobes of the lip; dorsal sepal ovate, acute or acuminate, 2–2.5 cm long, 1 cm wide; lateral sepals falcate-lanceolate, acuminate, 1.8–2.2 cm long, 1–1.1 cm wide, forming an obliquely conical, 8–10 mm long mentum; petals obovate, spathulate, apiculate, 3–3.5 cm long, 2–2.5 cm wide; lip obscurely trilobed towards apex, broadly obovate, 2.8 cm long, 3.1 cm wide, the lateral lobes erect, incurved, semicircular, larger than the mid-lobe, the mid-lobe transversely oblong, apiculate, the callus 3-ridged at base; column 2.5 mm long with a 7–9 mm long foot. (See plate 10B).

DISTRIBUTION: Bougainville. Also in New Guinea, New Britain and New Ireland.
HABITAT: Rain forest, 950 m.
COLLECTIONS: *Cribb & Morrison* 1913 & 1914 (K); *Kajewski* 2078 (K).

Dendrobium rhodostictum and *D. ruginosum* readily hybridize where found growing together (see *Slade* s.n. (K) from the Panguna Pass in Bougainville).

25. D. ruginosum *Ames* in Journ. Arn. Arb. 14: 109 (1933). Type: Santa Cruz Islands, Vanikoro, *Kajewski* 606 (holotype AMES!).
Sayeria ruginosa (Ames) S. Rauschert in Fedde, Repert. Spec. Nov. Regni Veg. 94(7–8) : 468 (1983).

An epiphytic herb. *Pseudobulbs* clustered, clavate, 4-angled in cross-section, 7–25 cm long, 5–8 mm wide, 4-noded below leaves, yellow to purple. *Leaves* 2–4 at apex, suberect, coriaceous, lanceolate, acute, 4–12 cm long, 1–2.5 cm wide. *Inflorescences* terminal, erect, up to 20 cm long, few- to several-flowered. *Flowers* white with a creamy-green lip with purplish-brown stripes; dorsal sepal lanceolate, acute, 1.5–1.8 cm long; lateral sepals obliquely lanceolate, acuminate, 1.7–2 cm long, 7 mm wide, forming an incurved, subconical, 6–7 mm long mentum; petals spathulate, 2.5–2.7 cm long, 1.2–1.7 cm wide; lip trilobed, 1.8–2.1 cm long, 1–1.2 cm wide, the lateral lobes rounded in front, the mid-lobe longer than the lateral lobes and oblong, the callus fleshy, obscurely 3-ridged at base; column 1.5 mm long, with an incurved, 6–7 mm long foot. (See plate 9E).

DISTRIBUTION: Bougainville, Guadalcanal, Kolombangara, Makira and New Georgia.
HABITAT: Montane rain forest, 600–1900 m.
COLLECTIONS: *Brass* s.n. & 3263 (AMES); *Chaplin* 843 (K); *Craven & Schodde* 309 (K, LAE); *Cribb* S27 & S69 (K);*Cribb & Campbell* 5122 (K); *Hunt*

in RSS 2321a & 2484 (K); *Kajewski* 606 (K); *Lavarack & Ridsdale* in NGF 31419 (K, L, LAE); *Mitchell* 8 (K); *Slade* s.n. (K); *Wickison* 33 & 229 (K).

Dendrobium ruginosum resembles *D. mooreanum* Lindley from Vanuatu, which differs in its relatively narrower petals and entire or very obscurely lobed rhombic lip with a triangular acute apex.

26. D. spectabile *(Blume) Miq.*, Fl. Ind. Bat. 3: 645 (1855). Type: New Guinea, *Latour-Leschenault* s.n. (holotype L!).
Latouria spectabilis Blume in Rumphia 4: 41, t. 195 fig. 1 & t. 199 C (1850).
Dendrobium tigrinum Rolfe ex Hemsley in Ann. Bot. 5: 507 (1891). Type: Solomon Islands, Makira, *Comins* 187 (holotype K!).

An epiphytic or lithophytic herb. *Pseudobulbs* clustered, cane-like, up to 40 cm high, 1–1.3 cm wide, 5- to 8-noded below leaves, swollen at base. *Leaves* 4–6 towards the apex, suberect, coriaceous, elliptic, obtuse, up to 23 cm long, 4–8 cm wide, shortly petiolate at base. *Inflorescences* emerging from just below leaf-bases, erect, 20–40 cm long, few- to many-flowered; peduncle terete. *Flowers* large, somewhat grotesque, yellow or greenish yellow, commonly heavily mottled with maroon on sepals and petals and lined with maroon on the lip, with a white callus, scented; sepals with somewhat undulate margins; dorsal sepal recurved, lanceolate, acuminate, 3.6 cm long, 5 mm wide; lateral sepals recurved, lanceolate-falcate, acuminate, 3.6 cm long, 1.3 cm wide, forming an obliquely conical, 1 cm long mentum; petals somewhat twisted, linear-lanceolate, acuminate, 4 cm long, 6 mm wide; lip trilobed, recurved, 4 cm long, 2.2 cm wide, the lateral lobes erect, subquadrate-semicircular, rounded in front, the mid-lobe much longer than the lateral lobes, lanceolate, acuminate, undulate-twisted, the callus 3-ridged, raised at base and apex; column short, 3 mm long with a 8.5 mm long foot. (See plate 9C).

DISTRIBUTION: Bougainville, Gizo, Guadalcanal, Kolombangara, Makira, Malaita and New Georgia. Also in New Guinea.
VERNACULAR NAME: Ubutu (New Georgia).
HABITAT: Lowland swampy forest, lower montane forest or on planted coconut or *Casuarina* trees, sea level to 1100 m.
COLLECTIONS: *Chaplin* 850 (K); *Gafui et al.* in BSIP 14708 (K); *Hunt* in RSS 2419 (K); *Im Thurn* 362 (K); *Mitchell* 18 (K); *Moore* s.n. (K); *Morrison & Cribb* 1908 (K); *Waterhouse* 114 (K); *Wickison* 55 (K).

Section MONANTHOS

27. D. bilobum *Lindley* in Hook., Journ. Bot., Lond. 2: 236 (1843). Type: New Guinea, *Hinds* s.n. (holotype K!).
Callista biloba (Lindley) Kuntze, Rev. Gen. Pl. 2: 654 (1891).
Cadetia biloba (Lindley) Blume, Mus. Bot. Lugd. Bat. 1: 30 (1849).
Monanthus biloba (Lindley) F.G. Brieger in Schltr., Die Orchideen ed. 3, 1(11–12): 660 (1981).

An epiphytic herb forming clumps. *Stems* clustered, erect to pendent, 20–40 cm high, slightly bilaterally compressed, leafy. *Leaves* distichous, suberect, thin-textured, linear, obtuse, 4–7 cm long, 4–8 mm wide, bilobed at apex, bright green, twisted at base to lie in one plane, articulated to a sheathing compressed leaf-base 1.4–1.6 cm long. *Inflorescences* lateral, 1-flowered. *Flower* non-resupinate, cream, pale yellow or yellowish green with a dark purple-marked lip; pedicel and ovary 4–5 mm long; dorsal sepal ovate, 4–5 mm long, 2–3 cm wide; lateral sepals obliquely ovate, 6.5 mm long, 4–5 mm wide, forming an obtuse, slightly upcurved, 4 mm long mentum; petals linear-lanceolate, 3.5–4 mm long, 1.5–2 mm wide; lip trilobed, 5–5.5 mm long, 3–4 mm wide, the lateral lobes erect, semicircular, fleshy at margins, the mid-lobe small, subcircular, fleshy; column 2–3 mm long, column-foot 9 mm long. (See fig. 42 A–H).

DISTRIBUTION: Bougainville, Choiseul, Guadalcanal, Kolombangara, Makira, Malaita and New Georgia. Also in New Guinea, Vanuatu and Fiji.
VERNACULAR NAME: Tuku (Bougainville).
HABITAT: Rain forest, sea level to 1350 m.
COLLECTIONS: *Brass* 2760 (G); *Chaplin* 832 & 833 (K); *Cribb* S54 (K); *Cribb, Dennis & Campbell* 5068 (K); *Cribb & Morrison* 1920 (K); *Dennis* in RSS 2028 (K); *Dickson* L8 (K); *Hunt* in RSS 2186a & 2553 (K); *Im Thurn* 334 (K); *Mitchell* 15 (K); *Waterhouse* 457 (K); *Wickison* 19 & 258 (K).

S. Wickison (pers. comm.) notes that at higher altitudes (above 320 m) the leaves of *D. bilobum* are smaller and more numerous compared to collections made on the coast.

Section OXYGLOSSUM

We have followed the Reeve & Woods (1989) revision of this section. Also see Warren & Woods (1990).

1. Leaves linear to narrowly elliptic, up to 3 mm wide 2
   Leaves oblong, lanceolate, ovate or elliptic, 4–35 mm wide 3
2. Flowers pale blue with an orange or red lip; ovary triangular in cross-section .... **34. D. violaceominiatum**
   Flowers white with a green base and an orange to golden-yellow apex of the lip; ovary indistinctly 5-ribbed .... **30. D. masarangense** subsp. **masarangense**
3. Dorsal sepal up to 2.5 cm long; lip boat-shaped, blunt at apex .... **29. D. laevifolium**
   Dorsal sepal up to 1.3 cm long; lip subtrilobed or oblong-linear, acute 4
4. Plant up to 25 cm high; pseudobulbs unifoliate, leaves distinctly petiolate; inflorescence many-flowered .... **31. D. petiolatum**
   Plant up to 8 cm high; pseudobulbs 1- to 3-leaved, leaves not petiolate; inflorescence 1- to 3-flowered 5

Fig. 42. Dendrobium bilobum A-H. A, habit x 2/3; **B**, flower x 6; **C**, dorsal sepal x 6; **D**, petal x 6; **E**, lateral sepal x 6; **F**, lip x 8; **G**, ventral and dorsal view of anther-cap x 8; **H**, column x 8. **D. melanostictum J-P. J**, inflorescence x 2; **K**, lip x 4; **L**, dorsal and ventral view of anther-cap x 4; **M**, column x 4; **N**, dorsal sepal x 3; **O**, petal x 3; **P**, lateral sepal x 3. **A-H** drawn from *Wickison* 19; **J-P** from *Dennis* 5. All drawn by Sue Wickison.

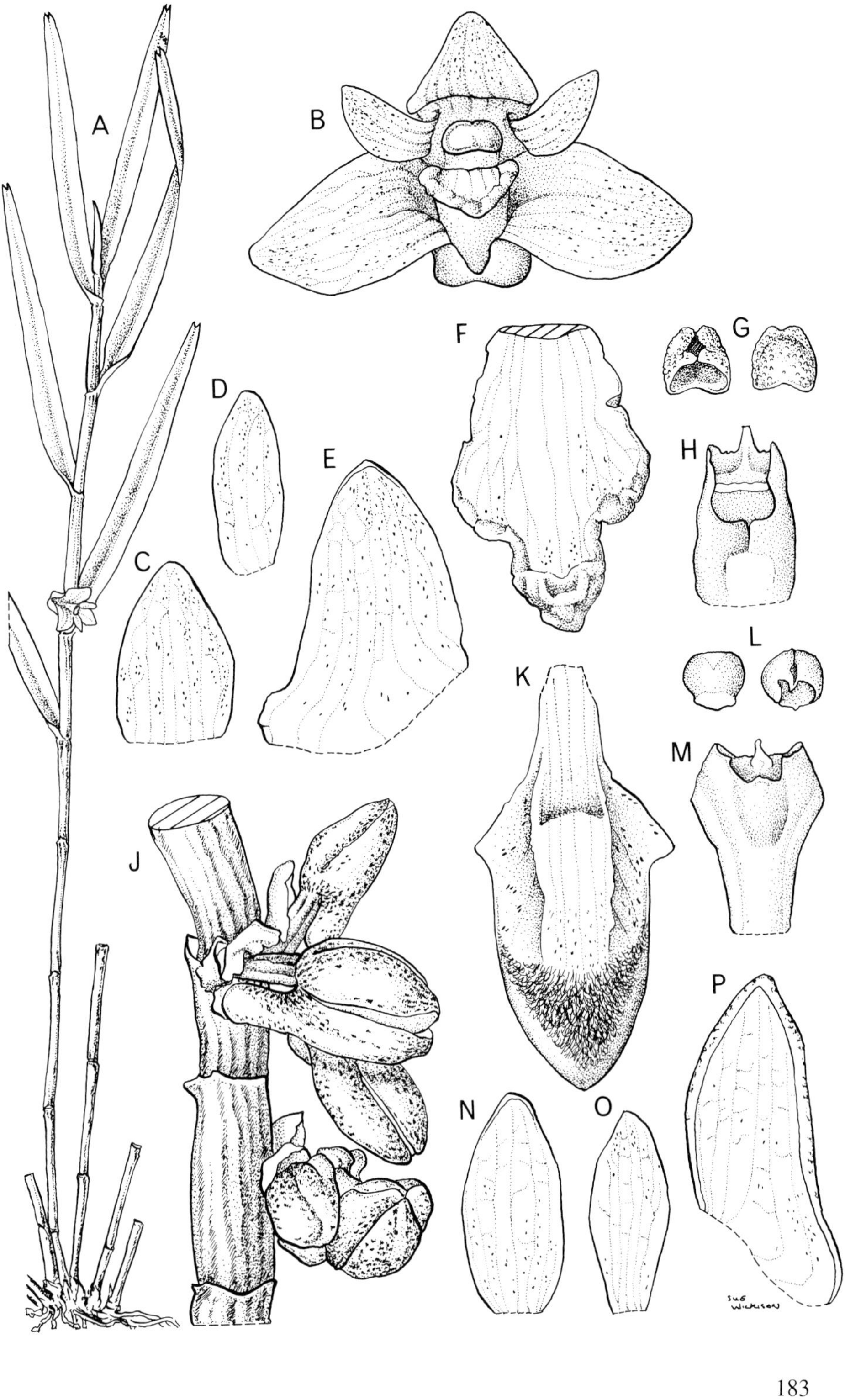
A
B
C
D
E
F
G
H
J
K
L
M
N
O
P
SUE WICKISON

5. Leaves lanceolate, 3–5 cm long; flowers orange-red with an orange lip and a yellow column .... **32. D. puniceum**
Leaves lanceolate or ovate, less than 2 cm long; flowers pink, purple or red 6
6. Leaves ovate to elliptic; flowers pink to purple; ovary distinctly 5-ribbed .... **28. D. delicatulum** subsp. **delicatulum**
Leaves lanceolate to elliptic; flowers red; ovary sharply 3-winged with dorsal wing often projecting up behind base of dorsal sepal .... **33. D. subacaule**

28. D. delicatulum Kraenzl. in Engler, Bot. Jahrb. 16: 17 (1892). Types: New Guinea, *Hellwig* 303 (holotype B†) & *A. Millar* in NGF 22862 (neotype LAE, isoneotype K!).

subspecies **delicatulum**

Dendrobium minutum Schltr. in Fedde, Repert. Spec. Nov. Regni Veg., Beih. 1: 531 (1912). Type: New Guinea, *Schlechter* 18754 (holotype B†, isotypes AMES!, BO!, E, K!, L).

For full synonymy see Reeve & Woods (1981 & 1989).

A dwarf epiphytic herb with a creeping habit forming dense mats on trees and branches, up to 3 cm high. *Pseudobulbs* ovoid-globose or sometimes clavate, up to 7 mm long, 7 mm wide, yellow-green, bearing 2 leaves at the apex. *Leaves* ovate to elliptic, acute, up to 2.2 cm long, 7 mm wide, glossy dark green. *Inflorescences* terminal, arising from both leafless and leafy stems (from between the leaves), 1- to 3-flowered. *Flowers* pink to purple; ovary distinctly 5-ribbed; dorsal sepal ovate-oblong, up to 7 mm long, 2.5 mm wide; lateral sepals obliquely ovate-elliptic, up to 1.2 cm long, 3 mm wide, forming a cylindrical mentum up to 5 mm long; petals oblong-oblanceolate, up to 5 mm long, 2 mm wide; lip oblong-linear, acute, up to 8 mm long, 2 mm wide, the apex deflexed; column 2–3 mm long with a 5 mm long column-foot. (See fig. 43 A–J).

DISTRIBUTION: Bougainville, Guadalcanal, Kolombangara and New Georgia. Also in New Guinea, Vanuatu, Fiji and the Caroline Islands (Ponape Island).

HABITAT: Montane forest, 480–1000 m.

COLLECTIONS: *Craven & Schodde* 149 (K); *Cribb & Campbell* 5133 (K); *Nicholls* 6 (K); *Wickison* 23 & 82 (K).

Subspecies *delicatulum* is distinguished from subsp. *huliorum* Reeve & Woods, from east New Guinea, by its smaller flowers which are less than 1.2 cm long, and

---

Fig. 43. Dendrobium delicatulum subsp. **delicatulum A-J. A**, habit life size; **B**, flower x 4; **C**, dorsal sepal x 6; **D**, petal x 6; **E**, lateral sepal x 6; **F**, lip x 6; **G**, column x 8; **H**, pollinia x 8; **J**, dorsal and ventral view of anther-cap x 8. **D. masarangense** subsp. **masarangense K-T. K**, habit x 3; **L**, flower x 4; **M**, dorsal view of anther-cap x 8; **N**, lip x 6; **O** ventral view of anther-cap x 8; **P**, pollinia x 8; **Q**, dorsal sepal x 6; **R**, petal x 6; **S**, lateral sepal x 6; **T**, column x 8. **A-J** drawn from *Wickison* 23; **K-T** from *Hunt* 2936. All drawn by Sue Wickison.

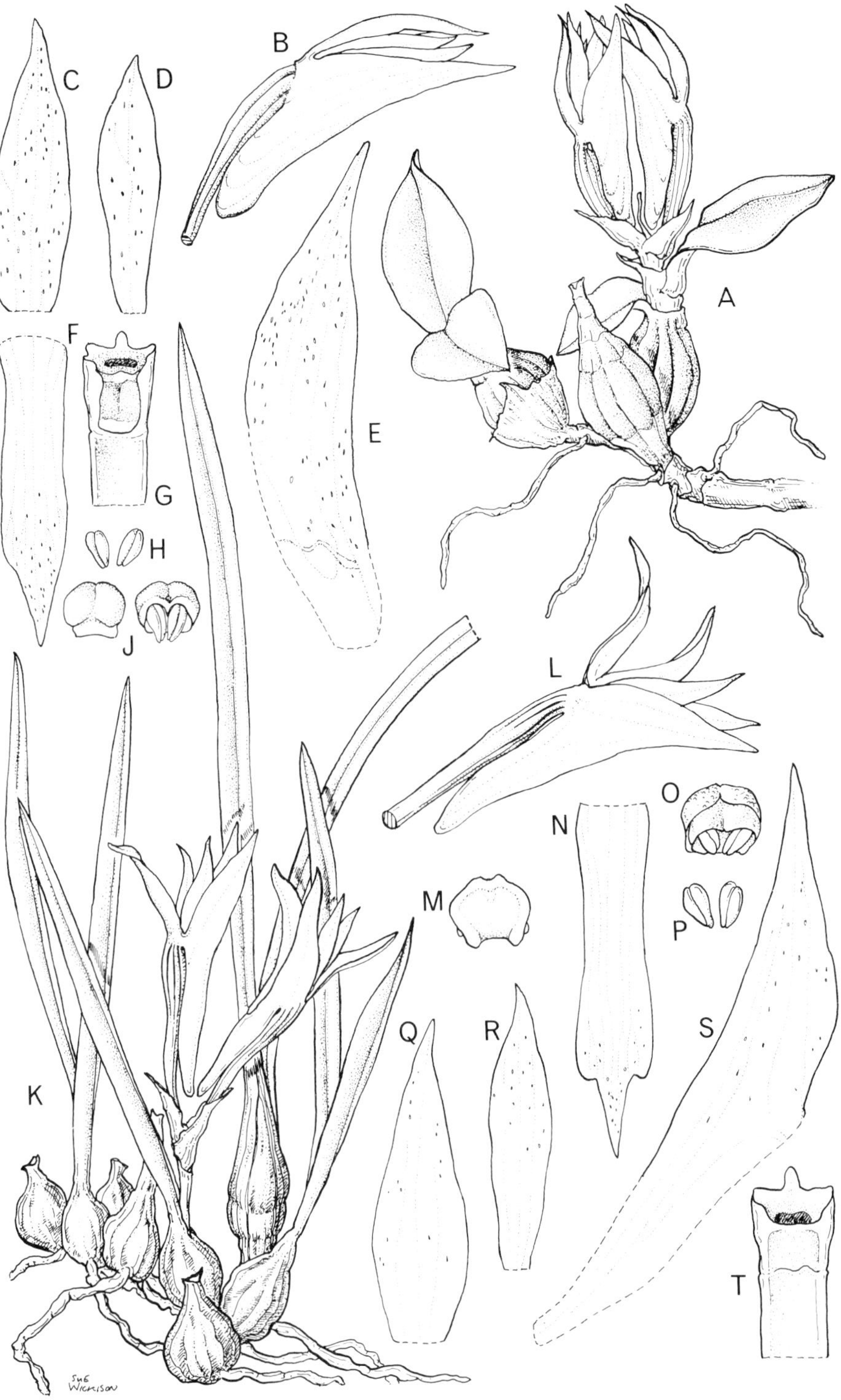
A
B
C
D
E
F
G
H
J
K
L
M
N
O
P
Q
R
S
T

from subsp. *parvulum* (Rolfe) Reeve & Woods, from Sulawesi and western New Guinea, by the colour of the flowers.

29. D. laevifolium *Stapf* in Curtis's Bot. Mag. 149: t. 9011 (1924). Type: New Guinea: Rossell Island, *J.C. Frost* s.n. (holotype K!).

Dendrobium occultum Ames in Journ. Arn. Arb. 14: 108 (1933). Type: Santa Cruz Islands, Vanikoro, *Kajewski* 604 (holotype AMES!).

Pedilonum occultum (Ames) S. Rauschert in Fedde, Repert. Spec. Nov. Regni Veg. 94(7–8) : 462 (1983).

A tufted epiphyte up to 15 cm high. *Pseudobulbs* clustered, ovate-oblong, some elongated up to 4.5 cm long, 1 cm wide, covered in fibres of decayed cataphylls, the apex 2-(3-) leaved. *Leaves* oblong-lanceolate to elliptic, acute, up to 13 cm long, 3 cm wide, bright green. *Inflorescences* terminal or from upper nodes on shorter pseudobulbs, 1– to 3-flowered. *Flowers* purple, pink or white often with a yellow or orange margin to the lip; ovary obscurely triangular, with rounded angles, shallowly 6-grooved; dorsal sepal ovate, up to 2.5 cm long, 1 cm wide; lateral sepals obliquely ovate, up to 3 cm long, 1 cm wide, forming a tubular to narrowly conical mentum up to 1.8 cm long; petals obovate, rotund, up to 1.8 cm long, 1 cm wide; lip naviculate, up to 2.1 cm long, 6 mm wide when flattened, with the margins incurved; column up to 4.5 mm long, column-foot up to 1.2 cm long. (See fig. 44 L–M and plate 11C).

DISTRIBUTION: Bougainville, Guadalcanal, the Santa Cruz Islands (Vanikoro) and Santa Isabel. Also in the Louisades Archipelago and Vanuatu.

HABITAT: Mossy forest, 1300–2340 m.

COLLECTIONS: *Brass* 3271 (AMES); *Brown* 5339 (BM); *Hunt* in RSS 2960 (K, L) & 2981 (K); *Whitmore & Womersley* in BSIP 1031 (LAE); *Wickison* 231 (K).

G. Dennis reports that 'in nature *D. laevifolium* occurs in deep growths of lichen and moss on the trunks or slightly sloping branches of treelets a few feet above the mossy elfin forest bed almost constantly saturated with cold mist or light rain with brief bursts of very hot sunshine' (Warren & Woods, 1990). Also see Wilson (1924) and Womersley (1967) for more information on this species.

30. D. masarangense *Schltr.* in Fedde, Repert. Spec. Nov. Regni Veg. 10: 78 (1911). Types: Celebes, *Schlechter* 20473 (holotype B†), and New Britain, *Ridsdale & Latik* in NGF 38046 (neotype LAE, isoneotypes K!, L).

subspecies **masarangense**

Fig. 44. Dendrobium erosum A-K. A, habit x 2/3; **B**, inflorescence x 2/3; **C**, flower x 2; **D**, dorsal and ventral view of anther-cap x 6; **E**, pollinia x 6; **F**, dorsal sepal x 4; **G**, petal x 4; **H**, lateral sepal x 4; **J**, column x 4; **K**, lip x 4. **D. laevifolium L-M. L**, habit x 2/3; **M**, lip x 3. **A** & **C-K** drawn from *Cribb & Wheatley* 21; **B** from *Mitchell* 6; **L-M** from *Cribb & Wheatley* 108. All drawn by Sue Wickison.

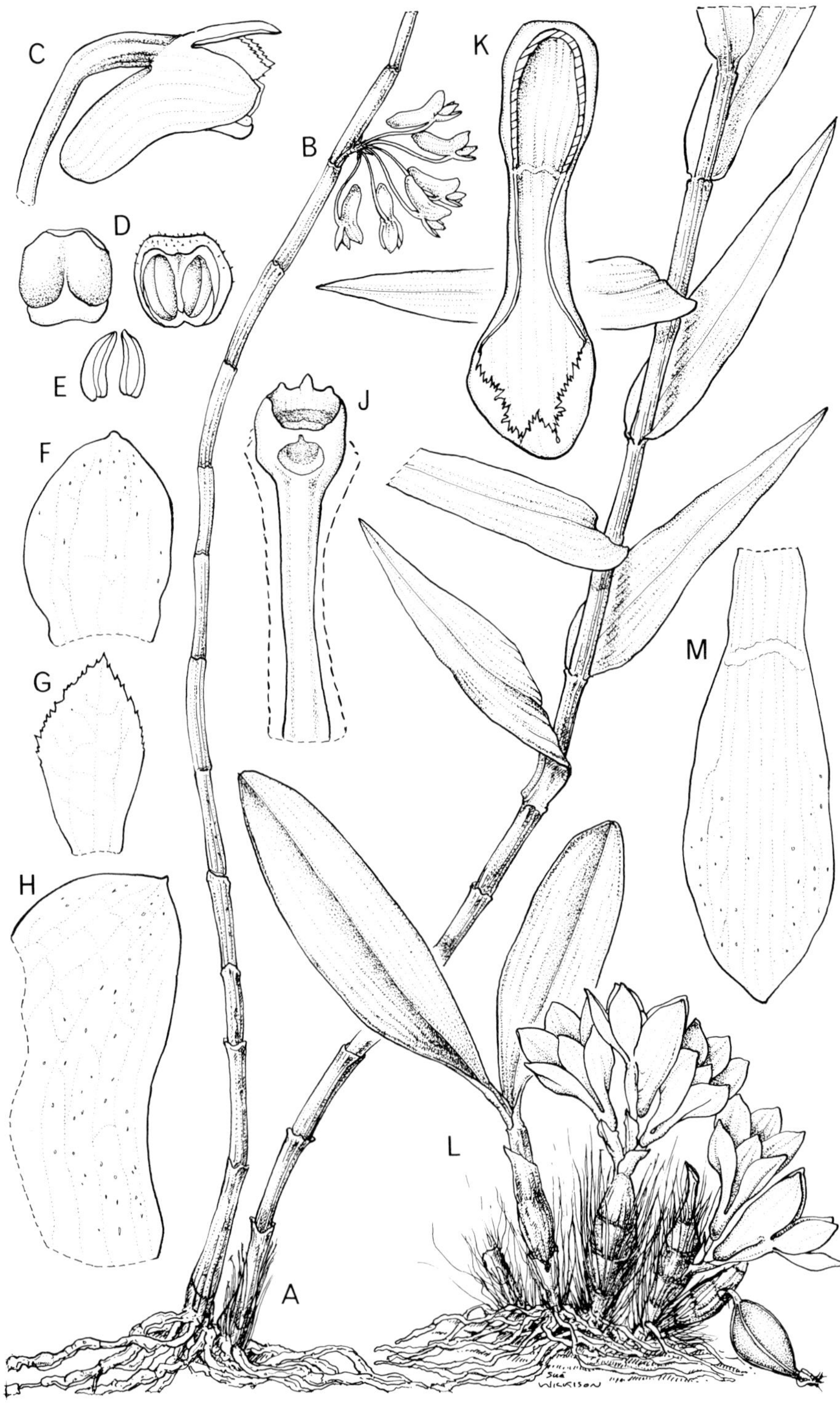
C
B
K
D
E
J
F
G
M
H
L
A

Dendrobium pumilio Schltr. in Fedde, Repert. Spec. Nov. Regni Veg., Beih. 1: 527 (1912). Types: New Guinea, *Schlechter* 16545 (syntype B†, isosyntypes BO!, E, K!, L, NSW) and *Schlechter* 17998 (syntype B†, isosyntypes BO!, E, K!, L, NSW).

An erect, tufted epiphyte up to 7 cm high. *Pseudobulbs* clustered, ovoid, up to 2 cm long, 2.5 mm wide, light green, 2– to 3(-4)-leaved at apex. *Leaves* erect, linear, acute, up to 5 cm long, 2 mm wide, occasionally semi-terete. *Inflorescence* terminal, arising mainly from leafy stems, 1– to 3-flowered. *Flowers* white with a green base and an orange to golden-yellow apex to the lip; ovary indistinctly 5-ribbed; dorsal sepal ovate-lanceolate, up to 7 mm long, 2.5 mm wide; lateral sepals obliquely triangular, up to 1.4 cm long, 3 mm wide, forming a narrowly conical, 5 mm long mentum; petals lanceolate, acuminate, up to 7 mm long, 1.5 mm wide; lip obscurely subtrilobed, linear-oblanceolate, acuminate, up to 11 mm long, 2 mm wide, deflexed at apex; column 2 mm long, with a 7 mm long column-foot. (See fig. 43 K–T).

DISTRIBUTION: Bougainville and Guadalcanal. Also in Sulawesi, New Guinea, New Britain, Vanuatu, Fiji and New Caledonia.
HABITAT: Montane rain forest, 500–1150 m.
COLLECTIONS: *Cribb* S30 (K); *Cribb, Dennis & Campbell* 5043 (K); *Hunt* in RSS 2936 (K); *Lavarack & Ridsdale* in NGF 31374 (K, LAE).

The typical subspecies is distinguished from subsp. *theionanthum* (Schltr.) Reeve & Woods, from New Guinea, mainly by its white flowers usually marked with orange or yellow below the apex of the lip.

31. D. petiolatum *Schltr.* in Fedde, Repert. Spec. Nov. Regni Veg., Beih. 1: 541 (1912). Type: New Guinea, *Schlechter* 18710 (holotype B†, isotypes AMES!, BO!, E, K!, L, NSW).
Dendrobium unifoliatum Schltr. in Fedde, Repert. Spec. Nov. Regni Veg. 17: 375 (1921). Type: New Guinea, *Kempter* s.n. (holotype B†).
Pedilonum petiolatum (Schltr.) S. Rauschert in Fedde, Repert. Spec. Nov. Regni Veg. 94(7–8): 462 (1983).

An erect to suberect tufted epiphyte up to 25 cm high. *Pseudobulbs* obclavate to shortly fusiform, 0.5–4.5 cm long, 0.5–1.6 cm wide, surrounded by fibers from decayed sheaths, unifoliate at apex. *Leaves* erect to spreading, oblong-elliptic, mucronate, 3–16 cm long, 1–3.5 cm wide, distinctly petiolate. *Inflorescences* arising terminally or laterally from leafless pseudobulbs, few- to many-flowered in dense racemes; peduncle distinct, 0.5–4 cm long (usually elongated when plant is embedded deep in moss). *Flowers* bright purple with a orange apex to the lip; ovary 5(-6)-winged; dorsal sepal oblong-lanceolate, acute, 1 cm long, 2–3 mm wide; lateral sepals carinate, obliquely lanceolate, strongly acuminate, 1.5 cm long, 3–4 mm wide, the basal part fused to form a 7 mm long mentum; petals linear-lanceolate, acute, 7 mm long, 2 mm wide; lip subtrilobed, linear, 1 cm long, 2–2.5 mm wide, with a central U- or V-shaped cross-ridge, the apex narrowly triangular, acute; column c. 2 mm long, column-foot 6–10 mm long.

DISTRIBUTION: Guadalcanal. Also in New Guinea and New Britain.
HABITAT: Montane forest, 800 m.
COLLECTION: *Metcalfe* s.n., cult. R.B.G. Sydney (K).

32. D. puniceum *Ridley* in Journ. Bot., Lond. 24: 324 (1886). Types: New Guinea, *Forbes* 425 & 598 (syntypes BM!).
Dendrobium scarlatinum Schltr. in Fedde, Repert. Spec. Nov. Regni Veg., Beih. 1: 535 (1912). Type: New Guinea, *Schlechter* 17966 (holotype B†, isotypes BO, E, K!, L).

For full synonymy see Reeve & Woods (1989).

A tufted epiphyte, up to 8 cm high. *Pseudobulbs* narrowly conical, up to 1 cm long, 3 mm wide, 1- to 2(-3)- leaved at apex. *Leaves* lanceolate, 3–5 cm long, 5 mm wide. *Inflorescences* terminal, arising from both leafy and leafless stems, 1- to 2-flowered. *Flowers* orange-red with an orange lip and a yellow column; ovary 5-winged, sometimes with 5 less distinct ribs alternating with main wings, the dorsal wings projecting distinctly and shortly upward at base of dorsal sepal; dorsal sepal ovate-lanceolate, 5–9 mm long, 2.5–3 mm wide; lateral sepals obliquely triangular, 1.3–2.1 cm long, 3–4.5 mm wide, forming a conical, obtuse, 8–12 mm long mentum; petals linear, 4.5–8.5 mm long, 1–2 mm wide; lip linear to narrowly lanceolate, 1–1.8 cm long, 2–2.5 mm wide, the apex subtrilobed, acute, recurved; column 2–3 mm long, column-foot 8–12 mm long. (See plate 11B).

DISTRIBUTION: Guadalcanal and Kolombangara. Also in New Guinea and New Britain.
HABITAT: Montane forest, 900–1000 m.
COLLECTIONS: *Cribb & Campbell* 5134 (K); *Dennis* in RSS 2104 (K).

33. D. subacaule *Reinw. ex Lindley* in Journ. Linn. Soc. Bot. 3: 11 (1858). Type: Moluccas, *Reinwardt* s.n. (holotype K!, isotype L, W).
Dendrobium tricostatum Schltr. in Fedde, Repert. Spec. Nov. Regni Veg., Beih. 1: 532 (1912). Type: New Guinea, *Schlechter* 20207 (holotype B†).
Dendrobium oreocharis Schltr. l.c.: 533 (1912). Type: New Guinea, *Schlechter* 18753 (holotype B†, isotypes E, K!, NSW).

For full synonymy see Reeve & Woods (1989).

A small erect to suberect tufted epiphyte, 1–5 cm high. *Pseudobulbs* clavate, globose, c. 5 mm long, 4 mm wide, yellow, apex 2- to 3(-4)-leaved. *Leaves* suberect to spreading, lanceolate to elliptic, up to 1.5 cm long, 4 mm wide, mucronate. *Inflorescences* terminal, 1– to 3-flowered. *Flowers* red; ovary sharply 3-winged with dorsal wing often projecting up behind base of dorsal sepal; dorsal sepal oblong-ovate, 5 mm long, 2 mm wide; lateral sepals obliquely ovate-triangular, 1.5 cm long, 2–3 mm wide, basal part fused to form an 8 mm long, cylindrical to conical mentum; petals obliquely linear to lanceolate, 5 mm long, 1–1.5 mm wide; lip subtrilobed, linear-oblanceolate, 1 cm long, 1.5–2 mm wide, adnate to column-foot at base, the upper margins incurved, the apex cuspidate; column 2 mm long, column-foot 8 mm long.

DISTRIBUTION: Guadalcanal. Also in the Moluccas and New Guinea.
HABITAT: Montane forest, 2300 m.
COLLECTION: *Hunt* in RSS 2963 (K).

34. D. violaceominiatum *Schltr.* in Engler, Bot. Jahrb. 56: 470 (1921). Type: Caroline Islands, Ponape, *Ledermann* 13415 (holotype B†), Ponape, *Glassman* 2482 (neotype US).

A small tufted epiphyte 3–5 cm high. *Pseudobulbs* crowded, obliquely fusiform, 0.4–1.8 cm long, 3–7.5 mm wide, apex 1- to 2-leaved. *Leaves* erect to spreading, linear to narrowly elliptic, up to 10 cm long, 3 mm wide, mucronate. *Inflorescence* terminal, 1- to 2-flowered. *Flowers* pale blue with an orange or red lip and a green anther-cap; ovary acutely triangular in cross-section; dorsal sepal lanceolate, 5–9.5 mm long, 2–3 mm wide; lateral sepals obliquely lanceolate, acuminate, 1.2–1.6 cm long, 3–3.5 mm wide, the basal part fused to form a broadly cylindrical 5–7 mm long mentum; petals linear, widest above the middle, acuminate, 6–8 mm long, 1–1.5 mm wide; lip subtrilobed, 7–10.5 mm long, 1.5–2 mm wide, narrowly triangular at the apex; column c. 2.5 mm long, column-foot 5–7 mm long.

DISTRIBUTION: Kolombangara and New Georgia. Also in the Caroline Islands.
HABITAT: Montane forest, 240–600 m.
COLLECTIONS: *Cribb & Campbell* 5108 (K); *Dickson* P1 (K); *Wickison* 8 (K).

Section OXYSTOPHYLLUM

See Seidenfaden (1985) for a discussion of this section.

35. D. tumoriferum *J.J. Smith* in Bull. Dép. Agric. Indes Néerl. 39: 11 (1910). Type: New Guinea, *Von Römer* 252 (holotype BO!).
Aporum tumoriferum (J.J. Smith) S. Rauschert in Fedde, Repert. Spec. Nov. Regni Veg. 94(7–8) : 443 (1983).

An epiphytic herb. *Stems* clustered, pendulous, up to 65 cm long, leafy. *Leaves* alternate, imbricate, bilaterally flattened, linear, acute, up to 3.4 cm long, 5 mm wide, mid-green paler at margins and base, sheathing at base. *Inflorescences* lateral or subterminal, 1-flowered, arising from the axil of the leaf; bracts imbricate. *Flowers* non-resupinate, not opening widely and possibly self-pollinating; dorsal sepal, petals and lip transparent with a tinge of maroon, lateral sepals pale transparent yellow; dorsal sepal ovate, acute, concave, 4 mm long, 2 mm wide; lateral sepals obliquely triangular, acute, 7 mm long, 5 mm wide, forming an obtuse mentum 4 mm long; petals ovate, 3.5 mm long, 1 mm wide, thickened at base; lip entire, oblong, 5 mm long, 2 mm wide, with raised, undulate margins, hard and fleshy at the base and on the mid-lobe; column 1 mm long. *Fruit* triangular in cross-section.

DISTRIBUTION: Bougainville, Guadalcanal and New Georgia. Also in New Guinea.
HABITAT: Damp rain forest, sea level to 600 m.
COLLECTIONS: *Maenu'u* in BSIP 6045 (K); *Waterhouse* 597 (K); *Wickison* 148 (K).

Section PEDILONUM

Wood (1985) suggests that these species are bird pollinated. The habit of producing short, rigid inflorescences on the mature, leafless stems enables pollinators, such as sunbirds, to gain an easy grip while feeding on the nectar contained within the flowers. The pollinia are also flattened and brown and easily adhere to a bird's bill, lying flush with its surface until contact is made with the next flower.

See Seidenfaden (1985) and Warren (1990c) for a discussion of this section.

1. Lip apex fleshy, rounded and concave .... **39. D. concavissimum**
   Lip apex, acute, thin-textured 2
2. Flowers orange .... **36. D. calcaratum**
   Flowers purple with or without white tips to the sepals or white tinged with green 3
3. Flowers many, densely clustered, white tinged with green .... **38. D. capituliflorum**
   Flowers few to many, loosely arranged, purple 4
4. Exterior of sepals covered in brown dots; petals glistening silvery-white in the light; pedicel and ovary scabrid .... **40. D. gnomus**
   Exterior of sepals not covered in brown dots; petals not glistening silvery-white in the light; pedicel and ovary not scabrid .... **37. D. caliculimentum**

36. D. calcaratum *A. Rich.*, Sert. Astrol.: 18, t. 7 (1834). Type: Santa Cruz Islands, Vanikoro, *A. Richard* s.n. (holotype P-lost).

D. triviale Kraenzl. in Engler, Bot. Jahrb. 25:604 (1898). Type: Samoa, *Reinecke* 422 (holotype B†).

D. separatum Ames in Journ. Arn. Arb. 13: 133 (1932). Type: Santa Cruz Islands, Vanikoro, *Kajewski* 503 (holotype AMES!).

Pedilonum separatum (Ames) S. Rauschert in Fedde, Repert. Spec. Nov. Regni Veg. 94(7–8): 463 (1983).

Pedilonum triviale (Kraenzl.) S. Rauschert, l.c. 464.

An epiphytic herb. *Stems* clustered, pendent, ribbed, up to 1.5 m long, slightly swollen at the base, yellow-green turning maroon with age. *Leaves* distichous, on upper part of stem only, oblong-lanceolate, acute, 9–16 cm long, 1.3–1.7 cm wide, unequally bilobed at apex. *Inflorescences* several, lateral, pendent, from nodes near apex of old leafless stems, 10- to 18-flowered, with the basal flowers developing first; peduncle up to 2 cm long, pale mauve tinged with olive-green. *Flowers* fleshy, bright orange becoming pink at the back of the spur, not opening widely; pedicel and ovary c. 1 cm long, green to mauve; dorsal sepal oblong, concave, 5–6 mm long, 3 mm wide; lateral sepals oblong, concave, 1.4–1.5 cm long, 3–4 mm wide; petals elliptic-lanceolate with an erose upper margin, 5–6 mm long, 2–3 mm wide; lip with the saccate basal portion fused to the column-foot, the free portion obovate, 9–10 mm long, 3–4 mm wide, with the lateral margins inrolled, and with an erose margin at the apex, the basal and apical parts of the lip divided by a transverse lamella, the spur cylindrical, obtuse, 8–9 mm long, appressed to the ovary; column 2–3 mm long with an 8 mm long column-foot. (See fig. 45).

DISTRIBUTION: Guadalcanal and the Santa Cruz Islands (Vanikoro). Also in New Britain, Vanuatu and Samoa.
HABITAT: Strand forest and lower montane forest, up to 850 m.
COLLECTIONS: *Dennis* in RSS 2042 (K); *Douglas* K/8 (K); *Wickison* 141 (K).

37. D. caliculimentum R.S. Rogers in Trans. Roy. Soc. Austr. 49: 261 (1925). Type: New Guinea, *C.E. Lane-Poole* 265 (holotype AD!).

An epiphytic herb. *Stems* pendent, slender, long. *Leaves* distichous, lanceolate, acute or mucronate, up to 3.5 cm long, 3.5 mm wide, sheathing at base. *Inflorescences* terminal or subterminal, short, few- to several-flowered. *Flowers* purple with white apices; dorsal sepal ovate, 6.5 mm long, 2.5 mm wide; lateral sepals obliquely ovate, 1.3 cm long, 3 mm wide, forming a conical mentum 7 mm long; petals ovate-linear, 6 mm long, 1.5 mm wide; lip with the saccate basal portion fused to the column-foot, the free part ovate-lanceolate, acute, 8.5 mm long, 3 mm wide, with the lateral margins inrolled, and with an erose margin at the apex, a transverse lamella dividing the basal and apical parts of the lip, the spur cylindrical, obtuse, 7 mm long, appressed to the ovary; column 4 mm long with an 8 mm long column-foot.

DISTRIBUTION: Guadalcanal. Also in New Guinea.
HABITAT: Mossy cloud forest, 1900 m.
COLLECTION: *Hunt* in RSS 2973 (K).

38. D. capituliflorum *Rolfe* in Bull. Misc. Inf., Kew 1901: 146 (1901). Type: New Guinea, *Messrs Sanders & Co.* (holotype K!).
Pedilonum capituliflorum (Rolfe) F.G. Brieger in Schltr., Die Orchideen ed. 3, 1(11–12): 679 (1981).

An epiphytic herb. *Stems* clustered, pendent, ribbed, up to 70 cm long, slightly swollen at the base, yellow-green turning purple with age. *Leaves* alternate, oblong-lanceolate, 7–14 cm long, 1.2–2 cm wide, unequally bilobed at apex, dark green. *Inflorescence* a globose head, lateral, at right-angles to the leafless stem, arising from nodes, many-flowered; raceme very condensed; peduncle 3–5 mm long. *Flowers* white tinged green, base of lip and column green, rarely yellow; pedicel and ovary c. 1 cm long; dorsal sepal oblong-lanceolate, 6.5–10 mm long, 2–3 mm wide; lateral sepals ovate, 1.3–1.5 cm long, 3–5 mm wide; petals lanceolate, with erose upper margins, 5–9 mm long, 1.5–3 mm wide; lip with the saccate basal portion fused with the column-foot, the free portion obovate, acute, 9.5–1.2 cm long, 3–5 mm wide, erose on the apical margins, slightly deflexed at apex, the spur cylindrical, obtuse, appressed to ovary, 5–7 mm long; column 2–3 mm long with a 6–8 mm long column-foot. (See plate 10E).

Fig. 45. Dendrobium calcaratum. **A**, habit x 2/3; **B**, flower x 3; **C**, lip x 4; **D**, lateral sepal x 4; **E**, dorsal sepal x 4; **F**, petal x 4; **G**, column x 6; **H**, anther-cap x 6; **J**, pollinia x 6. All drawn from *Wickison* 141 by Sue Wickison.

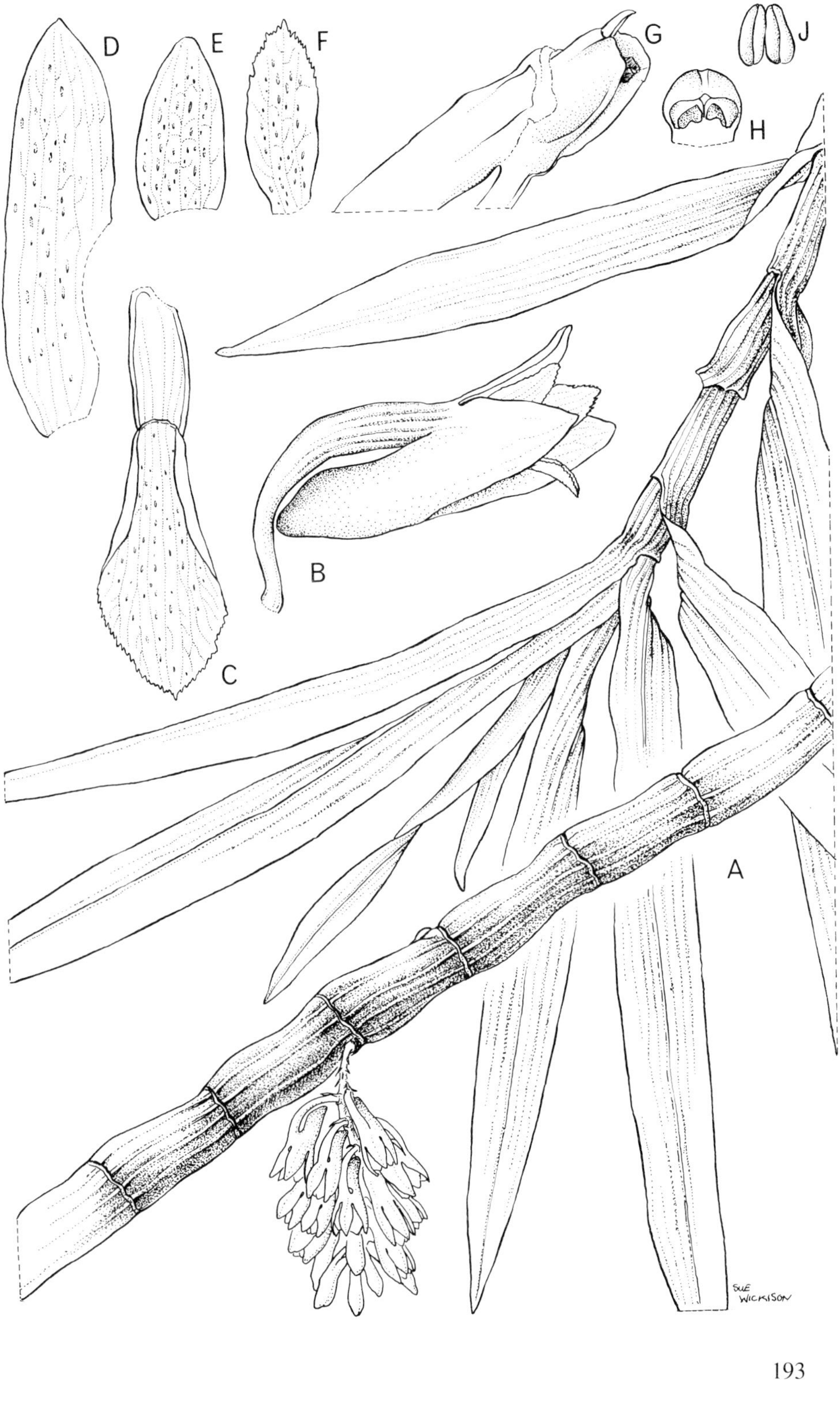
D
E
F
G
J
H
B
C
A
SUE WICKISON

DISTRIBUTION: Bougainville, Choiseul, Guadalcanal, Kolombangara, Makira and Malaita. Also in the Malay Archipelago, the Caroline Islands and New Guinea.

HABITAT: Lowland to montane forest, sea level up to 200 m.

COLLECTIONS: *Craven & Schodde* 511 (L); *Cribb, Dennis & Campbell* 5189 (K); *Cribb & Morrison* 1926 (K); *Dennis* in RSS 7935 (K, LAE, L); *Hunt* in RSS 2184, 2197, 2209 & 2412 (K); *Millar* in NGF 38402 (LAE); *Mitchell* 20 (K); *Peake* 2013 (K, L); *Schodde & Craven* 3927 (K, LAE, L); *Wickison* 21 & 203 (K).

Dendrobium capituliflorum has often been confused with *D. purpureum* Roxb., although the two species are quite distinct (see plate 10E & F).

39. D. concavissimum *J.J. Smith* in Bull. Jard. Bot. Buitenz., ser. 2, 2: 11 (1911). Type: New Guinea, *Koch* 203 (holotype BO!).

Dendrobium fornicatum Schltr. in Fedde, Repert. Spec. Nov. Regni Veg., Beih. 1: 509 (1912). Type: New Guinea, *Schlechter* 19539 (holotype B†, isotypes BO!, L!).

Pedilonum concavissimum (J.J. Smith) S. Rauschert in Fedde, Repert. Spec. Nov. Regni Veg. 94(7–8): 458 (1983).

Pedilonum fornicatum (Schltr.) S. Rauschert l.c.: 459.

An epiphytic herb. *Stems* up to 60 cm long, leafy in apical half. *Leaves* lanceolate, acute, up to 9 cm long, 1.5 cm wide, sheathing at base. *Inflorescences* lateral or subterminal, arising from older, leafless stems, many-flowered in a cluster. *Flowers* not opening widely, purple with an orange lip; dorsal sepal oblong, obtuse, 8 mm long, 4 mm wide; lateral sepals obliquely oblong, obtuse, 1.6 cm long, 5 mm wide, forming a conical, obtuse mentum 5 mm long; petals elliptic, obtuse, concave, 9 mm long, 4 mm wide; lip oblong, obtuse, 1.2 cm long, 4 mm wide, the apical lateral margins incurved, minutely lacerate, the spur obtuse, 6 mm long; column 4 mm long.

DISTRIBUTION: Bougainville. Also in New Guinea.

VERNACULAR NAME: Hita (Bougainville).

HABITAT: Montane and cloud forest, 1350–1850 m.

COLLECTIONS: *Craven & Schodde* 333 (K); *Lavarack & Ridsdale* in NGF 31291 (K); *Waterhouse* 614 (K).

40. D. gnomus *Ames* in Journ. Arn. Arb. 14: 106–7 (1933). Type: Santa Cruz Islands, Vanikoro, *Kajewski* 605 (holotype AMES-lost).

Pedilonum gnomus (Ames) S. Rauschert in Fedde, Repert. Spec. Nov. Regni Veg. 94(7–8) : 460 (1983).

Fig. 46. Dendrobium gnomus. **A**, habit x 2/3; **B**, flower x 4; **C**, dorsal and ventral view of anther-cap x 6; **D**, pollinia x 6; **E**, lip x 6; **F**, column x 6; **G**, petal x 6; **H**, dorsal sepal x 6; **J**, lateral sepal x 6. All drawn from *Mitchell* 68 by Sue Wickison.

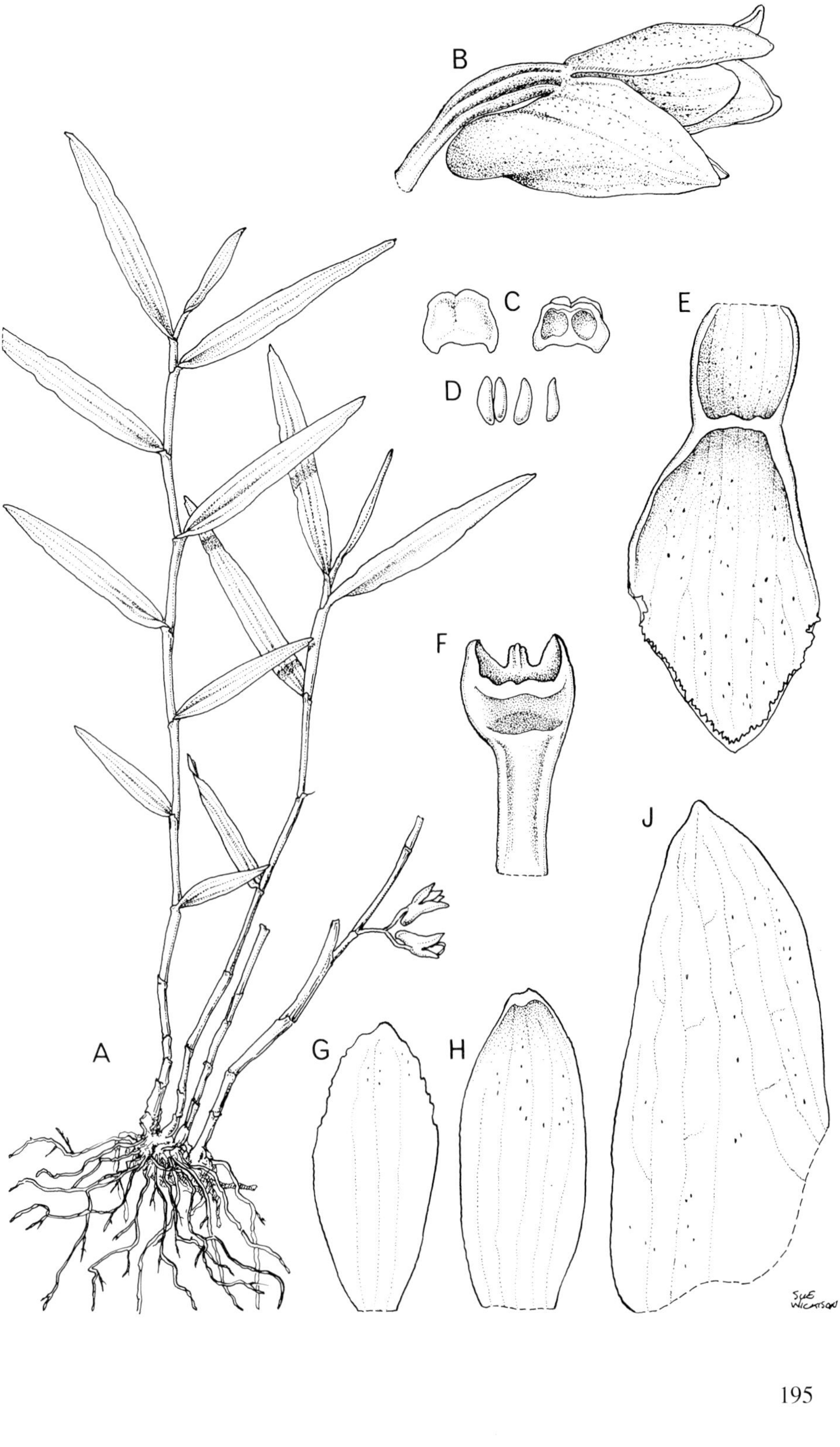
B
C
D
E
F
J
A
G
H
SUE WICKISON

An epiphytic herb. *Stems* clustered, slender, up to 30 cm long, leafy, turning dark purple-brown with age. *Leaves* distichous, lanceolate, 1.8–5 cm long, 5–8 mm wide, unequally bilobed at the apex, mid-green, sheathing at base, sheaths covered in brown spots. *Inflorescences* lateral, arising from the older stems, 2– to 4-flowered. *Flowers* pale mauve with the petals glistening silvery white in the light; pedicel and ovary scabrid, up to 1 cm long; exterior of sepals covered in brown dots; dorsal sepal elliptic-oblong, obtuse, 5–7 mm long, 2–3 mm wide; lateral sepals obliquely ovate, 0.9–1.2 cm long, 4–4.7 mm wide, forming a conical, rounded mentum 4–5 mm long; petals ovate, obtuse, 5–7 mm long, 2–3 mm wide; lip with a saccate basal portion fused to the column-foot, the free part obovate, acute, 0.7–1 cm long, 5 mm wide, with a minutely erose margin, the spur conical, obtuse, 5 mm long; column 2–2.5 mm long, with a 4–5 mm long column-foot. (See fig. 46).

DISTRIBUTION: Guadalcanal, Rannongga and the Santa Cruz Islands (Vanikoro).
HABITAT: Montane forest, 750–1000 m.
COLLECTIONS: *Cribb, Dennis & Campbell* 5031 (K); *Dennis* in RSS 2054 (K); *Douglas* K/14 (K); *Mitchell* 68 (K).

Section PLATYCAULON

41. D. platygastrium *Reichb. f.* in Otia Bot. Hamb. 1: 55 (1878). Type: Fiji, *U.S. Expl. Exped.* s.n. (holotype W!).
Dendrobium platycaulon Rolfe in Bull. Misc. Inf., Kew 1892: 139 (1892). Type: Philippines, *Messrs. Sander & Co.* s.n. (holotype K!).
Pedilonum platygastrium (Reichb. f.) S. Rauschert in Fedde, Repert. Spec. Nov. Regni Veg. 94(7–8): 462 (1983).
Pedilonum platycaulon (Rolfe) S. Rauschert, l.c.

An epiphytic plant with the young growths erect and the mature ones pendent. *Pseudobulbs* clustered, the basal part slender and terete, the upper leaf-bearing part enlarged and strongly bilaterally compressed, 12–30 cm long, 1.5–4 cm wide, acute at apex, dull yellow to olive brown; internodes 2–3 cm long. *Leaves* alternate, from nodes at narrow edge of the upper part of the stem, up to 8, elliptic, acute, 6.5–10 cm long, 2.5–4.5 cm wide, unequally bilobed at apex, dark green, sometimes tinged mauve on underside. *Inflorescences* lateral, 1–4, arising from near the apex of the pseudobulb, up to 2 cm long, with 2 to 7 flowers. *Flowers* usually not opening widely, pale yellow, white, pink or purple, with purple stripes at the base of the lip, the underside of column rose-pink; dorsal sepal ovate, up to 1.8 cm long, 9 mm wide; lateral sepals obliquely triangular, up to 2 cm long, 1.8 cm wide, forming a cylindrical mentum up to 1.3 cm long at 45–90 degrees to the ovary; petals ovate, up to 1.9 cm long, 9 mm wide; lip trilobed, 2.5–3.5 cm long, 1.4–1.7 cm wide, the lateral lobes erect, triangular, acute or truncate, the disc with 5–7 low longitudinal callus ridges, the mid-lobe obtuse or

---

Fig. 47. Dendrobium platygastrium A-G. A, habit x 2/3; **B**, lip x 3; **C**, dorsal and ventral view of anther-cap x 6; **D**, dorsal sepal x 3; **E**, column x 6; **F**, petal x 3; **G**, lateral sepal x 3. **D. pleianthum H-K. H**, habit x 2/3; **J**, lip x 8; **K**, inflorescence (in bud) x 2. **A-G** drawn from *Wickison* 151; **H-K** from *Hunt* in RSS 2947. All drawn by Sue Wickison.

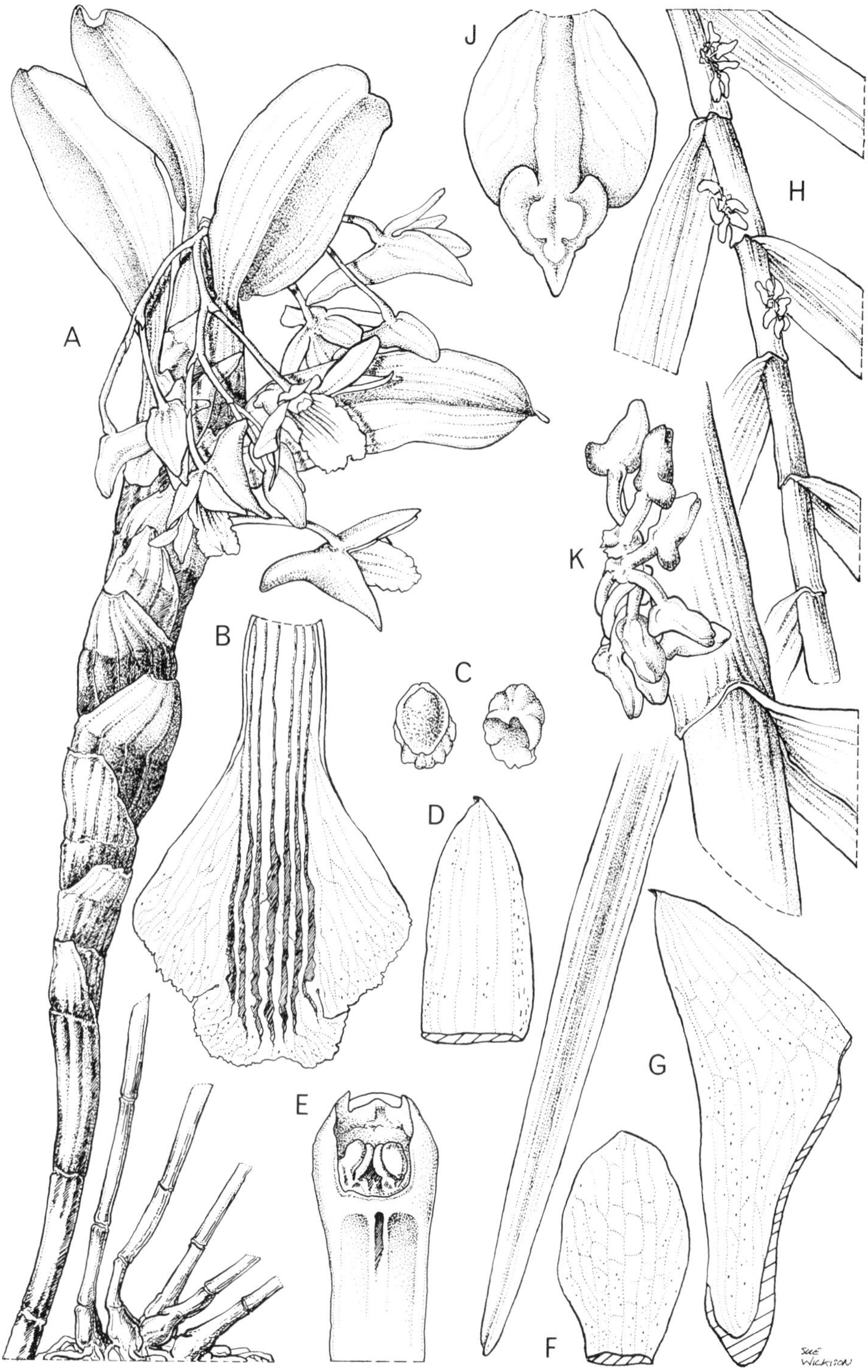

A
B
C
D
E
F
G
H
J
K

acute; column short, 2–4 mm long with a 1.8–2 cm long foot. (See fig. 47 A–G and plate 7F).

DISTRIBUTION: Bougainville, Guadalcanal, Makira and Santa Isabel. Also in the Philippines, Vanuatu, Fiji and Tonga.
HABITAT: Rain forest, 50–600 m.
COLLECTIONS: *Douglas* K/3 (K); *Hunt* in RSS 2062, 2257 & 2925 (K); *Mason* 1986 (K); *Millar* in NGF 38395 & 38427 (LAE); *Wickison* 151 (K).

Dendrobium platygastrium is similar to, and may prove to be conspecific with, *D. lamprocaulon* Schltr. (1905), from New Ireland. However, the type collection of that species has been destroyed. This species is also similar to *D. camptocentrum* Schltr. (1906) from New Caledonia.

Section PLEIANTHE

42. D. pleianthum *Schltr.* in Fedde, Repert. Spec. Nov. Regni Veg., Beih. 1: 625 (1912). Type: New Guinea, *Schlechter* 16303 (holotype B†, isotypes AMES, E, G, K!, L, S, Z).
Grastidium pleianthum (Schltr.) F.G. Brieger in Schltr., Die Orchideen ed. 3, 1(11–12) : 653 (1981).

An epiphytic herb. *Stems* pendulous, fleshy, up to 1.2 m long, clustered. *Leaves* oblong-lanceolate, up to 21 cm long, 1.5 cm wide, unequally bilobed at the apex, sheathing at base. *Inflorescences* lateral, arising from a small conical, laterally compressed protuberance on the side of the leaf sheath, 2 mm high, 6 mm long, 4- to 10-flowered. *Flowers* in rows, pale yellow with the lip faintly margined with red; dorsal sepal oblong-lanceolate, 1.1 cm long, 3.5 mm wide; lateral sepals falcate, keeled, 1.1 cm long, 5 mm wide, forming a 3 mm long obtuse mentum; petals obliquely linear-ligulate, 1 cm long, 2 mm wide; lip erect, trilobed, 6.5 mm long, 4.5 mm wide, the lateral lobes obliquely oblong, rounded at apex, the disc fleshy, the mid-lobe obovate, apiculate, recurved at apex; column short, 3 mm long, column-foot 3 mm long. (See fig. 47 H–K).

DISTRIBUTION: Guadalcanal. Also in New Guinea.
HABITAT: Ridge-top forest, 800 m.
COLLECTION: *Hunt* in RSS 2947 (K).

The collection from the Solomon Islands is in bud and, therefore, the description of the flowers has been taken from New Guinea material.

Section RHOPALANTHE

See Seidenfaden (1985) for a discussion of this section.

1. Leaves dorsiventrally flattened, oblong-ovate; apex of the lip entire .. .. **43. *D. crumenatum**
   Leaves bilaterally flattened, linear-lanceolate; apex of the lip cleft 2
2. Sepals acute at apex; petals linear, obtuse .. .. **44. D. goldfinchii**
   Sepals acuminate at apex; petals ovate, acuminate .. .. .. **45. D. hymenocentrum**

***43. D. crumenatum** *Sw.* in Schrader, Journ. Bot. 2: 237 (1799). Type: Java, collector unknown (holotype S).
Onychium crumenatum (Sw.) Blume, Bijdr. Fl. Ned. Ind.: 326 (1825).
Callista crumenata (Sw.) Kuntze, Rev. Gen. Pl. 2: 654 (1891).

For synonymy see Seidenfaden (1985) and Clements (1989).

An epiphytic herb. *Stems* erect, long, swollen near the base, terete above. *Leaves* distichous, ovate-oblong, thick and leathery, 4–7 cm long, 1–2 cm wide, reddish green, darkening with age, unequally bilobed at the apex, sheathing at base. *Inflorescences* terminal, up to 20 cm long, laxly many-flowered. *Flowers* cream; dorsal sepal oblong-ovate, acute, 1.2 cm long, 4 mm wide; lateral sepals obliquely ovate, 1.3 cm long, 8 mm wide, forming a conical mentum, 1.3 cm long, tapering at the apex; petals ovate-oblong, 1.1 cm long, 4 mm wide; lip trilobed, narrowed at base, 1.7 cm long, 1 cm wide, the lateral lobes erect, oblong, the disc with 5 raised longitudinal callus ridges, the apex oblong.

DISTRIBUTION: Bougainville. Widely distributed from India and China to Malaysia and the Philippines.
HABITAT: Cultivated on a leguminous tree at Arawa, 20 m.
COLLECTION: *Millar* in NGF 38400 (K, LAE).

This species, the well-known 'Pigeon orchid', is not native in Bougainville.

44. D. goldfinchii *F. Muell.* in Wing's South Sci. Rec. 3: 4 (1883). Type: Solomon Islands, *Goldfinch* s.n. (holotype MEL-lost?).
Aporum goldfinchii (F. Muell.) F.G. Brieger in Schltr., Die Orchideen ed. 3, 1(11–12): 673 (1981).

An epiphytic or lithophytic herb. *Stems* clustered, pendent or erect, strongly constricted below, swollen in the middle, tapering to a thin, wiry, leafless apical portion, zig-zag, up to 50 cm long, yellow-green often tinged red. *Leaves* alternate, forming two rows, stiff, thick, bilaterally flattened, linear-lanceolate, 3–7 cm long, 4–8 mm wide, usually tinged red, sheathing at base. *Inflorescences* terminal, up to 25 cm long, laxly many-flowered. *Flowers* pale yellow or white with a pale yellow or white lip with a green or yellow centre, variable in size; dorsal sepal lanceolate, acute, up to 1 cm long, 4 mm wide; lateral sepals obliquely triangular, acute, up to 1.3 cm long, 1. 3 cm wide, forming a conical, incurved mentum up to 1.4 cm long; petals linear, obtuse, up to 8 mm long, 2.5 mm wide; lip recurved, trilobed at apex, narrowed at base, up to 1.8 cm long, 1.2 cm wide, the lateral lobes rounded, the mid-lobe cleft into two roundish segments, the disc with three raised callus ridges becoming papillose at the base of the mid-lobe; column up to 4 mm long. (See fig. 48 and plate 10A).

DISTRIBUTION: Bougainville, Choiseul, the Florida Islands, Guadalcanal, Kolombangara, Makira, Malaita, New Georgia, Rendova, Rennell, Russell Islands and Santa Isabel. Also in New Guinea, Vanuatu, Fiji and Samoa.
VERNACULAR NAME: Tuku (Bougainville).
HABITAT: Strand forest and rain forest, sea level up to 800 m.

COLLECTIONS: *Chaplin* 799 (K); *Comins* 166 & 315 (K); *Cribb* 6025 & 6029 (K); *Cribb & Campbell* 5152 (K); *Dickson* M24, M25 & ECR3 (K); *Fa'arodo et al.* in BSIP 12334 (K); *Guppy* 182 (K); *Hunt* in RSS 2171, 2183, 2361, 2413 & 2643 (K); *Lever* July 1934 (K); *Mauriasi et al.* in BSIP 11841 (K); *Mitchell* 45 & 53 (K); *Noona Dan Expedition* 2879 (K); *Officers of HMS Pengiun* s.n. (K); *Ridsdale & Lavarack* in NGF 31025 (K); *Waterhouse* 320 (K); *Wickison* 10 & 11 (K).

There is a considerable range of variation in size of the plants and flowers of *D. goldfinchii* (see fig.48) and further study may possibly show that more than one taxon is included in this concept.

There is some confusion over the true identify of *D. goldfinchii* because of the loss of the type collection. Its relationship with *D. hymenocentrum* needs further study.

45. D. hymenocentrum *Schltr.* in Fedde, Repert. Spec. Nov. Regni Veg., Beih. 1: 567 (1912). Type: New Guinea, *Schlechter* 20000 (holotype B†, isotypes AMES, BOG, E, G, K!, L, NSW, S).

Aporum hymenocentrum (Schltr.) S. Rauschert in Fedde, Repert. Spec. Nov. Regni Veg. 94(7–8): 439 (1983).

An epiphytic or lithophytic herb. *Stems* clustered, pendent or erect, strongly constricted below, swollen above, tapering to a thin, wiry, leafless apical portion, zig-zag, up to 60 cm long. *Leaves* alternate, forming two rows, stiff, thick, bilaterally flattened, falcate, linear-lanceolate, 4–9 cm long, 4–8 mm wide, mid-green tinged purple, sheathing at base. *Inflorescences* terminal, c. 20 cm long, laxly many-flowered. *Flowers* creamy yellow tinged with red on the mentum; dorsal sepal lanceolate, acuminate, 6 mm long, 2.5 mm wide; lateral sepals obliquely ovate, acuminate, 1 cm long, 6 mm wide, forming a conical, incurved mentum 1.4 cm long; petals ovate, acuminate, 5 mm long, 1.5 mm wide; lip trilobed, narrowed at base, 1.4 cm long, 6 mm wide, the lateral lobes triangular, rounded, the mid-lobe cleft into two roundish segments, the disc with three longitudinal callus ridges; column 3 mm long.

DISTRIBUTION: Bougainville. Also in New Guinea.
HABITAT: Coastal forest, 20 m.
COLLECTION: *Millar & Vandenberg* in NGF 48504 (K).

This species is similar to *D. macfarlanei* F. Muell. from New Guinea and their relationship needs reassessment.

Fig. 48. Dendrobium goldfinchii. **A**, habit x 2/3; **B**, flower x 3; **C**, lip x 6; **D**, side view of lip x 6; **E**, dorsal and ventral view of anther-cap x 8; **F**, column x 6; **G**, dorsal sepal x 6; **H**, petal x 6; **J**, pollinia x 8; **K**, lateral sepal x 6; **L**, flower x 3; **M**, dorsal and ventral view of anther-cap x 6; **N**, pollinia x 6; **O**, column x 4; **P**, lip x 4; **Q**, petal x 4; **R**, side view of lip x 4; **S**, dorsal sepal x 4; **T**, lateral sepal x 4. **A-K** drawn from *Wickison* 11; **L-T** from *Wickison* 10. All drawn by Sue Wickison.

C
D
E
F
G
H
J
K
B
L
M
N
O
A
P
Q
R
S
T
SUE WICKISON

Section SPATULATA

See Cribb (1986) for a revision of this section. See also Warren (1990b) for a discussion of this section.

1. Mid-lobe of lip bearing one or three raised flap-like lamellae at apex of callus 2
   Mid-lobe of lip lacking raised lamellae 5
2. Mid-lobe of lip with a single raised flap-like lamella at apex of central ridge of callus; lateral lobes of lip spreading, rugulose; sepals and petals yellow to brown, recurved and twisted .... **47. D. conanthum**
   Mid-lobe of lip with three raised flap-like lamellae at apex of callus ridges; lateral lobes of lip porrect, not rugulose; sepals and petals yellow, brown, white, purple or mauve 3
3. Lip less than 1 cm wide when flattened; mid-lobe oblong; flower golden or brown without darker petals .... **54. D. sylvanum**
   Lip more than 1.5 cm wide when flattened; mid-lobe elliptic to obovate; flower not as above in colour 4
4. Mid-lobe of lip elliptic, half the length of the lip; flower white with purple veins on the lateral lobes of the lip .... **53. D. sancristobalense**
   Mid-lobe of lip obovate, a third the length of the lip; flower white, yellow or mauve ± with darker petals .... **18. D. gouldii**
5. Lip mid-lobe narrowly triangular, less than 4 mm wide; lateral lobes markedly erose; lip yellow or orange sometimes heavily veined with red .... **50. D. magistratus**
   Lip mid-lobe lanceolate or ovate, more than 5 mm wide; lateral lobes not as above 6
6. Floral bracts sheathing; sepals white; petals green; lip white veined with purple; leaves ovate, acute .... **46. D. antennatum**
   Floral bracts not sheathing; flowers green or white with purple veins on the lateral lobes of the lip; leaves oblong, elliptic to elliptic-lanceolate 7
7. Flowers ivory white veined with purple on lip .... **52. D. rennellii**
   Flowers yellow, yellow-brown or green 8
8. Flowers mustard yellow to brownish-yellow with deep orange veins and a pale yellow lip with purple veins and a pale yellow callus; callus of 5 erose ridges.................... **51. D. mirbelianum**
   Flowers yellow to yellow-green or green with a whitish, greenish or pale yellow lip; callus of 3 ridges .... **49. D. macranthum**

46. D. antennatum *Lindley* in Hook., Journ. Bot., Lond. 2: 236 (1843). Type: New Guinea, *Hinds* s.n. (holotype K!).
Dendrobium d'albertisii Reichb. f. in Gard. Chron. n.s. 9: 366 (1878). Type: New Guinea, *D'Albertis* (holotype W!).
Callista antennata (Lindley) Kuntze, Rev. Gen. Pl. 2: 654 (1891).

An epiphytic herb. *Stems* clustered, erect or pendulous, up to 75 cm tall. *Leaves* distichous, lanceolate, acute, 7–14 cm long, 1.5–4 cm wide, unequally bilobed at apex. *Inflorescences* 1-several, from the upper nodes of the stem, up to 25 cm long, 3- to 15-flowered; bracts tubular, 3–4 mm long. *Flowers* white with green or yellow-green petals with a white lip veined with mauve, scented; dorsal sepal oblong-lanceolate, 1.6–2.3 cm long, 6–7 mm wide, often spirally twisted; lateral sepals obliquely triangular, 1.6–2.5 cm long, 7 mm wide, forming a narrowly conical mentum, 8–10 mm long; petals linear to oblanceolate, acute, 2.5–5 cm long, 2–3.5 mm wide, once- to twice-twisted; lip trilobed, 1.5–2.3 cm long, 0.9–1.5 cm wide, the lateral lobes elliptic, rounded in front, the mid-lobe circular to ovate, acute, the callus of 5 longitudinal ridges, slightly dilated towards apex at the base of the mid-lobe; column 5–6 mm long. Flower turning green and persisting as fruit develops.

DISTRIBUTION: Guadalcanal, Malaita and Rennell. Also in New Guinea and Australia.
HABITAT: Coastal and lowland forest, sea level to 40 m.
COLLECTIONS: *Cribb* 6007 (K); *Hunt* in RSS 3079 (K); *Mauriasi et al.* in BSIP 13638 (K, LAE); *Wickison* 240 & 241 (K).

The Solomon Islands' collections of *D. antennatum* differ from those collected in northwestern New Guinea in having shorter and less twisted petals and are apparently self-pollinating. However, the continuous nature of this variation precludes the recognition of infraspecific taxa (Cribb, 1986).

47. D. conanthum *Schltr.* in Fedde, Repert. Spec. Nov. Regni Veg., Beih. 1: 550 (1912). Type: New Guinea, *Schlechter* 19996 (holotype B†).
Dendrobium kajewskii Ames in Journ. Arn. Arb. 13: 131 (1932). Type: Santa Cruz Islands, Vanikoro, *Kajewski* 638a (holotype AMES!).

An epiphytic herb. *Stems* clustered, 1–3 m tall. *Leaves* distichous, coriaceous, oblong, elliptic or ovate-elliptic, 4–12 cm long, 1.8–5.5 cm wide, unequally bilobed at obtuse or rounded apex, dark green; leaf bases sheathing. *Inflorescences* subterminal, spreading or suberect, 30–65 cm long, rather densely 6- to 20- (or more) flowered; peduncle 10–14 cm long. *Flowers* appearing contorted, yellow-green to yellow, more or less flushed with red-brown with a yellow lip veined with purple or brown; sepals and petals with strongly undulate margins; dorsal sepal reflexed and sometimes spirally twisted, narrowly oblong-lanceolate, obtuse or subacute, 2–2.3 cm long, 7 mm wide; lateral sepals reflexed, obliquely oblong-lanceolate, acute, 2.3–2.4 cm long, 1 cm wide, forming a conical, obtuse, 1 cm long mentum; petals linear-spathulate, acute or obtuse, 2.2–3 cm long, 7 mm wide, spirally twisted; lip trilobed, 2.1–2.35 cm long, 1.3–1.75 cm wide, the lateral lobes narrowly elliptic, rounded or obtuse and spreading at apex, with verrucose venation and erose margins, the mid-lobe recurved at apex, ovate, acute, with erose margins, the callus of 3 ridges, the middle one prominently raised at the apex on base of the mid-lobe; column 4–6 mm long. (See fig. 49 and plate 8F).

DISTRIBUTION: Bougainville, Guadalcanal, Makira, New Georgia, Olasana (reef island east of Gizo), Rendova, Santa Ana and the Santa Cruz Islands (Vanikoro). Also in New Guinea and Vanuatu.
HABITAT: Coastal and lowland forest.
COLLECTIONS: *Brass* 3205 (K); *Dennis* in RSS 7750 (K); *Dickson* M53 (K); *Kajewski* 638 (K) & 638a (AMES); *Wickison* 69 (K); *Woodward* s.n. (K); *Womersley* in NGF 24865 (K, LAE).

Where *D. macranthum* grows with *D. conanthum* it may hybridize with it; G. Dennis (pers. comm.) also notes that *D. conanthum* may hybridize with *D. sancristobalense*.

48. D. gouldii *Reichb. f.* in Gard. Chron. 1867: 901 (1867). Type: Solomon Islands, *Veitch* s.n. (holotype W!).

Dendrobium gouldii var. *acutum* Reichb. f., Xen. Orch. 2: 167, t.169 (1870). Type: Solomon Islands, *Veitch* s.n. (holotype W!).

Dendrobium undulatum var. *woodfordianum* Maid. in Proc. Linn. Soc. N.S.W. 24: 652 (1899). Type: Solomon Islands, *Woodford* s.n. (holotype NSW!).

Dendrobium woodfordianum (Maid.) Schltr. in Fedde, Repert. Spec. Nov. Regni Veg. 1: 545 (1912).

Dendrobium imthurnii Rolfe in Curtis's Bot. Mag. 138: t. 8452 (1912). Type: Vanuatu, *Im Thurn* s.n. (holotype K!).

An epiphytic herb. *Stems* clustered, 0.9–1.8 m tall, often slightly dilated in the middle. *Leaves* distichous, coriaceous, oblong-elliptic or elliptic, 12–17.5 cm long, 5–6.3 cm wide, obtuse or rounded at unequally bilobed apex; leaf bases sheathing. *Inflorescences* subterminal, 30–70 cm long, 7-to 40-flowered. *Flowers* very variable in colour and size, the sepals white, pale yellow or yellow, the petals pale yellow, brown or violet, the lip white to yellow veined with purple or violet; dorsal sepal recurved, linear-lanceolate, acute, 2–2.6 cm long, 4–5 mm wide; lateral sepals falcate, lanceolate, acute or acuminate, 2.2–2.6 cm long, 5–6 mm wide, forming a conical 8–10 mm long mentum deflexed slightly at the apex; petals linear-spathulate, obtuse, 2.7–4 cm long, 4–5 mm wide, one- to three-times twisted; lip trilobed, 2.2–2.4 cm long, 1.3–1.5 cm wide, the lateral lobes obliquely oblong, rounded in front, erose on margins, the mid-lobe subspathulate, obtuse, erose on margins, the callus of 5 ridges, the central 3 prominent and raised at the apex on the centre of the mid-lobe; column 6–8 mm long. (See plates 1B, 8A, B & C).

DISTRIBUTION: Bougainville, Florida Group (Small Nggela Island), Guadalcanal, Krihi Island (off San Jorge), Malaita, New Georgia, Rendova, Santa Isabel. Also in New Ireland and Vanuatu.

VERNACULAR NAME: Ubutu (New Georgia).

HABITAT: Strand forest and rain forest, sea level to 700 m.

COLLECTIONS: *Comins* 325 (K); *Cribb & Campbell* 5167 (K); *Crocker* 1 (K); *Dickson* ECR1 (K); *Fa'arodo et al.* in BSIP 13724 (K); *Gafui et al.* in BSIP 14769 (K, LAE); *Guppy* 267 (K); *Hunt* in RSS 2645, 2734 & 2779 (K); *Im Thurn* 249 (K); *Lever* s.n. (K); *Milne* 543 (K); *Officers of HMS Penguin* s.n. (K); *Ridsdale & Lavarack* in NGF 30557 & 31121 (K), 30605 (LAE); *Shiress* s.n. (SYD); *Veitch* 2/67 (K); *Waterhouse* 98 (K); *Womersley* in NGF 24866 (K, LAE).

Fig. 49. Dendrobium conanthum. A, habit reduced; **B**, leaves x 2/3; **C**, inflorescence x 2/3; **D**, flower x 5/3; **E**, column and lip x 2; **F**, pollinia x 4; **G**, dorsal and ventral view of anther-cap x 4; **H**, column x 4; **J**, lip x 2; **K**, dorsal sepal x 2; **L**, petal x 2; **M**, lateral sepal x 2. Drawn from *Wickison* 69 by Sue Wickison.

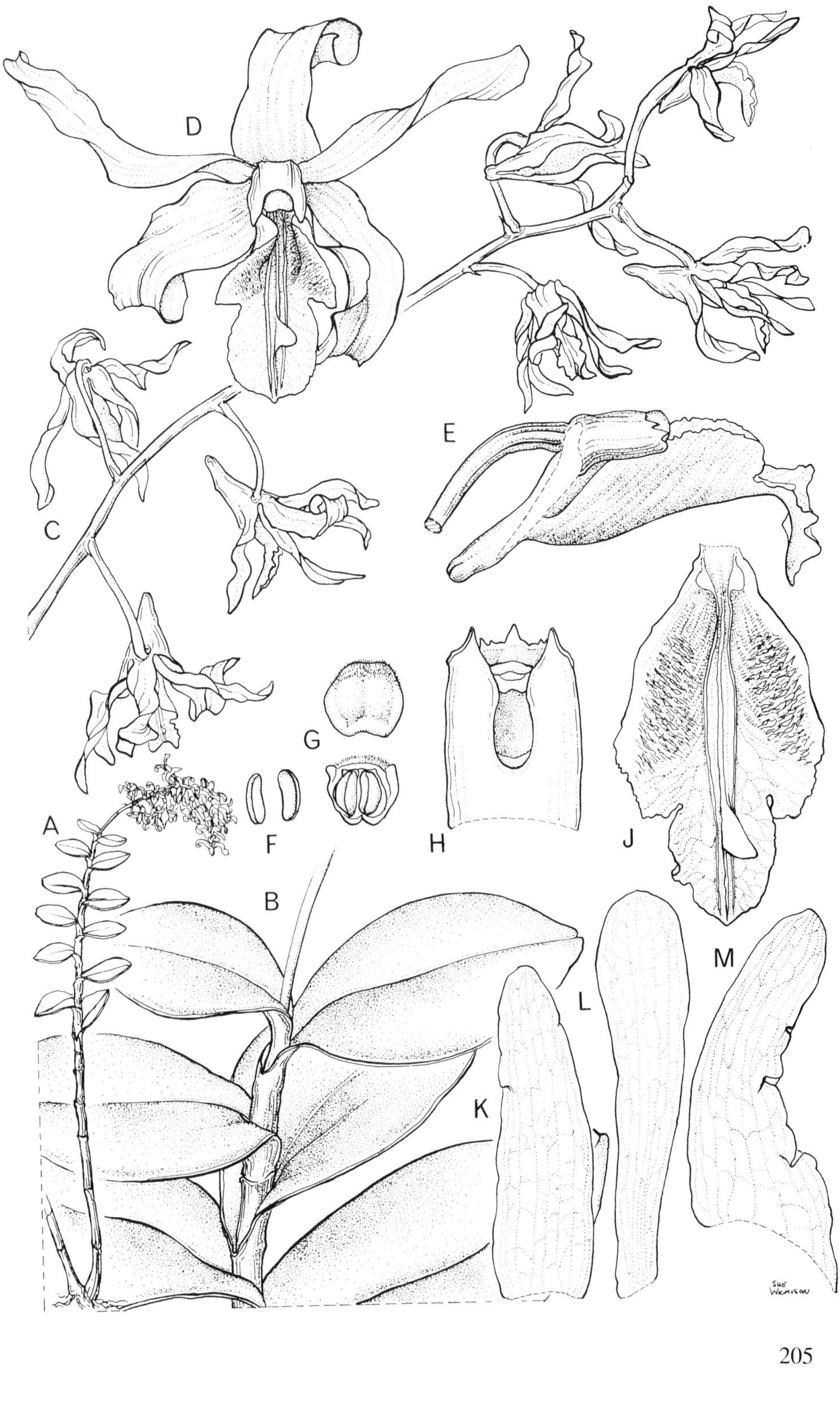
D
E
C
G
A
F
H
J
B
M
L
K
SUE WICKISON

Dendrobium gouldii is a variable species with a number of distinct geographical variants. White-flowered forms with a purple- or violet-veined lip are found in New Ireland and Bougainville and the easternmost Solomon Islands from Choiseul to Malaita (see plate 8C). On Kolombangara, New Georgia and adjacent islands in the western Solomons *D. gouldii* is represented by a cream-flowered form with a violet-veined lip. The two most striking variants are found on Guadalcanal and the neighbouring Florida Islands. Plants with golden yellow flowers, deep maroon petals and a maroon-veined lip are found on Guadalcanal (see plate 8B). These have been called 'Guadalcanal Gold' by growers. In the Florida Islands plants have white sepals, bluish violet petals and a violet-veined lip (see plate 8A). The variation patterns may be somewhat confused in places because this species has been taken from island to island and planted on trees around villages.

49. D. macranthum *A. Rich.*, Sert. Astrol.: 15, t.6 (1832). Type: Santa Cruz Islands, Vanikoro, *A. Richard* 39 (holotype P!).

Dendrobium arachnostachyum Reichb. f. in Gard. Chron. n.s. 7: 334 (1877). Type: 'New Guinea', but more likely from a Pacific Island, *P. Veitch* s.n. (holotype W!).

Dendrobium pseudotokai Kraenzl. in Bull. Soc. Bot. Fr. 76: 300 (1929). Type: Vanuatu, Efate, *Levat* s.n. (holotype P!).

Dendrobium tokai var. *crassinerve* Finet in Bull. Soc. Bot. Fr. 50: 381 (1903). Types: New Caledonia, *Thibault* s.n., *Germain* s.n. & *Balansa* 2387 (syntypes P!).

An epiphytic herb. *Stems* clustered, up to 60 cm tall, slightly swollen in lower half and at base. *Leaves* distichous, coriaceous, oblong to elliptic, 7–12 cm long, 2.5–4.5 cm wide, unequally bilobed at apex; leaf bases sheathing. *Inflorescences* subterminal, erect or suberect, 22–32 cm long, laxly 8- to 20-flowered. *Flowers* yellow, yellow-green or lime green with a whitish, greenish or pale yellow lip, with lilac callus ridges and purple side veins; dorsal sepal erect, linear-lanceolate, acute or acuminate, 2.4–3.8 cm long, 3–4 mm wide; lateral sepals recurved, falcate, obliquely linear-lanceolate, acuminate, 3–3.6 cm long, 7 mm wide, forming a narrowly conical, slightly upcurved or decurved, 1–1.2 cm long mentum; petals linear-subspathulate, acute, 3–4.5 cm long, 3 mm wide, not or once-twisted; lip trilobed in middle, porrect, 2.5–3.2 cm long, 1.2–1.3 cm wide, the lateral lobes obliquely oblong, truncate in front, erose on margins, the mid-lobe linear-lanceolate, acuminate, the callus of 3 ridges, all slightly raised at the apex on the basal half of the mid-lobe; column 6 mm long. (See plate 9B).

DISTRIBUTION: The Santa Cruz Islands. Also in Vanuatu and New Caledonia.

HABITAT: Coastal and lowland rain forest.

COLLECTION: *Mauriasi et al.* in BSIP 16687 (K).

50. D. magistratus *Cribb* in Orchadian 6(12): 274 (1981). Type: New Guinea, *Reeve* 393 (holotype K!, isotype LAE!).

A medium-sized to large epiphytic herb. *Stems* cane-like, clustered, tapering somewhat towards the apex, dilated somewhat at base, up to 60 cm long, 1 cm wide, covered by tubular sheaths and leaf bases, dark purple. *Leaves* distichous, coriaceous, flat, elliptic or oblong-elliptic, slightly bilobed at the rounded apex, up to 10 cm long, 3.5 cm wide. *Inflorescences* several, spreading to arcuate, lateral and emerging through the base of the leaf sheaths or subterminal, up to 14 cm long, densely few- to many-flowered; peduncle terete, much longer than the rhachis. *Flowers* subnutant, creamy-white with a yellow lip ± veined with red; dorsal sepal lanceolate, acute, 2.4 cm long, 5 mm wide; lateral sepals lanceolate, acute, 2.7 cm long, 7 mm wide, forming a conical, backward pointing, 6 mm long mentum; petals linear, acute, 3.2 cm long, 2 mm wide, 1 to 3 times twisted; lip trilobed in apical half, 2.4 cm long, 1.3 cm wide, the lateral lobes rounded, erect, verrucose on lateral veins and with erose margins, the mid-lobe narrowly triangular, acuminate, the callus of 5 longitudinal low somewhat erose lamellae; column 3.5 mm long, with a 6 mm long foot. (See plate 9D).

DISTRIBUTION: New Georgia. Also in New Guinea.
HABITAT: Lowland rain forest.
COLLECTION: *Wickison* 65 (K).

The Solomon Islands' collection differs from the type in lacking the red venation on the lip.

51. D. mirbelianum *Gaud.* in Freyc., Voy. Bot.: 423 (1826). Type: Moluccas, *Gaudichaud Dupré* s.n. (holotype P!).

For full synonymy see Cribb (1986).

An erect epiphytic herb. *Stems* 0.5–3.2 m tall, cane-like, swollen somewhat in basal half and at base. *Leaves* distichous, coriaceous, oblong-elliptic to elliptic-ovate, obtuse or rounded at obliquely bilobed apex, 8–12.5 cm long, 2.5–5.4 cm wide. *Inflorescences* lateral, suberect to spreading, 12–45 cm long, many-flowered. *Flowers* mustard-yellow or brownish-yellow with deep orange veins and a pale yellow lip with purple veins and a pale yellow callus; dorsal sepal erect, linear-lanceolate, 2.5 cm long, 6 mm wide; lateral sepals obliquely lanceolate, 2.5 cm long, 1.1 cm wide, forming an obliquely conical, 7–8 mm long mentum; petals spathulate, acute to subacute, 3.2 cm long, 7 mm wide, not twisted or half twisted; lip trilobed, 2.5 cm long, 1.8 cm wide, the lateral lobes elliptic, rounded in front, the mid-lobe obovate, acute, c. half length of lip, margins erose and undulate, the callus of 5 erose ridges, the central ridge longest and slightly dilated towards the apex on the middle of the mid-lobe; column 4 mm long. (See plate 9A).

DISTIBUTION: Bougainville, New Georgia and Rendova. Also in the Moluccas, New Guinea and Australia.
HABITAT: Coconut plantations, sea level.
COLLECTIONS: *Cribb* 1908 (K); *Dennis* (pers. comm.); *Dickson* ECR 2 & M23 (K).

52. D. rennellii *Cribb* in Kew Bull. 37(4): 583 (1983). Type: Solomon Islands, Rennell Island, *Noona Dan Expedition* 2864 (holotype K!).

An epiphytic or lithophytic herb. *Stems* cane-like, up to 70 cm or more long. *Leaves* distichous, coriaceous, narrowly elliptic or elliptic-lanceolate, unequally roundly bilobed at apex, up to 11 cm long, 3 cm wide. *Inflorescences* subterminal, arcuate to suberect, 5- to 15-flowered, up to 30 cm long. *Flowers* attractive, scented, white to creamy-white, veined with bright purple on the lip; dorsal sepal somewhat recurved, lanceolate, subacute to obtuse, 2.8–3 cm long, 0.7–1 cm wide, forming an obliquely conical mentum 7–9 mm long; petals spathulate, obtuse, not or half twisted, 3.1–3.7 cm long, 5–6 mm wide; lip trilobed, porrect, 2.5–2.8 cm long, 1.4 cm wide, the lateral lobes erect, oblong-elliptic, rounded in front, the mid-lobe oblong-elliptic, acute or apiculate, 1–1.2 cm long, the callus of three longitudinal ridges neither noticeably raised nor dilated towards apex; column 5 mm long. (See plate 8D).

DISTRIBUTION: Rennell and Bellona only.
HABITAT: Coastal forest and on raised coral islands in lake, sea level to 100 m.
COLLECTIONS: *Cribb* S38 (K); *Davis* 28882 (K); *Noona Dan Expedition* 2864 (K); *Wolff* in BSIP 4434 (K, LAE).

Dendrobium rennellii is endemic to the isolated raised coral islands of Rennell and Bellona. Hybrids of *D. rennellii* and *D. sylvanum* have been collected from the shores of Lake Tenggano, Rennell (*Dennis* 16).

53. D. sancristobalense *Cribb* in Kew Bull. 37(4): 585 (1983). Type: Solomon Islands, Makira, *Dennis* in RSS 2174 (holotype K!).

An epiphytic herb. *Stems* cane-like, up to c. 1 m tall. *Leaves* distichous, coriaceous, elliptic to oblong-elliptic, roundly or obtusely unequally bilobed at apex, up to 14 cm long, 7 cm wide. *Inflorescences* subterminal, erect or suberect, up to 36 cm long, laxly 2- to many-flowered. *Flowers* cream or white with purple or violet veins on the lip; dorsal sepal oblong-ovate, subacute, 2–2.8 cm long, 4–7 mm wide; lateral sepals obliquely narrowly oblong, obtuse, 2.5–3 cm long, 7–9 mm wide, with a backward pointing mentum, narrowly conical but narrowly cylindrical towards apex, 1–1.3 cm long; lip trilobed in middle, porrect, 2.3–2.5 cm long, 1.4–1.5 cm wide, the lateral lobes obliquely oblong, rounded in front, the mid-lobe elliptic to obovate, rounded or obtuse, the callus of 3 ridges raised at apex on basal half of the mid-lobe; column 4–6 mm long.

DISTRIBUTION: Makira.
HABITAT: Coastal forest and rain forest, up to 850 m.
COLLECTIONS: *Comins* 326 (K); *Hunt* in RSS 2174 & 2321 (K); *Stewart* s.n. (K).

Dendrobium sancristobalense is closely related to the variable *D. gouldii* but it differs in its longer more prominent mentum and lip in which the elliptic or obovate mid-lobe is half the total length and bears three lamellae that are semi-elliptic rather than obliquely triangular and acute.

54. D. sylvanum *Reichb. f.* in Linnaea 41: 91 (1877). Type: New Caledonia, *Veillard* 1304 (holotype W!).
Dendrobium warianum Schltr. in Fedde, Repert. Spec. Nov. Regni Veg., Beih. 1: 548 (1912). Type: New Guinea, *Schlechter* 19951 (holotype B†).

For full synonymy see Cribb (1986).

An epiphytic herb. *Stems* cane-like, 50–120 cm tall, slightly dilated in middle. *Leaves* distichous, coriaceous, oblong to narrowly oblong-ovate or elliptic, obtuse or rounded at the unequally bilobed apex, 8–16 cm long, 2.5–5.5 cm wide. *Inflorescences* apparently terminal, 20–55 cm long, densely up to 40-flowered. *Flowers* greenish yellow to yellow-ochre with purple venation on the lip and a white callus margined with violet; dorsal sepal narrowly oblong, obtuse, apiculate, 1.8–2.2 cm long, 5–6 mm wide, recurved; lateral sepals obliquely oblong, obtuse, 2.1–2.7 cm long, 7–9 mm wide, forming a conical mentum, 6–7 mm long; petals spathulate, rounded or emarginate, 1.9–2.6 cm long, 5–6 mm wide, not or half-twisted; lip narrow, trilobed, 1.8–2.4 cm long, 8–11 mm wide, the lateral lobes narrowly oblong, rounded and erose in front, the mid-lobe recurved, oblong, obtuse with undulate margins, the callus of three undulate ridges raised at apex into oblong flaps on the middle of the mid-lobe, the mid-ridge being slightly longer; column 4 mm long. (See plate 8E).

DISTRIBUTION: Rennell, Makira and Savo Island (north of Guadalcanal). Also in New Guinea, New Ireland and New Caledonia.
HABITAT: Strand forest, sea level to 70 m.
COLLECTIONS: *Comins* 327 (K); *Dennis* in RSS 2267 (K); *Hunt* in RSS 2924 (K); *Wickison* 52 (K).

Section TRACHYRHIZUM

55. D. prostheciglossum *Schltr.* in Fedde, Repert. Spec. Nov. Regni Veg., Beih. 1: 555 (1912). Type: New Guinea, *Schlechter* 18174 (holotype B†, isotype K!).
Trachyrhizum prostheciglossum (Schltr.) F.G. Brieger in Schltr., Die Orchideen ed. 3, 1(11–12): 687 (1981).

An epiphytic herb. *Stems* clustered, erect, ribbed, 20–90 cm tall. *Leaves* distichous, ovate, 4–7 cm long, 1.5–2.2 cm wide, unequally bilobed at apex, the leaf bases sheathing (in dried specimens they may appear darker giving the stem a banded appearance). *Inflorescences* lateral, from base of leaf sheath, 4- to 9-flowered; peduncle up to 10 cm long. *Flowers* yellow-green tinged on the outside with purple or brown with a white lip; dorsal sepal broadly ovate, 5 mm long, 5 mm wide; lateral sepals obliquely triangular, 7 mm long, 1.2 cm wide, forming a conical, obtuse, 9 mm long mentum; petals linear-elliptic, 7 mm long, 3 mm wide; lip trilobed, 1.1 cm long, 9 mm wide, the lateral lobes erect, large, falcate, the disc with a 3-ridged central callus extending to the base of the 2-tailed mid-lobe; column 2 mm long, column-foot 6 mm long, upturned at apex.

DISTRIBUTION: Choiseul. Also in New Guinea, New Britain and Vanuatu.
HABITAT: Rain forest, 150 m.
COLLECTION: *Mitchell* 56 (K).

56. DIPLOCAULOBIUM (Reichenbach f.) Kraenzlin

Epiphytic herbs. *Pseudobulbs* clustered, of one internode. *Leaves* solitary, terminal, subcoriaceous, erect, linear to elliptic, lacking a petiole. *Inflorescences* 1 to several, one-flowered, usually borne in succession. *Flowers* lasting less than a day, borne on pedicels from within a large bract at the apex of the pseudobulb; sepals and petals slender, often filiform for the most part but the bases of the lateral sepals rather broad and forming with the column-foot a distinct mentum; lip attached to the apex only of the column-foot, trilobed with rather small lateral lobes or almost entire, the mid-lobe usually rather long, the callus of longitudinal keels which may be undulate; column with a distinct but not very long foot; pollinia 4.

A genus of about one hundred species centred on New Guinea and extending to Malaysia in the west and Australia and the Pacific Islands in the east, seven species being recorded.

1. Leaves elliptic, 1–1.5 cm long .... **2. D. cyclobulbon**
   Leaves oblong-lanceolate, more than 3.5 cm long 2
2. Leaves more than 2 cm wide 3
   Leaves less than 2 cm wide 4
3. Flowers white to lemon yellow at the base becoming pink or pale yellow to white at the tips, with a pink lip with a pale green callus; lip 7.5–9 mm long, 4–4.5 mm wide .... **6. D. solomonense**
   Flowers cream, sepals and petals turning dark yellow at the apices, becoming pink-purple as the flowers go over; lip white with dark purple markings; lip 18 mm long, 5 mm wide .... **3. D. nitidissimum**
4. Flowers dull yellow with a white lip and wine-red lateral lobes .... **1. D. bicolor**
   Flowers pale green, cream or lemon yellow; lip not as above 5
5. Lip mid-lobe spathulate, with an oblong claw and undulate callus ridges on the claw .... **6. D. mekynosepalum**
   Lip mid-lobe without a claw and lacking undulate callus ridges 6
6. Plant up to 22 cm tall; flowers with lemon yellow sepals and petals becoming darker yellow above, the lip yellow with purple spots down the centre, the flowers maturing pink-salmon; lip trilobed, 22 mm long, 11 mm wide; lateral sepals more than 30 mm long .... **5. D. magnilabre**
   Plant small, up to 15 cm tall; flowers light green at base becoming pale yellow to off-white at the tips of the sepals and petals, with a yellow lip with purple markings on the lateral lobes; lip entire to obscurely trilobed, 2–7(9.5) mm long, 1–3(4) mm wide; lateral sepals 13–25 mm long .... **7. D. subintegrum**

1. D. bicolor *Cribb & B. Lewis* in Lindleyana 6(1):28 (1991). Type: Solomon Islands, Malaita, *P.F. Hunt* 3082 (holotype K!).

Fig. 50. Diplocaulobium bicolor. A, habit x 1; **B**, flower x 1; **C**, column x 6; **D**, lip x 4; **E**, dorsal and ventral view of anther-cap x 6; **F**, pollinia x 6; **G**, dorsal sepal x 3; **H**, petal x 3; **J**, lateral sepal x 3. Drawn from *Wickison 2* by Sue Wickison.

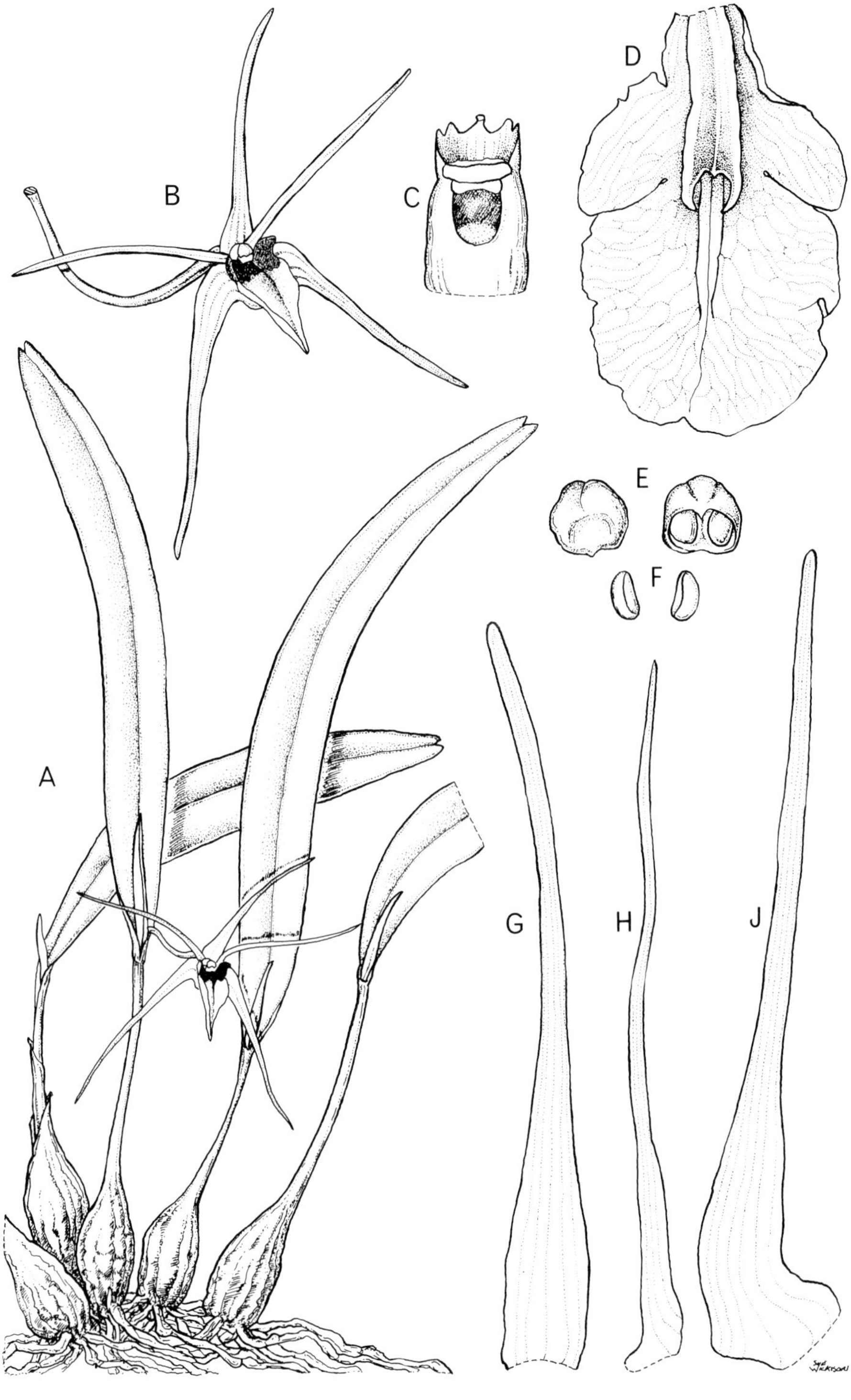

A
B
C
D
E
F
G
H
J

An epiphytic herb up to 29 cm tall. *Pseudobulbs* yellow-green, swollen at the base, up to 15 cm tall, 8 mm wide at the base, 1.5–2 mm wide above, unifoliate. *Leaf* erect, oblong, 10–15 cm long, 1–2 cm wide, minutely bilobed at apex. *Inflorescence* terminal, 1-flowered; bract lanceolate, 1.8–3.2 cm long. *Flower* with dirty yellow sepals and petals and a white lip with wine-red lateral lobes; pedicel and ovary 4.5–5 cm long; sepals and petals apparently filiform with apical margins inrolled; dorsal sepal linear, lanceolate, acuminate, 2.2–3.5 cm long, 1–2 mm wide; lateral sepals obliquely triangular at base, acuminate above, 2.2–3.5 cm long, 6 mm wide at base, forming an obtuse mentum with the column-foot; petals linear, 2–2.8 cm long, 1 mm wide; lip trilobed, 1–1.4 cm long, 8–12 mm wide, the lateral lobes ovate, the mid-lobe oblong, obtuse, the callus of 2 raised ridges extending to the mid-lobe; column 4 mm long with a maroon-red blotch at the base and an 8 mm long column-foot; anther-cap maroon. (See fig. 50 and plate 12C).

DISTRIBUTION: Choiseul, Guadalcanal, New Georgia and Malaita. Also in New Guinea.
HABITAT: Coastal and lowland mountain forest, sea level to 200 m.
COLLECTIONS: *Balodis* 3 (K); *Cribb, Campbell & Morrison* 1870 (K); *Cribb, Dennis & Campbell* 5181 (K); *Douglas* s.n. (K); *Hunt* in RSS 3082 (K); *Mitchell* 59 (K); *Robinson* K171 (K); *Wickison* 2, 4 & 60A (K); *Womersley & Whitmore* in BSIP 1135 (K).

This species is easily recognised by the lip shape, callus structure and the wine-red colour of the lateral lobes of the lip. *Diplocaulobium bicolor* is perhaps closest to *D. iboense* (Schltr.) A.D. Hawkes, from New Guinea, but this species is distinguished by having a clawed mid-lobe and light brown lateral lobes of the lip.

2. D. cyclobulbon *(Schltr.) A.D. Hawkes* in Lloydia 20: 128 (1957). Type: New Guinea, *Schlechter* 18650 (holotype B†).
Dendrobium cyclobulbon Schltr. in Fedde, Repert. Spec. Nov. Regni Veg., Beih. 1: 461 (1912).

A small creeping herb up to 4 cm tall. *Pseudobulbs* closely overlapping, quadrangular, 5–9 mm tall, 2.5–8 mm wide, slightly ridged, especially when old, unifoliate at apex. *Leaf* elliptic, minutely emarginate, 1–1.5 cm long, 6–7 mm wide. *Inflorescence* terminal, 1-flowered; bract lanceolate, 5–7 mm long. *Flower* pale yellow to white with faint purple on the margins of the lateral lobes of the lip and a bright lemon yellow mid-lobe and anther-cap; pedicel and ovary 10 mm long; dorsal sepal lanceolate, acuminate, 6.5–8 mm long, 1.5–3 mm wide; lateral sepals obliquely triangular, acute, 1.1–1.2 cm long, 5 mm wide at base, forming an obtuse mentum with the column-foot; petals linear, acute, 5.5–9 mm long, 0.5–1 mm wide; lip trilobed, 8–11.5 mm long, 2.5–3 mm wide, the lateral lobes small,

---

Fig. 51. Diplocaulobium cyclobulbon A-H. A, habit x 2/3; **B**, flower x 4; **C**, dorsal sepal x 6; **D**, petal x 6; **E**, lateral sepal x 6; **F**, column x 8; **G**, dorsal and ventral view of anther-cap x 8; **H**, lip x 6. **D. solomonense J-K. J**, lateral sepal x 2; **K**, lip x 4. **D. mekynosepalum L-N. L**, lip x 4; **M**, lip x 4; **N**, lateral sepal x 2. **D. nitidissimum O-P. O**, lip x 4; **P**, lateral sepal x 2. **A-H** drawn from *Mitchell 48*; **J-P** drawn from illustrations by M. Dickson. All drawn by Sue Wickison.

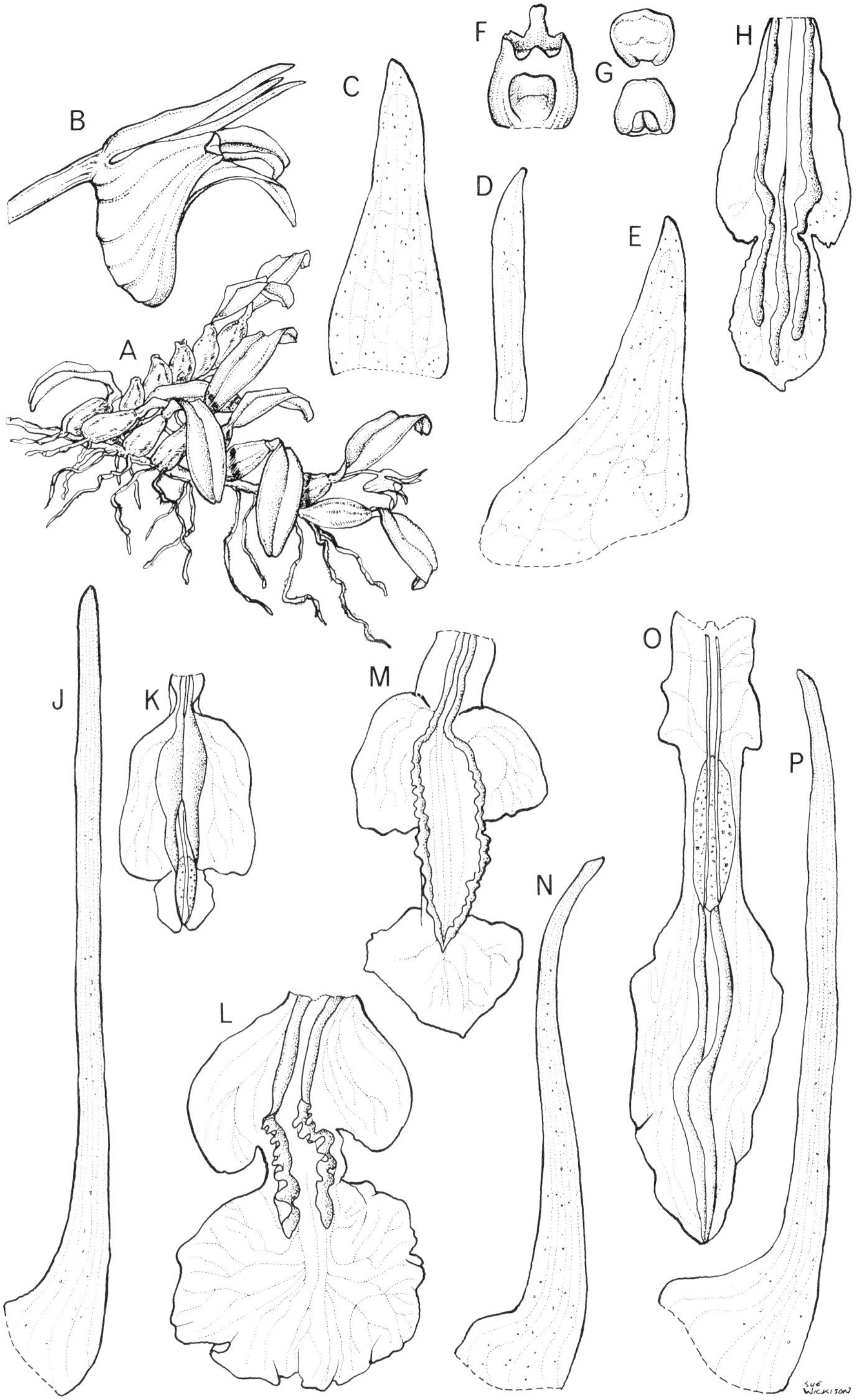
A
B
C
D
E
F
G
H
J
K
L
M
N
O
P
SUE WICKISON

rounded, the mid-lobe oblong-ovate, obtuse, the callus with 3 undulate ridges which extend almost to the end of the mid-lobe with only the 2 outer ones raised at the base; column 2 mm long, with a 5 mm long foot. (See fig. 51 A–H).

DISTRIBUTION: Choiseul and Guadalcanal. Also in New Guinea.
HABITAT: Lowland forest to lower montane forest, 70–550 m.
COLLECTIONS: *Campbell* s.n. (K); *Cribb & Morrison* 1891 (K); *Mitchell* 48 (K); *Robinson* K/6 (K).

3. D. magnilabre *Cribb & B. Lewis* in Lindleyana 6(1):31 (1991). Type: Solomon Islands, New Georgia, *S. Wickison* 34 (holotype K!).

An epiphytic herb up to 22 cm tall. *Pseudobulbs* swollen near the base, 4–15 cm long, 7–8 mm wide at the base, 1 mm wide above, light yellow-green, unifoliate at apex. *Leaf* erect, linear, minutely and unequally bilobed at the apex, 6–10 cm long, 7–9 mm wide, with a well defined midrib. *Inflorescence* terminal, 1-flowered; bract linear-lanceolate, 1.4 cm long. *Flower* lemon-yellow becoming darker yellow at tips, with a yellow lip with purple spots down the centre and a purple-marked column, the flowers maturing pink-salmon; pedicel and ovary up to 8 cm long; sepals and petals apparently filiform with margins inrolled; dorsal sepal linear-lanceolate, acuminate, 29–32 mm long, 1 mm wide; lateral sepals obliquely triangular at base, acuminate above, 31–33 mm long, 11–15 mm wide at base, forming an obtuse mentum with the column-foot; petals linear, 26–32 mm long, 1–2 mm wide; lip trilobed in basal half, 20–22 mm long, 8–11 mm wide, with small, crescent-shaped lateral lobes, the mid-lobe ovate, acuminate, the disc with 2 raised basal callus ridges extending half of the way down the lip; column 4 mm long, with apical teeth, column-foot 6 mm long. (See fig. 52 and plate 12A).

DISTRIBUTION: Bougainville, Choiseul, New Georgia and Rendova.
HABITAT: Primary montane rain forest, 370–900 m.
COLLECTIONS: *Mason* 1935 (K); *Mitchell* 29 (K); *Shelford* 1 (K);*Wickison* 34, 39 & 248 (K).

This species is readily distinguished by its large and distinctively coloured flowers and the lip which has a broadly ovate mid-lobe and bicarinate basal callus. It is perhaps closest to *D. tentaculatum* (Schltr.) Kraenzl., from New Guinea, but that differs in having a narrower mid-lobe and a 3-ridged callus on the lip.

4. D. mekynosepalum *(Schltr.) Kraenzl.* in Engler, Planzenr. Orch.- Mon.- Dendr.: 343 (1910). Type: New Guinea, *Schlechter* 14061 (holotype B†).
Dendrobium mekynosepalum Schltr. in K. Schum. & Laut., Nachtr. Fl. Deutsch. Südsee: 154 (1905).

Fig. 52. Diplocaulobium magnilabre. **A**, habit x 2/3; **B**, flower x 1; **C**, dorsal and ventral view of anther-cap x 6; **D**, lip x 3; **E**, pollinia x 6; **F**, column x 6; **G**, dorsal sepal x 3; **H**, petal x 3; **J**, lateral sepal x 3. Drawn from *Wickison 34* by Sue Wickison.

B
C
D
E
F
A
G
H
J
Sue Wickison

An epiphytic herb up to 25 cm tall. *Pseudobulb* up to 8 cm long, swollen at the base, 5–8 mm wide, 1–1.5 mm wide above, green-yellow, unifoliate. *Leaf* erect, oblong-lanceolate, 4.5–10 cm long, 0.6–1 cm wide, deep green. *Inflorescence* terminal, 1-flowered; bract lanceolate, 1.8–2 cm long. *Flower* cream to pale green with the sepals tipped yellow to reddish-orange, and with a white to cream lip with a yellow callus and brown to red marks on the lateral lobes; pedicel and ovary 3.4–4.6 cm long; sepals and petals apparently filiform; dorsal sepal linear-lanceolate, acuminate, 2.7–3.2 cm long, 2–4 mm wide; lateral sepals obliquely triangular at base, acuminate above, 2.7–3.2 cm long, 4–6 mm wide, forming an obtuse mentum with the column-foot; petals linear, 2.5–2.7 cm long, 1–2 mm wide; lip trilobed, 6–12 mm long, 3–6 mm wide, the lateral lobes semicircular, clawed, the mid-lobe reniform, orbicular or triangular, the callus with 2 raised undulate ridges which extend onto the beginning of the mid-lobe; column 4 mm long with a 6 mm long foot. (See fig. 51 L–N and plate 12B).

DISTRIBUTION: Guadalcanal and New Georgia. Also in New Guinea.
VERNACULAR NAME: Fi'i Adi (Kwara'ae name).
HABITAT: Lowland forest, sea level to 500 m.
COLLECTIONS: *Chaplin* 814B (K); *Cribb & Campbell* 5100 (K); *Hunt* in RSS 2932 (K); *Wickison* 9 & 53 (K).

Schlechter (1912) published illustrations of two rather distinct forms of *D. mekynosepalum* with rather different lip shapes. Both of these are found in the Solomon Islands and Bougainville and further study may show that this variation requires taxonomic recognition.

Hancock and Henderson (1988) record the same Kwara'ae name for *Cadetia hispida*, *Diplocaulobium meckynosepalum* and *Diplocaulobium solomonense*.

5. D. nitidissimum *(Reichb. f.) Kraenzl.* in Engler, Pflanzenr. Orch.-Mon.-Dendr.: 337 (1910). Type: Admiralty Islands, *Moseley* s.n., March 1875 (holotype W, isotype K!).
Dendrobium nitidissimum Reichb. f. in Journ. Linn. Soc. Bot. 15: 112 (1876).
Dendrobium mettkeanum Kraenzl. in Gard. Chron. ser. 3, 16: 306 (1894). Type: ex Messrs Sanders & Co., *Micholitz* s.n. (holotype B†).

A large robust epiphyte growing in clumps up to 55 cm tall. *Pseudobulb* up to 34 cm long, swollen at the base 0.9–1.2 cm wide, 2–4 mm wide above, dark olive-green, covered basally in sheathing bracts, unifoliate. *Leaf* erect, oblong, up to 20 cm long, 3.2 cm wide, dark green. *Inflorescence* terminal, 1-flowered. *Flower* cream-coloured becoming dark yellow at the tips of the sepals and petals with a white lip with dark purple markings, the flower becoming pink-purple as it goes over; pedicel and ovary pale green, 3 cm long; sepals and petals apparently filiform with margins inrolled; dorsal sepal linear-lanceolate, acuminate, 4–4.6 cm long, 2.5–3 mm wide; lateral sepals obliquely triangular at base, acuminate above, 4–4.5 cm long, 2.5–3 mm wide, forming an obtuse mentum with the column-foot; petals linear, 4–4.2 cm long, 2.5–3 mm wide; lip obscurely trilobed in basal half, 1.8 cm long, 5 mm wide, the lateral lobes small, semi-circular, the mid-lobe obscurely clawed becoming narrowly ovate above, the callus with 2 central ridges extending for the length of the lip and an oblong callus on the claw; column 4 mm long, with an upcurved foot 8 mm long. (See fig. 51 O–P).

DISTRIBUTION: Bougainville. Also in New Guinea, the Admiralty Islands and New Ireland.
HABITAT: Plantation areas, sea level.
COLLECTION: *Millar* in NGF 38386 (K).

The description is taken from the Bougainville specimen cited above. It agrees well in its floral morphology with *D. nitidissimum*, from the Admiralty Islands, but it differs in being more robust and in having a broader leaf. The Bougainville collection is also similar to *D. ischnopetalum* (Schltr.) Kraenzl., but differs in being more robust and in having a papillose callus on the claw of the mid-lobe. More material is needed of this species before the identificiation can be confirmed.

6. D. solomonense *Carr* in Bull. Misc. Inf., Kew 1934: 380 (1934). Type: Bougainville, *J.H.L. Waterhouse* 339 (holotype K!).

A large robust epiphyte up to 80 cm tall growing in clumps. *Pseudobulbs* up to 60 cm tall, 1–2 cm wide at the swollen base, 2–3 mm wide above, covered by bracts at the base, bright yellow, unifoliate. *Leaf* erect, linear-oblong, up to 27 cm long, 3.1 cm wide, with a minutely bilobed apex, bright green. *Inflorescence* terminal, 1-flowered; bract lanceolate, 2.5–3.2 cm long, straw-coloured. *Flower* lemon-yellow at the base becoming pink or pale yellow to white at the tips, with a pink lip and a pale green callus on the mid-lobe, scented; pedicel and ovary 3–3.5 cm long; sepals and petals apparently filiform; dorsal sepal linear-lanceolate, acuminate, 3.3–5.1 mm long, 2 mm wide; lateral sepals obliquely triangular at base, acuminate above, 2.9–5.1 cm long, 2 mm wide, forming an obtuse mentum with the column-foot; petals linear, 3.1–4.8 mm long, 1.5 mm wide; lip trilobed in basal half, 7.5–9 mm long, 4–4.5 mm wide, the lateral lobes oblong-semicircular, the mid-lobe oblong, the callus with 2 fleshy ridges, extending up to the mid-lobe with an elliptic callus on the mid-lobe; column 3.5 mm long, with a 7 mm long column-foot. (See fig. 51 J–K).

DISTRIBUTION: Bougainville, Choiseul, Guadalcanal, Kolombangara, Malaita and New Georgia.
VERNACULAR NAMES: Nah (Bougainville), Fi'i Adi (Kwara'ae name).
USES: The Solomon Islanders use the bright yellow stems of *D. solomonense* to make plaited armlets and to weave baskets.
HABITAT: Montane forest, 1600–1730 m.
COLLECTIONS: *Craven & Schodde* 313 (K); *Cribb & Campbell* 5118 (K); *Gafui et al.* in BSIP 16279 (K); *Hunt* in RSS 2045 & 3002 (K); *Lavarack & Ridsdale* in NGF 31322 (K); *Waterhouse* 339 (K); *Whitmore* in BSIP 2078 (K); *Wickison* 29 (K).

Hancock and Henderson (1988) record the same Kwara'ae name for *Cadetia hispida*, *Diplocaulobium meckynosepalum* and *Diplocaulobium solomonense*.

7. D. subintegrum *Cribb & B. Lewis* in Lindleyana 6(1):33 (1991). Type: Solomon Islands, New Georgia, *S. Wickison* 30 (holotype K!).

An epiphytic herb up to 30 cm tall. *Pseudobulbs* clustered, 1.5–9.5 cm long, swollen to 5–7 mm wide at base, 0.5–0.8 mm wide above, light greenish yellow with a purple tinge, often sheathed basally by brown bracts, unifoliate. *Leaf* erect, linear, minutely bilobed at apex, 3.5–8 cm long, 2.5–8 mm wide, glossy green with a well defined midrib. *Inflorescence* terminal; bract linear-lanceolate, 5–8 mm long. *Flower* light green-yellow at base becoming pale yellow to off-white at the tips of the sepals and petals, with a hint of purple, or pale green-yellow throughout, with a yellow lip with purple markings on the lateral lobes and a purple anther-cap; pedicel and ovary 1.4–4 cm long; sepals and petals apparently filiform with margins inrolled; dorsal sepal linear, lanceolate, acuminate, (0.8-)1.3–2.6 cm long, 1–2 mm wide; lateral sepals obliquely triangular at base, acuminate above, (0.8-)1.3–2.5 cm long, 2.2–3 mm wide at base, forming an obtuse mentum with the column-foot; petals linear, (0.7-)1.3–2.4 cm long, 1–1.2 mm wide; lip entire to obscurely trilobed, 2–7(-9) mm long, 1–3(- 4.5) mm wide, oblong-ovate with an acute or truncated apex, with undulate margins, the callus of 2 broad fleshy ridges which extend half way along the lip; column 1–2 mm long with a 2–6 mm long, gently upcurved foot. (See fig. 53 and plate 12D).

DISTRIBUTION: Bougainville, Guadalcanal, Kolombangara, New Georgia, Ranongga, Rendova, and the Santa Cruz Islands.
HABITAT: Primary rain forest and exposed mountain ridges, sea level to 1600 m.
COLLECTIONS: *Chaplin* 804, 809, 810, 811, 812 & 826 (K); *Cribb* 5037 (K); *Cribb & Campbell* 5136 (K); *Cribb, Dennis & Campbell* 5188 (K); *Lavarack & Ridsdale* in NGF 31185 (K); *Metcalfe* s.n. (K); *Mitchell* 71 (K); *Smith & Newmark* s.n. (K); *Wickison* 30 & 47 (K).

This species is closest to *D. pentanema* (Schltr.) Kraenzl., from New Ireland, but differs in being a smaller plant with flowers which have shorter sepals and a different callus of 2 parallel ridges.

57. FLICKINGERIA A.D. Hawkes

Epiphytic herbs with a creeping, rooting rhizome from which erect or pendent branching stems arise. *Stems* of a few to about eight internodes with sheathing scale leaves, the uppermost internode swollen to form a pseudobulb with a solitary leaf at the apex, branching usually from the node below the pseudobulb. *Leaf* coriaceous, suberect. *Inflorescences* terminal or subterminal, several, each one-flowered. *Flowers* ephemeral, lasting less than a day; dorsal sepal and petals free; lateral sepals united at base to column-foot to form a chin-like mentum; lip trilobed, the mid-lobe clawed, the callus of 2 or 3 ridges; pollinia 4.

A genus of about sixty-five species distributed from India and China through SE. Asia to the Pacific Islands and Australia, two species being recorded. The species are arranged in 2 sections according to Seidenfaden (1980).

Fig. 53. Diplocaulobium subintegrum. **A**, habit x 2/3; **B**, flower x 5/3; **C**, ventral and dorsal view of anther-cap x 8; **D**, lip x 8; **E**, pollinia x 8; **F**, column x 8; **G**, dorsal sepal x 4; **H** petal x 4; **J**, lateral sepal x 4. Drawn from *Wickison 30* by Sue Wickison.

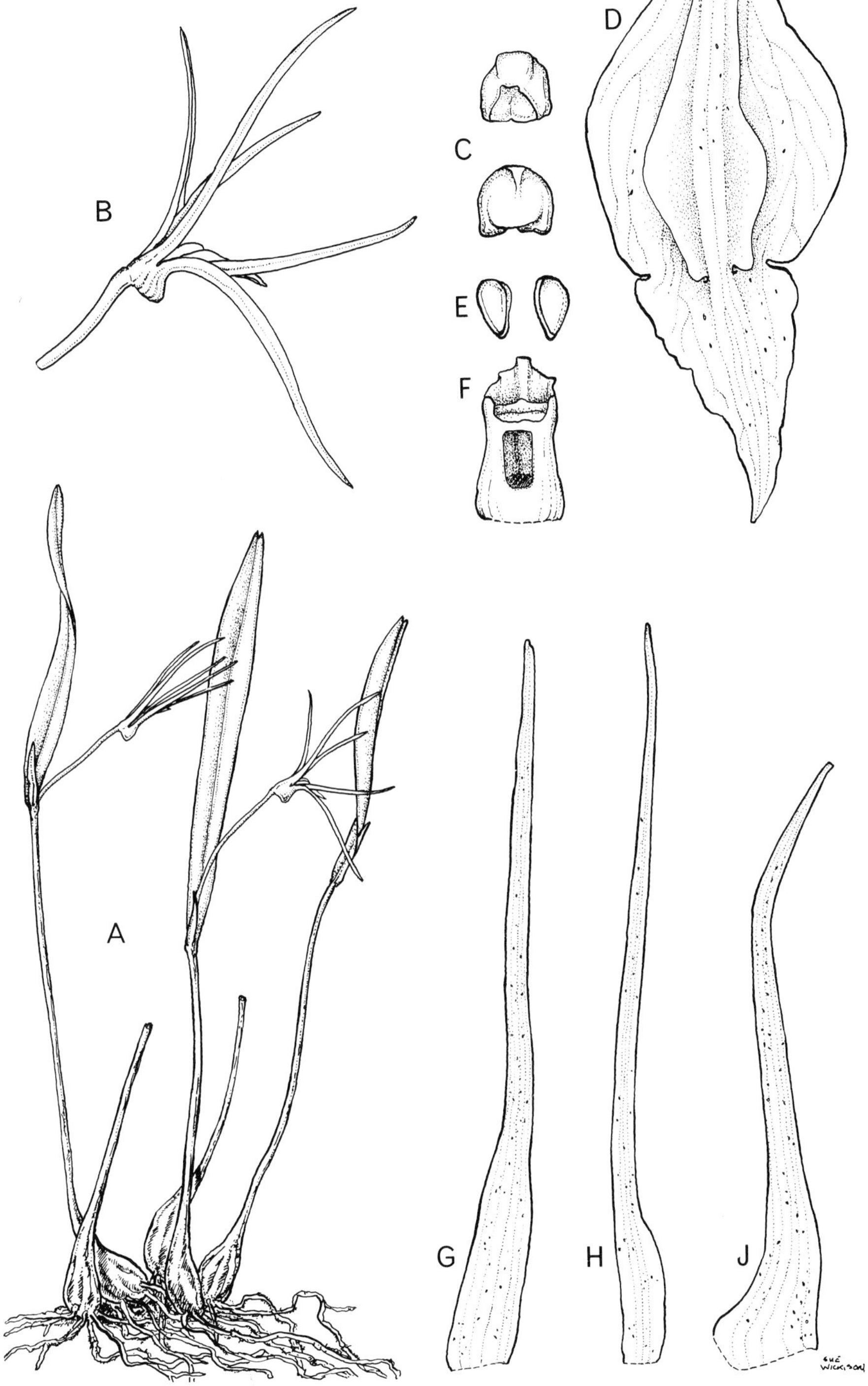
B
C
D
E
F
A
G
H
J
SUE WICKISON

Lip mid-lobe lacerate, divided into many fine segments; lip white or cream-coloured, sepals spotted with red; leaves up to 22 cm long, 10 cm wide .... **1. F. comata**
Lip mid-lobe not lacerate, retuse at apex; lip red in basal half, dull orange above, sepals off-white with purple veins; leaves up to 8 cm long, 2 cm wide .... **2. F. convexa**

Section PLICATILES

1. F. comata *(Blume) A.D. Hawkes* in Orch. Weekly 2: 453 (1961). Type: Java, *Kuhl & van Hasselt* s.n. (holotype L, isotype P!).
Desmotrichum comatum Blume, Bijdr. Fl. Ned. Ind.: 230 (1825).
Dendrobium comatum (Blume) Lindley, Gen. Sp. Orch. Pl.: 76 (1830).
Dendrobium criniferum Lindley in Bot. Reg. 30: misc. 41 (1844). Type: Sri Lanka, *MacRae* s.n. (holotype K!).
Dendrobium thysanochilum Schltr. in K. Schum. & Laut., Nachtr. Fl. Deutsch. Südsee: 152 (1905). Type: New Britain, *Schlechter* 13720 (holotype B†).
Dendrobium amesianum Schltr. in Engler, Bot. Jahrb. 56: 468 (1921). Type: Micronesia, Ponape, *Ledermann* 13460 (holotype B†, isotype K!).
Ephemerantha comata (Blume) P.F. Hunt & Summerhayes in Taxon 10: 102 (1961).

For full synonymy see Clements (1989).

An epiphytic or lithophytic plant, erect or pendent, up to 1 m long. *Stems* branching, the uppermost internode swollen to form a pseudobulb. *Pseudobulbs* clavate, becoming progressively smaller the further they are from the base of the branching stem, up to 17 cm long, 1 cm in diameter, yellow, unifoliate at apex. *Leaves* similarly becoming progressively smaller the further they are from the base of the stem, elliptic, 6–22 cm long, 3–10 cm wide. *Inflorescences* terminal, 1-flowered. *Flowers* opening widely, lasting less than a day, faintly scented, cream to pale yellow with or without pale purple spots, with a cream to yellow lip; dorsal sepal oblong, acute, 1–1.5 cm long, 3–5 mm wide; lateral sepals narrowly triangular, acute, 1–1.5 cm long, 4–6 mm wide; petals linear, 1–1.5 cm long, 1–1.5 mm wide; lip trilobed, longer than the sepals, 1.3–1.9 cm long, 6–8 mm wide, the lateral lobes triangular with erose margins, the mid-lobe shortly clawed, oblong, obtuse, fringed with thick, 5 mm long hairs, the disc with three parallel, crenate keels which extend on to the base of the mid-lobe; column 4.5 mm long, foot 6 mm long. (See fig. 54 A–J and plate 11E).

DISTRIBUTION: Bougainville, Choiseul, Makira and New Georgia. Widely distributed from Sri Lanka, Indonesia and the Malay Peninsula to Sumatra, New Guinea, Vanuatu, New Caledonia, Fiji, Samoa and Australia.
HABITAT: On trunks and branches of trees and rarely on rocks in strand and riverine forest, sea level to 480 m.

Fig. 54. Flickingeria comata A-J. A, habit x 2/3; **B**, flower x 5/3; **C**, lip, column and left lateral sepal x 3; **D**, lip x 3; **E**, column x 4; **F**, dorsal and ventral view of anther-cap x 4; **G**, dorsal sepal x 3; **H**, petal x 3; **J**, lateral sepal x 3. **F. convexa K-L. K**, flower x 2; **L**, lip x 3. **A-J** drawn from *Wickison* 42; **K-L** from photo taken by Gosta Kjellsson. All drawn by Sue Wickison.

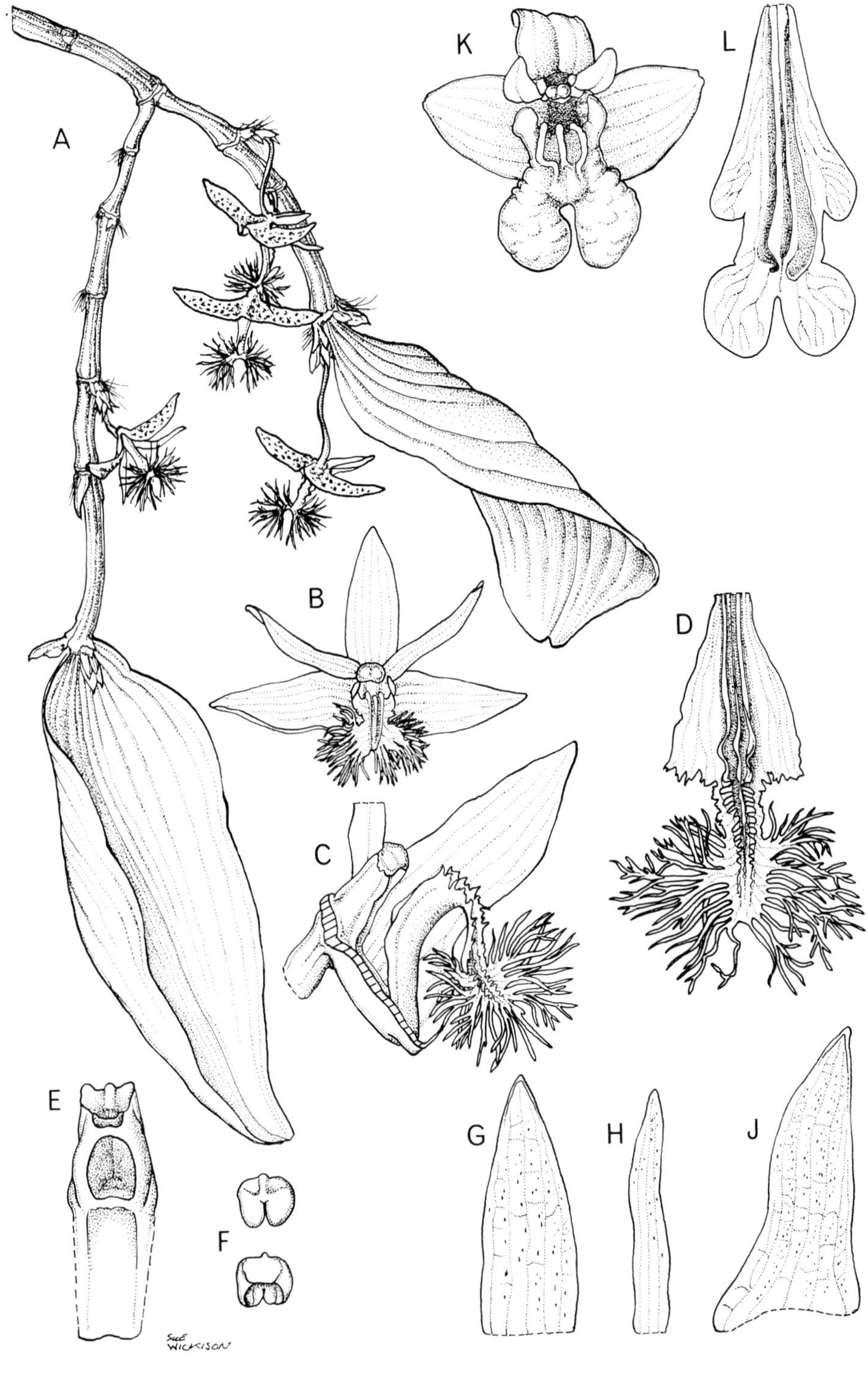
A
B
C
D
E
F
G
H
J
K
L
SUE WICKISON

COLLECTIONS: *Chaplin* 798, 805 & 807 (K); *Dickson* M11 (K); *Hunt* in RSS 2365 (K); *Millar & Vandenberg* in NGF 48521 (K); *Mitchell* 57 (K); *Waterhouse* s.n., 490 & 627 (K); *Wickison* 42 (K).

Section BILOBULATAE

2. F. convexa *(Blume) A. D. Hawkes* in Orch. Weekly 2: 46, 454 (1961). Type: Java, *Blume* s.n. (holotype L).
Desmotrichum convexum Blume, Bijdr. Fl. Ned. Ind.: 330 (1825).

For full synonymy see Seidenfaden (1980).

An epiphytic plant branching into a creeping network. *Stems* erect or pendent, uppermost internode swollen to form a pseudobulb. *Pseudobulbs* clavate, 1–5 cm long, up to 7 mm across, unifoliate at the apex. *Leaves* ovate-elliptic, emarginate, 2.5–8 cm long, 1–2 cm wide. *Inflorescence* terminal, 1-flowered. *Flowers* off-white veined with red-purple, the lateral lobes of the lip, disc and callus red with an orange mid-lobe, lasting less than a day; dorsal sepal ovate, 6 mm long, 4 mm wide; lateral sepals broadly triangular, 7 mm long, 5 mm wide; petals linear, 6 mm long, 2 mm wide; lip trilobed, 11 mm long, 5 mm wide, the lateral lobes narrow, obliquely oblong, rounded, the mid-lobe clawed, retuse at apex, the callus of 3 ridges, the middle one obscure, the outer two lamellate and raised at apex; column 5 mm long, column-foot 5 mm long. (See fig. 54 K–L and plate 11F).

DISTRIBUTION: Guadalcanal and New Georgia. Also in Thailand and the Malay Peninsula to New Guinea and Australia.
HABITAT: Riverine forest and mangroves, sea level to 800 m.
COLLECTIONS: *Metcalfe* s.n. (K); *Nicholls* JN 8 (K).

58. PSEUDERIA Schlechter

Large climbing plants. *Stems* branched, leafy near apex, sheathed in tubular leaf bases. *Leaves* distichous. *Inflorescences* lateral, short, emerging at the internodes through the leaf sheaths, few-flowered; peduncle sheathed in imbricate bracts at the base. *Flowers* with sepals and petals free; lateral sepals usually falcate; lip entire, often with 2 raised callus ridges in the basal half, converging in the middle of the lip; column slender, without a column-foot; pollinia 4.
A genus of about eighteen species from the Moluccas and New Guinea to the SW. Pacific, a single species being recorded.

P. similis *(Schltr.) Schltr.* in Fedde, Repert. Spec. Nov. Regni Veg., Beih. 1: 644 (1912). Type: New Guinea, *Schlechter* 14491 (holotype B†, isotypes AMES!, BR, K!, P).
Dendrobium simile Schltr. in K. Schum. & Laut., Nachtr. Fl. Deutsch. Südsee: 175 (1905).
Pseuderia vanikorensis Ames in Journ. Arn. Arb. 13: 130 (1933), **synon. nov.** Type: Santa Cruz Islands, Vanikoro, *Kajewski* 669 (holotype AMES!).

A large climbing plant up to 10 m long. *Stems* branched, leafy near apex, sheathed in tubular leaf bases. *Leaves* distichous, lanceolate, acute, up to 20 cm long, 2.8 cm wide, green with a glossy upper surface, paler below. *Inflorescences* lateral, emerging at the internodes through the leaf sheath, up to 3.5 cm long, bearing 3–4 flowers, sheathed in imbricate bracts at the base. *Flowers* pale yellow-green, with or without purple spots, purple with yellow patches or white; pedicel and ovary slender, up to 1.2 cm long; dorsal sepal oblong, acute, 10–11 mm long, 2–4 mm wide; lateral sepals falcate, 8–11 mm long, 3–4 mm wide; petals oblong, acute, 8–10 mm long, 1.5–4 mm wide; lip entire, obovate, 5–7 mm long, 4 mm wide, finely pubescent, with 2 raised callus ridges in the basal half, converging in the middle of the lip and sometimes extending almost to the recurved apex; column slender, curved, 5–6 mm long.

DISTRIBUTION: Bougainville, Duke Island (near New Georgia), Guadalcanal, Makira, Ranongga, the Santa Cruz Islands (Vanikoro) and Santa Isabel. Also in New Guinea.
VERNACULAR NAME: Totoa (Duke Island).
HABITAT: Rain forest, 150–750 m.
COLLECTIONS: *Beer et al.* in BSIP 7827 (K); *Dennis* in RSS 2025 & 7853 (K); *Hunt* in RSS 2251 (K); *Kajewski* 669 (AMES); *Mauriasi et al.* in BSIP 16682 (K); *Mitchell* 74 (K); *Schodde & Craven* 3640 (K); *Waterhouse* 253 (K).

The type of *P. similis* has two raised callus ridges in the basal half of the lip which converge in the middle of the lip. The type of *P. vanikorensis* is similar but the convergent callus ridges extend almost to the lip apex. Ames notes in his type description of *A. vanikorensis* that the lip margin is ciliate but we have not found that to be the case. Specimens from Bougainville, (e.g. *Schodde & Craven* 3640) are most similar to material from New Guinea but specimens from the Solomon Islands, e.g. Ranongga (*Mitchell* 74), are intermediate between the two extremes. We have therefore concluded that we are dealing with one variable species with a distinct cline from north to south.

59. BULBOPHYLLUM Thouars

Epiphytic herbs. *Pseudobulbs* clustered or spaced on a creeping or pendent rhizome, unifoliate at the apex (in the Solomon Islands and Bougainville). *Leaf* usually erect, thin-textured or fleshy. *Inflorescences* arising from the base of the pseudobulb or from nodes on the rhizome, short or long, one- to many-flowered. *Flowers* small to large; sepals larger than the petals; dorsal sepal and petals usually free; lateral sepals united at base to column-foot to form a short mentum and sometimes connate above; lip fleshy, entire to trilobed, usually hinged to column-foot so that it is mobile; column short, usually with a column-foot; pollinia 2 or 4.
A large cosmopolitan genus of about one thousand species.
Bulbophyllum is the second largest genus in the Solomon Islands and Bougainville after *Dendrobium*. Forty-one species have been recorded (but *B. medusae* is not native), and they are arranged alphabetically in the following sections after Schlechter (1913) and Vermeulen (pers. comm.): *Antennisepalum* (1), *Aphanobulbon* (1), *Brachyostele* (1), *Brachypus* (2), *Cirrhopetalum* (3), *Dialeipanthe* (2), *Ephippium* (6), *Fruticicola* (2), *Hapalochilus* (4),

Harpobrachium (2), *Leptopus* (2), *Macrouris* (5), *Oxysepala* (2), *Pelma* (2), *Polyblepharon* (3) and *Sestochilus* (3).

KEY TO SECTIONS OF BULBOPHYLLUM

This key refers only to the genus in the Solomon Islands and Bougainville.

1. Flowers two or more 2
   Flowers solitary 9
2. Pseudobulbs very small and inconspicuous compared to the size of the plant 3
   Pseudobulbs conspicuously present 4
3. Sepals ending in a thickened knob .... section **ANTENNISEPALUM**
   Sepals obtuse to acute .... section **APHANOBULBON**
4. Infloresence umbellate .... section **CIRRHOPETALUM**
   Inflorescence an elongated raceme, or inflorescence two-flowered 5
5. Lip more than 5 mm long 6
   Lip less than 5 mm long 7
6. Flowers opening simultaneously .... section **BRACHYOSTELE**
   Flowers opening one at a time .... section **DIALEIPANTHE**
7. Stelidia at apex of column with a semi-elliptic, downward pointing tooth on the lower margin, close to the tip .... section **HARPOBRACHIUM** (in part)
   Stelidia at apex of column without a tooth, or with a deltoid tooth on the lower margin 8
8. Lip with a distinct cavity at its base, or sepals with fine glandular hairs along the margins .... section **PELMA**
   Lip without a cavity at its base, and sepals without glandular hairs along their margins .... section **MACROURIS**
9. Lip immovably fused to the column base .... section **HAPALOCHILUS**
   Lip easily movable, attached to the column-foot by a thin ligament 10
10. Dorsal sepal more than 1 cm long, or dorsal sepal less than 1 cm long but then lateral sepals twice as long as the dorsal sepal or longer 11
    Dorsal sepal less than 1 cm long and lateral sepals less than twice the length of the dorsal sepal 13
11. Petals more than half the length of the dorsal sepal .... section **SESTOCHILUS**
    Petals half the length of the dorsal sepal or shorter 12
12. Lateral sepals 1.3–1.8 cm long .... section **BRACHYPUS**
    Lateral sepals 4–8 cm long .... section **EPHIPPIUM**
13. Stelidia at apex of column with a semi-elliptic, downward pointing tooth on the lower margin close to the tip .... section **HARPOBRACHIUM** (in part)
    Stelidia at apex of column without a tooth, or with a deltoid tooth on the lower margin 14
14. Lateral sepals connate or held parallel to one another for most of their length .... section **POLYBLEPHARON**
    Lateral sepals spreading 15

15. Lip papillose over most of its lower surface .. .. .. section **FRUTICICOLA**
    Lip glabrous or densely hairy over most of its surface 16
16. Margins of petals ciliate, or rhizome creeping .. .. section **LEPTOPUS**
    Margins of petals glabrous to erose; rhizome hanging or pendent .. .. section **OXYSEPALUM**

Section ANTENNISEPALUM

1. B. bulliferum *J.J. Smith* in Bull. Dép. Agric. Indes Néerl. 19: 5 (1908). Type: New Guinea, *Versteeg* 1449 (holotype BO).

An epiphytic herb. *Rhizome* creeping. *Pseudobulbs* 1–4 cm apart, orbicular to ovoid, 2–4 mm long, 1–3 mm wide, unifoliate at the apex. *Leaf* elliptic, obtuse, 1.3–7 cm long, 0.8–2.2 cm wide; petiole 1–9 mm long. *Inflorescence* arising from the base of the pseudobulb, up to 22 cm long, bearing 5 to 12 flowers; rhachis zig-zag. *Flowers* maroon and white with a yellow anther-cap; dorsal sepal ovate to elliptic, 3–6 mm long, 1.4–2.8 mm wide, ending in a thickened knob, margins ciliate; lateral sepals obliquely ovate, 3–4.5 mm long, 1.7–2.2 mm wide, ending in a thickened knob, upper margin erose or sparsely ciliate; petals obovate to elliptic, retuse to acute, 1–2 mm long, 0.4–0.8 mm wide, margins erose to denticulate; lip trilobed to obovate-rectangular, retuse to obtuse, 2–2.5 mm long, 0.7–1.2 mm wide, thick, papillose on surface; column 0.5–0.7 mm long, the column-foot with a distinct central tooth close to the attachment of the lip.

DISTRIBUTION: Guadalcanal, New Georgia, San Cristobal. Also in Irian Jaya and New Guinea.
HABITAT: Rain forest, 50–500 m.
COLLECTIONS: *Hunt* in RSS 2236 (K); *Nicholls* 3 (K); *Wickison* 134a (K).

Section APHANOBULBON

2. B. ebulbe *Schltr.* in K. Schum. & Laut., Nachtr. Fl. Deutsch. Südsee.: 200 (1905), non *B. ebulbum* King & Pantling (1895). Type: New Ireland, *Schlechter* 14639 (holotype B†).
Bulbophyllum polypodioides Schltr. in Engler, Bot. Jahrb. 39: 86 (1906). Type: New Caledonia, *Schlechter* 15422 (holotype B†, isotypes K!, P!, Z).
Bulbophyllum nigroscapum Ames in Orchid. 7: 86 (1922). Type: Samoa, *Setchell* 383 (holotype AMES!).

An epiphytic herb. *Rhizome* creeping, covered in adventitious roots. *Pseudobulbs* very reduced, spaced on a creeping rhizome, unifoliate at the apex. *Leaf* erect, oblong, 5–16 cm long, 1.5–2.5 cm wide, pale green; petiole 2–5 cm long. *Inflorescence* arising from the base of the pseudobulb; rhachis up to 15 cm long, many-flowered; peduncle sometimes covered in bracts, 4–11 cm long (often black in dried specimens). *Flowers* yellow to pale green with a yellow lip; dorsal sepal oblong-lanceolate, 4.5 mm long, 1.5 mm wide; lateral sepals obliquely triangular, 5.5 mm long, 3.5 mm wide, forming a 1.5 mm long, obtuse mentum; petals lanceolate-oblong, 2.5 mm long, 1 mm wide; lip tongue-shaped, obtuse, 2.5–3 mm long, 1.5 mm wide, recurved at apex; column 1 mm long, the column-foot 2.5 mm long.

DISTRIBUTION: Kolombangara, Makira and Rennell. Also in Irian Jaya, New Guinea, New Ireland, Vanuatu, New Caledonia, Fiji and Samoa.
HABITAT: Rain forest, sea level to 670 m.
COLLECTIONS: *Cribb* 6028 (K); *Dennis* in RSS 2352 (K); *Hunt* in RSS 2551 (K); *Officers of HMS Penguin* s.n. (K).

Section BRACHYOSTELE

3. B. pachyglossum *Schltr.* in Fedde, Repert. Spec. Nov. Regni Veg. 16: 125 (1920). Type: New Guinea, *Kempter* s.n. (holotype B†).

A robust epiphytic herb up to 30 cm tall. *Rhizome* creeping. *Pseudobulbs* well-spaced, conical, 5–6 cm long, 5–7 mm wide, unifoliate at the apex. *Leaf* elliptic, 18–22 cm long, 6.5–8.5 cm wide; petiole 4–5 cm long. *Inflorescence* arising from the base of the rhizome, spreading, up to 20 cm long, bearing c. 7 flowers; peduncle sheathed in bracts; bracts large, ovate, up to 2.5 cm long, 1.2 cm wide, concealing the flowers. *Flowers* not opening widely, pale green spotted purple with a purple lip; sepals papillose on the exterior surface; dorsal sepal ovate, 1.1 cm long, 6 mm wide; lateral sepals fleshy, obliquely ovate, 1.6 cm long, 1 cm wide; petals elliptic, acute, 9 mm long, 6 mm wide; lip tongue-shaped, recurved, fleshy and papillose on the upper surface, 7 mm long, 5 mm wide; column 4 mm long, the column-foot 8 mm long. (See fig. 55).

DISTRIBUTION: Kolombangara. Also in New Guinea.
HABITAT: In moss cushion in montane forest on ridge and at summit, 1500–1600 m.
COLLECTION: *Cribb & Campbell* 5127 (K).

Section BRACHYPUS

Leaves more than 1.5 cm wide; sepals and petals without ciliate margins .... **4. B . cerinum**
Leaves less than 1.5 cm wide; dorsal sepal and petals with ciliate margins .... **5. B. samoanum**

4. B. cerinum *Schltr.* in Fedde, Repert. Spec. Nov. Regni Veg., Beih. 1: 765 (1913). Types: New Guinea, *Schlechter* 17669 (syntype B†, isosyntype K!), *Schlechter* 18249 (syntype B†).

An epiphytic herb. *Rhizome* creeping. *Pseudobulbs* clustered, well developed, ovoid, up to 3 cm long, 2 cm wide, yellow to light green, unifoliate at the apex.

Fig. 55. Bulbophyllum pachyglossum. **A**, habit x 2/3; **B**, flower x 3; **C**, dorsal and ventral view of anther-cap x 4; **D**, side view of lip x 4; **E**, lip x 4; **F**, dorsal sepal x 4; **G**, column x 4; **H**, petal x 3; **J**, lateral sepal x 3. Drawn from *Cribb & Campbell* 5127 by Sue Wickison.

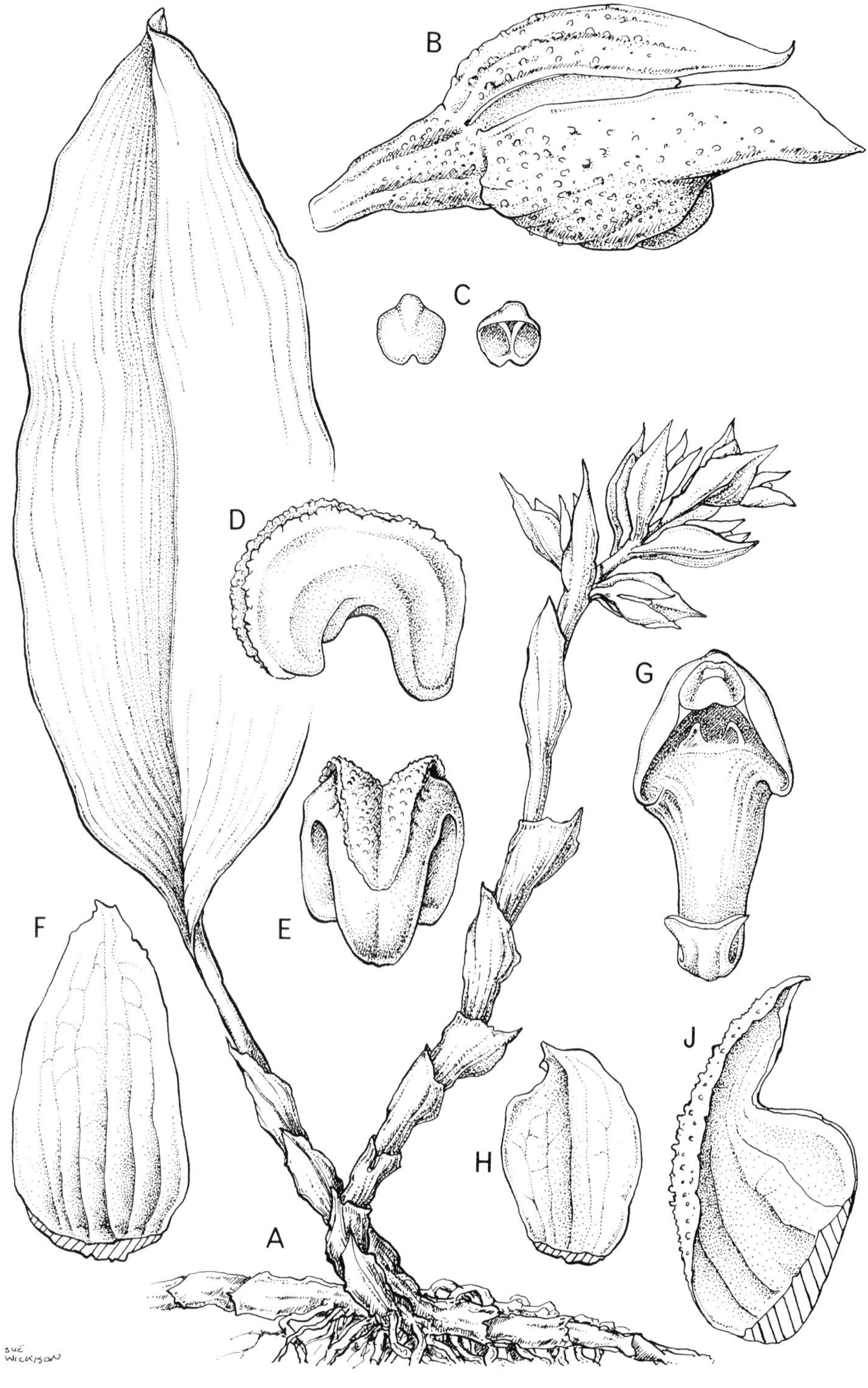
B
C
D
G
E
F
J
H
A
SUE WICKISON

Leaf elliptic, acute, 8–19 cm long, 1.5–5 cm wide, light to mid-green; petiole up to 5 cm long. *Inflorescence* arising from the base of the pseudobulb, 2–11 cm long, one-flowered at apex; peduncle sheathed by 2 bracts. *Flowers* light green, pale yellow or white sometimes veined and spotted red on the inside of the sepals, the petals light green, pale yellow or white, the lip yellow underneath and scarlet above, the column pale green; dorsal sepal ovate, acute, 1.4–1.6 cm long, 4–6 mm wide; lateral sepals ovate, acute, 1.6–1.8 cm long, 6 mm wide; petals narrowly ovate, 3.5–6 mm long, 1.5 mm wide; lip tongue-shaped, 4–5.5 mm long, 2 mm wide, thick and fleshy, with 2 ridged keels; column 3 mm long, the column-foot 3.5 mm long.

DISTRIBUTION: Guadalcanal and New Georgia. Also in New Guinea.
HABITAT: Montane and moss forest, 440–1900 m.
COLLECTIONS: *Eadley* 2 (K); *Hunt* in RSS 2976 (K); *Wickison* 81, 117 & 131 (K).

This species is similar to *B. iboense* Schltr., from New Guinea and the two may be conspecific.

5. B. samoanum *Schltr.* in Fedde, Repert. Spec. Nov. Regni Veg. 9: 107 (1911). Type: Samoa, *Vaupel* 546 (holotype B†).
B. christophersenii L.O. Williams in Bot. Mus. Leafl. Harv. Univ. 7: 143 (1939). Type: Samoa, *E. Hume* 2297 (holotype BISH, isotypes AMES!, K!, P!).

An epiphytic herb. *Rhizome* creeping. *Pseudobulbs* clustered, 7 mm high, 4 mm wide, unifoliate at the apex. *Leaf* oblong, subacute, 5–15 cm long, 0.8–1.5 cm wide, narrowed at the base. *Inflorescence* arising from the base of the pseudobulb; peduncle 7–10 cm long, bearing a solitary flower. *Flower* yellow-green tinged with purple or red,with purple petals and a white to pale pink lip, and a yellow anther-cap; dorsal sepal ovate, acute, 1 cm long, 4 mm wide, with ciliate margins; lateral sepals ovate-oblong, 1.3 cm long, 5 mm wide; petals minute, linear, 3 mm long, 1 mm wide, with ciliate margins; lip tongue-shaped, 3–4 mm long, 1 mm wide, with 2 ridges; column 3 mm long, the column-foot 1 mm long, upcurved.

DISTRIBUTION: Guadalcanal. Also in Vanuatu, Fiji, Samoa and New Caledonia.
HABITAT: Rain forest, 540 m.
COLLECTION: *Wickison* 139 (K).

Section CIRRHOPETALUM

1. Lateral sepals up to 12 cm long .. .. ***8. B. medusae**
   Lateral sepals less than 4 cm long 2
2. Flowers maroon, usually red-black at base; lateral sepals filiform .. .. **6. B. gracillimum**
   Flowers creamy yellow, blotched and speckled maroon-red with purple petals and purple lip; lateral sepals oblong, acute .. .. **7. B. longiflorum**

6. B. gracillimum (Rolfe) Rolfe in Bull. Misc. Inf., Kew 1907: 412 (1907), non Hayata (1912). Type: Burma, *Jarvis, O'Brien & Horsman* s.n. (holotype K!).
Cirrhopetalum gracillimum Rolfe in Bull. Misc. Inf., Kew 1895: 34 (1895).
Cirrhopetalum leratii Schltr. in Fedde, Repert. Spec. Nov. Regni Veg. 9: 216 (1911), non *C. leratiae* Kraenzl. (1920). Type: New Caledonia, *Le Rat* s.n. (holotype B†).
Cirrhopetalum warianum Schltr. in Fedde, Repert. Spec. Nov. Regni Veg., Beih. 1: 890 (1913). Type: New Guinea, *Schlechter* 19215 (holotype B†, isotype K!).

For full synonymy see Seidenfaden (1973).

An epiphytic herb. *Rhizome* creeping. *Pseudobulbs* spaced 0.5–2 cm apart, ovoid, 5–10 mm long, 5–10 mm wide, unifoliate at the apex. *Leaf* elliptic, 4–9 cm long, 1.3–3 cm wide; petiole up to 2 cm long. *Inflorescence* umbellate, arising from the base of the pseudobulb, up to 30 cm long. *Flowers* maroon, usually red-black at base; dorsal sepal ovate, filiform, 6–13 mm long, 2 mm wide, with hairy margins; lateral sepals lanceolate, filiform, up to 3.5 cm long, 2–3 mm wide; petals obliquely ovate, 6 mm long, 1 mm wide, with hairy margins; lip tongue-shaped, 1.6–2.5 mm long, 1–2 mm wide, recurved; column 1.4 mm long, the column-foot 1 mm long, upcurved at apex.

DISTRIBUTION: Choiseul, Guadalcanal, Kolombangara, Rendova and Santa Isabel. A widespread species from SE. Asia and Borneo to New Guinea, New Caledonia and Fiji.
HABITAT: Lowland rain forest and coconut plantations, sea level to 500 m.
COLLECTIONS: *Chaplin* 853, 854 & 856 (K); *Cribb & Campbell* 5103 & 5151 (K); *Hunt* in RSS 2414 & 2644 (K); *Mitchell* 1 & 24 (K); *Woodford* 3 (K).

7. B. longiflorum *Thouars*, Orch. Iles Austr. Afr.: 98 (1822). Type: Madagascar, *Thouars* s.n. (holotype P!, isotype L).
Epidendrum umbellatum Forst. f., Ins. Austr. Prodr.: 60 (1786). Type: Society Islands, *Forster* s.n. (holotype BM!, isotype K!).
Cirrhopetalum thouarsii Lindley in Bot. Reg 10: sub t. 832 (1824). Type: Fiji, *Seemann* 598 (holotype K!).
Cirrhopetalum clavigerum Fitzg. in Journ. Bot. 21: 204 (1883). Type: N. Australia, *Fitzgerald* s.n. (holotype BM!).
Cirrhopetalum kenejianum Schltr. in Fedde, Repert. Spec. Nov. Regni Veg., Beih. 1: 889 (1913). Type: New Guinea, *Schlechter* 18462 (holotype B†).
Cirrhopetalum longiflorum (Thou.) Schltr. in Beih. Bot. Centralbl. 33, 2: 420 (1915).

For full synonymy see Clements (1989).

An epiphytic herb. *Rhizome* creeping, thick and woody. *Pseudobulbs* usually 4–6 cm apart but may be closer, conical, 1.5–3.5 cm high, 0.8–1.2 cm wide, often covered with hair-like remains of sheathing bracts, unifoliate at the apex. *Leaf* oblong, 9–16 cm long, 2–3.5 cm wide, thick; petiole 1.5–3 cm long. *Inflorescence* arising from base of pseudobulb, with 6- to 8-flowers in a umbel; peduncle up to 30 cm long. *Flowers* cream-yellow, blotched and speckled deep maroon-red with purple petals and a purple lip; dorsal sepal elliptic, concave with an apical hair, 8–9 mm long, 6–6.5 mm wide; lateral sepals oblong, acute, connate in apical half,

2.5–3.5 cm long, 4 mm wide; petals obliquely triangular, 5.5–9 mm long, 3 mm wide, hairy; lip tongue-shaped, 4–5 mm long, 1.5 mm wide, recurved; column 3 mm long, the column-foot 4 mm long, upcurved at apex.

DISTRIBUTION: Widespread from Africa and Madagascar to the Malay Archipelago, New Guinea, Vanuatu, Fiji, Tahiti, Guam, the Austral Islands, the Society Islands and Australia.
HABITAT: Rain forest, growing in plenty of light, 200 m.
COLLECTIONS: none seen.

Although we have not seen any collections of *B. longiflorum* from the Solomon Islands and Bougainville we suspect that it might be found there as it occurs in all the neighbouring islands.

***8. B. medusae** *(Lindley) Reichb. f.*, Walp. Ann. Bot. 6: 262 (1861). Type: Singapore, *Loddiges* s.n. (holotype K!).
Cirrhopetalum medusae Lindley in Bot. Reg. 28: 12 (1842).

For full synonymy see Seidenfaden (1973).

An epiphytic herb. *Rhizome* creeping, rather stout. *Pseudobulbs* about 3 cm apart, ovoid, 2–3.5 cm long, 1–1.5 cm wide, unifoliate at the apex. *Leaf* oblong, 10–20 cm long, 1–2.5 cm wide, the apex obtuse and slightly cleft; petiole up to 3 cm long. *Inflorescence* arising from the base of the pseudobulb, pendent, 9–20 cm long; flowers in an umbel like a mop; bracts conspicuous, cream. *Flowers* white or white with pink spots, with a yellowish lip; dorsal sepal ovate, acuminate, 3 cm long, 3–5 mm wide; lateral sepals lanceolate, filiform, up to 12 cm long, 4 mm wide; petals obliquely triangular, filiform, 5 mm long, 1–2 mm wide; lip tongue-shaped, 1.5–2 mm long, 1–1.5 mm wide; column 3 mm long, the column-foot 2 mm long, recurved at apex.

DISTRIBUTION: Bougainville. Distributed from Thailand to Borneo.
HABITAT: Established on trees at Arawa.
COLLECTIONS: *Cribb* (sight record).

This species is not native but it is found on trees at Arawa, Bougainville (see Slade, 1965).

Section DIALEIPANTHE

Roots smooth, glabrous; lip tongue-shaped, 2.3–3 cm long; petals minute, mucronate, 4 mm long .... 9. **B. longiscapum**
Roots rough, papillose; lip ovate to tongue-shaped, 1.2 cm long; petals ovate, 1.5 cm long .... 10. **B. pachyanthum**

9. B. longiscapum *Rolfe* in Bull. Misc. Inf., Kew 1896: 45 (1896). Type: Fiji, *Yeoward* s.n. (holotype K!).

Macrolepis longiscapa A. Rich., Sert. Astrol.: 25, t. 10. (1833). Type: Vanuatu, *Richard* s.n. (holotype P?).
Bulbophyllum praealtum Kraenzl. in Notizbl. Bot. Gart. Berlin 5: 109 (1909). Type: Samoa, *Vaupel* 322 (holotype B†, isotype AMES!).
Bulbophyllum macrolepis L.O. Williams in Bot. Mus. Leafl. Harv. Univ. 5: 135 (1938). Type as for *Macrolepis longiscapa*.

An epiphytic herb. *Rhizome* creeping, thick and woody; roots smooth. *Pseudobulbs* well spaced on the rhizome, ovoid, up to 2.5 cm high, 0.8–1.5 cm wide, unifoliate at the apex. *Leaf* erect, oblong-lanceolate, 11–22 cm long, 1.5–3 cm wide, mid-green with a sheen; petiole 2–4 cm long. *Inflorescence* arising from the base of the pseudobulb; peduncle 28–60 cm long; bracts large and imbricate. *Flowers* opening one at a time, lemon-green with purple lines at the base, with yellow-green to purple petals, a maroon lip with a yellow apex and a yellow-white column; scent unpleasant; dorsal sepal lanceolate, 3 cm long, 7–8 mm wide; lateral sepals obliquely lanceolate, 3.5 cm long, 1.3 cm wide, forming a short mentum; petals minute, mucronate, 4 mm long, 4 mm wide; lip very fleshy, tongue-shaped, 2.3–3 cm long, 5 mm wide; column 7–10 mm long, the column-foot 7 mm long. (See fig. 56 and plate 13C).

DISTRIBUTION: Guadalcanal, Kolombangara, New Georgia and San Cristobal. Also in Vanuatu, Fiji, Samoa, Tonga, Niue, the Wallis Islands, the Mariana Islands and the Horn Islands.
HABITAT: Rain forest, 320–1000 m.
COLLECTIONS: *Dennis* in RSS 2032 & 2283 (K); *Hunt* in RSS 2564 (K); *Mitchell* 4 (K); *Wickison* 51 (K).

10. B. pachyanthum *Schltr.* in Engler, Bot. Jahrb. 39: 85 (1906). Type: New Caledonia, *Schlechter* 15678 (holotype B†).

An epiphytic herb. *Rhizome* creeping with papillose roots. *Pseudobulbs* usually clustered, ovoid, 4 cm long, up to 1.5 cm wide, unifoliate at the apex. *Leaf* erect, oblong-elliptic, up to 22 cm long, 5.2 cm wide; petiole up to 7 cm long. *Inflorescence* arising from the base of the pseudobulb, up to 27 cm long. *Flowers* opening one at a time, tawny-yellow; dorsal sepal ovate, keeled, 3 cm long, 1.2 cm wide; lateral sepals lanceolate, 4.2 cm long, 6 mm wide, with a fleshy central keel; petals ovate, 1.5 cm long, 7 mm wide; lip fleshy, ovate to tongue-shaped, 1.2 cm long, 5 mm wide, the lateral margins erect, the disc minutely papillose; column 7 mm long with 2.5 mm long stelidia, the column-foot 7 mm long, upcurved at apex.

DISTRIBUTION: Guadalcanal. Also in New Caledonia, Fiji, Tonga and Samoa.
HABITAT: Montane forest, 700–800 m.
COLLECTION: *Cribb, Dennis & Campbell* 5164 (K).

This species is similar to *B. exasperatum* Schltr., from New Guinea, but differs in having smaller flowers.

Section EPHIPPIUM

1. Dorsal sepal elliptic, more than 4 cm long, 3 cm wide; sepals lime-yellow with a fine net-like pattern of fine cinnamon to purplish-brown dotting .... **14. B. grandiflorum**
   Dorsal sepal lanceolate to ovate, ± acuminate, less than 6.5 cm long, 1.5 cm wide; flower colour not as above 2
2. Leaves 2–3 cm long, 1–1.4 cm wide; petals small, trifid at apex .... **15. B. streptosepalum**
   Leaves (6-)7.5–17 cm long, 1–3 cm wide; petals not as above 3
3. Petals elliptic, acuminate, clavate at apex .... **13. B. dennisii**
   Petals not clavate at apex 4
4. Lateral sepals lanceolate, filiform above but swollen at apex, twisted together in apical third to half .... **12. B. blumei**
   Lateral sepals lanceolate, not swollen at apex, held parallel to one another 5
5. Flowers with a light brown or reddish dorsal sepal, paler and shortly white-hairy on the inside; lateral sepals cream; petals lanceolate, acuminate, 1 cm long, 3 mm wide .... **11. B. antenniferum**
   Flowers with a purple dorsal sepal with transparent, translucent spots or blotches; lateral sepals off-white speckled purple near the base; petals filiform, antenna-like, 1.5 cm long, 0.5 mm wide .. .. **14. B. trachyanthum**

11. B. antenniferum *(Lindley) Reichb. f.*, Walp. Ann. Bot. 6: 250 (1861). Type: Philippines, *Cuming* (holotype K!).
Cirrhopetalum antenniferum Lindley in Bot. Reg. 29: sub t. 49 (1843).
Phyllorchis antennifera (Lindley) Kuntze, Rev. Gen. Pl. 2: 675 (1891).

An epiphytic herb. *Rhizome* creeping. *Pseudobulbs* spaced c. 1.5 cm apart on the creeping rhizome, ovoid, up to 2.5 cm long, 1 cm wide, unifoliate at the apex. *Leaf* erect, oblong-lanceolate, obtuse, up to 13 cm long, 3 cm wide; petiole 1–3 cm long. *Inflorescence* arising from the rhizome or the base of the pseudobulb, shorter than the leaves, up to 10 cm long, bearing a single flower. *Flower* not opening widely, the dorsal sepal light brown or reddish on the back with a greenish apex, paler and shortly white-hairy on the inside, the lateral sepals cream, petals pale with three green veins and pink apices, with a pale brown lip with pink spots; dorsal sepal lanceolate, acuminate, up to 6 cm long, 1.5 cm wide, concave over the column; lateral sepals lanceolate, 4 cm long, 4.5 mm wide; petals lanceolate, acuminate, 1 cm long, 3 mm wide; lip tongue-shaped, 4.5 mm long, 2 mm wide; column 2 mm long, the column-foot 6 mm long, upcurved.

DISTRIBUTION: Guadalcanal. Also in Java, Borneo, the Philippines and New Guinea.
HABITAT: No information from the Solomon Islands.
COLLECTION: *Metcalfe* s.n., cult. R.B.G. Sydney no. 840824 (K).

Fig. 56. Bulbophyllum longiscapum. A, habit x 2/3; **B**, dorsal sepal x 2; **C**, petal x 4; **D**, lip, column and petal x 3; **E**, ventral and dorsal view of anther-cap x 6; **F**, flower x 5/3; **G**, pollinia x 6; **H**, lateral sepal x 2. Drawn from *Bregulla* 10 by Sue Wickison

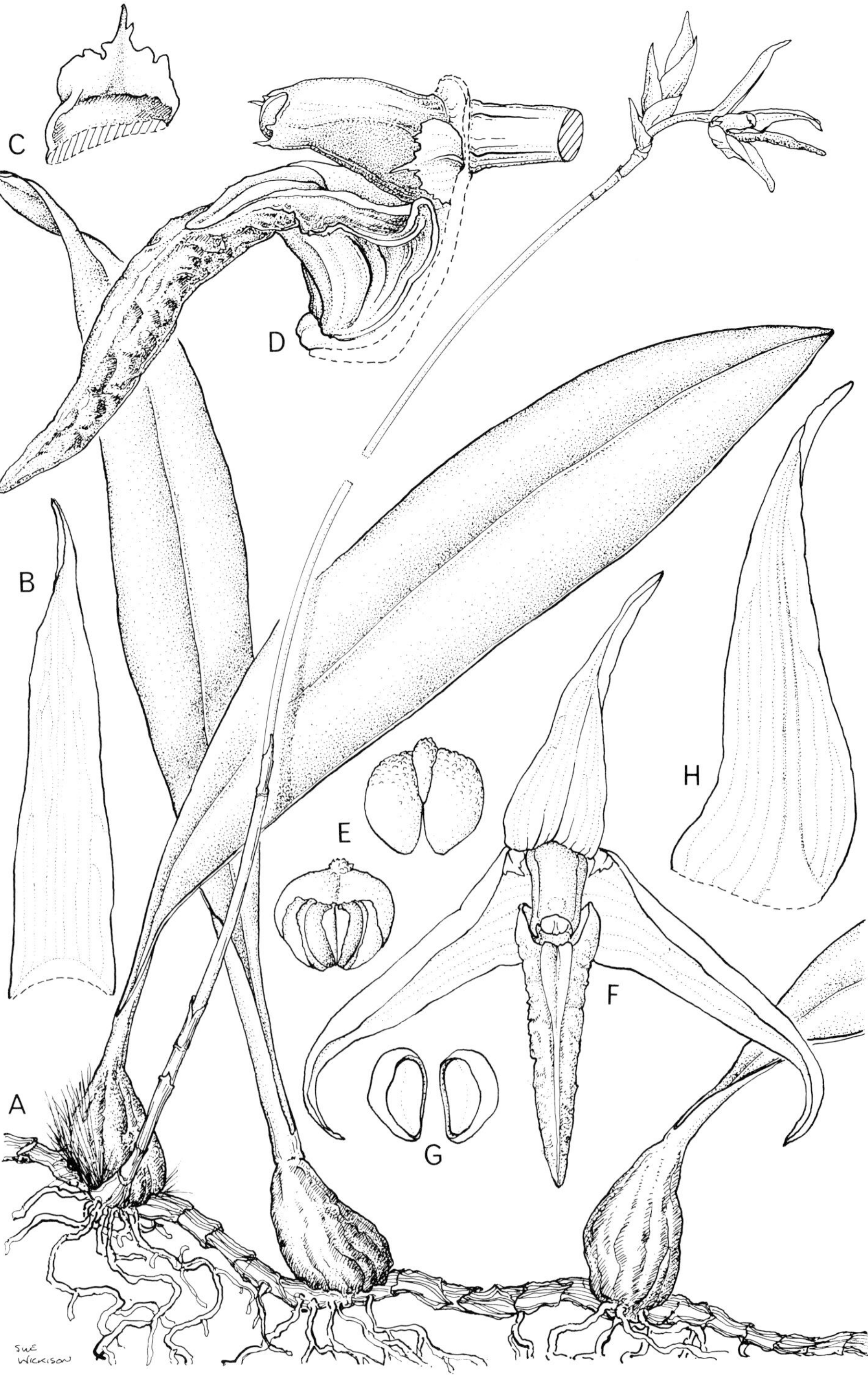
C
D
B
E
H
F
A
G
SUE WICKISON

12. B. blumei *(Lindley) J.J. Smith*, Orch. Java: 459 (1905). Type: Java, *Blume* s.n. (holotype L).
Cirrhopetalum blumei Lindley, Gen. Sp. Orch. Pl.: 59 (1830).
Bulbophyllum masdevalliaceum Kraenzl. in Engler, Bot. Jahrb. 34: 251 (1905). Type: New Guinea, *Beccari* 486 (holotype FI).

An epiphytic herb. *Rhizome* creeping. *Pseudobulbs* spaced 1.5–4 cm apart on rhizome, ovoid, up to 2 cm long, 1.5 cm wide, unifoliate at the apex. *Leaf* erect, oblong-lanceolate, up to 17 cm long, 2 cm wide; petiole 3 cm long. *Inflorescence* arising from the rhizome or the base of the pseudobulb, up to 14 cm long, bearing a single flower. *Flower* with a white dorsal sepal lined with purple, the lateral sepals and lip purple with a yellow apex; dorsal sepal ovate, acuminate, 2.1 cm long, 4.5 mm wide, with finely crenellate margins, concave over the column; lateral sepals lanceolate, filiform, swollen at apex, 4.3–4.8 cm long, 7 mm wide, the apical third to half twisted together; petals falcate, acute, 4 mm long, 2 mm wide; lip obscurely trilobed, filiform above, 6 mm long, 1.5 mm wide, fleshy at base with erect, rounded, sparsely hairy lateral lobes; column 2 mm long with 1 mm long, erect stelidia, the column-foot 2 mm long. (See plate 13D).

DISTRIBUTION: Kolombangara, Malaita, New Georgia and Rendova. Also in the Malay Peninsula, Java, Sumatra, Borneo and New Guinea.
HABITAT: Coastal forest and lowland rain forest, sea level to 100 m.
COLLECTIONS: *Balodis & Dickson* AB1 (K); *Chaplin* 857 (K); *Cribb* 5107 (K); *Cribb & Campbell* 5146 (K); *Hunt* in RSS 3072 (K).

13. B. dennisii *J.J. Wood* in Kew Bull. 38(1): 71 (1983). Type: Solomon Islands, Guadalcanal, *Morrison, Cribb & Dennis* 1834 (holotype K!).

An erect epiphytic herb. *Rhizome* creeping. *Pseudobulbs* up to 1.5 cm apart on the rhizome, narrowly conical to ± cylindrical, 2–4 cm long, 1.2 cm wide, unifoliate at the apex. *Leaf* narrowly oblong-elliptic, ligulate, (6-)7.5–16 cm long, 1–2.2 cm wide, thickly coriaceous, shortly petiolate. *Inflorescence* arising from the base of pseudobulbs, up to 10.5 cm long, green flecked pale purple near the apex, bearing a solitary flower. *Flower* with the dorsal sepal maroon-purple, orange at base, white at apex, the lateral sepals white, becoming light purple at base, the petals dirty white with yellow apices, the lip off-white flecked purple-red; dorsal sepal lanceolate-ligulate, obtuse, slightly conduplicate above, 6.2 cm long, 1 cm wide, inner surface covered with minutely pubescent yellow hairs; lateral sepals lanceolate, connate above and below, 4.5–4.7 cm long, 8 mm wide, sparsely minutely hairy at base of inner surface; petals elliptic, filiform in middle but clavate at the apex, 1.5 cm long, 2 mm wide; lip ovate, fleshy, obtuse, recurved, 5 mm long, 3 mm wide, the disc hairy; column 4 mm long, the column-foot 5–6 mm long, incurved at apex. (See plate 14D).

DISTRIBUTION: Bougainville, Guadalcanal, Kolombangara and Makira.
HABITAT: Rain forest, 70–500 m.
COLLECTIONS: *Chaplin* 855 (K); *Cribb & Campbell* 5106 (K); *Herklots* s.n. (K); *Hunt* in RSS 2204 (K); *Morrison, Cribb & Dennis* 1834 (K); *Wickison* 144 (K).

14. B. grandiflorum *Blume* in Rumphia 4: 42 t. 195, fig. 3 (1850). Type: New Guinea, *Zippelius* (holotype L).
Bulbophyllum cominsii Rolfe in Bull. Misc. Inf., Kew 1895: 138 (1895). Type: Solomon Islands, Florida group, *Comins* 289 (holotype K!).

A robust epiphyte often forming a mat on the trunk of the host tree. *Rhizome* creeping, rooting from nodes. *Pseudobulbs* well-spaced on the rhizome, ovoid-quadrangular with rounded edges, 1.8 cm long, 1.3 cm wide, unifoliate at the apex. *Leaf* elliptic, 7–14 cm long, 1.5–3.3 cm wide, minutely bilobed at apex; petiole 1.5–2 cm long. *Inflorescence* arising from the rhizome near the base of the pseudobulb, 10–20 cm long, bearing a solitary flower. *Flower* lime-yellow with a fine net-like pattern of minute cinnamon to purplish-brown dotting on the sepals, the petals lime-green, the lip white with purple-magenta spotting; dorsal sepal elliptic, 4–8 cm long, 1.8–3.2 cm wide, concave over the column; lateral sepals falcate, incurved, joined at apices, 4–8 cm long, 1.8–3 cm wide; petals obliquely oblong, obtuse, 3–5 mm long, 2–3 mm wide; lip fleshy, tongue-shaped, 6 mm long, 2.5 mm wide, with ciliate margins, recurved in basal half; column 3 mm long, the column-foot 6 mm long. (See plate 14B).

DISTRIBUTION: Bougainville, Choiseul, Kolombangara, Makira, Malaita, New Georgia, Rendova, Santa Ana Island and Tanabulo Island (off Santa Isabel). Also in New Guinea and Indonesia.

HABITAT: Coastal forest, mangrove swamps and coconut plantations, sea level to 50 m.

COLLECTIONS: *Chaplin* 800, 821B, 851 & 852 (K); *Cribb & Campbell* 5150 (K); cult. *Dennis* s.n. & in RSS 8176 (K); *Dickson* M25a (K); *Dockrill et al.* in NGF 34177 (K); *Hunt* in RSS 2229, 2799 & 3071 (K); *Mitchell* 19 (K); *Woodford* s.n. (K).

15. B. streptosepalum *Schltr.* in Fedde, Repert. Spec. Nov. Regni Veg., Beih. 1: 779 (1913). Type: New Guinea, *Schlechter* 19326 (holotype B†, isotype AMES).

An epiphytic herb. *Rhizome* creeping, ascending. *Pseudobulbs* clustered, ovoid to conical, 7–10 mm high, 3–6 mm wide, unifoliate at the apex. *Leaf* oblong-elliptic, obtuse, 2–3 cm long, 9–11 mm wide. *Inflorescence* arising from the rhizome, bearing a solitary flower; peduncle 14 cm long. *Flower* with a cream dorsal sepal heavily reticulated with scarlet, pale greenish-cream lateral sepals heavily veined and flushed with maroon at the base, a pale green column and petals and a scarlet lip; dorsal sepal ovate, concave, twisted to one side, 8 mm long, 3 mm wide; lateral sepals lanceolate, acuminate, twisted at apex, 5.5 cm long, 2.5 cm wide; petals small, trifid at apex, 2.5 mm long, 0.5 mm wide; lip fleshy, tongue-shaped, 2 mm long, 1 mm wide, with hairy margins; column 2 mm long, the column-foot 2.5 mm long.

DISTRIBUTION: Guadalcanal. Also in New Guinea and Vanuatu.

HABITAT: Epiphytic in forest on ultra-basic soil, 500 m.

COLLECTION: *Hunt* in RSS 2937 (K).

16. B. trachyanthum *Kraenzl.* in Oest. Bot. Zeit. 44: 336 (1894). Type: New Ireland, *Micholitz* s.n. (holotype B†).

An epiphytic herb. *Rhizome* creeping. *Pseudobulbs* well spaced on the rhizome, conical, up to 3 cm long, 2.2 cm wide, unifoliate at the apex. *Leaf* erect, oblong-elliptic, 2–9 cm long, 1.5–3.5 cm wide, shortly petiolate. *Inflorescence* arising from the rhizome, up to 24 cm long, bearing a solitary flower. *Flower* with the dorsal sepal purple with transparent, translucent spots or blotches, the lateral sepals off-white speckled purple, heaviest near the base, the petals white with maroon apices, the lip pale green speckled with maroon; dorsal sepal lanceolate, acuminate, 5.5–8 cm long, 1 cm wide; lateral sepals lanceolate, acuminate, 5–6 cm long, 6 mm wide; petals elliptic, filiform, antenna-like, 1.5–2 cm long, 5 mm wide; lip ovate, recurved, 3 mm long, 1 mm wide; column 3 mm long, the column-foot 3 mm long. (See plate 14A).

DISTRIBUTION: New Guinea, New Ireland and Fiji.
HABITAT: Rain forest, sea level to 800 m.
COLLECTIONS: none seen.

Although we have not seen any collections of *B. trachyanthum* from the Solomon Islands or Bougainville, it may well be found there as it grows both in New Ireland and Fiji.

Section FRUTICICOLA

Plant creeping; lateral sepals 5 mm long .. .. **17. B. betchei**
Plant pendent; lateral sepals 10 mm long .. .. **18. B. epibulbon**

17. B. betchei *F. Muell.* in Wing's South Sci. Rec. 1: 173. Type: Samoa, *Betche* 261 (holotype MEL).
Bulbophyllum finetianum Schltr. in Engler, Bot. Jahrb. 39: 83 (1906). Type: New Caledonia, *Schlechter* 15416 (holotype B†, isotypes K!, P!, Z).
Bulbophyllum atroviolaceum H. Fleischm. & Rechinger in Denkschr. Akad. Wiss. Wien 85: 261 (1910). Types: Samoa, *Rechinger* 105, 441, 1617 & 1824 (syntypes W).
Bulbophyllum ponapense Schltr. in Engler, Bot. Jahrb. 56: 484 (1921). Type: Caroline Islands, *Ledermann* 13447 (holotype B†).

An epiphytic herb. *Rhizome* creeping, covered in adventitious roots and cataphylls. *Pseudobulbs* 3–15 mm apart, more or less appressed to the rhizome,

---

Fig. 57. Bulbophyllum betchei A-K. A, habit x 2/3; **B**, flower x 6; **C**, lip x 10; **D**, column x 14; **E**, ventral and dorsal view of anther-cap x 14; **F**, pollinia x 14; **G**, dorsal sepal x 6; **H**, lateral sepal x 6; **J**, petal x 6; **K**, column and lip x 10. **B. citrinilabre L-U. L**, habit x 2/3; **M**, flower and pseudobulb x 4; **N**, lateral sepal x 6; **O**, petal x 6; **P**, dorsal sepal x 6; **Q**, side view of flower with dorsal sepal and right petal and lateral sepal removed x 6; **R**, anther-cap x 14; **S**, pollinia x 14; **T** side view of lip x 14; **U**, lip x 14. **A** drawn from *MacKee* 33669; **B-K** from *Sprunger* 206; **L-U** from *Wickison* 252. All drawn by Sue Wickison.

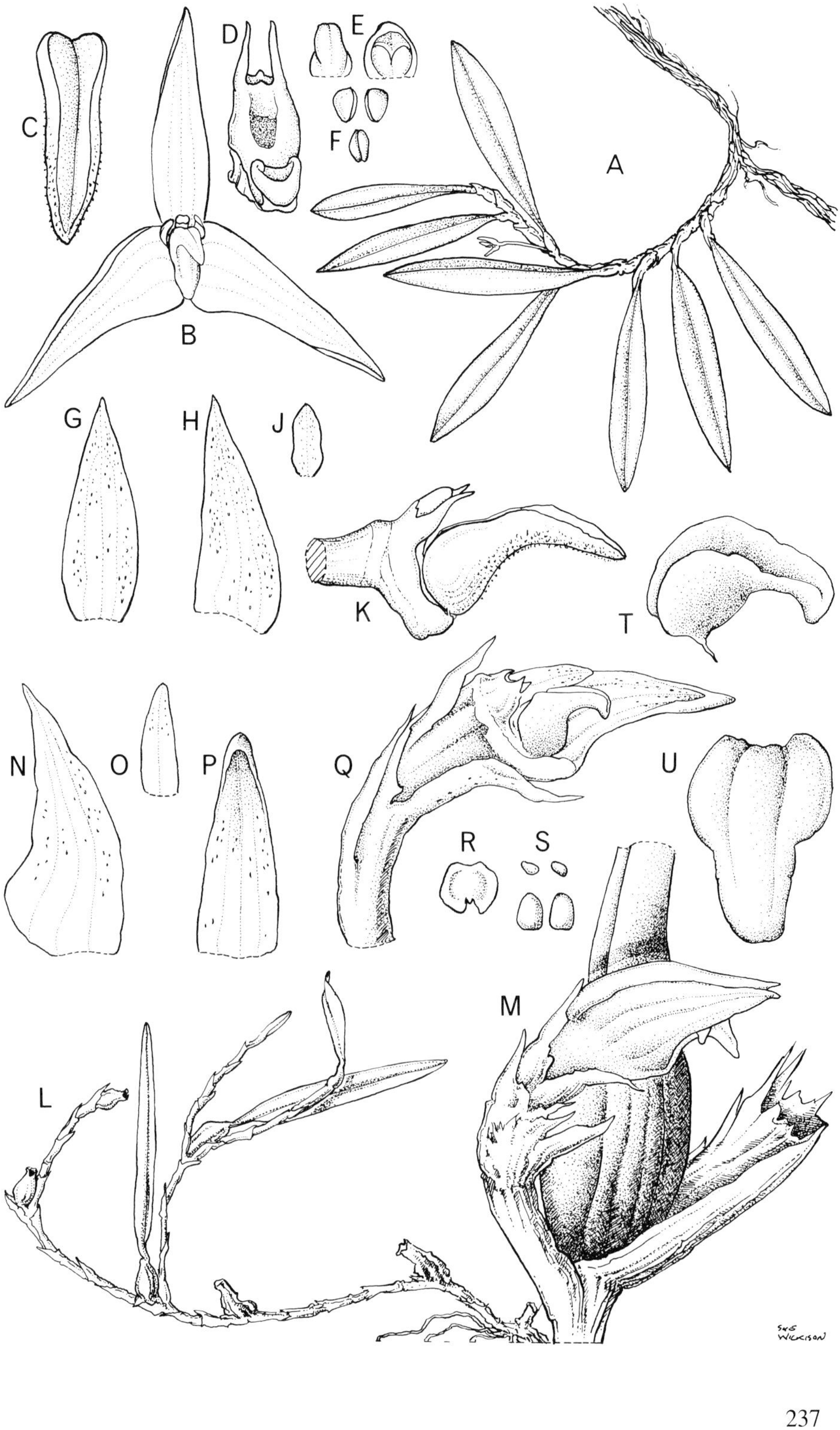
A
B
C
D
E
F
G
H
J
K
T
N
O
P
Q
U
R
S
M
L

cylindrical, 4–6 mm long, 1–2 mm wide, unifoliate at the apex. *Leaf* oblong-lanceolate, 3.5–5 cm long, 0.5–1.0 cm wide, yellow-green. *Inflorescence* arising from the base of the pseudobulb, thin, 0.5–1 cm long, bearing a solitary flower. *Flower* purple with green apices and with a purple lip with a yellow apex; dorsal sepal ovate, acute, 5 mm long, 2 mm wide; lateral sepals ovate, acute, 5 mm long, 2 mm wide; petals oblong-ovate, 1.5 mm long, 0.5 mm wide; lip tongue-shaped, 2.5–3 mm long, 0.9 mm wide, thick, covered below with papillae but glabrous above, fleshy; column 1.5 mm long, the column-foot 1 mm long. (See fig. 57 A–K).

DISTRIBUTION: Guadalcanal. Also in the Caroline Islands, Vanuatu, New Caledonia, Fiji and Samoa.
HABITAT: Lowland and montane rain forest, up to 800 m.
COLLECTIONS: *Cribb, Dennis & Campbell* 5049 & 5073 (K).

18. B. epibulbon *Schltr.* in Fedde, Repert. Spec. Nov. Regni Veg., Beih. 1: 847 (1912). Type: New Guinea, *Schlechter* 19763 (holotype B†, isotype AMES).

A pendent epiphytic herb, c. 20 cm long. *Rhizome* hanging, appearing zig-zag. *Pseudobulbs* 1.5–2 cm apart, appressed to the rhizome, ovoid, up to 5 mm high, 2 mm wide, pale yellow-green, unifoliate at the apex. *Leaf* suberect, oblong-lanceolate, 4–6 cm long, 7–9 mm wide, green, paler on reverse. *Inflorescence* arising from the base of the pseudobulb, up to 1 cm long, peduncle often covered by a sheathing bract, bearing a solitary flower. *Flower* pale translucent straw-colour with a yellow lip with a red mid-line, the column yellow, red at the base; dorsal sepal lanceolate, 9 mm long, 1.5 mm wide; lateral sepals lanceolate, 1 cm long, 2 mm wide; petals oblong, 2 mm long, 1.2 mm wide; lip obovate, 5 mm long, 1.5 mm wide, papillose below, finely papillose above at base; column 1 mm long, the column-foot 1 mm long.

DISTRIBUTION: Makira. Also in New Guinea.
HABITAT: Lowland rain forest, 70 m.
COLLECTIONS: *Hunt* in RSS 2191 & 2243 (K).

Section HAPALOCHILUS

Section *Hapalochilus* was established as a segregate genus from *Bulbophyllum* by Senghas (1978) because of the long, arcuate, footless column and the lip which is immovably adnate to it. Unfortunately, Senghas only transferred the type of the genus, but more species were transferred by Garay and Kittredge (1985). We have chosen, however, to keep *Hapalochilus* as a section of *Bulbophyllum* until the entire genus *Bulbophyllum* has been critically studied.

1. Leaves 4–7.5 cm long, 0.7–1.8 cm wide 2
   Leaves 9–17 cm long, 0.9–2 cm wide 3
2. Leaves lanceolate, acute, 0.7–1.1 cm wide .... **20. B. novahiberniae**
   Leaves elliptic-oblong, obtuse, 1.2–1.8 cm wide .... **19. B. microrhombos**
3. Inflorescences up to 3 cm long .... **21. B. stenophyllum**
   Inflorescences 5–12 cm long .... **22. B. trachyglossum**

19. B. microrhombos *Schltr.* in Fedde, Repert. Spec. Nov. Regni Veg., Beih. 1: 719 (1913). Type: New Guinea, *Schlechter* 19757 (holotype B†).

An epiphytic herb. *Rhizome* creeping. *Pseudobulbs* spaced 4–6 mm apart on the rhizome, ovoid, 1 cm long, 7 mm wide, unifoliate at the apex. *Leaf* elliptic-oblong, obtuse, 5–7.5 cm long, 1.2–1.8 cm wide, with a prominent mid-rib; petiole 1–2 cm long, tinged purple. *Inflorescence* arising from the base of the rhizome, up to 5 cm long, bearing a solitary flower. *Flower* pale yellow-green with purple stripes, the petals translucent with a central purple stripe, the lip bright yellow with purple spotting on the back, the column pale yellow with a streak of purple; dorsal sepal ovate, 8.5–11 mm long, 2–2.5 mm wide; lateral sepals ovate, 9–11 mm long, 3.5 mm wide; petals rhombic, 1.3–1–2 mm long, 1 mm wide; lip convex, elliptic, 9 mm long, 3 mm wide, papillose, thickened at apex; column 3 mm long, nearly footless. (Plate 12E).

DISTRIBUTION: New Georgia and Ranongga. Also in New Guinea and Vanuatu.
HABITAT: Montane rain forest, 700–900 m.
COLLECTIONS: *Mitchell* 73 (K); *Peach* 1 (K).

This species gives the impression of a small *B. trachyglossum*, but differs in the shorter leaves and small rhombic petals. Vermeulen (pers. comm.) notes that this species is similar to *B. aureoapex* Schltr., from New Guinea, but that has a longer inflorescence.

20. B. novahiberniae *Schltr.* in K. Schum. & Laut., Nachtr. Fl. Deutsch. Südsee: 211 (1905). Type: New Ireland, *Schlechter* 14641 (holotype B†).

An epiphytic herb. *Rhizome* creeping. *Pseudobulbs* clustered, ovoid, 8–12 mm long, 5–8 mm wide, dirty light green, unifoliate at the apex. *Leaf* lanceolate, acute, 4–7.5 cm long, 7–11 mm wide, light green. *Inflorescence* arising from the base of the pseudobulb, short, up to 1 cm long, bearing a solitary flower. *Flower* not opening widely, pale yellow with maroon stripes, the petals white with a maroon spot, the lip orange, the column pale yellow; dorsal sepal ovate, 9 mm long, 1.5 mm wide; lateral sepals ovate, acute, 9 mm long, 2.5 mm wide; petals obliquely ovate, acute, 1.5 mm long, 0.8 mm wide; lip elliptic, 9 mm long, 1.2 mm wide, concave at base, convex and papillose at apex; column 1 mm long, nearly footless. (See fig. 58 A–J).

DISTRIBUTION: Guadalcanal and Makira. Also in New Ireland.
HABITAT: Rain forest, 50–600 m.
COLLECTIONS: *Hunt* in RSS 2237 (K); *Wickison* 136 (K).

21. B. stenophyllum *Schltr.* in Fedde, Repert. Spec. Nov. Regni Veg., Beih. 1: 719 (1912). Type: New Guinea, *Schlechter* 18521 (holotype B†).

An epiphytic herb. *Rhizome* creeping. *Pseudobulbs* spaced 1–4 cm apart, ovoid, 1.5–3 cm long, 4–7 mm wide, dull maroon, unifoliate at the apex. *Leaf* oblong-lanceolate, acute, 9–17 cm long, 1.5–2 cm wide, blue-green; petiole up to 2 cm long. *Inflorescence* arising from the base of the pseudobulb, up to 3 cm long, bearing a solitary flower. *Flower* purple edged with white and with a bright yellow or purple lip; dorsal sepal narrowly ovate, 1.1 cm long, 2.5 mm wide; lateral sepals oblong-ovate, 1.1 cm long, 3.5 mm wide; petals falcate, 3 mm long, 0.7 mm wide; lip convex, elliptic, 11 mm long, 2.5 mm wide; column bent over the lip, 2 mm long, nearly footless. (See plate 12E).

DISTRIBUTION: Guadalcanal. Also in New Guinea and Vanuatu.
HABITAT: Montane forest, 800–1300 m.
COLLECTIONS: *Cribb, Dennis & Campbell* 5028 (K); *Hunt* in RSS 2952 (K).

This species is closely allied to *B. trachyglossum* Schltr. but is distinguished by its shorter inflorescence.

22. B. trachyglossum *Schltr.* in K. Schum. & Laut., Nachtr. Fl. Deutsch. Südsee: 217 (1905). Type: New Ireland, *Schlechter* 14685 (holotype B†).

An epiphytic herb. *Rhizome* creeping. *Pseudobulbs* spaced 1–3 cm apart, ovoid, 1–3 cm long, 4–8 mm wide, green, unifoliate at the apex. *Leaf* oblong-lanceolate, 9–16 cm long, 0.9–2 cm wide, mid-green; petiole up to 3 cm long. *Inflorescence* arising from the base of the pseudobulb, 5–12 cm long, bearing a solitary flower. *Flower* purple edged with white and with a bright orange-yellow or purple lip; dorsal sepal narrowly ovate, 1.4–2.2 cm long, 2.5–3 mm wide; lateral sepals obliquely ovate, 1.1–2.5 cm long, 2.5–3 mm wide; petals obliquely lanceolate, 3.5–4 mm long, 1 mm wide; lip convex, elliptic, 9–18 mm long, 2–5 mm wide, papillose; column 3.5 mm long. (See fig. 58 K–S).

DISTRIBUTION: Bougainville, Guadalcanal and Kolombangara. Also in New Ireland.
HABITAT: Montane rain forest, 830–1350 m.
COLLECTIONS: *Chaplin* 841 (K); *Craven & Schodde* 152 (K); *Cribb & Campbell* 5123 (K); *Cribb, Dennis & Campbell* 5044 (K); *Wickison* 74 (K).

Section HARPOBRACHIUM

Inflorescences few-flowered; lip recurved at the apex .. .. **23. B. cylindrobulbum**
Inflorescences 1-(2-)flowered (open one at a time if 2 present); lip strongly curved .. .. **24. B. manobulbum**

Fig. 58. Bulbophyllum novaehiberniae A-J. A, habit x 2/3; **B**, flower x 4; **C**, column x 8; **D**, anther-cap x 8; **E**, pollinia x 8; **F**, petal x 6; **G**, dorsal sepal x 6; **H**, lateral sepal x 6; **J**, lip x 6. **B. trachyglossum K-S. K**, habit x 2/3; **L**, flower x 4; **M**, petal x 4; **N**, dorsal sepal x 4; **O**, lateral sepal x 4; **P**, pollinia x 8; **Q**, column x 8; **R**, ventral view of lip x 4; **S**, dorsal view of lip x 4. **A-J** drawn from *Wickison* 156; **K-S** from *Wickison* 74. All drawn by Sue Wickison.

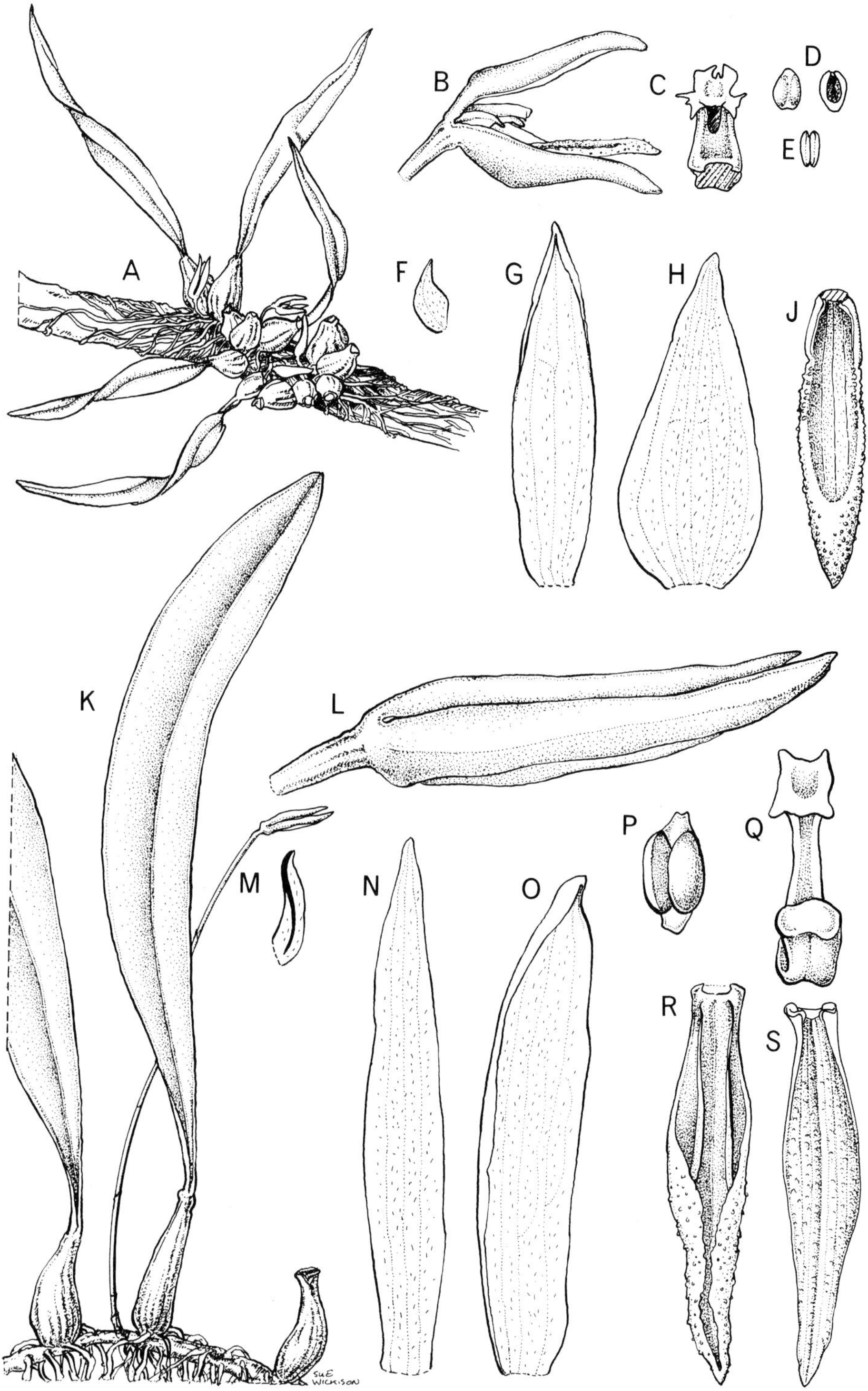
A
B
C
D
E
F
G
H
J
K
L
M
N
O
P
Q
R
S
SUE WICKISON

23. B. cylindrobulbum *Schltr.* in K. Schum. & Laut., Nachtr. Fl. Deutsch. Südsee: 200 (1905). Type: New Guinea, *Schlechter* 13999 (holotype B†).

An epiphytic herb. *Rhizome* creeping, roots present along the entire rhizome. *Pseudobulbs* well spaced on the rhizome, 5–10 cm apart, orbicular, ovoid, ellipsoid, or cylindrical, 0.3–6 mm long, 0.2–1.2 cm wide, unifoliate at the apex. *Leaf* lanceolate-oblong, rounded to acute at the apex, 3–13 cm long, 0.5–2 cm wide; petiole 3–8 mm long. *Inflorescences* single or a few together on a short sympodium, 4–10 cm long, few-flowered; rhachis erect or arching; peduncle and rhachis turning red when the flowers have died; bracts ovate to elliptic, red. *Flowers* cream or yellow; dorsal sepal ovate to elliptic, acute, 4–14 mm long, 1.5–4 mm wide; lateral sepals obliquely ovate, 4.2–15 mm long, 1.8–3.6 mm wide; petals obliquely ovate to elliptic, apex rounded or acute, 1.6–5 mm long, 1.2–2.8 mm wide, occasionally somewhat papillose towards the tip; lip ovate, elliptic, slightly constricted half-way, obtuse, 1.5–4 mm long, 0.8–2 mm wide, recurved, with 2 inconspicuous to distinct knobs near the base; column 0.8–2.2 mm long.

DISTRIBUTION: Makira and New Georgia. Also in New Guinea.
HABITAT: Lowland rain forest, sea level to 380 m.
COLLECTIONS: *Hunt* in RSS 2208 (K); *Wickison* s.n. & 64a (K).

24. B. manobulbum *Schltr.* in K. Schum. & Laut., Nachtr. Fl. Deutsch. Südsee: 207 (1905). Type: New Guinea, *Schlechter* 15743 (holotype B†), *Schlechter* 19005 (neotype BM!, isoneotype K!).

An epiphytic herb. *Rhizome* creeping. *Pseudobulbs* 1.5–21 cm apart, ovoid to ellipsoid, 0.8–3 cm long, 0.3–1.5 cm wide, somewhat flattened, unifoliate at the apex. *Leaf* ovate to elliptic, obtuse to subacute, 2–7 cm long, 0.5–2 cm wide; petiole 1–6 mm long. *Inflorescences* usually single, sometimes a few together on a short sympodium, 1-(2-)flowered, (opening one at a time if 2 present); peduncle 0.6–1 cm long; bracts ovate, acute. *Flowers* pale yellow sometimes suffused with pale purple, with a yellow lip, purplish towards the base; pedicel and ovary 4.5–10 mm long; dorsal sepal ovate to elliptic, acute, 5.5–8 mm long, 1.7–2.2 mm wide, thin-textured; lateral sepals slightly oblique, ovate, acute, 6–7.5 mm long, 1.7–2.2 mm wide, thin-textured; petals ovate to elliptic, 3–3.8 mm long, 1–1.6 mm wide; lip strongly curved, ovate, sometimes slightly constricted near the middle, 1–1.4 mm long, 1–2 mm wide, thick, glabrous, the ventral surface moderately concave with two inconspicuous knobs near the base, with two inconspicuous ridges converging in front; column 0.8–1.2 mm long. (See fig. 59).

DISTRIBUTION: New Georgia. Also in New Guinea.
HABITAT: Rain forest, 480 m.
COLLECTION: *Wickison* 64b (K).

Fig. 59. Bulbophyllum manobulbon. **A**, habit; **B**, flower; **C**, dissection of flower; **D**, dorsal and ventral view of lip; **E**, side view of lip and column; **F**, pollinia and pollen sac; **G**, ventral view of anther-cap; **H**, dorsal view of anther-cap. Drawn from *Jongejan* 332 by Jaap Vermeulen.

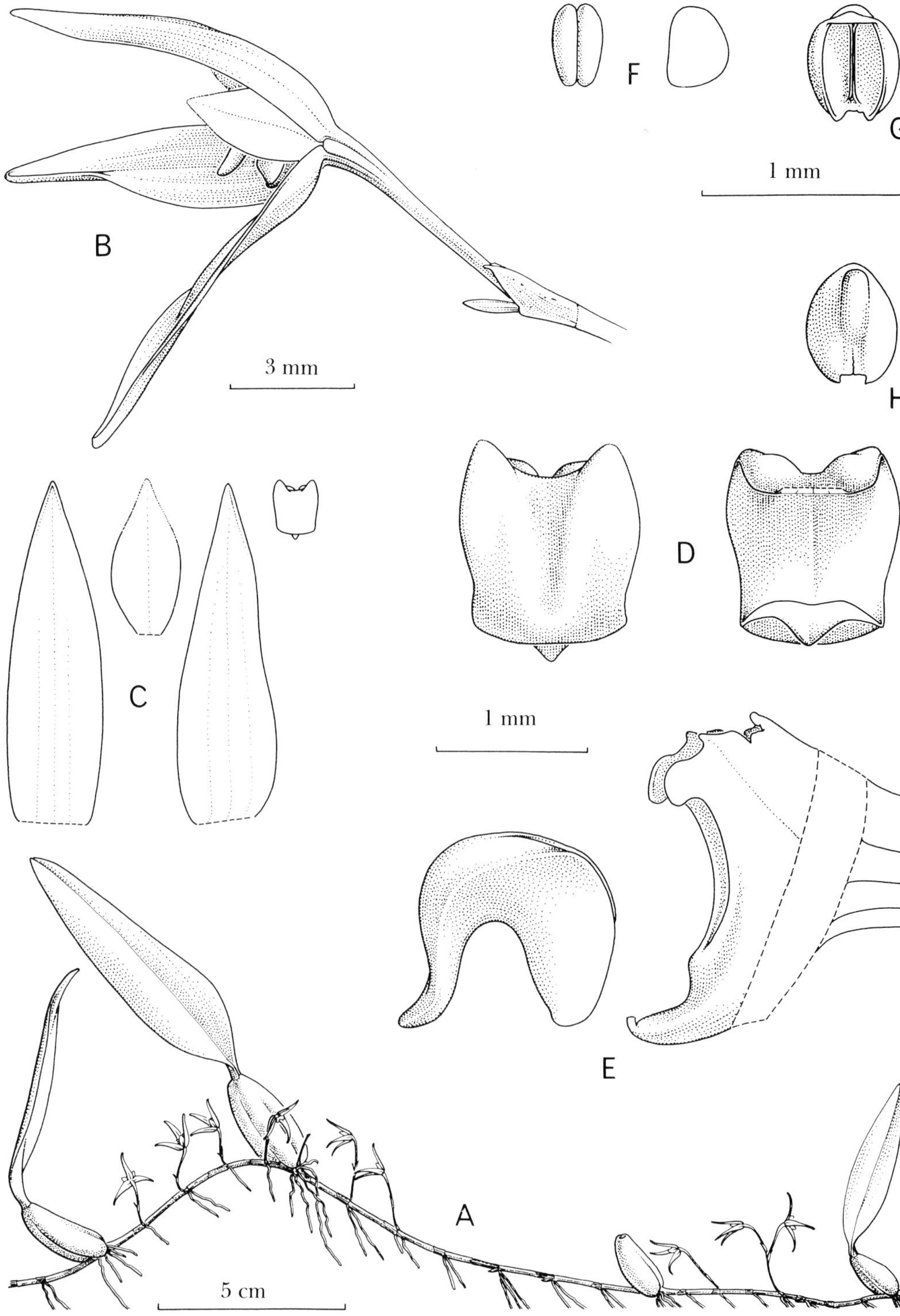
F
G
1 mm
B
H
3 mm
D
C
1 mm
E
A
5 cm

Wickison 64b is sterile but Vermeulen (pers. comm.) considers that it is probably this species. The description is from New Guinea material.

Section LEPTOPUS

Plant creeping; rhizome rarely dichotomising at pseudobulbs; leaves elliptic; lip glabrous .... **25. B. hassallii**
Plant pendent; rhizome often dichotomising at pseudobulbs; leaves ovate, acute; lip hairy .... **6. B. piestobulbon**

25. B. hassallii *Kores* in Allertonia (5)1: 148 (1989). Type: Fiji, *Hassall* 108154 (holotype K!)

An epiphytic herb. *Pseudobulbs* clustered or spaced up to 1.1 cm apart on the creeping rhizome, ovoid, 5 mm long, 5 mm wide, unifoliate at the apex. *Leaf* elliptic, 2–4 cm long, 0.9–1.2 cm wide, purple-green, paler underneath. *Inflorescence* arising from the base of the pseudobulb; peduncle 0.5–5.5 cm long, bearing a solitary flower. *Flower* pale yellow to orange-yellow speckled with scarlet or dark purple, the petals yellow, the lip crimson, the column yellow; dorsal sepal lanceolate, 7–9 mm long, 1.5 mm wide; lateral sepals lanceolate, 7–9 mm long, 1.5–2 mm wide; petals minute, triangular, acute, 0.7–1 mm long, 0.3–0.6 mm wide; lip fleshy, obovate, with 2 basal ridges, 1.5–5 mm long, 1–2 mm wide, the mid-lobe variable in size, recurved or straight; column 2 mm long, with a short, upcurved column-foot. (See fig. 60 and plate 13F).

DISTRIBUTION: Guadalcanal, Makira, Malaita and New Georgia. Also in Fiji.
HABITAT: Rain forest, 600 m.
COLLECTIONS: *Cribb* 6003 (K); *Cribb, Dennis & Campbell* 5072 (K); *Dickson* 29A (K); *Hunt* in RSS 2268 & 2948 (K); *Wickison* 251 (K).

Bulbophyllum hassallii is very variable in the length of the inflorescence and the size of the mid-lobe of the lip. Specimens with a short, broad mid-lobe are identical to the type of *B. hassallii*, whilst specimens with a longer mid-lobe (eg. *Cribb, Dennis & Campbell* 5072) approach *B. filamentosum* Schltr., from New Guinea. Apparent intermediates occur on the Solomon Islands (eg. *Cribb* 6003), and they may all belong to one variable species.

A sterile specimen from Niue Island (*Sykes* 347) may be referrable to this species.

Fig. 60. Bulbophyllum hassallii. A, habit x 2/3; **B**, flower x 6; **C**, dorsal sepal x 6; **D**, lateral sepal x 6; **E**, column and petals x 12; **F**, side view of lip x 14; **G**, lip x 14; **H**, habit x 2/3; **J**, flower x 6; **K**, side view of lip x 14; **L**, lip x 14; **M**, column and petals x 12; **N**, dorsal sepal x 6; **O**, lateral sepal x 6. **A-G** drawn from *Wickison* 251; **H** from *Dickson* 29A; **J-O** from *Cribb* 6003. All drawn by Sue Wickison.

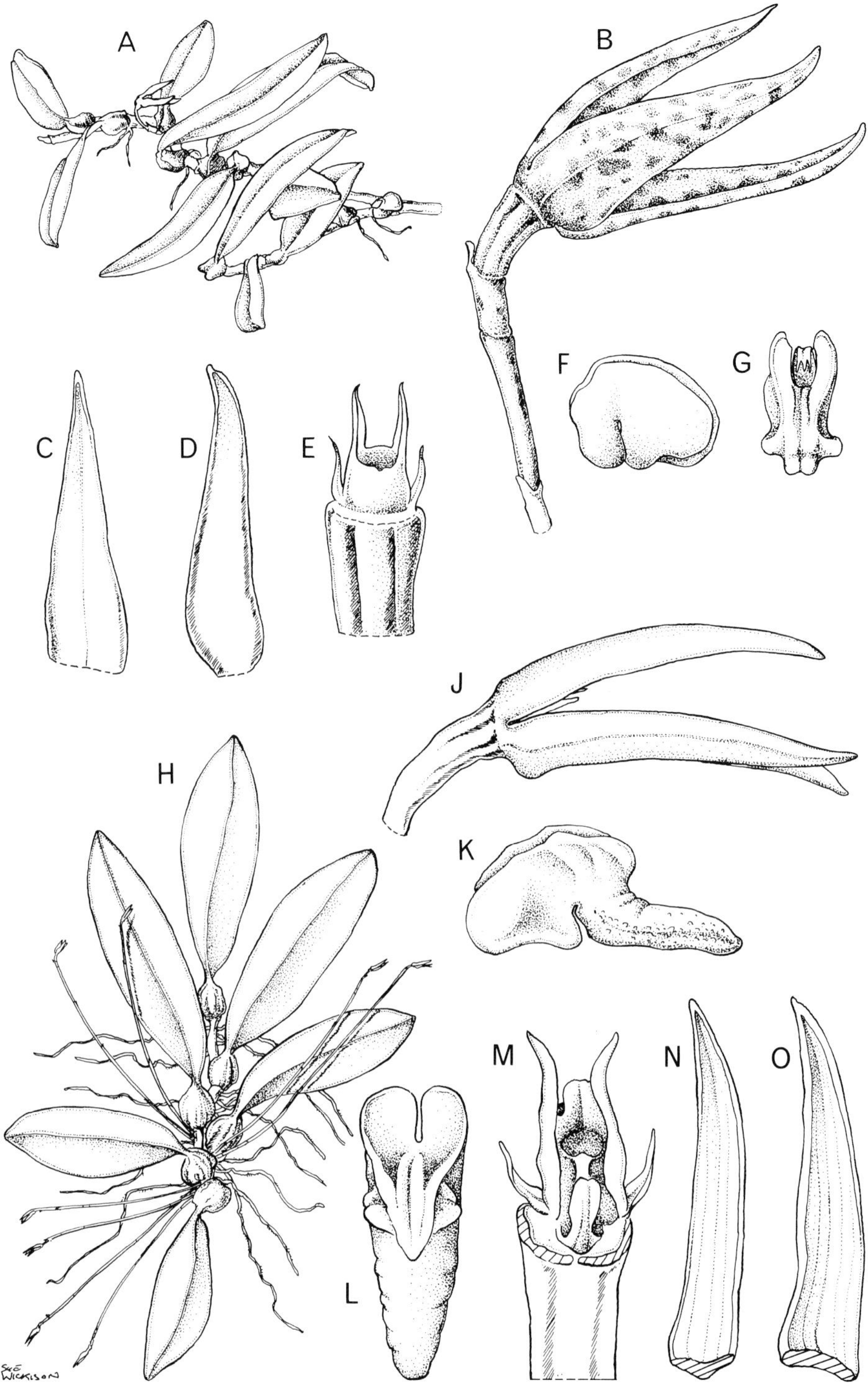
A
B
C
D
E
F
G
H
J
K
L
M
N
O
WICKISON

26. B. piestobulbon *Schltr.* in Engler, Bot. Jarhb. 38: 140 (1923). Type: New Guinea, *Schlechter* 8081 (holotype B†, isotype K!).
Bulbophyllum bambusifolium J.J. Smith in Nova Guinea 18: 63 (1935). Type: New Guinea, *Docters van Leeuwen* 9459 (holotype BO!).

A pendent epiphytic herb. *Rhizome* woody, dichotomising at nodes. *Pseudobulbs* spaced 3–9 cm apart, ovoid, up to 1.5 cm long, 7 mm wide, unifoliate at the apex. *Leaf* ovate, acute, 3–6.5 cm long, 1.5–2.2 cm wide. *Inflorescences* arising from the rhizome; peduncle up to 1.5 cm long, sheathed at base by a brown papery bract, bearing a solitary flower. *Flower* white striped with purple and with a maroon lip with a yellow apex; dorsal sepal ovate, sparsely hairy at margins, 4.5–5 mm long, 1.5 mm wide; lateral sepals obliquely ovate, 5–5.5 mm long, 2.5 mm wide; petals lanceolate, hairy at margins, 1.5 mm long, 0.5 mm wide; lip oblong, rounded at apex, 4 mm long, 1 mm wide, the basal half concave, the remainder densely hairy; column 1–1.5 mm long, the column-foot 0.5 mm long. (See fig. 61).

DISTRIBUTION: Bougainville, Choiseul, Guadalcanal, Kolombangara, Makira, Malaita and New Georgia. Also in New Guinea.
HABITAT: Rain forest, sea level to 1200 m.
COLLECTIONS: *Chaplin* 791 & 792 (K); *Craven & Schodde* 75 (K); *Dickson* L3 (K); *Hunt* in RSS 2187, 2549, 3074 & 3078 (K); *Kajewski* 2663 (BM, G); *Mitchell* 22 (K); *Wickison* s.n. (K).

Section MACROURIS

1. Rhizome pendent; inflorescences generally two-flowered; lip straight .... **28. B. dichotomum**
   Rhizome creeping; inflorescences two- or more-flowered; lip recurved 2
2. Lip with a hairy margin .... **31. B. orbiculare**
   Lip without a hairy margin 3
3. Lateral sepals less than 5 mm long .... **29. B. graciliscapum**
   Lateral sepals 1 cm long or more 4
4. Lateral sepals 1.8–4.6 cm long .... **27. B. choranthum**
   Lateral sepals 1–1.2 cm long .... **30. B. levatii**

27. B. choranthum *Schltr.* in K. Schum. & Laut., Nachtr. Fl. Deutsch. Südsee: 197 (1905). Type: New Guinea, *Schlechter* 14642 (holotype B†, isotypes BR, K!, P).

An epiphytic herb. *Rhizome* creeping, rooting in tufts below each pseudobulb. *Pseudobulbs* 3–15 cm apart, ovoid, 0.7–2.2 cm long, 0.5–2.5 cm wide, unifoliate at the apex. *Leaf* ovate to elliptic, acute, 4–14 cm long, 0.9–4 cm wide; petiole 1–3 cm long. *Inflorescence* arising from the base of the pseudobulb, 10–39 cm long,

---

Fig. 61. Bulbophyllum piestobulbon. **A**, habit x 2/3; **B**, flower x 6; **C**, lip x 8; **D**, dorsal and ventral view of anther-cap x 10; **E**, pollinia x 10; **F**, side view of lip x 8; **G**, dorsal sepal x 8; **H**, column x 10; **J**, petal x 8; **K**, lateral sepal x 8. **A-C** & **F-K** drawn from *Mitchell* 22; **D-E** from *Hunt* in RSS 3074. All drawn by Sue Wickison.

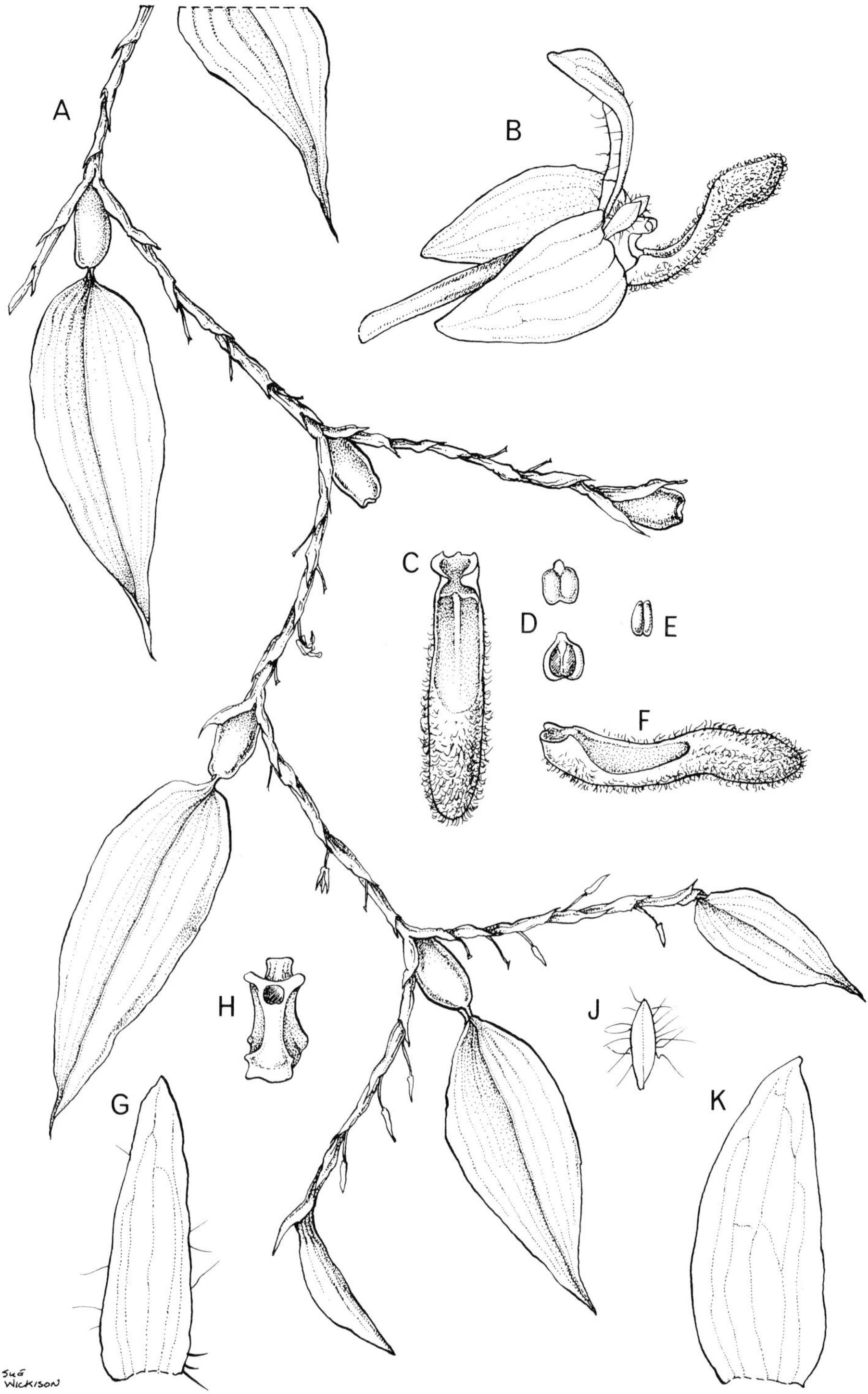
A
B
C
D
E
F
G
H
J
K
SUE WICKISON

*ff*2- to 40-flowered; rhachis pendulous; bracts triangular, acute. *Flowers* pale yellowish to yellowish green, the lateral sepals sometimes suffused with purple or entirely purple; dorsal sepal ovate to triangular, acute to acuminate, 1.2–2.6 cm long, 3–6 mm wide; lateral sepals connate or free, obliquely ovate to triangular, acute to acuminate, 1.8–4.6 cm long, 3–8 mm wide; petals ovate to elliptic, obtuse to acute, 2.5–7 mm long, 1.5–3.5 mm wide; lip curved, ovate to subpanduriform, rounded at apex, 1.4–3 mm long, 0.8–1.8 mm wide, the margins glabrous or very finely papillose, thick, the ventral surface with 2 ridges; column 1–2 mm long.

DISTRIBUTION: New Georgia. Also in New Guinea.
HABITAT: Rain forest and rubber plantations, sea level to 2000 m.
COLLECTION: *Chaplin* 815 (K).

Unfortunately the specimen from the Solomon Islands is sterile and the description of the flowers has been provided by Vermeulen (unpublished manuscript).

Bulbophyllum chloranthum differs from *B. levatii* mainly by its larger flowers and wider sepals.

28. B. dichotomum *J.J. Smith* in Bull. Dép. Agric. Indes Néerl. 19: 5 (1908). Type: New Guinea, *Versteeg* 1537 (holotype L).

A pendent epiphytic herb. *Rhizome* dichotomising at nodes, up to 2 m long. *Pseudobulbs* 3–10 cm apart on the rhizome, ellipsoid to cylindrical, 0.5–1.8 cm long, 1–4 mm wide, unifoliate at the apex. *Leaf* narrowly ovate, acute, 6.5–8 cm long, 1–1.3 cm wide. *Inflorescences* arising from the rhizome; peduncle 1–3 mm long, (1-)2-flowered, the flowers opening one at a time. *Flowers* cream-coloured or white; pedicel and ovary 1.2–2.5 mm long; dorsal sepal ovate, acuminate, glabrous, 3.5–6 mm long, 0.8–1.6 mm wide; lateral sepals free, ovate to triangular, acuminate, glabrous, 4.5–6.5 mm long, 0.8–1.5 mm wide; petals obovate to elliptic, obtuse, glabrous, 1.2–2.1 mm long, 0.5–1.3 mm wide; lip fleshy, approximately straight, ovate, obtuse, 1–1.7 mm long, 0.3–0.7 mm wide, glabrous; column 0.5–1 mm long.

DISTRIBUTION: Bougainville. Also in Indonesia, New Guinea and Vanuatu.
HABITAT: Montane rain forest, 750 m.
COLLECTION: *Craven & Schodde* 282 (K).

Bulbophyllum dichotomum differs from the other species in this section by the generally two-flowered inflorescences, the straight lip and the presence of two pollinia only (the other species have four).

29. B. graciliscapum *Schltr.* in K. Schum. & Laut., Nachtr. Fl. Deutsch. Südsee: 203 (1905). Type: New Guinea, *Schlechter* 13924 (holotype B†).
Bulbophyllum setipes Schltr. in Fedde, Repert. Spec. Nov. Regni Veg. Beih. 1: 869 (1913). Type: New Guinea, *Schlechter* 18684 (holotype B†).

Fig. 62. Bulbophyllum levatii. **A**, habit; **B**, flower; **C**, dissection of flower; **D**, dorsal and ventral view of lip; **E**, side view of lip and column; **F**, ventral and dorsal view of anther-cap; **G**, pollinia and pollen sac. Drawn from *Bowden* s.n. and *Green* 1057 by Jaap Vermeulen.

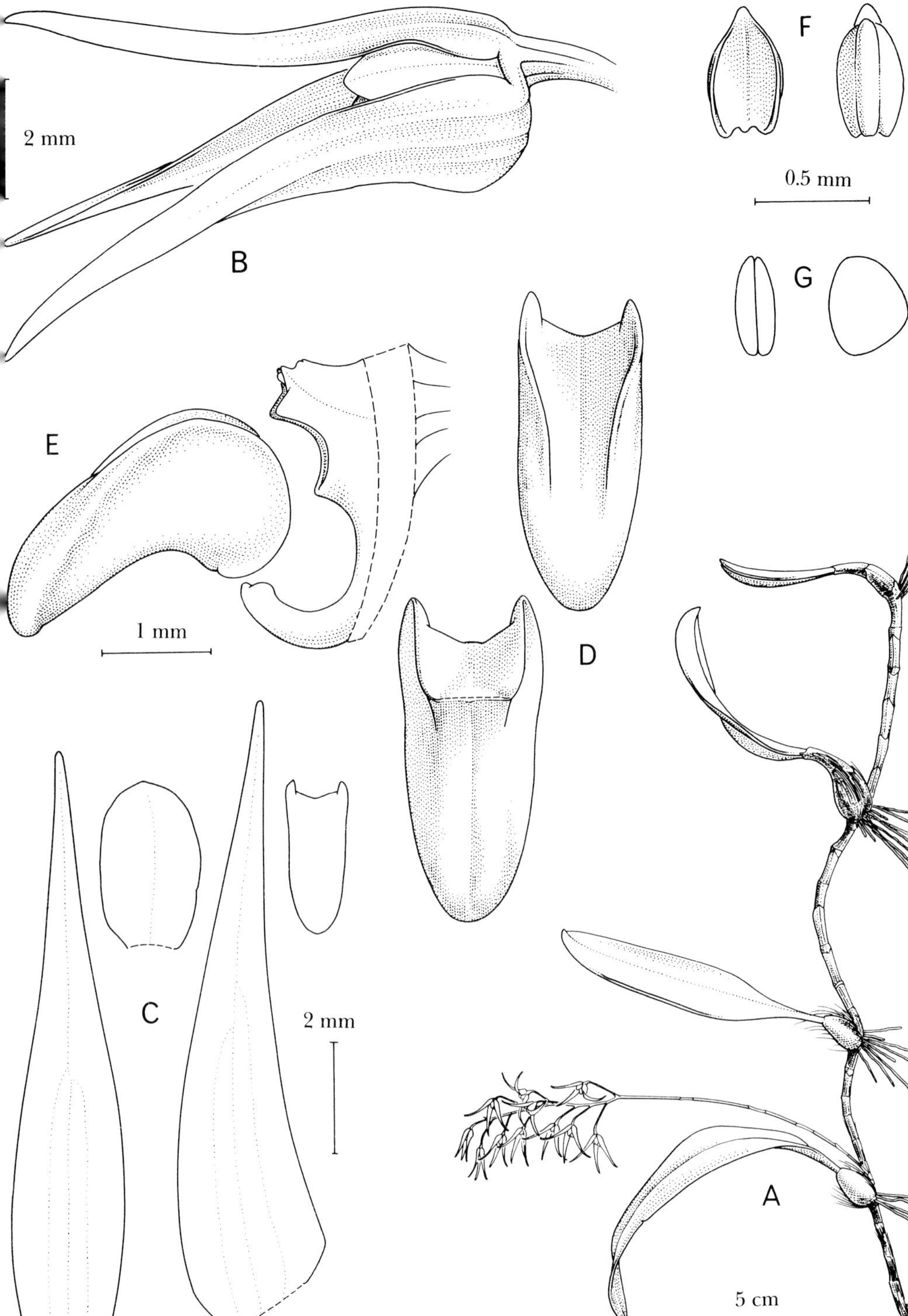
B
2 mm
F
0.5 mm
G
E
1 mm
D
C
2 mm
A
5 cm

An epiphytic herb. *Rhizome* creeping, woody. *Pseudobulbs* spaced 2–5 cm apart, ovate, 5–6 mm high, 3–4 mm wide, unifoliate at the apex. *Leaf* erect, oblong-lanceolate, subacute, (1.5-) 2–6 cm long, 5–8 mm wide. *Inflorescence* arising from the base of the pseudobulb; peduncle 10–18 cm long; rhachis zig-zag when in flower, straight in fruit, 4- to 6-flowered. *Flowers* yellow with maroon stripes; dorsal sepal ovate, acute, 4 mm long, 2 mm wide; lateral sepals ovate, acute, 4 mm long, 2 mm wide; petals oblong, obtuse, 2 mm long, 1.5 mm wide; lip thick, tongue-shaped with 2 ridged keels, 2–3 mm long, 1 mm wide, the surface finely papillose; column 1.5 mm long, the column-foot 1–1.5 mm long, upcurved at the apex. (See plate 13B).

DISTRIBUTION: Malaita and Rendova. Also in New Guinea and Vanuatu.
HABITAT: Rain forest, 50 m.
COLLECTIONS: *Brown* W/261 (BM); *Cribb & Campbell* 5157 (K).

30. B. levatii *Kraenzl.* in Bull. Soc. Bot. Fr. 76: 300 (1929). Type: Vanuatu, *Levat* s.n. (holotype MPU?).

An epiphytic herb. *Rhizome* creeping, woody. *Pseudobulbs* well spaced on the rhizome, 2–4 cm apart, ovate, 7 mm high, 7 mm wide, yellow-green to green, unifoliate at the apex. *Leaf* erect, oblong, obtuse to subacute, 3–7 cm long, 0.8–1.2 cm wide; petiole up to 1 cm long. *Inflorescence* arising from the base of pseudobulb; peduncle 10–12 cm long; rhachis purple basally, paler to white apically, 4- to 11-flowered. *Flowers* white; dorsal sepal lanceolate, acuminate, 1–1.2 cm long, 2.5 mm wide; lateral sepals lanceolate-acuminate, 1–1.2 cm long, 3 mm wide; petals oblong, obtuse, 3–3.5 mm long, 2 mm wide; lip thick, tongue-shaped, 2–3 mm long, 1.5 mm wide, recurved with 2 ridged keels; column 1–2 mm long, the column-foot 1 mm long, upcurved. (See fig. 62 and plate 13A).

DISTRIBUTION: Guadalcanal. Also in Vanuatu.
HABITAT: Montane forest, 1250 m.
COLLECTION: *Cribb, Dennis & Campbell* 5047 (K).

We originally considered *B. levatii* to be conspecific with *B. pensile* Schltr. (see Lewis & Cribb, 1989). However, Vermeulen (pers. comm.) has established that *B. pensile* is a distinct species confined to New Guinea.

31. B. orbiculare *J.J. Smith* in Bull. Jard. Bot. Buitenz. ser. 2, 3: 75 (1912). Type: New Guinea, *Gjellerup* 483 (holotype BO!).

An epiphytic herb. *Rhizome* creeping. *Pseudobulbs* up to 3 cm apart on rhizome but may be clustered in young growth, ovoid, 4–10 mm long, 5–8 mm wide,

Fig. 63. Bulbophyllum orbiculare. **A**, habit x 2/3; **B**, part of inflorescence x 6; **C**, flower x 6; **D**, dorsal sepal x 12; **E**, petal x 12; **F**, lateral sepal x 12; **G**, anther-cap x 14; **H**, pollinia x 14; **J**, column x 14; **K**, lip x 14; **L**, side view of lip x 14. Drawn from *Wickison* 250 by Sue Wickison

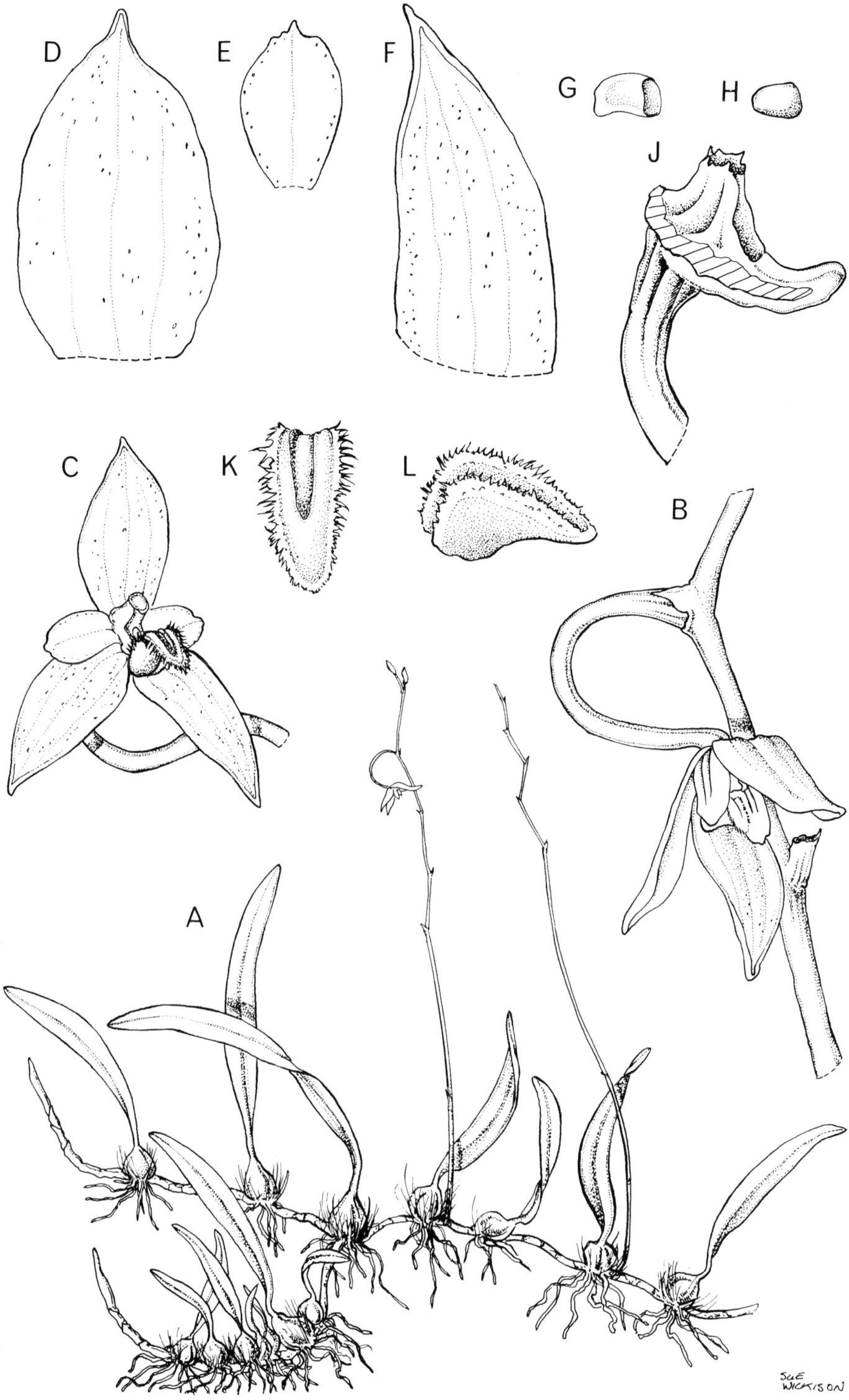
D
E
F
G
H
J
C
K
L
B
A
SUE WICKISON

unifoliate at the apex. *Leaf* oblong-lanceolate, acute, 3.5–6.5 cm long, 5–12 mm wide. *Inflorescence* arising from the base of the pseudobulb, up to 30 cm long, laxly few-flowered, with the flowers at the base developing first and the pedicel and ovary recurving as the flower matures. *Flowers* greenish-yellow with a green lip or purple with transparent white margins and a purple lip with a white hairy margin; dorsal sepal ovate, acute, 3.5–6.5 mm long, 2–4.5 mm wide; lateral sepals obliquely ovate, acute, 4–7 mm long, 2–3.5 mm wide; petals elliptic, apiculate, 1.5–2.8 mm long, 1–2.8 mm wide; lip fleshy, tongue-shaped, 1.5–3 mm long, 0.7–1.6 mm wide, with a hairy upper margin; column 1–1.2 mm long, the column-foot 1 mm long. (See fig. 63 and plate 13E).

DISTRIBUTION: Bougainville, Guadalcanal and Kolombangara. Also in New Guinea.

HABITAT: Montane forest, 1100–1800 m.

COLLECTIONS: *Chaplin* 831 (K); *Cribb et al.* 5121 (K); *Cribb & Morrison* 1977 (K); *Hunt* in RSS 2984 (K); *Mitchell* 7 (K); *Lavarack & Ridsdale* in NGF 31329 (K) & 31399 (LAE); *Wickison* 76, 225 & 250 (K).

Section OXYSEPALA

Pseudobulbs distinct; leaves oblong-lanceolate; lip fleshy, tongue-shaped .... **32. B. citrinilabre**

Pseudobulbs inconspicuous; leaves narrowly elliptic to elliptic; lip thin-textured, oblong to pandurate .... **33. B. sessile**

32. B. citrinilabre *J.J. Smith* in Fedde, Repert. Spec. Nov. Regni Veg. 12: 403 (1912). Type: New Guinea, *Lorentz* 530 (holotype BO).

An epiphytic herb. *Rhizome* hanging. *Pseudobulbs* spaced 1.5–3 cm apart, cylindrical, appressed to rhizome, 5- 10 mm long, 5 mm wide, unifoliate at the apex. *Leaf* erect, oblong-lanceolate, 4.5–5 cm long, 4–6 mm wide. *Inflorescences* arising from the base of the rhizome, 5 mm long, sheathed in bracts, bearing a solitary flower. *Flower* not opening widely, yellow; dorsal sepal ovate, acute, 5 mm long, 2 mm wide, thickened at apex; lateral sepals obliquely ovate, acute, 6 mm long, 3 mm wide; petals oblong-ovate, 2.5 mm long, 1 mm wide; lip fleshy, tongue-shaped to obscurely trilobed, recurved at apex, 2 mm long, 1.5 mm wide, with the lateral margins erect; column 1 mm long, the column-foot 1.5–2 mm long. (See fig. 57 L–U and plate 12F).

Fig. 64. Bulbophyllum fractiflexum subsp. **solomonense A-K. A**, habit x 2/3; **B**, flower and bract x 8; **C**, column x 14; **D**, dorsal and ventral view of anther-cap x 14; **E**, pollinia x 14; **F**, side view of lip x 14; **G**, dorsal sepal x 10; **H**, petal x 10; **J**, lateral sepal x 10; **K**, lip x 14. **B. sessile L-T. L**, habit x 2/3; **M**, leaf x 2/3; **N**, flower and bract x 8; **O**, column x 14; **P**, dorsal and ventral view of anther-cap x 14; **Q**, lip x 14; **R**, dorsal sepal x 10; **S**, petal x 10; **T**, lateral sepal x 10. **A-K** drawn from *Wickison* 84; **L-T** from *Wickison* 13A. All drawn by Sue Wickison.

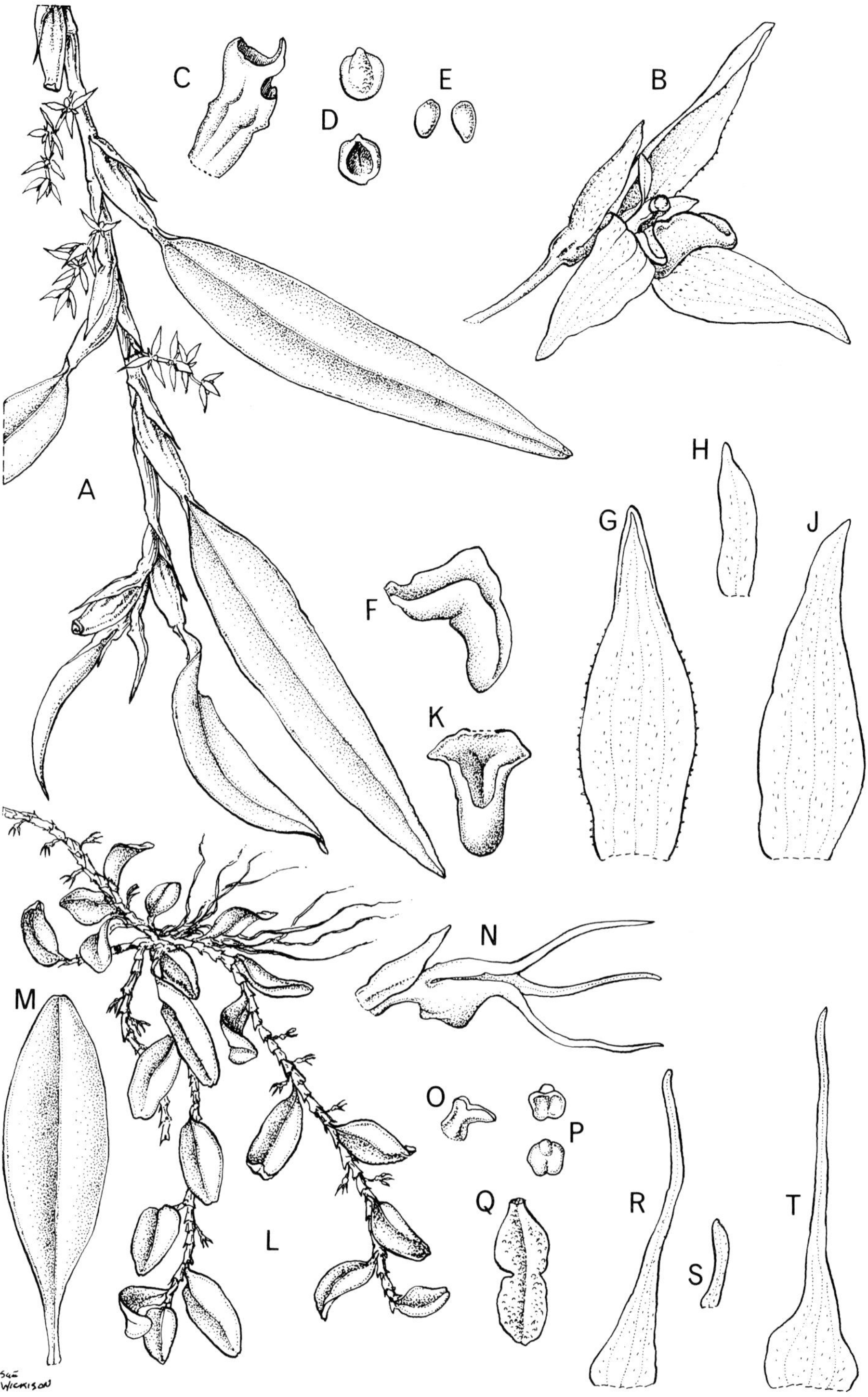
A
B
C
D
E
F
G
H
J
K
L
M
N
O
P
Q
R
S
T
SUE WICKISON

DISTRIBUTION: Malaita. Also in New Guinea.
HABITAT: Rain forest, 300–600 m.
COLLECTIONS: *Wickison* 202 & 252 (K).

33. B. sessile *(Koen.) J.J. Smith*, Orch. Java: 448 (1905). Type: Thailand, *Koenig* s.n. (holotype K!).
Epidendrum sessile Koen. in Retz., Obs. Bot. 6: 60 (1791).
Phyllorchis sessile Kuntze, Rev. Gen. Pl. 2: 676 (1891).

For full synonymy see Seidenfaden (1979).

A small creeping epiphytic herb forming dense clumps. *Rhizome* pendent, woody, branching, covered in bracts. *Pseudobulbs* 1–2 cm apart and appressed to the rhizome, minute, conical, up to 4 mm long, 2 mm wide, green, unifoliate at the apex. *Leaf* variable in size, narrowly elliptic to elliptic, 1–7 cm long, 0.5–2 cm wide, fleshy, light green. *Inflorescences* arising from the rhizome, 1-flowered; peduncle abbreviated, 0.5–1 mm long. *Flower* creamy-white to yellow with a yellow lip; dorsal sepal lanceolate, acuminate, 5–7 mm long, 0.8 mm wide; lateral sepals lanceolate, acuminate, 5–8 mm long, 1.2 mm wide; petals narrowly oblong, 1–1.5 mm long, 0.3–0.8 mm wide; lip oblong to pandurate, 1.5–1.7 mm long, 0.8 mm wide, recurved at apex, thin-textured; column 0.2 mm long, the column-foot 0.3 mm long. (See fig. 64 L–T).

DISTRIBUTION: Bougainville, Choiseul, Guadalcanal and New Georgia. A widespread species from SE. Asia to New Guinea, New Ireland and Fiji.
HABITAT: Rain forest, sea level to 1430 m.
COLLECTIONS: *Chaplin* 858 (K); *Cribb et al.* 5089 (K); *Im Thurn* 371 (K); *Lazarus* 1 (K); *Mitchell* 21 (K); *Waterhouse* 380 (K); *Wickison* 13A & 13B (K).

Vermeulen (pers. comm.) has noted that two different species may have been confused in this concept. The plants found in the east of its distribution differ in having a narrower lip than those found in the west and the former may be better referred to *B. volkensii* Schltr., originally described from the Caroline Islands (Palau). A critical revision of this group is needed.

Section PELMA

Leaves lanceolate, 5–7 cm long, 0.8–1 cm wide; lip obscurely trilobed and without a closed slit in the ventral basal half
.... **34. B. fractiflexum** subsp. **solomonense**
Leaves ovate to elliptic, acute, 5.5–8.5 cm long, 1–2.3 cm wide; lip elliptic, the ventral surface with a closed slit in the basal half .. ..**35. B. melanoxanthum**

34. B. fractiflexum *J.J. Smith* in Bull. Dép. Agric. Indes Néerl. 19: 6 (1908), non Kraenzl. (1912). Type: New Guinea, *Versteeg* 1295 (holotype BO, isotypes K!, L, U).

subsp. **solomonense** *J. Vermeulen & B. Lewis* **subsp. nov.** a subsp. typ. sepalis latioribus et labello longiore distinguendo. Typus: Bougainville, *Schodde & Craven* 3858 (holotypus CANB, isotypi L, LAE!).

A pendent epiphyte. *Rhizome* woody, covered in sheathing bracts. *Pseudobulbs* appressed to the rhizome, 1.5–2 cm apart, conical, up to 2 cm long, 8 mm wide, unifoliate at the apex. *Leaf* lanceolate, acute, 5–7 cm long, 0.8–1 cm wide. *Inflorescences* arising from the rhizome, 3- to 5-flowered; peduncle up to 1.7 cm long. *Flowers* creamy-white; dorsal sepal ovate, acute, 4.5 mm long, 1.5 mm wide, with a finely ciliate margin; lateral sepals obliquely ovate, acute, 4.5 mm long, 1.5 mm wide; petals linear, acute, 2 mm long, 0.5 mm wide; lip fleshy, obscurely trilobed in basal half, 1 mm long, 1 mm wide, the lateral lobes erect, the mid-lobe oblong, rounded, reflexed; column 1 mm long, the column-foot 1 mm long. (See fig. 64 A–K).

DISTRIBUTION: Bougainville, Guadalacanal and Makira.
HABITAT: Lowland rain forest, 50–600 m.
COLLECTIONS: *Hunt* in RSS 2203 & 2225 (K); *Wickison* 84 (K); *Schodde & Craven* 3858 (LAE).

Subsp. *solomonense* can be distinguished from the typical subspecies by the broader sepals and longer lip.

35. B. melanoxanthum *J. Vermeulen & B. Lewis* **sp. nov.** affine *B. bacillifero* J.J. Smith sed dentibus basalibus labelli marginibus integri contingentibus, pseudobulbis flavis et foliis fere nigris in horto sicco satis differt. Type: Bougainville, *Craven & Schodde* 137 (holotypus CANB!, isotypus L!).

An epiphytic herb. *Rhizome* pendulous, up to 50 cm long. *Pseudobulbs* 2.5–4.5 cm apart, ovoid, 2–3 cm long, 6–8 mm wide, somewhat flattened, unifoliate at the apex. *Leaf* ovate to elliptic, acute, 5.8–8.5 cm long, 1–2.3 cm wide; petiole 4–7 mm long. *Inflorescence* arising from the rhizome, 4- to 8-flowered; peduncle 1–1.7 cm long; bracts triangular, acute. *Flowers* scarcely opening, creamy-white with a creamy-yellow lip; pedicel and ovary 2 mm long; dorsal sepal rather thick, elliptic, obtuse, 4 mm long, 1.8 mm wide; lateral sepals fleshy, free, elliptic, obtuse, 4.3 mm long, 1.7 mm wide; petals thin-textured, obovate, rounded at the tip, 3 mm long, 1.3 mm wide, the margins finely papillose towards the tip; lip fleshy, slightly curved towards the tip, elliptic, rounded at apex, 2.4 mm long, 1 mm wide, the ventral surface with a distinct basal concavity, the basal teeth converging, touching along their entire proximal margins, leaving only an approximate closed slit between, the basal part of the lip concave, gradually passing into the convex top part of the lip; column 0.7 mm long, the rostellum distinctly protruding, the column-foot with a rounded central knob above the attachment of the lip. (See fig. 65).

DISTRIBUTION: Bougainville and New Georgia.
HABITAT: Rain forest, 400–800 m.
COLLECTIONS: *Craven & Schodde* 137 (CANB, L); *Wickison* s.n. (K).

This species differs from *B. bacilliferum* J.J. Smith in the structure of the basal teeth on the lip; in *B. melanoxanthum* these are touching approximately along their entire front margin, whereas in *B. bacilliferum* they only touch at the tip, leaving a distinct channel between towards the base.

The herbarium specimens are of a distinctive colour with yellow pseudobulbs and almost black leaves. The name has been derived from Greek and refers to this feature, 'melanos' meaning black, and 'xanthos', yellow.

Section POLYBLEPHARON

1. Leaves 3–5 mm wide .... **37. B. stenochilum**
   Leaves 7–15 mm wide 2
2. Leaves (3.5-)4–7 cm long; flowers crimson with black scales on the outside of the sepals and the ovary .... **38. B. ulcerosum**
   Leaves 2–4 cm long; flowers yellow-green to purple, the petals pale translucent yellow and the lip purple, without scales .... **36. B. membranaceum**

36. B. membranaceum *Teijsm. & Binnend.* in Natuurk. Tijdschr. Ned. Ind. 3: 397 (1855). Type: Java, *Teijsmann & Binnendijk* s.n. (holotype BO!).
Bulbophyllum nuruanum Schltr. in K. Schum. & Laut., Nachtr. Fl. Deutsch. Südsee: 212 (1905). Type: New Guinea, *Schlechter* 13800 (holotype B†).

An epiphytic herb. *Rhizome* creeping. *Pseudobulbs* 2.5–6 cm apart, ovoid, 6–7 mm long, 5 mm wide, unifoliate at the apex. *Leaf* ovate, acute, 2–4 cm long, (0.7-)1–1.5 cm wide. *Inflorescences* arising from the base of the pseudobulb or from the rhizome, 2 mm long, bearing a solitary flower. *Flower* not opening widely, yellow-green to purple, the petals pale translucent yellow, the lip purple; dorsal sepal ovate, acute, 3.5 mm long, 1.5 mm wide; lateral sepals joined together and held parallel to one another, obliquely ovate, acute, 4 mm long, 2.5 mm wide; petals oblong, apiculate, 1.5 mm long, 0.8 mm wide; lip oblong to tongue-shaped, 1.5 mm long, 0.8 mm wide, papillose, recurved; column 1 mm long, the column-foot 1.2 mm long.

DISTRIBUTION: Guadalcanal. A wide spread species from Thailand, the Malay Peninsula, Java, Borneo and Sulawesi to New Guinea, Vanuatu and Samoa.
HABITAT: Lowland and montane rain forest.
COLLECTION: *Cribb, Campbell & Dennis* 5057 (K).

This species is similar to *B. gibbonianum* Schltr. from the Caroline Islands (Palau).

Fig. 65. Bulbophyllum melanoxanthum. **A**, habit; **B**, part of inflorescence; **C**, dissection of flower; **D**, dorsal and ventral view of lip; **E**, side view of lip and column; **F**, dorsal and ventral view of anther-cap. Drawn from *Craven & Schodde* 137 by Jaap Vermeulen.

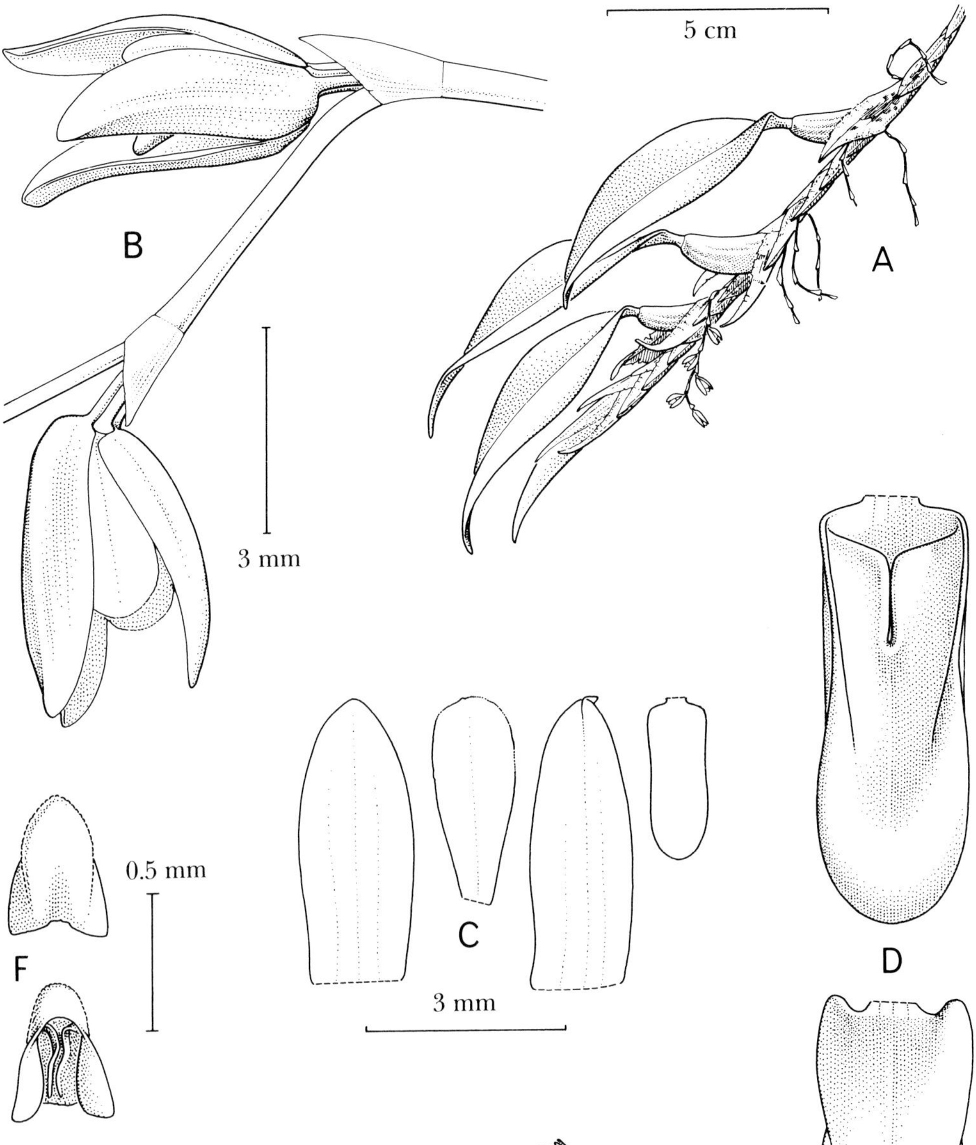

5 cm
A
B
3 mm
C
3 mm
D
0.5 mm
F
E
1 mm

37. B. stenochilum *Schltr.* in Fedde, Repert. Spec. Nov. Regni Veg., Beih. 1: 823 (1913). Type: New Guinea, *Schlechter* 19136 (holotype B†).

A delicate, pendent epiphytic herb. *Pseudobulbs* appressed to the rhizome, 7–12 mm apart, conical, up to 8 mm long, 1–2 mm wide, unifoliate at the apex. *Leaf* lanceolate, 2–3.5 cm long, 3–5 mm wide. *Inflorescence* arising from the base of the pseudobulb, 3 mm long, few-flowered, the flowers surrounded by bracts. *Flowers* rarely opening, purple; sepals papillose on exterior surface; dorsal sepal ovate, 4.2 mm long, 2.4 mm wide; lateral sepals connate almost to the apices, 4 mm long, 2.3 mm wide; petals lanceolate, obliquely acute, 1.1 mm long, 0.7 mm wide; lip narrowly oblong, 1 mm long, 0.2 mm wide, shortly hairy; column 1 mm long, the column-foot 8 mm long.

DISTRIBUTION: Guadalcanal and Kolombangara. Also in New Guinea.
HABITAT: Montane moss forest, 1600–1900 m.
COLLECTIONS: *Cribb & Campbell* 5139 (K); *Hunt* in RSS 2979 (K).

This species is similar to *B. bisepalum* Schltr. from New Guinea.

38. B. ulcerosum *J.J. Smith* in Bull. Dép. Agric. Indes Néerl. 39: 5 (1910). Type: New Guinea, *von Romer* 809 (holotype BO!).

An epiphytic pendent herb. *Pseudobulbs* appressed to the rhizome, c. 1 cm apart, conical, up to 2 cm long, 1 cm wide, dull green, unifoliate at the apex. *Leaf* lanceolate, acute, (3.5-)4–7 cm long, 0.7–1.5 cm wide, dull green spotted red on the undersurface towards the base. *Inflorescence* arising from the base of the pseudobulb, 5–10 mm long, few-flowered, surrounded by bracts. *Flowers* crimson with black scales on the outside of the sepals and the ovary; dorsal sepal oblong-ovate, obtuse, 3.5 mm long, 1.5 mm wide; lateral sepals connate for a half to two-thirds of their length, 6 mm long, 1–1.5 mm wide; petals rotund, obtuse, thin, 0.7 mm long, 0.9 mm wide; lip obovate, 1.5 mm long, 0.8 mm wide, shortly hairy, thin, recurved in the centre; column 2.5 mm long, the column-foot 2.2 mm long.

DISTRIBUTION: Makira. Also in New Guinea.
HABITAT: Ridge-top forest, 560 m.
COLLECTION: *Hunt* in RSS 2221 (K).

This species is similar to *B. oligochaete* Schltr., from New Guinea.

Section SESTOCHILUS

1. Lip 2 cm long, acuminate at the apex .. .. **41. B. tollenoniferum**
   Lip less than 1 cm long, not acuminate at the apex 2
2. Flowers yellow; sepals pustular on outer surface .. .. **39. B. hahlianum**
   Flowers cream with purple blotches inside with a yellow lip with purple markings; sepals not pustular on outer surface .. .. **40. B. macranthum**

39. B. hahlianum *Schltr.* in K. Schum. & Laut., Nachtr. Fl. Deutsch. Südsee: 204 (1905). Type: New Ireland, *Schlechter* 13714 (holotype B†).

A robust epiphytic herb. *Rhizome* creeping, woody, covered in thick roots, branching. *Pseudobulbs* well-spaced on the rhizome, ovoid, up to 2 cm long, 1 cm wide, unifoliate at the apex. *Leaf* oblong-elliptic, obtuse, 15–21 cm long, 3.5–4.5 cm wide, dark green; petiole up to 4.5 cm long. *Inflorescences* arising from nodes on the rhizome, 1.5–2 cm long, bearing a solitary flower. *Flower* non-resupinate, yellow; sepals pustular on outer surface; dorsal sepal ovate, 1.5 cm long, 6 mm wide; lateral sepals obliquely ovate, acute, 1.2 cm long, 8 mm wide; petals narrowly ovate, 1.1 cm long, 5 mm wide; lip absent; column 4 mm long, the column-foot 5 mm long.

DISTRIBUTION: New Georgia. Also in New Guinea and New Ireland.
HABITAT: Lowland rain forest.
COLLECTION: *Nicholls* 1 (K).

Unfortunately the lip is absent from the cited collection but *B. hahlianum* Schltr. has a small, recurved, tongue-shaped lip. Our specimen differs from the type description of *B. hahlianum* in having somewhat smaller flowers. More material is needed of this species to confirm the identification.

40. B. macranthum *Lindley*, Bot. Reg. 30: t. 13 (1844). Type: Singapore, *Loddiges* s.n. (holotype K!).
Sarcopodium macranthum (Lindley) Lindley in Paxt., Fl. Gard. 1: 155 (1850).

For full synonymy see Seidenfaden (1979).

A robust epiphytic herb. *Rhizome* creeping, covered in fibers and woody roots. *Pseudobulbs* 3–13 cm apart on the rhizome, ovoid, 2.5 cm high, 1 cm wide, yellow, unifoliate at the apex. *Leaf* erect, elliptic, up to 28 cm long, 7 cm wide, minutely bilobed at apex, thick and leathery, yellow-green; petiole up to 5 cm long. *Inflorescences* arising from the rhizome close to a pseudobulb; peduncle up to 10 cm long, bearing a solitary flower. *Flower* non-resupinate, cream with purple blotches inside, a yellow lip with purple markings and a yellow column, sweetly fragrant; dorsal sepal obliquely ovate, 2.2–2.5 cm long, 5.5–8 mm wide; lateral sepals falcate, 2.1 cm long, 7.5–10 mm wide, usually touching at apex; petals lanceolate, 1.9–2.2 cm long, 5–6 mm wide; lip obovate, fleshy, 7–9 mm long, 4–5 mm wide, recurved; column 4–7 mm long, the column-foot 3 mm long. (See fig. 66 A–J).

DISTRIBUTION: Bougainville, Choiseul, Guadalcanal, Kolombangara, Makira, Malaita, New Georgia and Rendova. Also in the Malay Peninsula and Archipelago.
HABITAT: Lowland rain forest, 15–150 m.
COLLECTIONS: *Chaplin* 756 & 758 (K); *Comins* 365 (K); *Dennis et al.* in BSIP 8172 (K); *Dockrill* in NGF 34181 (K); *Hunt* in RSS 2249 (K); *Metcalfe* s.n. & cult. R.B.G. Sydney (K); *Mitchell* 55 (K); *Wickison & Billy* 214 (K); *Woodford* 9 (K).

Slade (1965) notes that *B. macranthum* '...sets out to attract fruit flies by emitting a fruity odour like the first flush of steam-distilled aroma from raspberry jam; so powerful is it that one is led by the nose to its rather lovely, cathedral-window-like inverted flower from easily 20 yards away. The fruit flies behave as if possessed. They scarcely notice any outside object and wander over the waxlike sepals and petals constantly applying their disc proboscis to the flower, finally crawling onto the inverted, balanced labellum which suddenly precipitates them onto the column below. Few, if any, flowers retain their pollen more than a few hours, so successfully this design and specific odorous attraction works!'.

41. B. tollenoniferum *J.J. Smith* in Fedde, Repert. Spec. Nov. Regni Veg. 11: 139 (1912). Type: New Guinea, *K. Gjellerup* 997 (holotype BO!).

A robust epiphytic herb. *Rhizome* creeping, covered in fibers and woody roots. *Pseudobulbs* spaced 4–8 cm apart, ovoid, 1.5 cm high, 1 cm wide, unifoliate at the apex. *Leaf* erect, elliptic, up to 16 cm long, 5 cm wide, minutely bilobed at apex, thick and leathery; petiole up to 3 cm long. *Inflorescences* arising from the rhizome close to a pseudobulb; peduncle up to 7 cm long, bearing a solitary flower. *Flower* non-resupinate, orange with the lip and sepals purple-edged; dorsal sepal lanceolate, acute, 2.7 cm long, 5 mm wide, minutely papillose on outer surface; lateral sepals lanceolate, acute, 2.7 cm long, 1.2 cm wide; petals lanceolate, acute, 2.7 cm long, 4 mm wide; lip fleshy, obovate, acuminate, 2 cm long, 4 mm wide, recurved in basal half; column 6 mm long, the column-foot 6 mm long. (See fig. 66 K–L).

DISTRIBUTION: Rendova. Also in New Guinea.
HABITAT: Lowland rain forest, 50 m.
COLLECTION: *Cribb & Campbell* 5149 (K).

This plant differs from the type description in having a slightly longer lip.

60. PEDILOCHILUS Schlechter

Small, epiphytic herbs. *Pseudobulbs* clustered or spaced on creeping rhizome, unifoliate at the apex. *Leaf* erect. *Inflorescence* arising from the base of the pseudobulb, erect, 1-flowered. *Flower* small to medium-sized; sepals and petals free; petals curved slightly into the shape of the letter S; lip attached to the column-foot, immobile, saccate, the base clawed, with 2 basal auricles; column short, usually with a short column-foot; pollinia 2.

A genus of about fifteen species from New Guinea to Vanuatu, a single species being recorded.

Fig. 66. Bulbophyllum macranthum A-J. A, habit x 2/3; **B**, flower x 5/3; **C**, lip x 4; **D**, side view of lip x 4; **E**, dorsal sepal x 2; **F**, petal x 2; **G**, lateral sepal x 2; **H**, column x 4; **J**, dorsal and ventral view of anther-cap x 6. **B. tollenoniferum K-L. K**, flower x 5/3; **L**, lip x 2. **A-J** drawn from *Wickison & Billy* 214; **K-L** from *Cribb & Campbell* 5149. All drawn by Sue Wickison.

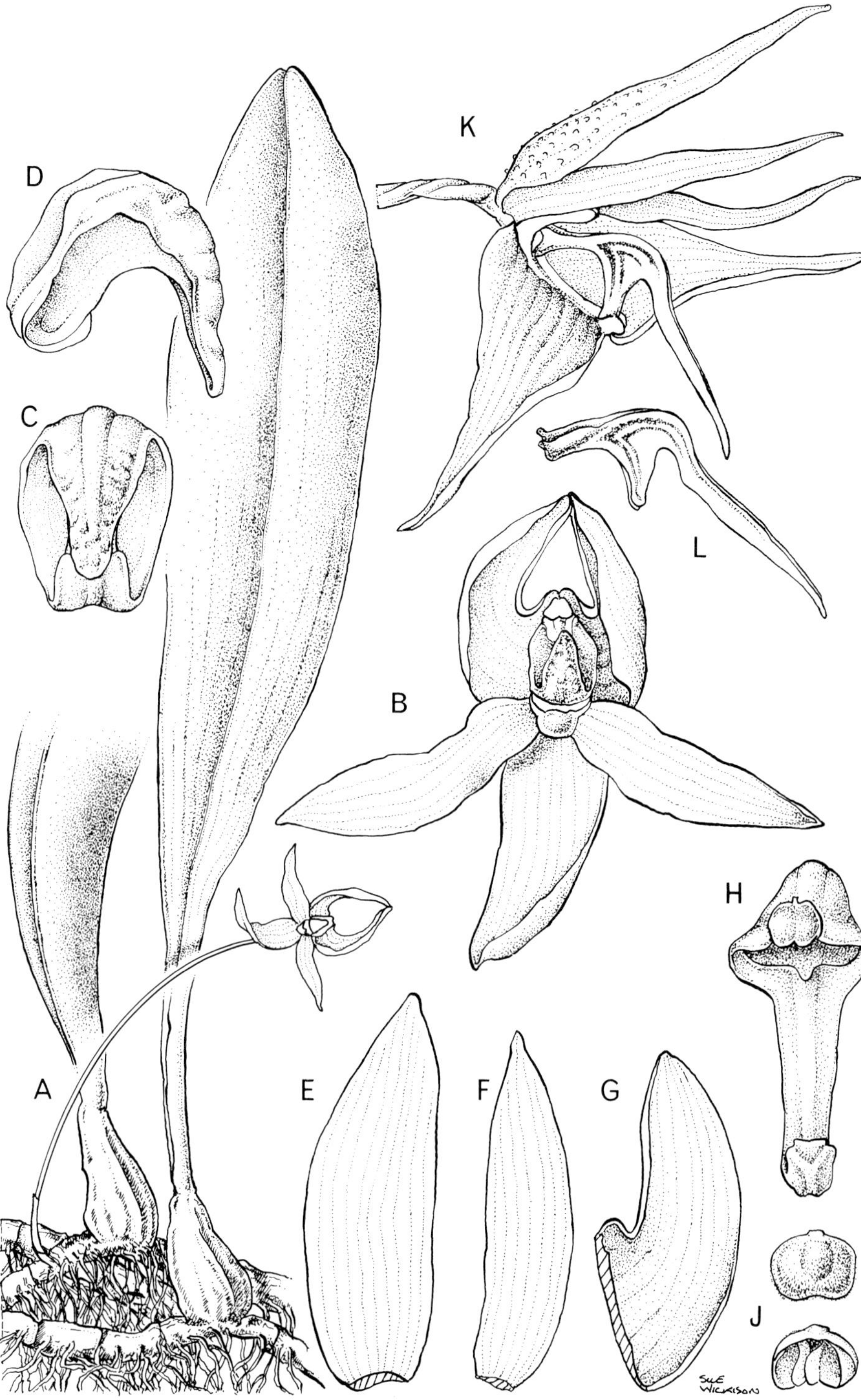
K
D
C
L
B
H
A
E
F
G
J
SUE WICKISON

P. ciliolatum *Schltr.* in Fedde, Repert. Spec. Nov. Regni Veg., Beih. 1: 688 (1912). Type: New Guinea, *Schlechter* 19750 (holotype B†).

A small epiphyte with a short creeping rhizome. *Pseudobulbs* erect, ovoid or obliquely ovoid, 5 mm tall, 3–5 mm wide, unifoliate at the apex and sheathed by papery bracts at the base. *Leaf* erect, coriaceous, linear to narrowly oblong-elliptic, obtuse, 5–8.5 cm long, 4–6 mm wide; petiole 0.5–1.5 cm long. *Inflorescence* arising from the base of the pseudobulb, 4 cm long, 1-flowered. *Flower* mid-yellow; dorsal sepal elliptic to obovate, 8–10 mm long, 4–5 mm wide, ciliolate; lateral sepals obliquely oblong-elliptic, 9–11 mm long, 5.5–6.5 mm wide, obscurely ciliolate; petals shortly clawed, obliquely ovate, 5–6 mm long, 3 mm wide; lip clawed, urceolate like a small slipper, acute at reflexed apex, 8–10 mm long, 4–5 mm wide, with recurved margins, with small recurved auricles at either side at the apex of the claw and a low fleshy callus between them; column 2 mm long, with long attenuate apical stelidia.

DISTRIBUTION: Guadalcanal. Also in New Guinea.
HABITAT: Montane mossy forest, 3000 m.
COLLECTION: *Hunt* in RSS 2971 (K).

61. CALYMMANTHERA Schlechter

Small epiphytic monopodial herbs. *Stems* short, leafy. *Leaves* succulent, distichous, linear-oblong to oblong-lanceolate, sheathing at base. *Inflorescences* lateral, erect, paniculate, many-flowered; bracts small. *Flowers* small, not opening widely; sepals and petals free, lanceolate, about equal in length; lip trilobed, with a relatively long, subfleshy mid-lobe; column and foot short; pollinia 4.

A small genus of about five species in New Guinea and the Pacific Islands, a single species being recorded.

C. major *Schltr.* in Fedde, Repert. Spec. Nov. Regni Veg., Beih. 1: 957 (1913). Type: New Guinea, *Schlechter* 16333 (holotype B†).

An epiphytic herb. *Stems* erect, sturdy, up to 17 cm tall, 6–9 mm wide. *Leaves* 6–9, distichous, oblong-lanceolate, 11.5–17 cm long, 1.5–2 cm wide, stiff, succulent, unequally bilobed at apex; leaf bases sheathing the stem. *Inflorescences* lateral, 2–4, pyramidal, many-flowered, up to 40 cm long, branched; branches 4–8 cm long, becoming shorter towards the apex; branchlets 0.5–4 cm long; bracts small. *Flowers* yellow becoming pale yellow at the base with a white lip; pedicel and ovary 1–2.5 mm long, pale yellow; sepals and petals equal, lanceolate, 2.5–3 mm long, not opening widely; lip trilobed, 1.5 mm long, 0.5 mm wide at base, the lateral lobes erect and rounded, the mid-lobe fleshy,

---

Fig. 67. Calymmanthera major. A, habit x 2/3; **B**, flower x 8; **C**, dorsal and ventral view of anther-cap x 14; **D**, column and lip x 14; **E**, pollinia x 14; **F**, dorsal sepal x 8; **G**, petal x 8; **H**, lateral sepal x 8; **J**, column x 14; **K**, lip x 14. Drawn from *Wickison* 137 by Sue Wickison.

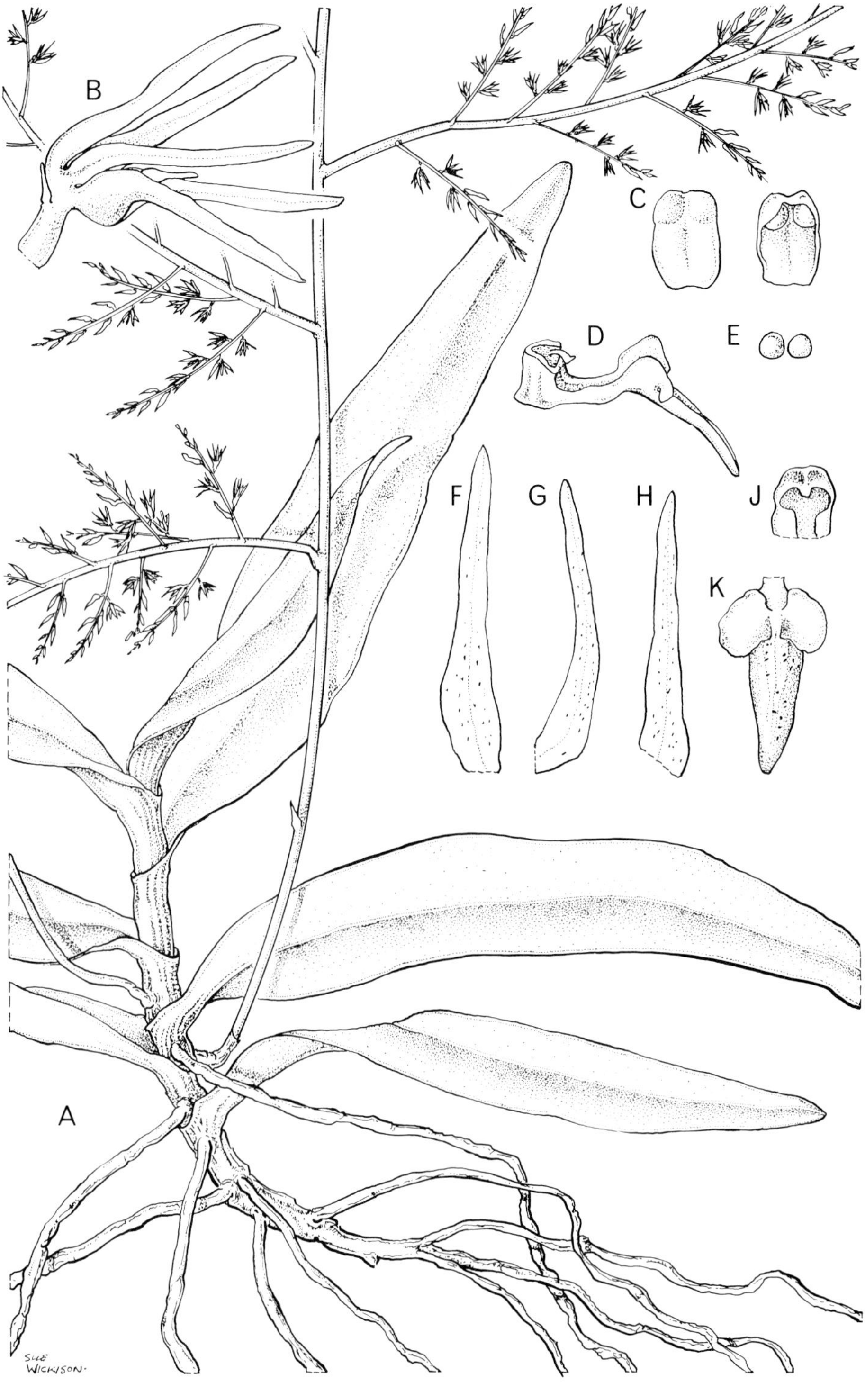
A
B
C
D
E
F
G
H
J
K
SUE WICKISON

obovate-lanceolate; column 0.7 mm long, column-foot 1 mm long. *Fruit* cylindrical, 1.8–2 cm long, 1–1.5 mm wide. (See fig. 67).

DISTRIBUTION: Guadalcanal, Kolombangara and Makira. Also in New Guinea and Fiji.
HABITAT: Rain forest, sea level to 580 m.
COLLECTIONS: *Cribb & Morrison* 1884 (K); *Dennis* 13 (K); *Hunt* in RSS 2235 & 2581 (K); *Wickison* 137 (K).

62. GUNNARELLA Senghas

Small epiphytic monopodial herbs. *Stems* short, leafy. *Leaves* distichous, often twisted at base to lie in one plane, sheathing at base. *Inflorescences* lateral, simple, pendent, many-flowered. *Flowers* not opening widely, diaphanous; sepals and petals free, lanceolate, about equal in length; lip trilobed, ecallose, lacking a spur, movably hinged to the column-foot, the lateral lobes erect; column short with a short foot; pollinia 4.
A small genus of about six species in New Guinea and the SW. Pacific Islands, a single species being recorded.

G. robertsii *(Schltr.) Senghas* in Die Orchidee 39(2): 73 (1988). Type: New Caledonia, *Roberts* s.n. (holotype B†).
Sarcochilus robertsii Schltr. in Fedde, Repert. Spec. Nov. Regni Veg. 3: 320 (1907).
Chamaenanthus robertsii (Schltr.) Schltr. in Fedde, Repert. Spec. Nov. Regni Veg., Beih 1: 957 (1913).

A small epiphytic herb, often pendent. *Stems* 1–2 cm long, leafy. *Leaves* up to 10, distichous, linear, 6–13 cm long, 0.7–1 cm wide, obliquely bilobed at apex, green sometimes with a reddish-mauve tinge, the leaf bases sheathing the stem. *Inflorescences* lateral, up to 28 cm long, simple, pendent, many-flowered. *Flowers* transluscent, white; sepals and petals linear-lanceolate, 4–5 mm long, 1–1.5 mm wide; lip obscurely trilobed, obovate, obtuse, 1–1.5 mm long, 0.5–1 mm wide, fleshy at the base, the lateral lobes small and erect. (See fig. 68 A–K and plate 14C).

DISTRIBUTION: Guadalcanal, Makira and Malaita. Also in New Guinea, Vanuatu and New Caledonia.
HABITAT: Lowland rain forest, 50–300 m.
COLLECTIONS: *Cribb* 6006 (K); *Cribb & Morrison* 1853 (K); *Hunt* in RSS 2207 (K).

Fig. 68. Gunnarella robertsii A-K. A, habit x 2/3; **B**, flower x 6; **C**, dorsal and ventral view of anther-cap x 18; **D**, pollinia x 18; **E**, side view of lip x 14; **F**, column x 18; **G**, lip x 14; **H**, dorsal sepal x 12; **J**, petal x 12; **K**, lateral sepal x 12. Drawn from *Cribb & Wheatley* 5 by Sue Wickison.

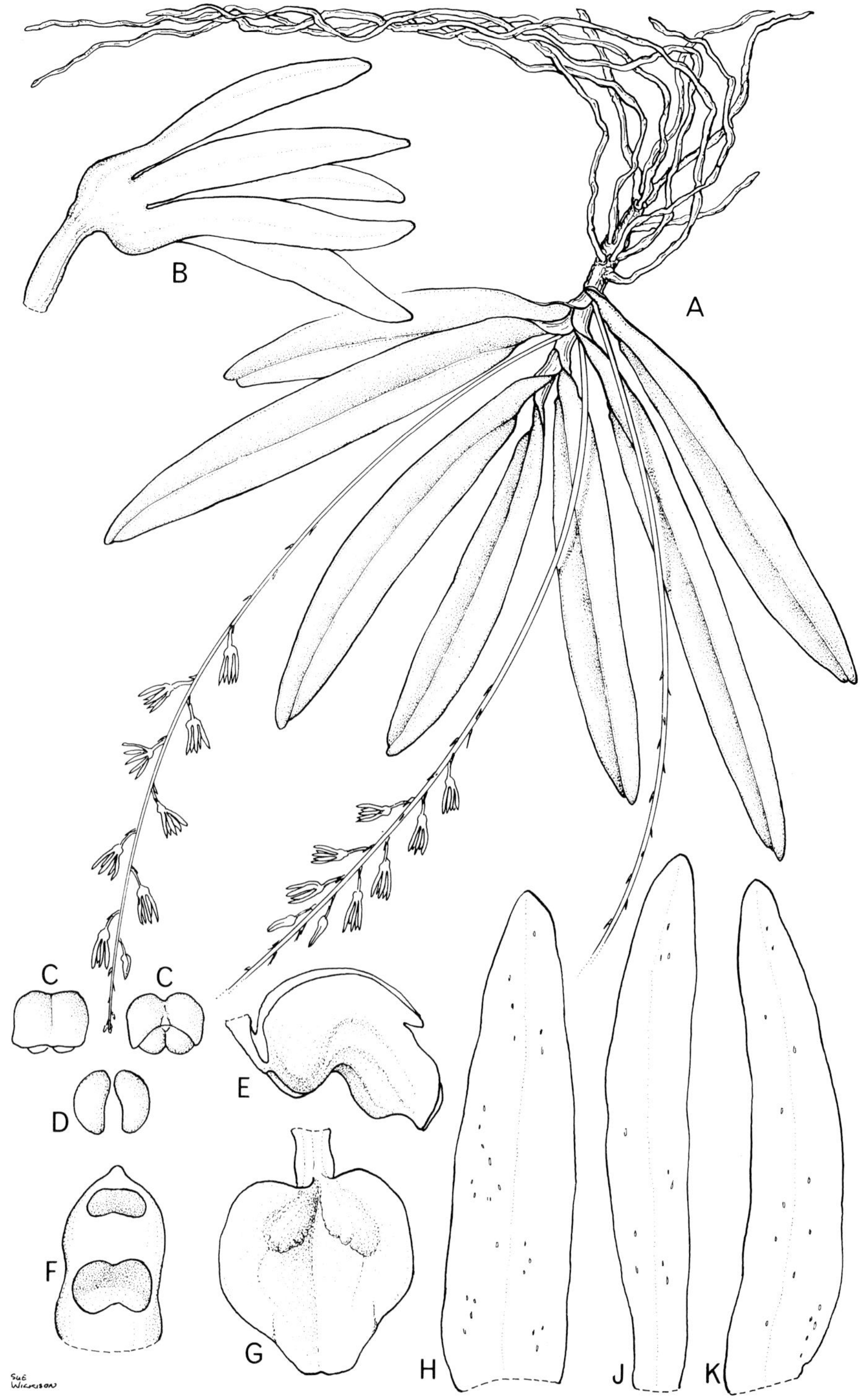
B
A
C
C
D
E
F
G
H
J
K

We originally considered *G. laxus* (Schltr.) Senghas, from New Guinea, to be conspecific with *G. robertsii* (Lewis & Cribb, 1989). Unfortunately, the type of *G. laxus* has been destroyed but recent observations by M. Clements (pers. comm.) have suggested that the two species are distinct.

63. RHINERRHIZA Rupp

Epiphytic monopodial plants. *Stems* short with robust, cylindrical roots. *Leaves* alternate, sheathing at base. *Inflorescences* lateral, simple, long, many-flowered. *Flowers* lasting 1–2 days; sepals and petals free, spathulate (in the Solomon Islands and Bougainville), similar; lip articulated on the apex of the column-foot, the lateral lobes large and erect, the mid-lobe small, situated in front of the spur, the disc with a central callus; column short with a long column-foot set at an acute angle to it; pollinia 4.

A small genus of two species from New Guinea, the Solomon Islands and Australia, a single species being recorded.

R. moorei *(Reichb. f.) M. Clements, B. Wallace & D. Jones* in Clements, Aust. Orchid Research 1: 128 (1989). Type: cult. London ex New Britain, *Moore* s.n. (holotype W).

Thrixspermum moorei Reichb. f. in Gard. Chron. n.s. 13: 104 (1880).

Thrixspermum beccarii Reichb. f. in Bot. Centralbl. 28: 343 (1886). Type: New Guinea, *Beccari* s.n. (holotype FI, isotype W).

Sarcochilus papuanus Kraenzl., Fl. Kaiser Wilh. Land: 34 (1889). Type: New Guinea, *Hollrung* 416 (holotype not traced).

Sarcochilus engleriana Kraenzl. in K. Schum. & Laut., Nachtr. Fl. Deutsch. Südsee: 252 (1901). Types: New Guinea, *Lauterbach* 2477 & 3355 (syntypes B†).

Sarcochilus solomonensis Rolfe in Bull. Misc. Inf., Kew 1908: 72 (1908). Type: Solomon Islands, Tulagi, *Woodford* 6 (holotype K!).

Sarcochilus moorei (Reichb. f.) Schltr. in Fedde, Repert. Spec. Nov. Regni Veg., Beih. 1: 967 (1912).

For full synonymy see Clements (1989).

A medium-sized epiphyte. *Stems* pendulous or spreading, short, up to 11 cm long. *Leaves* distichous, coriaceous, oblong or obliquely oblong-elliptic, obliquely obtuse at apex, up to 25 cm long, 5 cm wide, green tinged with purple, sheathing at base. *Inflorescence* lateral, pendent, up to 60 cm long, laxly many-flowered; peduncle up to 12 cm long. *Flowers* pale yellow spotted with red; pedicel and ovary 1.3–1.5 cm long; sepals and petals prominently narrowed at base; dorsal sepal spathulate, 1.2 cm long, 5 mm wide; lateral sepals elliptic-spathulate, 1.1 cm long, 5 mm wide; petals obovate-spathulate, 9 mm long, 4 mm

---

Fig. 69. Rhinerrhiza moorei. **A**, habit x 2/3; **B**, flower x 2; **C**, column x 8; **D**, dorsal and ventral view of anther-cap x 8; **E**, pollinia x 8; **F**, side view of lip x 6; **G**, lip x 6; **H**, transverse section through the lip x 6; **J**, dorsal sepal x 4; **K**, petal x 4; **L**, lateral sepal x 4. Drawn from *Wickison* 68 by Sue Wickison.

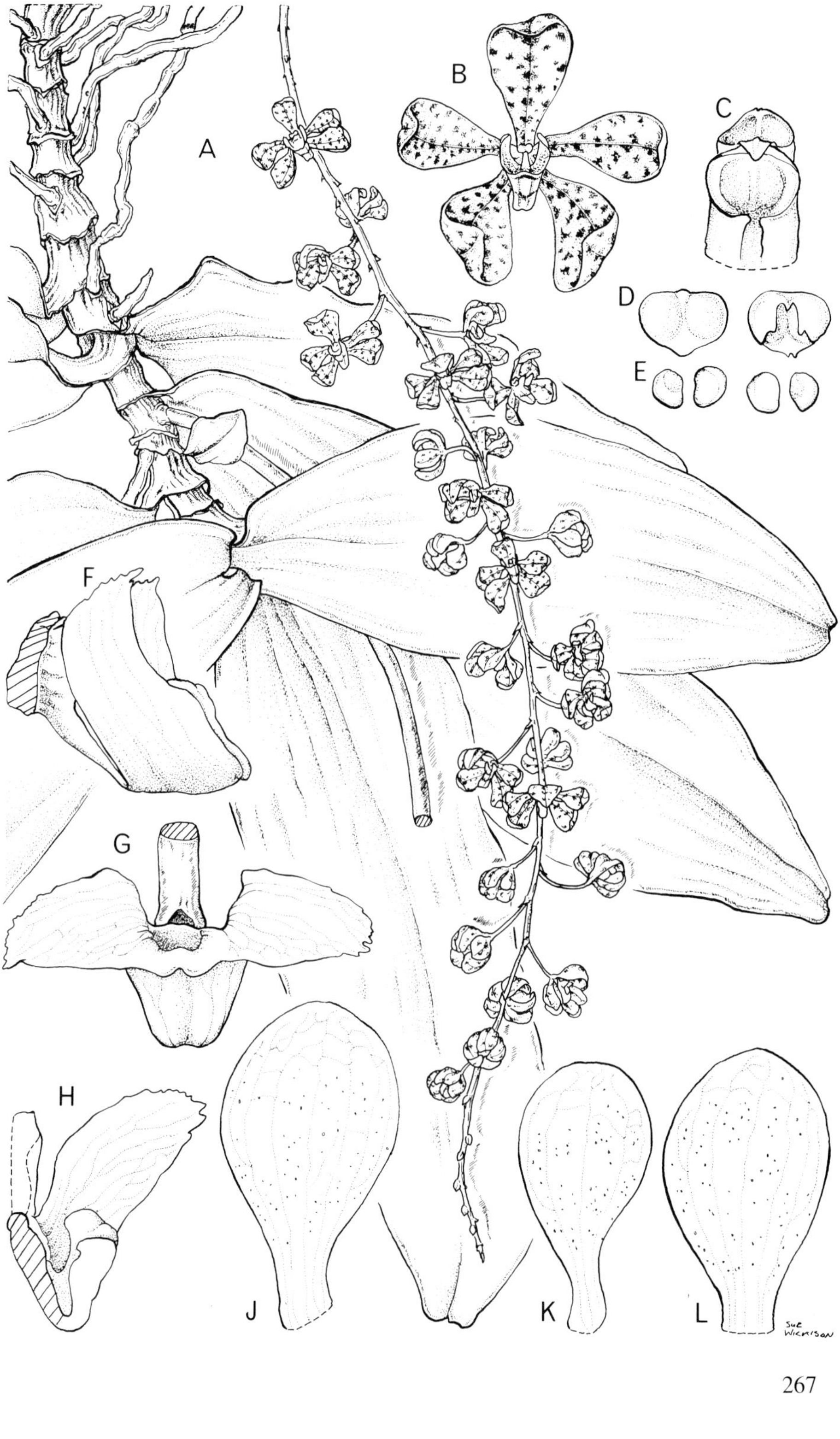
A
B
C
D
E
F
G
H
J
K
L
SUE WICKISON

wide; lip trilobed, 6 mm long, 8.5 mm wide, the lateral lobes erect, oblong, large with erose margins, the mid-lobe short with a central callus, the spur short and saccate; column short, 3 mm long with a 3 mm long column-foot set at an acute angle to it. (See fig. 69 and plate 16B).

DISTRIBUTION: Gizo, Makira, New Georgia, Olasaa (reef island east of Gizo Island) and Rennell. Also in New Guinea, New Britain and Australia.
HABITAT: Strand forest, sea level.
COLLECTIONS: *Bradley* 15 (BM); *Dickson* BT2 (K); *Guppy* 218 (K); *Hunt* in RSS 2173 (K); *Im Thurn* 347, 348 & s.n. (K); *Noona Dan Expedition* 2834 & 2896 (K); *Wickison* 68 (K); *Woodford* 6 (K).

64. SARCANTHOPSIS Garay

Medium-sized to large epiphytic or lithophytic monopodial plants. *Stems* elongate, erect, stout, leafy. *Leaves* distichous, ligulate, coriaceous, sheathing at base. *Inflorescences* lateral, erect or spreading, branched, stout, with few to many flowers. *Flowers* small to medium-sized, fleshy; sepals and petals free, similar and widely spreading; lip trilobed, the lateral lobes erect, the mid-lobe a fleshy flap which arches over the basal keel of the lip; column short, fleshy; pollinia 4.
A small genus of about five species in the Moluccas, New Guinea and the Pacific Islands, a single species being recorded. See Kores (1989) for a discussion on the status of this genus.

S. nagarensis *(Reichb. f.) Garay* in Bot. Mus. Leafl. Harv. Univ. 23(4): 199 (1972). Type: Fiji, *Seemann* 594 (holotype W, isotype K!).
Sarcanthus nagarensis Reichb. f. in Seem., Fl. Vit.: 298 (1868).
Stauropsis woodfordii Rolfe in Bull. Misc. Inf., Kew 1908: 72 (1908). Types: Solomon Islands, New Georgia, *Officers of H.M.S. Penguin* s.n. (syntype K!) and Santa Isabel, *Woodford* s.n. (syntype K!).
Stauropsis nagarensis (Reichb. f.) Rolfe, l.c.: 64 (1909).
Stauropsis quaifei Rolfe, l.c.: 64 (1909). Type: Vanuatu, Espiritu Santo, *Quaife* 265 (holotype K!).
Vandopsis quaifei (Rolfe) Schltr. in Fedde, Repert. Spec. Nov. Regni Veg. 10: 196 (1911).
Vandopsis nagarensis (Rolfe) Schltr. in Fedde, Repert. Spec. Nov. Regni Veg., Beih. 1: 972 (1913).
Vandopsis raymundii Schltr. in Engler, Bot. Jahrb. 58: 494 (1921). Type: Palau Islands, *Ledermann* 14066 (holotype B†, isotype K!).
Sarcanthopsis quaifei (Rolfe) Garay in Bot. Mus. Leafl. Harv. Univ. 23(4): 199 (1972).
Sarcanthopsis woodfordii (Rolfe) Garay l.c.

Fig. 70. Sarcanthopsis nagarensis. A, habit x 2/3; **B**, lip x 3; **C**, flower with the right petal and lateral sepal removed x 3; **D**, dorsal sepal x 3; **E**, longitudinal section through the lip x 4; **F**, lateral sepal x 3; **G**, petal x 3; **H**, flower x 3; **J**, dorsal and ventral view of anther-cap x 6; **K**, pollinia x 10; **L**, stipe x 10. Drawn from *Wickison* 140 by Sue Wickison.

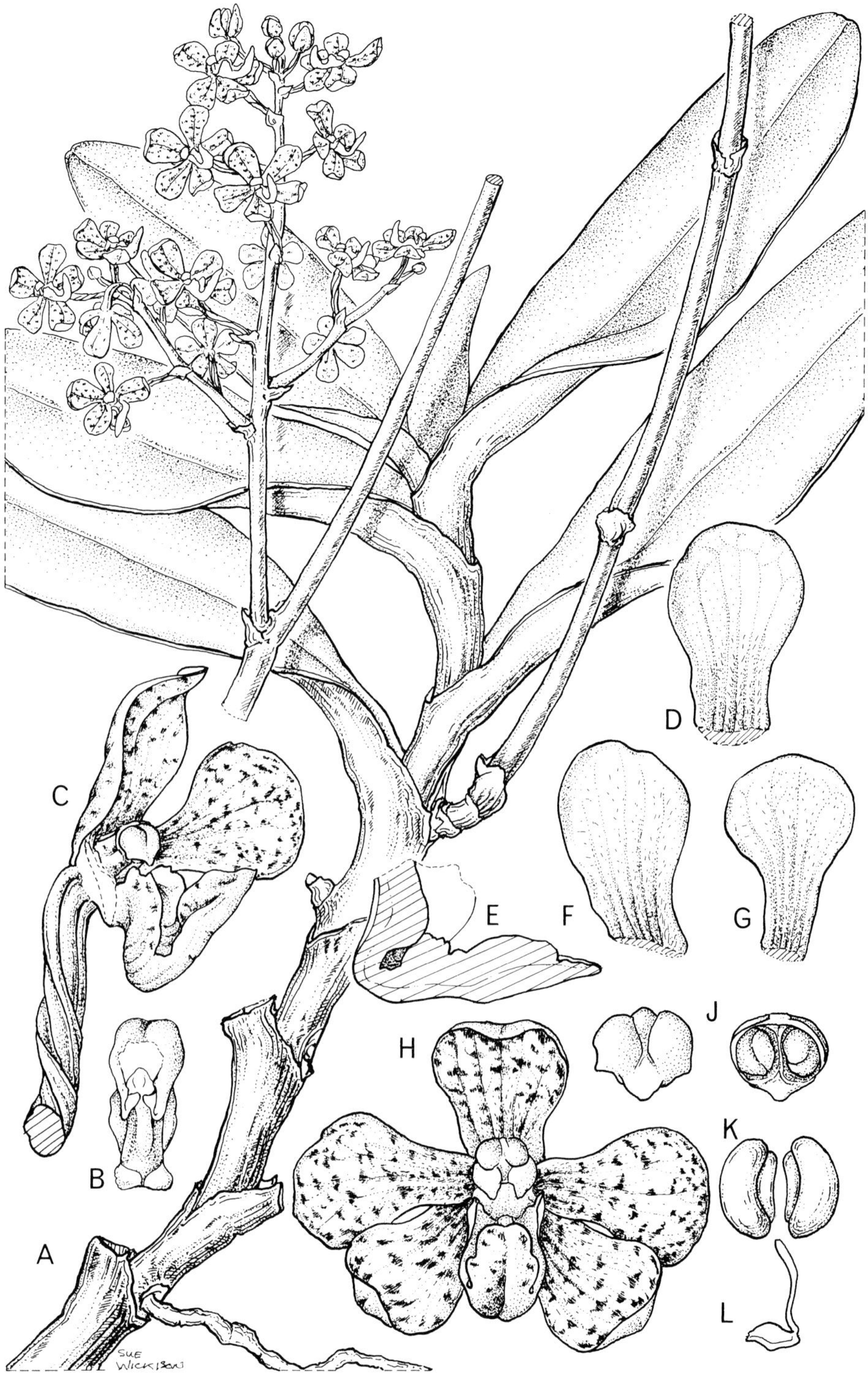

C
D
E
F
G
B
H
J
K
A
L
SUE WICKISON

An epiphytic or lithophytic plant. *Stems* erect or decurved with an erect apex, up to 3.5 m long, 2.5 cm wide, with long, aerial roots. *Leaves* fleshy, alternate, strap-like, up to 35 cm long, 6 cm wide, rounded at apex, dark green; internodes c. 5 cm long. *Inflorescence* lateral, from opposite a leaf axil, up to 45 cm long, branched, many-flowered. *Flowers* scented, fleshy, yellow spotted with reddish-brown, with the lip and column rather paler with purplish markings and lines; sepals and petals similar, oblong-spathulate, 8–9 mm long, 6–7 mm wide; lip trilobed, 8–9 mm long, 4 mm wide, the lateral lobes erect, the mid-lobe a fleshy flap which arches over the base of the mid-lobe, narrow, with a central fleshy ridge; column short, fleshy, 3 mm long. (See fig. 70).

DISTRIBUTION: Bougainville, Faro Island, Guadalcanal, Kolombangara, Makira, New Georgia, Rennell, the Santa Cruz Islands, Santa Isabel and the Shortland Islands (Faro Island). Also in the Caroline Islands (Palau), Vanuatu and Fiji.

HABITAT: Lowland and coastal forest, up to 300 m.

COLLECTIONS: *Comins* 205 (K); *Cribb* 6009 (K); *Cribb & Campbell* 5115 (K); *Dickson* BT4 (K); *Douglas* K11 (K); *Guppy* 255 (K); *Mauriasi et al.* in BSIP 16676 & 16685 (K); *Millar* in NGF 38413 (K); *Millar & Vandenberg* in NGF 48515 (K); *Noona Dan Expedition* 2888 (K); *Officers of HMS Penguin* s.n. (K); *Ridsdale & Lavarack* in NGF 31119 (K); *Waterhouse* 629 (K); *Wickison* 140 (K); *Woodford* 5 (K).

Douglas (pers. comm.) notes that the flowers tend to face the light.

65. THRIXSPERMUM Loureiro

Mostly epiphytic monopodial plants but occasionally terrestrial. *Stems* long or short. *Leaves* few to many, distichous. *Inflorescences* lateral, the apical part of the raceme often being flattened and the flowers borne either in two distinct ranks or spirally; bracts persistent. *Flowers* short-lived, produced in succession; sepals and petals free, more or less equal; lip immovably joined to the column-foot, trilobed, saccate at base, the lateral lobes usually erect, the mid-lobe fleshy; column short with a distinct broad foot; pollinia 4.

A genus of about one hundred species from tropical Asia to Australia and the Pacific Islands, four species being recorded from the Solomon Islands and Bougainville.

The species are arranged in two sections following Smith (1905): those with their flowers and bracts in two ranks (section *Thrixspermum*) and those where the flowers and bracts are spirally arranged (section *Dendrocolla*). See Kores (1989) for a discussion of the sections.

1. Flowers and bracts spirally arranged .... **1. T. congestum**
   Flowers and bracts in two ranks 2
2. Plant delicate, with stem less than 10 cm long; apex of inflorescence less than 4 mm wide across bracts .... **3. T. graeffei**
   Plant robust, with stem 30–150 cm long; apex of inflorescence more than 7 mm wide across bracts 3

3. Plants terrestrial climbers; leaves ovate, 2.5–4 cm long, 1–1.5 cm wide
.... **2. T. amplexicaule**
Plants epiphytic; leaves strap-like, up to 22 cm long, 4 cm wide
.... **4. T. platystachys**

Section DENDROCOLLA

1. T. congestum *(Bailey) Dockr.*, Aust. Sarcanth.: 27 (1967). Type: Australia, *L.T. Nugent* (holotype BRI).
Cleisostoma congestum Bailey in Proc. Roy. Soc. Queensl. 11: 17 (1895).
Thrixspermum adenotrichum Schltr. in Fedde, Repert. Spec. Nov. Regni Veg., Beih. 1: 962 (1913). Type: New Guinea, *Schlechter* 18923 (holotype B†).

For full synonymy see Clements (1989).

An epiphytic herb. *Stems* up to 3.5 cm long. *Leaves* distichous, oblong, up to 9 cm long, 8 mm wide, with the apex unequally notched, pale yellow-green, sheathing at base. *Inflorescence* lateral, up to 18 cm long; raceme up to 2 cm long; bracts spirally arranged. *Flowers* white to pale yellow-orange, the lip usually white with brown dots or the margins sometimes pale pink; dorsal sepal ovate-spathulate, 7 mm long, 3 mm wide; lateral sepals ovate, concave, 7 mm long, 3 mm wide; petals ovate-spathulate, concave, 6 mm long, 2.5 mm wide; lip trilobed, 6–7 mm long, 9–10 mm wide, hairy, saccate at base, the lateral lobes obliquely oblong, obtuse, erect and column-embracing, the mid-lobe oblong, obtuse, with a basal oblong callus, the apical margin hairy; column 2.5 mm long, column-foot 1 mm long. (See fig. 71 A–K).

DISTRIBUTION: Guadalcanal, Makira and Rendova. Also in New Guinea, Vanuatu, New Caledonia and Australia.
HABITAT: Lowland forest, sea level to 100 m.
COLLECTIONS: *Cribb & Campbell* 5156 & 5171 (K); *Cribb & Dennis* 5171 (K); cult. *Dennis* s.n. (K); *Hunt* in RSS 2232 & 3076 (K).

Section THRIXSPERMUM

2. T. amplexicaule *(Blume) Reichb. f.*, Xenia Orchid. 2: 121 (1867). Type: Java, *Blume* s.n. (holotype L).
Dendrocolla amplexicaulis Blume, Bijdr. Fl. Ned. Ind.: 288 (1825).

For full synonymy see Seidenfaden (1988).

A terrestrial climber reaching 1.5 m long. *Leaves* distichous, ovate, 2.5–4 cm long, 1–1.5 cm wide, becoming smaller towards the apex of the stem, with the bases clasping (amplexicaul) and sheathing the stem. *Inflorescences* lateral, up to 40 cm long; rhachis up to 14 cm long, with 1 to 2 flowers open at a time from the apical bracts; bracts alternate, falcate, 6 mm long, 3 mm wide. *Flowers* non-resupinate, pale purple and white; dorsal sepal ovate, 8 mm long, 6 mm wide; lateral sepals ovate, 8 mm long, 6 mm wide; petals ovate, 6.5 mm long, 5 mm wide; lip trilobed, 8 mm long, 8 mm wide, saccate at base, with a small callus inside and a band of hairs below it, the mid-lobe obtuse, fleshy, the lateral lobes small and pointed, erect and embracing the column; column 4 mm long.

DISTRIBUTION: Guadalcanal. Widespread in SE. Asia from Vietnam to New Guinea.
HABITAT: Shallow peat swamp, growing with *Phragmites*, *Hanguana* and *Stenochlaena* in full sunlight, 10 m.
COLLECTION: *Leach* in BSIP 19284 (K).

3. T. graeffei *Reichb. f.* in Seem., Fl. Vit.: 298 (1868). Type: Samoa, *Graeffe* s.n. (holotype W!).
Sarchilus graeffei (Reichb. f.) Benth. & Hook. f. ex Drake, Ill. Fl. Ins. Mar. Pacif.: 310 (1892).
Thrixspermum oreadum Schltr. in Fedde, Repert. Spec. Nov. Regni Veg., Beih. 1: 962 (1913). Type: New Guinea, *Schlechter* 16927 (holotype B†).

An epiphytic herb. *Stems* up to 8 cm high. *Leaves* 3–5, distichous, oblong, narrowing at apex, 3–6 cm long, 6–8 mm wide, purple-mauve, sheathing at base. *Inflorescences* lateral, up to 9 per stem, racemose, bilaterally flattened, up to 5 cm long, the flowers in 2 alternating rows; peduncle up to 10 cm long; bracts alternate. *Flowers* pale yellow with a creamy-white lip with orange-ochre blotches; dorsal sepal oblong, concave, 5 mm long, 1.3 mm wide; lateral sepals ovate, concave, 4.7 mm long, 1.5 mm wide; petals ovate-lanceolate, 4.7 mm long, 1.4 mm wide; lip obscurely trilobed, semicircular, widest at base, 5 mm long, 5.2 mm wide, fleshy and hairy at base, apex and along the mid-vein, the disc with 2 horse-shoe shaped calli, the basal one small and glabrous, the other one, towards the apex of the lip, larger and hairy; column 1 mm long. (See fig. 71 L–R).

DISTRIBUTION: New Georgia. Also in New Guinea, Vanuatu, Samoa and Fiji.
HABITAT: Lower montane forest, 250 m.
COLLECTION: *Nicholls* 5 (K).

4. T. platystachys *(Bailey) Schltr.* in Orchis 5: 55 (1911). Type: Australia, *T.L. Bancroft* s.n. (holotype BRI, isotypes K!, MEL).
Sarcochilus platystachys Bailey, Syn. Qd. Fl. Suppl. 1: 56 (1886).
Thrixspermum novaehiberniae Schltr. in K. Schum. & Laut., Nachtr. Fl. Deutsch. Südsee: 232 (1905). Type: New Ireland, *Schlechter* 13617 (holotype B†).

An epiphytic herb forming large pendulous clumps, often attached by a few roots. *Stems* up to 30 cm long, leafy. *Leaves* alternate, well-spaced, strap-like, coriaceous, oblong, up to 22 cm long, 4 cm wide, unequally bilobed at apex, folded

---

Fig. 71. Thrixspermum congestum A-K. A, habit x 2/3; **B**, bracts x 4; **C**, flower x 3; **D**, dorsal sepal x 6; **E**, petal x 6; **F**, lateral sepal x 6; **G**, column x 6; **H**, dorsal and ventral view of anther- cap x 8; **J**, pollinia x 8; **K**, lip x 6. **T. graeffei L-R. L**, flower; **M**, lip x 10; **N**, dorsal sepal x 8; **O**, petal x 8; **P**, lateral sepal x 8; **Q**, column x 14; **R**, bracts x 4. **A** drawn from *Cribb & Campbell* 5156; **B** from *Hunt* 2232; **C-K** from *Cribb & Dennis* 5171; **L-R** from *Nicholls* 5. All drawn by Sue Wickison.

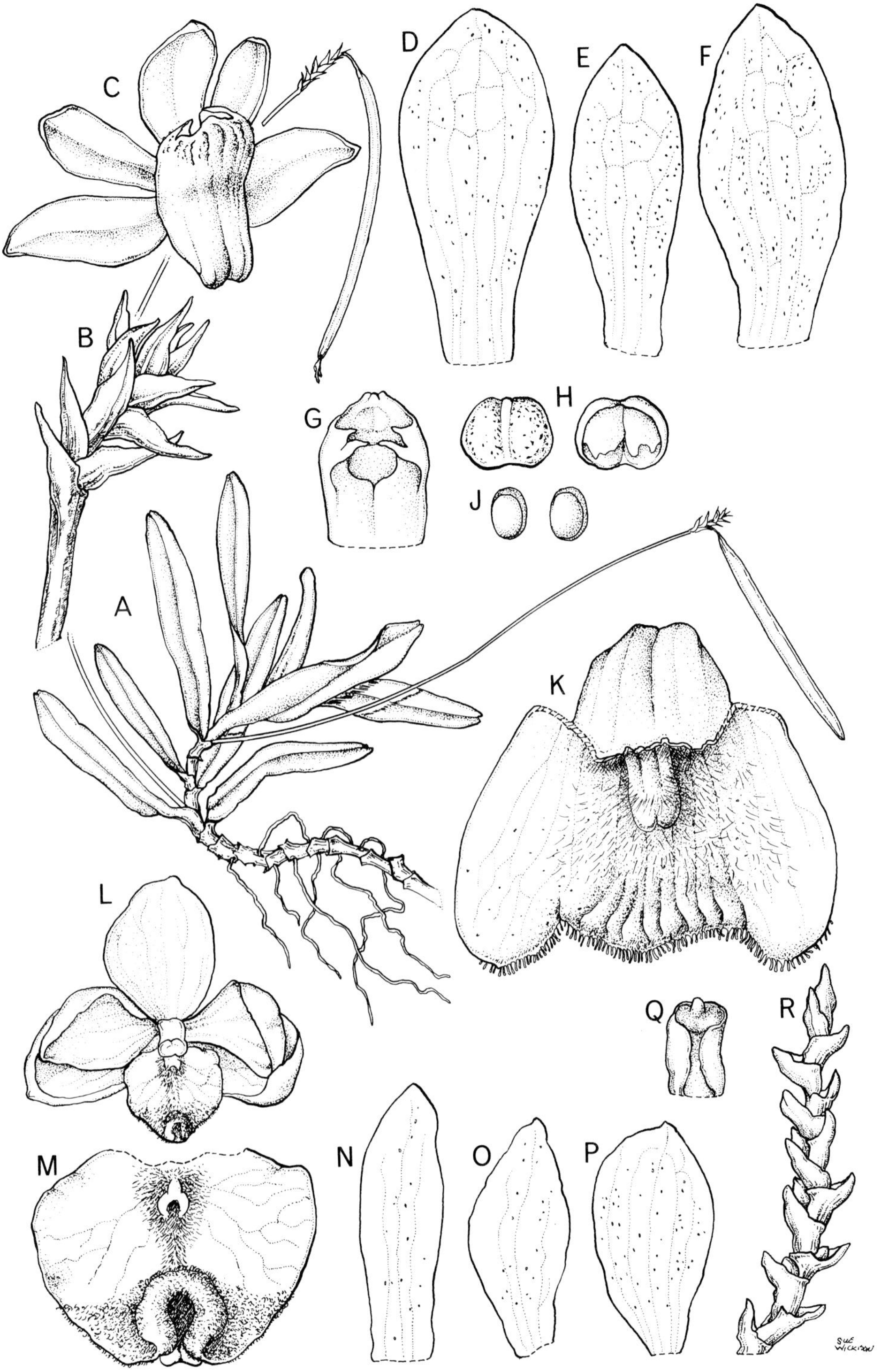
A
B
C
D
E
F
G
H
J
K
L
M
N
O
P
Q
R

at base with the bases sheathing stem. *Inflorescences* lateral, up to 12 per stem, up to 50 cm long; rhachis up to 25 cm long, flattened, with 1 or 2 flowers open at a time, developing in succession from the bracts; bracts alternate, falcate, imbricate, up to 1 cm long, 4 mm wide. *Flowers* opening for 1 day, non-resupinate, yellow tinged with orange, with a white lip with purple markings, sweetly scented; dorsal sepal lanceolate, spreading, 2 cm long, 4 mm wide; lateral sepals lanceolate, 2.3 cm long, 4 mm wide; petals lanceolate, 2 cm long, 3 mm wide; lip trilobed, 1.5 cm long, 7 mm wide, saccate at base with a few short hairs, the lateral lobes erect and embracing the column, obliquely oblong with rounded apices, the mid-lobe ovate, cucullate, fleshy at the apex and hairy, the disc with a central rounded callus; column 2.5 mm long. (See fig. 72 and plate 16E).

DISTRIBUTION: Bougainville, Choiseul, Guadalcanal, Kolombangara, Makira, New Georgia, Rendova and Santa Isabel. Also in New Guinea, New Ireland and Australia.
VERNACULAR NAME: Nuo-wacki (Bougainville).
HABITAT: Lowland forest, 50–650 m.
COLLECTIONS: *Dickson* 26 (K); *Hunt* in RSS 2227 & 2865 (K); *Kajewski* 1951 (G); *Lavarack & Ridsdale* in NGF 31169 (K); *Millar* in NGF 38398 (K); *Mitchell* 58 (K); *Wickison* 63 & 63B (K); *Womersley* in NGF 24826 (K).

66. ASCOGLOSSUM Schlechter

Epiphytic monopodial herbs. *Stems* stout, leafy in apical part. *Leaves* alternate, oblong, coriaceous, sheathing at base. *Inflorescence* lateral, branched, laxly many-flowered. *Flowers* showy, brightly coloured, with sepals and petals free and reflexed; petals slightly smaller than sepals; lip small, trilobed, the lateral lobes erect, the mid-lobe porrect, with a rounded saccate spur; column short; pollinia 4.
A monotypic genus from the Malay Archipelago to New Guinea and the Solomon Islands.

A. calopterum *(Reichb. f.) Schltr.* in Fedde, Repert. Spec. Nov. Regni Veg., Beih. 1: 975 (1913). Type: New Guinea, *MacFarlane* s.n. (holotype W).
Saccolabium calopterum Reichb. f. in Gard. Chron. n.s. 18: 520 (1882).

For full synonymy see Schlechter (1913).

A large epiphytic herb. *Stems* erect or pendent, often horizontal in the apical part, up to 50 cm long, 1.5 cm wide, leafy in the apical part. *Leaves* coriaceous, alternate, strap-like, up to 25 cm long, 2.5 cm wide, unequally bilobed at the apex, dark green and sometimes mottled purple on the under side, articulated to a sheathing base 4 cm long. *Inflorescence* lateral, branching, erect, 35–60 cm long,

---

Fig. 72. Thrixspermum platystachys. **A**, habit x 1/3; **B**, fruit and rhachis x 2/3; **C**, inflorescence x 1; **D**, flower x 5/3; **E**, lip x 3; **F**, side view of lip x 3; **G**, column x 4; **H**, dorsal and ventral view of anther-cap x 4; **J**, pollinia x 4; **K**, dorsal sepal x 3; **L**, petal x 3; **M**, lateral sepal x 3. Drawn from *Wickison* 63 by Sue Wickison.

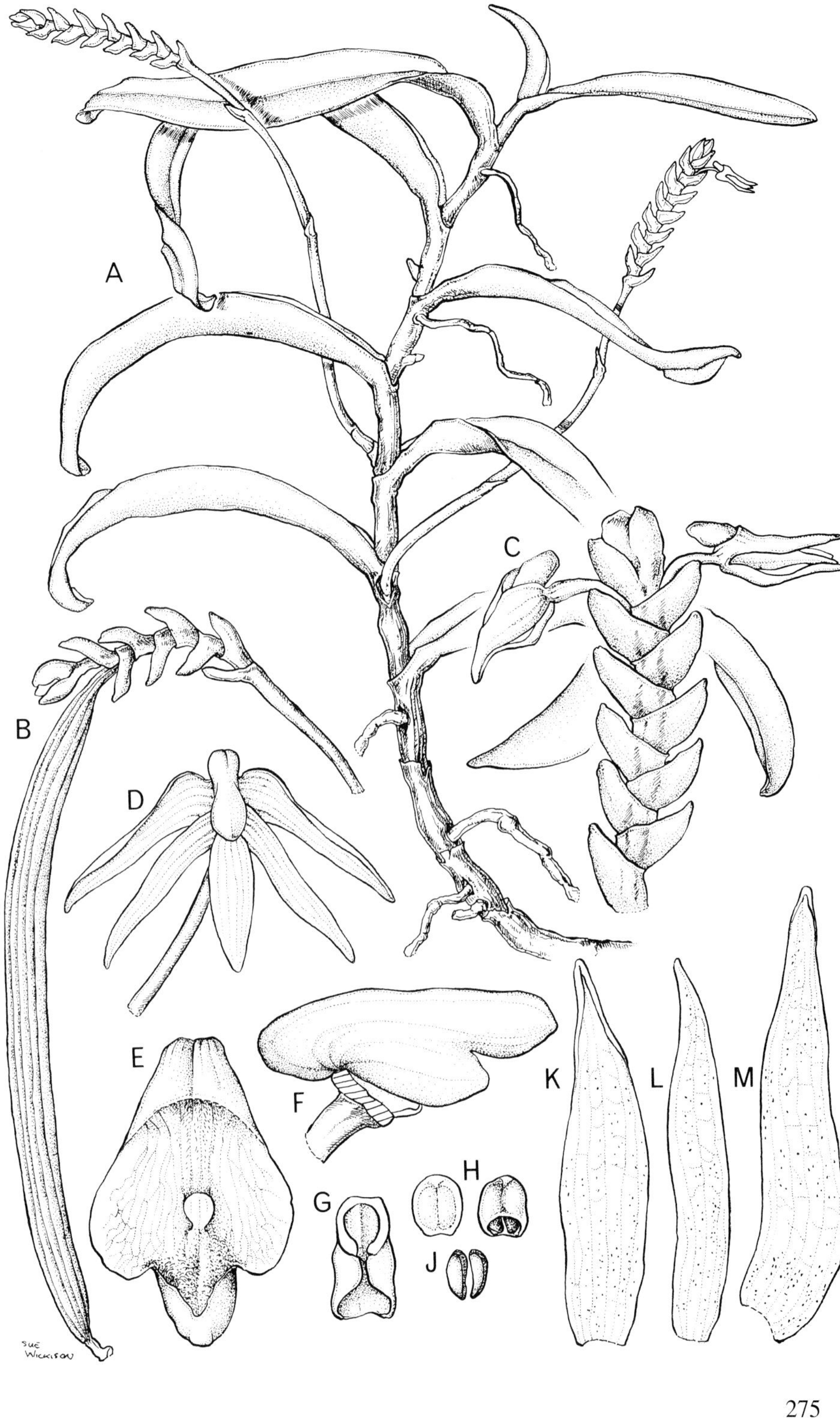
A
B
C
D
E
F
G
H
J
K
L
M
SUE WICKISON

the branches 4- to 7-flowered, the flowers opening in succession, with 2 or 3 open at a time; peduncle terete, 20–35 cm long, green flushed with purple; bracts ovate-triangular, subacute, 2–3 mm long. *Flowers* vivid purple occasionally with an orange or pink lip and spur; pedicel and ovary 1.3–1.8 cm long; sepals and petals reflexed; dorsal sepal elliptic-spathulate, obtuse, 1.4 cm long, 5 mm wide; lateral sepals spathulate, obtuse, 1.6 cm long, 6–9 mm wide, concealing the spur; petals elliptic, obtuse, 1.2 cm long, 5 mm wide; lip hinged to the base of the column, trilobed, 1.2 cm long, the lateral lobes erect, oblong-rounded, 2 mm long, 2 mm wide, attached to the column, the mid-lobe short, ligulate, porrect, spurred, the spur cylindrical-globose, 6–9 mm long; column 4–5 mm long. (see fig. 73 and plate 16F).

DISTRIBUTION: Bougainville, Guadalcanal and Santa Isabel. Also in the Philippines and New Guinea.
VERNACULAR NAME: Papino Poku (Bougainville).
HABITAT: Lowland rain forest, 100–200 m.
COLLECTIONS: *Cribb* s.n. (K); *Hunt* in RSS 2163 (K); *Ridsdale & Lavarack* in NGF 30593 & 31024 (LAE); *Robinson* K/5 & K185 (K); *Waterhouse* 673 (K).

Waterhouse records the same native name for *Renanthera endelfeldtii* F. Muell. & Kraenzl. and *A. calopterum* although the two species are quite distinct (Carr, 1934).

67. **ARACHNIS** Blume

Large lithophytic or epiphytic, monopodial plants. *Stems* long or short, leafy. *Leaves* distichous, oblong or gradually narrowing towards the apex, sheathing at the base. *Inflorescences* lateral, rigid, short or long, simple or branched, with few to many flowers. *Flowers* often fragrant, with sepals and petals free, almost equal, spreading, usually widening at the tips; lip articulated to the column-foot by a short strap of flexible tissue, fleshy, trilobed, the lateral lobes broad, quadrangular to triangular, erect, the mid-lobe fleshy; column short, thick, with a short foot; pollinia 4. *Fruit* a long, cylindrical capsule.
A small genus of about ten species from Indochina, Indonesia, the Malay Peninsula, Borneo, the Philippines, Sulawesi, New Guinea and the Solomon Islands, a single species being recorded.

A. beccarii *Reichb. f.* in Bot. Centralbl. 28: 343 (1886). Type: New Guinea, *Beccari* s.n. (lectotype W).

For full synonymy see Tan (1974).

Fig. 73. Ascoglossum calopterum. A, habit x 2/3; **B**, apex of leaf x 2/3; **C**, flower x 2; **D**, dorsal and ventral view of anther-cap x 4; **E**, pollinia x 4; **F**, side view of lip, spur and column x 3; **G**, lip and spur x 3; **H**, dorsal sepal x 3; **J**, petal x 3; **K**, lateral sepal x 3. Drawn from *Cribb* s.n. (Kew spirit no. 48241) by Sue Wickison.

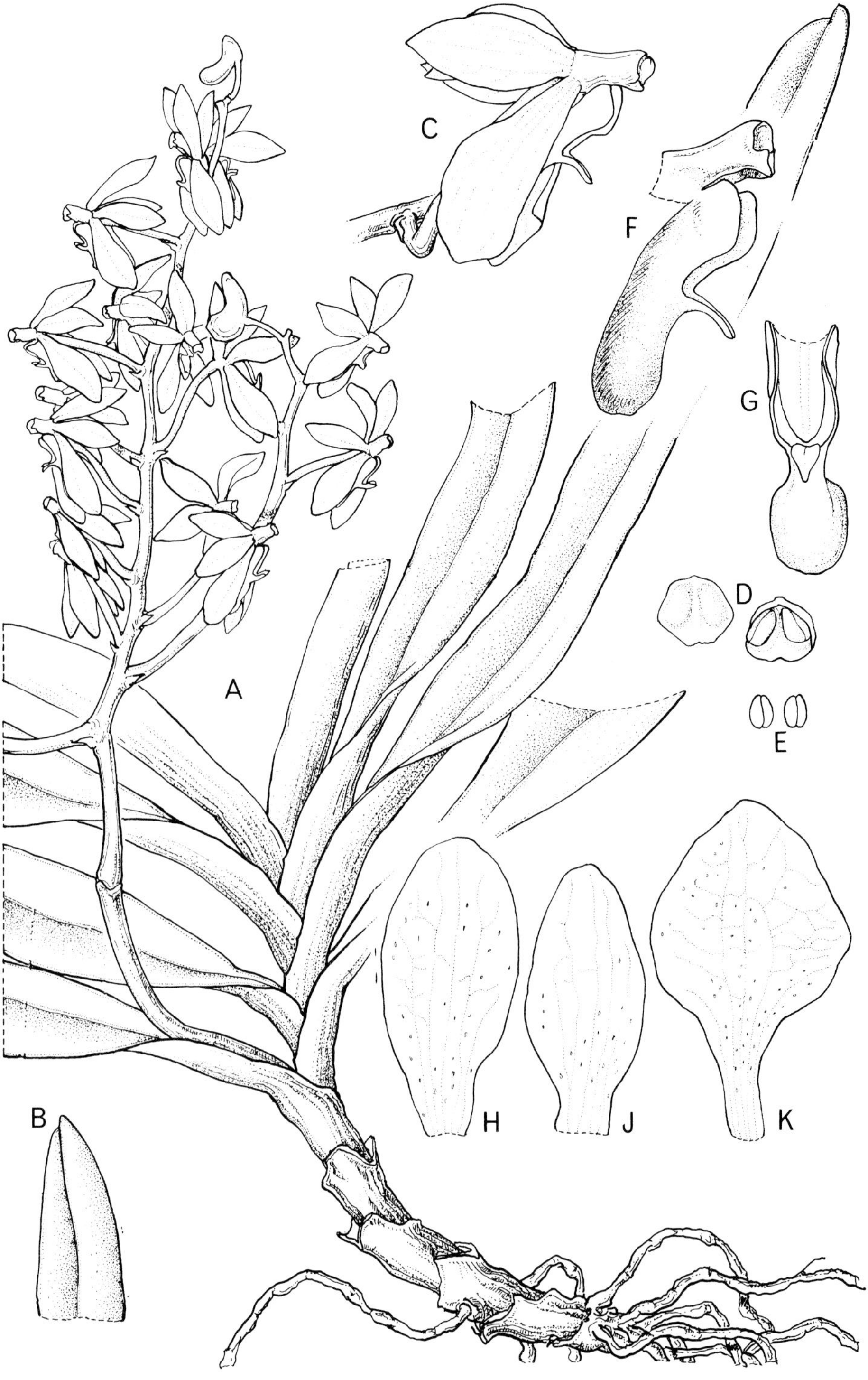
A
B
C
D
E
F
G
H
J
K

var. **imthurnii** *(Rolfe) Tan* in Taxon 23: 82 (1974). Type: Solomon Islands, Malaita, *Im Thurn* s.n. (holotype K!).
Stauropsis imthurnii Rolfe in Curtis's Bot. Mag. 143: 8714 (1917).
Arachnis imthurnii (Rolfe) L.O. Williams in Bot. Mus. Leafl. Harv. Univ. 5(4): 58 (1937).
Vandopsis imthurnii (Rolfe) P.F. Hunt in Kew Bull. 24: 98 (1970).

An epiphytic plant. *Stems* short, stout, erect, leafy. *Leaves* distichous, imbricate, arcuate, firmly coriaceous, elongate-liguate, acute, up to 2 m long, 7 cm wide, the margins sharp, dilated, sheathing the stem at the base. *Inflorescence* lateral from the base of the leaves, woody, longer than leaves, laxly branched near the apex; bracts short. *Flowers* somewhat fleshy, fragrant, white becoming yellow towards the apices with a yellow lip marked at the base with violet-blue; pedicel and ovary 1.5 cm long; sepals spreading, spathulate-oblong, obtuse, up to 3 cm long, 1 cm wide near the apex, the lateral sepals with a well defined median ridge on the back; petals spathulate, up to 3 cm long, 8 mm wide; lip with a short claw, fleshy, trilobed, 6 mm long, 7 mm across when flattened, saccate in the middle, the lateral lobes erect, orbicular-oblong, the mid-lobe inflexed, broadly oblong, obtuse; column broad, 8 mm long. (See fig. 74).

DISTRIBUTION: Bougainville, Guadalcanal, Malaita and Santa Isabel.
HABITAT: No information.
COLLECTIONS: *Davis* in BSIP s.n. (K); *Im Thurn* s.n. (K); *Schodde & Craven* 4105 (LAE); *Woodford* s.n. (K).

Var. *imthurnii* differs from the typical variety in lacking brown blotches on the white petals and sepals while the lip is marked with violet blue. The typical variety is found in New Guinea.

68. SCHOENORCHIS Reinwardt

Small epiphytic, monopodial herbs, often growing in tangled clumps. *Stems* erect, arcuate or pendulous, often branched near base. *Leaves* distichous, linear to terete, sheathing at base. *Inflorescences* lateral, simple, with many small flowers; peduncle usually shorter than the rhachis. *Flowers* minute; sepals and petals free, but not spreading widely so that the flowers appear bell-shaped; lip trilobed, spurred; column short, footless; pollinia 4.

A genus of about ten species from SE. Asia to Australia and the Pacific Islands, a single species being recorded.

S. micrantha *Reinw. ex* Blume, Bijdr. Fl. Ned. Ind. 1(8): 362 (1825). Type: Java, *Blume* s.n. (holotype L).
Saccolabium chionanthum Lindley in Journ. Linn. Soc. Bot. 3: 35 (1859). Type: Java, *Reinwardt* s.n. (drawing K!).

Fig. 74. Arachnis beccarii var. **imthurnii. A**, habit reduced; **B**, flower x 1; **C**, inflorescence; **D**, lip x 6; **E**, dorsal and ventral view of anther-cap x 6; **F**, pollinia x 6; **G**, transverse section through the lip x 6; **H**, dorsal sepal x 2; **J**, column x 4; **K**, petal x 2; **L**, outer surface of lateral sepal x 2; **M**, stipe x 3. Drawn from *Davis* in BSIP s.n. (Kew spirit no. 30763) by Sue Wickison.

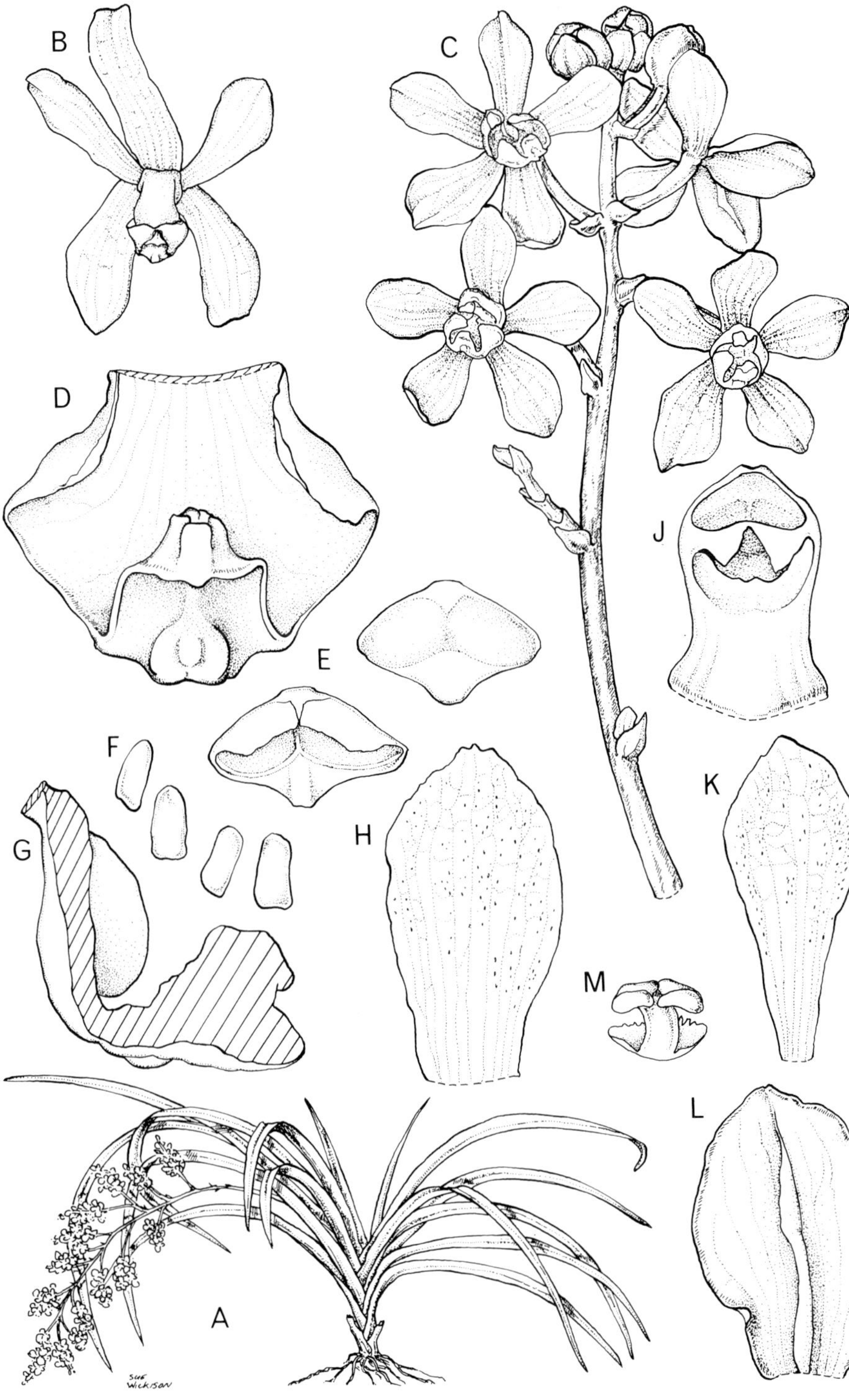
B
C
D
E
F
G
H
J
K
M
L
A
SUE WICKISON

Schoenorchis densiflora Schltr. in Fedde, Repert. Spec. Nov. Regni Veg., Beih. 1: 986 (1912). Type: New Guinea, *Schlechter* 17316 (holotype B†, isotype K!).
Saccolabium plebejum J.J. Smith in Bull. Jard. Bot. Buitenz. ser. 2, 3: 77 (1912). Type: New Guinea, *Gjellerup* 21 (holotype BO).
Ascocentrum micranthum (Reinw. ex Blume) Holtt. in Gard. Bull. Sing. 11: 275 (1947).

An epiphytic herb growing in small tangled clumps, due to progressive branching of the stem. *Stems* 2–15 cm long, leafy throughout. *Leaves* very thick, distichous, linear, sulcate above, 3–6 cm long, 1 mm wide, sheathing at base. *Inflorescences* lateral, 1–4, 1–2.5 cm long, many-flowered. *Flowers* off-white, bell-shaped; dorsal sepal lanceolate, 2 mm long, 1 mm wide; lateral sepals oblong, 2 mm long, 1 mm wide; petals shorter than sepals, oblong-obovate, 1.5–2 mm long, 0.5–1 mm wide; lip trilobed, 2.5 mm long, 1.5 mm wide, the lateral lobes erect and triangular, the mid-lobe short, ovate and recurved at the apex, the spur globose to ellipsoidal, less than 1 mm long. (See plate 14E).

DISTRIBUTION: Bougainville, Choiseul, Guadalcanal, Rendova and Santa Isabel. Widely distributed from Asia and the Malay Archipelago to New Guinea, Vanuatu, New Caledonia, Fiji and Australia.
HABITAT: Rain forest, 50–500 m.
COLLECTIONS: *Cribb & Campbell* 5154 (K); *Dennis* in RSS 7882 (K); *Hunt* in RSS 2923 (K); *Kajewski* 2270 (BM); *Schodde & Craven* 4108 (K); *Waterhouse* 777 (K); *Wickison* 114A (K).

69. LUISIA Gaudichaud

Epiphytic monopodial herbs. *Stems* long, erect or semi-pendulous, leafy. *Leaves* distichous, stiff, terete, the leaf bases sheathing. *Inflorescences* lateral, sessile, very short, thickened near the apex, with one or a few flowers opening at a time. *Flowers* small, not opening widely; sepals and petals free; petals often longer and narrower than the sepals; lip fleshy, relatively large, divided into 2 distinct portions, the basal part narrower than the apical part that is expanded into a somewhat broad lamina which is thicker than the basal part; column short with a short foot; pollinia 4.
A genus of about forty species from tropical Asia to the Pacific Islands, a single species being recorded.

L. teretifolia *Gaud.* in Freyc., Voy. Bot.: 427, t.37 (1826). Type: Mariana Islands, Guam, *Gaudichaud* 37 (holotype P!).

An epiphytic herb forming straggly clumps. *Stems* erect or semi-pendulous, woody, 10–30 cm long, leafy. *Leaves* distichous, rigid, terete, 5–15 cm long, 2–5 mm wide, dark green, sheathing at base. *Inflorescences* lateral, 5–15 mm long, erect, thickened near apex, the flowers opening 1 to 3 at a time. *Flowers* yellow to green sometimes with purple mottling inside, with a purple or rarely a green apex to the lip; pedicel and ovary 6–8 mm long; dorsal sepal oblong, concave, 5–7 mm long, 3–4 mm wide; lateral sepals obliquely ovate, 7–8 mm long, 3–4 mm wide;

petals narrowly oblong, 8–9 mm long, 1.5–2 mm wide; lip fleshy, 6–7 mm long, 6–7 mm wide, divided into two distinct parts, narrowed and concave in basal half, expanded into a fleshy, cordate apical lamina above; column 4 mm long, the column-foot 2 mm long.

DISTRIBUTION: Rendova. Widely distributed from SE. Asia to the Malay Peninsula, New Guinea, Vanuatu, the Mariana Islands (Guam), New Caledonia and Australia.
HABITAT: Rain forest, 50 m.
COLLECTION: *Cribb & Campbell* 5147 (K).

70. **RENANTHERA** Loureiro

Epiphytic, often climbing, monopodial plants. *Stems* with long internodes, straggling, leafy in apical half. *Leaves* alternate, somewhat coriaceous, relatively short, with a bilobed apex, sheathing at base. *Inflorescences* lateral, long, usually branching, many-flowered. *Flowers* usually predominently red but sometimes yellow; sepals and petals free, usually spreading widely; lateral sepals broader than the dorsal sepal, with wavy margins; lip trilobed, spurred, shorter than the sepals and petals, firmly attached to the column, with erect lateral lobes, the mid-lobe small, recurved, with a pair of calli at the base, the spur conical; column short and stout; pollinia 4.
A genus of about fifteen species from Assam and China to the Philippines, Malaysia, New Guinea and the Solomon Islands, a single species being recorded.

R. endsfeldtii *F. Muell. & Kraenzl.* in Oest. Bot. Zeit. 1894: 460 (1894). Type: New Guinea, *J. Chalmers* (holotype MEL).

A straggling epiphyte up to 10 m long. *Stems* rooting from the nodes, up to 50 cm long, 6 mm in diameter, leafy in apical section. *Leaves* alternate, oblong, 9–12 cm long, 2.2–2.6 cm wide, unequally roundly bilobed at apex, articulated to a sheathing base c. 3 cm long. *Inflorescence* lateral, from near the apex of the stem, emerging through the leaf sheath base, branched, erect, up to 60 cm long, each branch laxly 10–20-flowered, up to 13 cm long; peduncle up to 30 cm long, stout; bracts oblong, obtuse, 3–4 mm long, 1 mm wide. *Flowers* deep tangerine in colour or reddish-orange, with darker or red spots, a bright red lip with a creamy-yellow base, and a red column blotched with yellow; pedicel and ovary 1.5 cm long; sepals and petals spreading widely; dorsal sepal oblong-spathulate, 1–1.3 cm long, 3 mm wide; lateral sepals obliquely spathulate, 1–1.3 cm long, 4 mm wide, slightly twisted; petals linear-oblong, 8–9 mm long, 2–3 mm wide; lip trilobed, the lateral lobes erect, obliquely triangular, 2 mm long, 2 mm wide, the mid-lobe oblong, reflexed, 1.5 mm long, 1.5 mm wide, with two ovoid, fleshy calli at the base, the spur conical, obtuse, 3–4 mm long, 2–2.5 mm wide; column 3 mm long.

DISTRIBUTION: Bougainville, Guadalcanal and Ranongga. Also in the Moluccas, New Guinea and New Ireland.
VERNACULAR NAME: Papino Poku (Bougainville).
HABITAT: Rain forest and secondary scrub, sea level to 200 m.

COLLECTIONS: *Dennis* in RSS 7749 (K): *Mitchell* 78 (K); *Waterhouse* 616 (K).

Waterhouse records the same native name for *Ascoglossum calopterum* (Reichb. f.) Schltr. and *Renanthera endsfeldtii* although the two species are quite distinct (Carr, 1934).

71. VANDA W. Jones ex R. Brown

Large, coarse monopodial epiphytes or lithophytes. *Stems* long, usually erect, with thick cord-like roots. *Leaves* distichous, strap-like. *Inflorescences* lateral, racemose, rather short, several-flowered. *Flowers* fairly large, showy, opening widely, stiff or fleshy; sepals and petals free, similar, narrowed in the basal half, often with wavy margins; lip immovably fixed to the apex of the short column-foot, trilobed with the mid-lobe projecting as a landing platform, spurred; column short, thick, dilated near the base and with a very small column-foot; pollinia 4.

A genus of about sixty species from China, India, Malaysia, New Guinea and Australia, a single species being recorded.

V. hindsii *Lindley* in Journ. Bot., Lond. 2: 237 (1843). Type: New Guinea, *Hinds* s.n. (holotype K!).

For full synonymy see Clements (1989).

An epiphytic or lithophytic plant forming large, coarse clumps. *Stems* erect or semi-erect, up to 1 m long, with numerous thick cord-like roots. *Leaves* coriaceous, numerous, alternate in 2 ranks, crowded, strap-like, arcuate, up to 40 cm long, 4 cm wide, unequally bilobed at the apex, the apical teeth acute or acuminate. *Inflorescences* lateral, stiff, up to 20 cm long, bearing up to 7 flowers. *Flowers* shiny chestnut-brown with yellow margins and a white column; sepals and petals widely spreading, narrowed at the base, the margins wavy; dorsal sepal ovate-elliptic, 1.5 cm long, 1.3 cm wide; lateral sepals ovate-elliptic, 1.5 cm long, 1.3 cm wide; petals ovate, 1.4 cm long, 1.2 cm wide; lip trilobed, 1.6 cm long, 1.3 cm wide, the lateral lobes crescent-shaped, erect, the mid-lobe oblong, projecting forwards like a landing platform, slightly narrowed at base with a notched apex, spurred, the spur globose, 4 mm long; column 5 mm long, dilated near the base. (See plate 14F).

DISTRIBUTION: Bougainville and Guadalcanal. Also in New Guinea and Australia.

HABITAT: Rain forest.

COLLECTIONS: *Dennis* s.n. (K); *Waterhouse* s.n. (K).

72. CLEISOSTOMA Blume

Epiphytic monopodial herbs. *Stems* stout, elongate, leafy. *Leaves* distichous, oblong, sheathing at base. *Inflorescences* lateral, usually branched, many-flowered. *Flowers* small, with the sepals and petals free; petals narrower than sepals; lip trilobed with a rounded saccate spur, the front wall of the spur bearing an erect tongue; column short; pollinia 4.

A genus of eighty to a hundred species from tropical Asia to the Pacific Islands, two species being recorded.

Stems 2–4 mm wide; leaves 8–10 cm long, 5–7 mm wide .. .. **1. C. bicorne**
Stems stout, 7–14 mm wide; leaves 18–33 cm long, 15–22 mm wide
.... **2. C. pacificum**

1. C. bicorne *(J.J. Smith) Garay* in Bot. Mus. Leafl. Harv. Univ. 23(4): 169 (1972). Types: New Guinea, *Versteeg* 1077 & *Djibdja* 843 (syntypes BO).
Sarcanthus bicorne J.J. Smith in Bull. Dép. Agric. Indes Néerl. 19: 35 (1908).

An epiphytic herb. *Stem* up to 50 cm long, 2–3 mm wide, sheathed in leaf bases. *Leaves* distichous, oblong-lanceolate, 8–10 cm long, 5–7 mm wide, minutely bilobed at apex. *Inflorescence* lateral, 2–4 cm long, unbranched. *Flowers* white with purple markings on the lip; dorsal sepal oblong, acute, 3.5–4 mm long, 2 mm wide; lateral sepals obliquely ovate, acute, 4 mm long, 2 mm wide; petals linear, 1.5–2 mm long, 0.5 mm wide; lip very fleshy, united on basal margins to the column, trilobed, 5 mm long, 2–2.5 mm wide, the spur shortly conical-cylindrical, 1.5–2 mm long, with a transverse callus at the base of the lip almost blocking the mouth of the spur; column fleshy, 1.5–2 mm long.

DISTRIBUTION: New Georgia. Also in New Guinea and New Ireland.
HABITAT: Rain forest, prolific, 360 m.
COLLECTION: *Wickison* s.n. (K).

Unfortunately this specimen is sterile and therefore the identification is provisional. However, its habit agrees well with specimens of *C. bicorne* from New Guinea and New Ireland but it differs in having somewhat shorter, thicker peduncles. Further material is needed to confirm this identification. The description of the flowers is taken from the type description and other New Guinea material.

2. C. pacificum *Cribb & B. Lewis* in Orchid Review 97, 1150: 251 (1989). Type: Vanuatu, Pentecost, *Wheatley* 302 (holotype K!).
Cleisostoma longipaniculatum Kores in Allertonia 5, 1: 202 (1989), **synon. nov.** Type: Fiji, *Smith* 7416 (holotype US, isotypes AMES, BISH, K!, L, NY, UC).

An epiphytic plant. *Stems* stout, erect or pendent-arcuate, leafy, 10–30 cm long, 0.7–1.4 cm wide, covered by sheathing leaf bases. *Leaves* distichous, spreading, coriaceous, linear, unequally roundly bilobed at the apex, 18–33 cm long, 1.5–2.2

cm wide. *Inflorescences* lateral, emerging through the leaf sheaths opposite the leaves, paniculate or rarely unbranched; peduncle terete, 17–25 cm long; branches 2.5–7 cm long; bracts triangular, spreading or reflexed, 1–2 mm long. *Flowers* small, whitish, pale-green or yellow with a white lip; pedicel and ovary 4–5 mm long; dorsal sepal elliptic, obtuse or rounded, 3–4 mm long, 2 mm wide; lateral sepals obliquely oblong-elliptic, obtuse, 4–5 mm long, 2–2.5 mm wide; petals oblong, obtuse, 3 mm long, 1 mm wide; lip very fleshy, united on basal margins to the column, trilobed, obtuse, 2–2.5 mm long and wide, the spur nectariferous, shortly conical-cylindrical, obtuse, 3.5–4 mm long, with a high transverse callus at the base of the lip almost blocking the mouth of the spur; column fleshy, 1–1.3 mm long. (See fig. 75).

DISTRIBUTION: Guadalcanal, Makira and New Georgia. Also in Vanuatu, New Caledonia and Fiji.
HABITAT: Strand forest and lowland rain forest, sea level to 200 m.
COLLECTIONS: *Dennis* in RSS 2258, 2327 & 7936 (K); *Dickson* 30 (K); *Tayloroe* 3 (K); *Wickison* 49 (K).

Unfortunately, *C. pacificum* and *C. longipaniculatum* were both described in the same month. However, *C. pacificum* was published on 1 August 1989 whilst *C. longipaniculatum* was published on 4 August 1989. The former, therefore, has priority.

73. **MALLEOLA** J.J. Smith & Schlechter

Small epiphytic monopodial herbs. *Stems* short to elongate, leafy. *Leaves* alternate, unequally bilobed at the apex, twisted at the base to lie in one plane. *Inflorescence* lateral, unbranched, densely to laxly many-flowered. *Flowers* small; sepals and petals free, subsimilar, but petals usually narrower; lip trilobed, usually wider than long, with or without a callus, spurred, the spur with a wide mouth, elongate, tapering at apex; column short, fleshy; pollinia 2.
A genus of about twenty-five species from Indonesia and the Philippines to New Guinea and the Solomon Islands, two species being recorded.

Spur clavate from the front, 2.5 mm long, slightly recurved at the tip .... **1. M. pallida**
Spur conical and obtuse from the front, 1.5 mm long, straight .... **2. M. rara**

1. M. pallida *(Schltr.) Schltr.*, in Fedde, Repert. Spec. Nov. Regni Veg., Beih. 1: 980 (1913). Type: New Guinea, *Schlechter* 14578 (holotype B†).
Saccolabium pallidum Schltr. in K. Schum & Laut., Nachtr. Fl. Deutsch. Südsee: 229 (1905).

Fig. 75. Cleisostoma pacificum. **A**, habit x 2/3; **B**, apex of leaf x 2/3; **C**, flower x 4; **D**, side view of flower x 4; **E**, column x 8; **F**, dorsal and ventral view of anther-cap x 8; **G**, pollinia x 8; **H**, transverse section through the lip x 6; **J**, lip x 6; **K**, dorsal sepal x 6; **L**, petal x 6; **M**, lateral sepal x 6. Drawn from *Wickison* 49 by Sue Wickison.

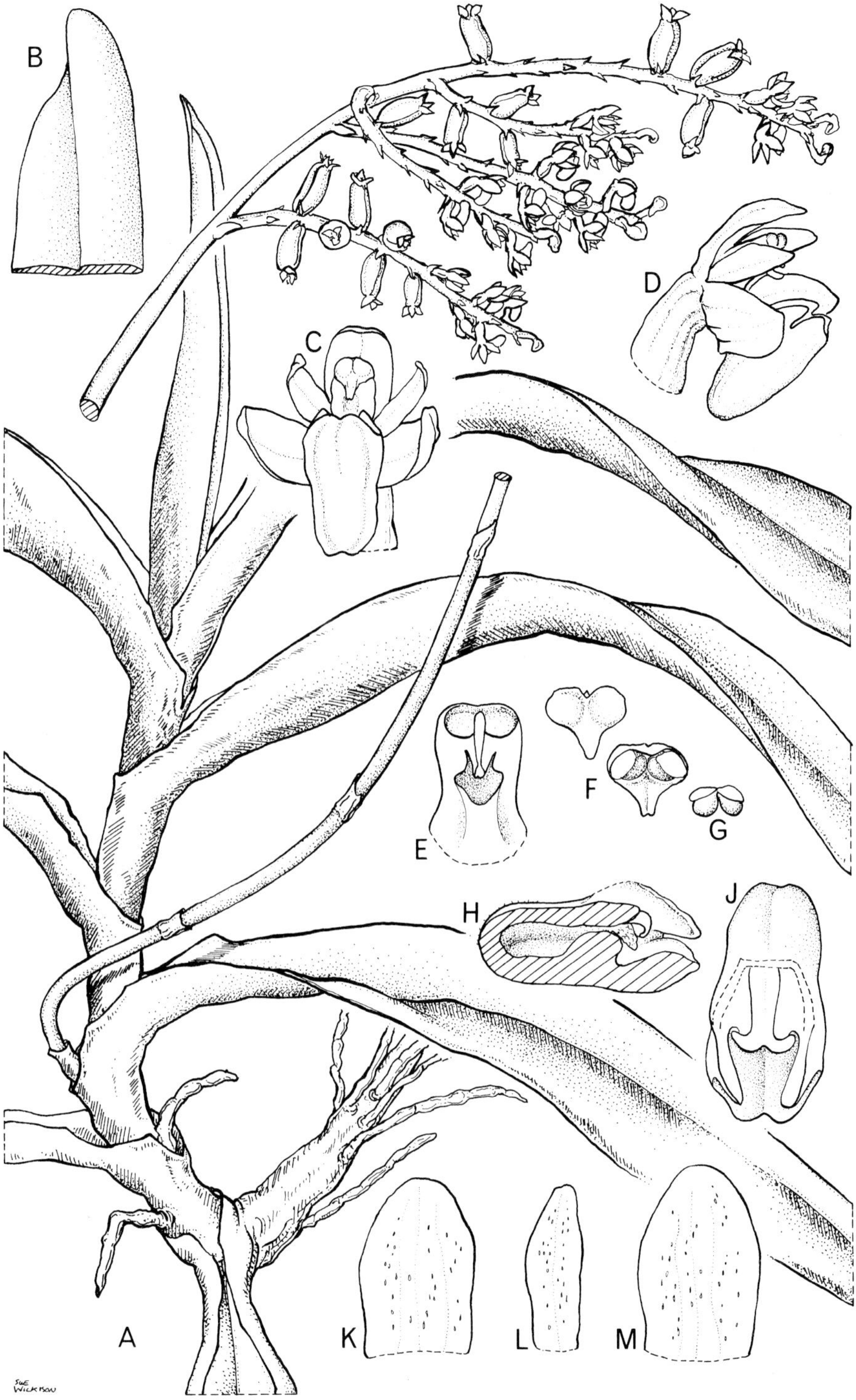
B
D
C
E
F
G
H
J
A
K
L
M

A dwarf epiphytic herb. *Stems* usually pendent, very short, up to 1 cm long. *Leaves* 2–7, alternate, ligulate, 2.5–5 cm long, 3–6 mm wide, unequally obtusely bilobed at apex, articulated at base to a short sheath. *Inflorescence* lateral, up to 6 cm long, laxly up to 20-flowered; peduncle very slender, wiry; bracts narrowly triangular, acute, 1 mm long. *Flowers* pale green with a white lip, very small, probably self-pollinating or cleistogamous; pedicel and ovary 3–4 mm long, the ovary minutely verruculose; sepals and petals oblong, obtuse, 1.5 mm long, 0.8 mm wide; lip trilobed, 1 mm long, 1.5 mm wide, the lobes semicircular, rounded, the lateral lobes erect, the spur tapering to apex in side view but clavate from the front, 2.5 mm long, slightly recurved at tip; column 0.8 mm long. (See plate 15B).

DISTRIBUTION: Guadalcanal and Makira. Also in New Guinea and New Ireland.
HABITAT: Rain forest, 40–800 m.
COLLECTIONS: *Corner* in RSS 2162 (K); *Hunt* in RSS 2239 (K); *Wickison* 238 (K).

2. M. rara *Schltr.* in Fedde, Repert. Spec. Nov. Regni Veg., Beih. 1: 980 (1913). Type: New Guinea, *Schlechter* 20365 (holotype B†).

A dwarf epiphytic herb. *Stems* usually pendent, very short, up to 1 cm long. *Leaves* 3, alternate, ligulate, 3.5–4.5 cm long, 8–10 mm wide, unequally obtusely bilobed at apex, articulated at base to a short sheath. *Inflorescence* lateral, up to 13 cm long, laxly up to 20-flowered, peduncle slender; bracts narrowly triangular, acute, 1 mm long. *Flowers* pale green with a white lip, very small, probably self-pollinating or cleistogamous; pedicel and ovary up to 2.5 mm long, elongating after fertilization; sepals and petals oblong, obtuse, 1.5 mm long, 0.8 mm wide; lip trilobed at apex, 1 mm long, 1.5 mm wide, the lobes rounded, spreading, the spur gradually tapering to apex in side view but broadly conical and obtuse from the front, 1.5 mm long; column 0.8 mm long.

DISTRIBUTION: Guadalcanal. Also in New Guinea.
HABITAT: Primary forest, 650 m.
COLLECTION: *Wickison* 143 (K).

74. MICROPERA Lindley

Large monopodial epiphytes. *Stems* long, erect or scrambling, leafy. *Leaves* distichous, well-spaced, coriaceous. *Inflorescence* lateral, racemose, erect or drooping, laxly few- to many-flowered. *Flowers* cream or dingy white; sepals and petals free, subsimilar, narrow, widely spreading or deflexed; lip immobile, fixed to the apex of the column-foot, trilobed, spurred, the spur with a prominent lobed appendage in its upper part; column short with a very short foot; pollinia 4.

A genus of about twelve species from SE. Asia, Malaysia, New Guinea, the Solomon Islands and Australia, a single species being recorded.

M. fasciculata *(Lindley) Garay* in Bot. Mus. Leafl. Harv. Univ. 23(4): 186 (1972). Type: New Guinea, *Hinds* s.n. (lectotype K!).
Saccolabium fasciculatum Lindley in Journ. Bot., Lond. 2: 238 (1843).
Sarcanthus papuanus J.J. Smith in Bull. Dép. Agric. Indes Néerl. 39: 19 (1910). Type: New Guinea, *von Roemer* 200 (holotype BO!).
Camarotis keffordii (Bailey) J.J. Smith in Natuurk. Tijdschr. Ned. Ind. 72: 97 (1913). Type: Australia, *Kefford* (holotype BRI).

For full synonymy see Clements (1989).

A large epiphytic plant forming coarse, straggling clumps. *Stems* erect, stout, often over 1 m long, 5–6 mm in diameter, sparsely branched with numerous thick cord-like roots, leafy in the apical half. *Leaves* alternate, oblong, up to 16 cm long, 2 cm wide, stiff, coriaceous, with a minutely bilobed apex, sheathing at base. *Inflorescence* lateral, up to 28 cm long, erect, stiff and slender, laxly 10- to 15-flowered; peduncle c. 10 cm long. *Flowers* cream becoming darker with age, scented, usually non-resupinate; dorsal sepal stiffly erect, broad, lanceolate, 6–8 mm long, 3 mm wide; lateral sepals reflexed behind the lip, oblong, 6–8 mm long, 3.5 mm wide; petals spreading, oblong, obtuse, 6–8 mm long, 2.5 mm wide; lip shoe-shaped, 9 mm long, 2.5 mm wide, the lateral lobes triangular, erect or incurved, joined to the base of the column, the mid-lobe pointed, thickened near base, the spur conical, tapering at base, 3 mm long, curved, the entrance blocked by a conspicuous toothed callus; column 4 mm long with a vestigial column-foot.

DISTRIBUTION: Bougainville, New Georgia and Santa Isabel. Also in New Guinea, New Caledonia and Australia.
VERNACULAR NAME: Gora Goraha Ubutu (New Georgia).
HABITAT: Ridge-top primary forest.
COLLECTIONS: *Beer et al.* in BSIP 7667 (K); *Kajewski* 2271 (K, BM); *Waterhouse* 173 (K).

75. POMATOCALPA Breda

Epiphytic monopodial plants. *Stems* elongate and stout or reduced, leafy. *Leaves* distichous, oblong, sheathing at base. *Inflorescence* lateral, branched or not, densely many-flowered. *Flowers* small; sepals and petals free and somewhat spreading; lip trilobed, the lateral lobes small, joined at the back to the base of the column, the front edges incurved, the spur rounded and saccate, the back wall of the spur bearing an erect tongue which reaches to the mouth of the spur; column short; pollinia 4.

A genus of about thirty species from SE. Asia to Australia and the Pacific Islands, 2 species being recorded.

Stem elongate, stout, up to 1.3 m long; leaves up to 38 cm long, 3–4 cm wide; inflorescence branched, up to 80 cm long .... **1. P. marsupiale**
Stem short, 2 cm long; leaves up to 22 cm long, 2 cm wide; inflorescence unbranched, c. 8 cm long .... **2. P. potamophilum**

1. P. marsupiale *(Kraenzl.) J.J. Smith* in Nova Guinea 7: 101 (1913). Type: New Guinea, *Hollrung* 743 (holotype B†).
Cleisostoma marsupiale Kraenzl. in K. Schum. & Hollr., Fl. Kais. Wilh. Land: 34 (1889).

For full synonymy see Schlechter (1913).

An epiphytic herb, forming robust, erect clumps. *Stems* stout, up to 1.3 m long, with thick cord-like roots arising from the nodes, leafy. *Leaves* numerous, coriaceous, alternate, strap-like, up to 38 cm long, 3–4 cm wide, unequally bilobed at apex, closely sheathing at base. *Inflorescences* lateral, branched in upper third, held stiffly erect, up to 80 cm long, the branches 3–8 cm long, densely many-flowered, the flowers upward-facing, with only a few open at once; bracts deciduous, elliptic, 1.5 mm long. *Flowers* yellowish-green with a cream or orange-yellow lip; sepals and petals widely spreading, thick and fleshy; dorsal sepal erect, oblong, 6–7 mm long, 2 mm wide; lateral sepals divergent, oblong, 6–7 mm long, 2–3 mm wide; petals obliquely erect, oblong-lanceolate, 6 mm long, 2–3 mm wide; lip trilobed, 5 mm long, 4 mm wide, the lateral lobes erect and incurved, the mid-lobe short, thick, fleshy, acute, recurved, the spur pear-shaped, 4–5 mm long, the callus a thick appendage which almost closes the entrance to the spur; column 2–3 mm long. (See plate 16A).

DISTRIBUTION: Bougainville, Guadalcanal and Malaita. Also in New Guinea and Australia.
HABITAT: Rain forest, 40–520 m.
COLLECTIONS: *Cribb* S16 (K); *Millar & Vandenberg* in NGF 48514 (K); *Nakisi* in BSIP 7349 (K); *Wickison* 155, 201 & 257 (K).

2. P. potamophilum Schltr. in Fedde, Repert. Spec. Nov. Regni Veg., Beih. 1: 988 (1913). Types: New Guinea, *Schlechter* 16312 & 18287 (syntypes B†, isosyntypes K!).

An epiphytic herb forming clumps. *Stems* short, 2 cm long, with thick cord-like roots arising from the base, leafy. *Leaves* numerous, alternate, oblong, coriaceous, up to 22 cm long, 2 cm wide, unequally bilobed at apex, sheathing at base. *Inflorescence* lateral, up to 8 cm long, unbranched, densely-flowered, the flowers upward-facing; bracts triangular, 1 mm long. *Flowers* with sepals and petals not opening widely, thick and fleshy; dorsal sepal oblong, 4 mm long, 3 mm wide; lateral sepals oblong, 4 mm long, 3 mm wide; petals oblong, 4 mm long, 1.5–2 mm wide; lip trilobed, 3–4 mm long, 3 mm wide, the lateral lobes erect and incurved, the mid-lobe short, thick, fleshy, acute, recurved, spurred, the spur conical, 3.5 mm long, 2.5 mm wide, with the callus a thick appendage which almost closes the entrance to the spur; column 2 mm long.

DISTRIBUTION: New Georgia. Also in New Guinea.
HABITAT: Cultivated in a garden at Munda but collected in New Georgia.
COLLECTION: *Chaplin* 814A (K).

There is no information on the colour of the flowers from the Solomon Islands' specimen. In New Guinea the flowers are yellow with brown spots and a white lip.

76. ROBIQUETIA Gaudichaud

Epiphytic monopodial plants. *Stems* stout, pendulous, leafy, sheathed in leaf bases, with cord-like roots. *Leaves* alternate, coriaceous, oblong, unequally bilobed at the apex. *Inflorescences* lateral, usually elongate on a long peduncle (except *R. woodfordii* which has a short peduncle and inflorescence), erect or pendulous, unbranched, with many crowded small flowers rather like a bottle brush. *Flowers* non-resupinate, often brightly coloured, fleshy; sepals and petals free, subsimilar, not opening widely, concave; lip small, trilobed, fleshy, with a large spur; column short and broad, without a foot; pollinia 4.

A genus of about twenty-five species from SE. Asia to Australia and the Pacific Islands, four species being recorded (but *R. wassellii* is not native).

1. Inflorescence very short, less than 7 cm long; peduncle 1.5–2 cm long .... **4. R. woodfordii**
   Inflorescence more than 10 cm long; peduncle elongate, more than 7 cm long 2
2. Flowers yellow with crimson spots and a golden spur, the lip with a green tip; the spur less than 4 mm long, subglobose .... **1. R. gracilistipes**
   Flowers white, pink, rose-purple or maroon, usually with green tips to the segments; the spur more than 4 mm long, clavate 3
3. Flowers maroon with a white lip, tipped with green; spur about 10 mm long; peduncle maroon .... ***3. R. wassellii**
   Flowers pink, rose-purple or white, tipped with green; spur 5–7 mm long; peduncle green .... **2. R. mooreana**

1. R. gracilistipes *(Schltr.) J.J. Smith* in Natuurk. Tijdschr. Ned. Ind. 72: 43 (1912). Type: New Guinea, *Schlechter* 13948 (holotype B†).
Saccolabium gracilistipes Schltr. in K. Schum. & Laut., Nachtr. Fl. Deutsch. Südsee: 227 (1905).
Saccolabium squamulosum J.J. Smith in Bull. Dépt. Agric. Indes Néerl. 19: 35 (1908). Types: New Guinea, *Versteeg* 1779 & *Djibdja* 873 (syntypes L).
Robiquetia squamulosa (J.J. Smith) J.J. Smith in Natuurk. Tijdschr. Ned. Ind. 72: 45 (1912).

For full synonymy see Clements (1989).

An epiphytic plant forming large clumps. *Stems* sturdy, up to 40 cm long, 1 cm wide, leafy, with thick, cord-like roots. *Leaves* alternate, sturdy, coriaceous, oblong-elliptic, up to 14 cm long, 4 cm wide, unequally bilobed at the apex, dark green above, light green below speckled with red-brown, sheathing at base. *Inflorescence* lateral, up to 18 cm long, densely many-flowered; peduncle up to 11 cm long. *Flowers* yellow with crimson spots, with or without a green tip to the lip, and with a golden spur, fleshy, finely pubescent; sepals and petals incurved, fairly widely spreading; dorsal sepal ovate, 4 mm long, 4 mm wide; lateral sepals ovate, 4.5 mm long, 4 mm wide; petals ovate, 3.5–4 mm long, 2.5 mm wide; lip trilobed, 2.5 mm long, 3 mm wide, the lateral lobes erect, oblong, the mid-lobe projected forwards, triangular, obtuse, fleshy, the spur subglobose, 3.5 mm long; column 3 mm long. (See fig. 76 L–M)

DISTRIBUTION: Bougainville, Guadalcanal and New Georgia. Also in New Guinea and Australia.
HABITAT: Rain forest, sea level to 530 m.
COLLECTIONS: *Dickson* 32 (K); *Millar* in NGF 38415 (K); *Wickison* 147 (K).

2. R. mooreana *(Rolfe) J.J. Smith* in Natuurk. Tidschr. Ned. Ind. 72: 44 (1912). Types: New Guinea, cult. *Moore* s.n., cult. *Lawrence* s.n. & cult. *Lucas* s.n. (syntypes K!).
Saccolabium mooreanum Rolfe in Bull. Misc. Inf., Kew 1893: 64 (1893).

An epiphytic plant forming large clumps. *Stems* sturdy, up to 30 cm long, 1 cm wide, with cord-like roots, leafy. *Leaves* alternate, coriaceous, oblong-lanceolate, up to 26 cm long, 4 cm wide, unequally bilobed at the apex, deep green to purple green, sheathing at base. *Inflorescence* lateral, up to 24 cm long, densely many-flowered; peduncle up to 16 cm long. *Flowers* white, pink or rose-purple, the sepals and lip with green tips; sepals and petals incurved, not opening widely, fleshy; dorsal sepal oblong, 6 mm long, 2.5 mm wide; lateral sepals obliquely ovate, 6 mm long, 3 mm wide; petals ovate, 6 mm long, 3 mm wide; lip obscurely trilobed, 3–4 mm long, 3 mm wide, the lateral lobes erect, with undulate margins, the mid-lobe projected forwards, triangular, obtuse, fleshy, the spur slightly inflated at the tip, 5–7 mm long; column 3 mm long. (See fig. 76 N and plate 16C).

DISTRIBUTION: Bougainville, Guadalcanal, Malaita, New Georgia and Rendova. Also in New Guinea.
HABITAT: Rain forest, sea level to 700 m.
COLLECTIONS: *Chaplin* s.n. (K); *Dennis* in RSS 2003, 2024, 7851 & 7883 (K); *Dickson* MD33 (K); *Mason* 1631 (K); *Millar & Vandenberg* in NGF 48507 (K); *Waterhouse* 374 (K); *Wickison* 152 & 212 (K).

Robiquetia mooreana is very similar in appearance to *R. bertholdii* (Reichb. f.) Schltr., from Vanuatu, Fiji and Tonga but differs in having larger bracts and slightly larger flowers. Kores (1989) notes that *R. bertholdii* occurs in the Santa Cruz Islands but we have not been able to examine any specimens from there.

Fig. 76. Robiquetia woodfordii A-K. A, habit x 2/3; **B**, flower x 4; **C**, lip, spur and column x 6; **D**, column x 8; **E**, anther-cap x 8; **F**, pollinia x 8; **G**, dorsal sepal x 6; **H**, petal x 6; **J**, lateral sepal x 6; **K**, lip and spur x 6. **R. gracilistipes L-M. L**, flower x 4; **M**, flower x 4. **R. mooreana. N**, flower x 4. **A-K** drawn from *Hunt* 3034; **L-M** from *Wickison* 147; **N** from *Wickison* 152. All drawn by Sue Wickison.

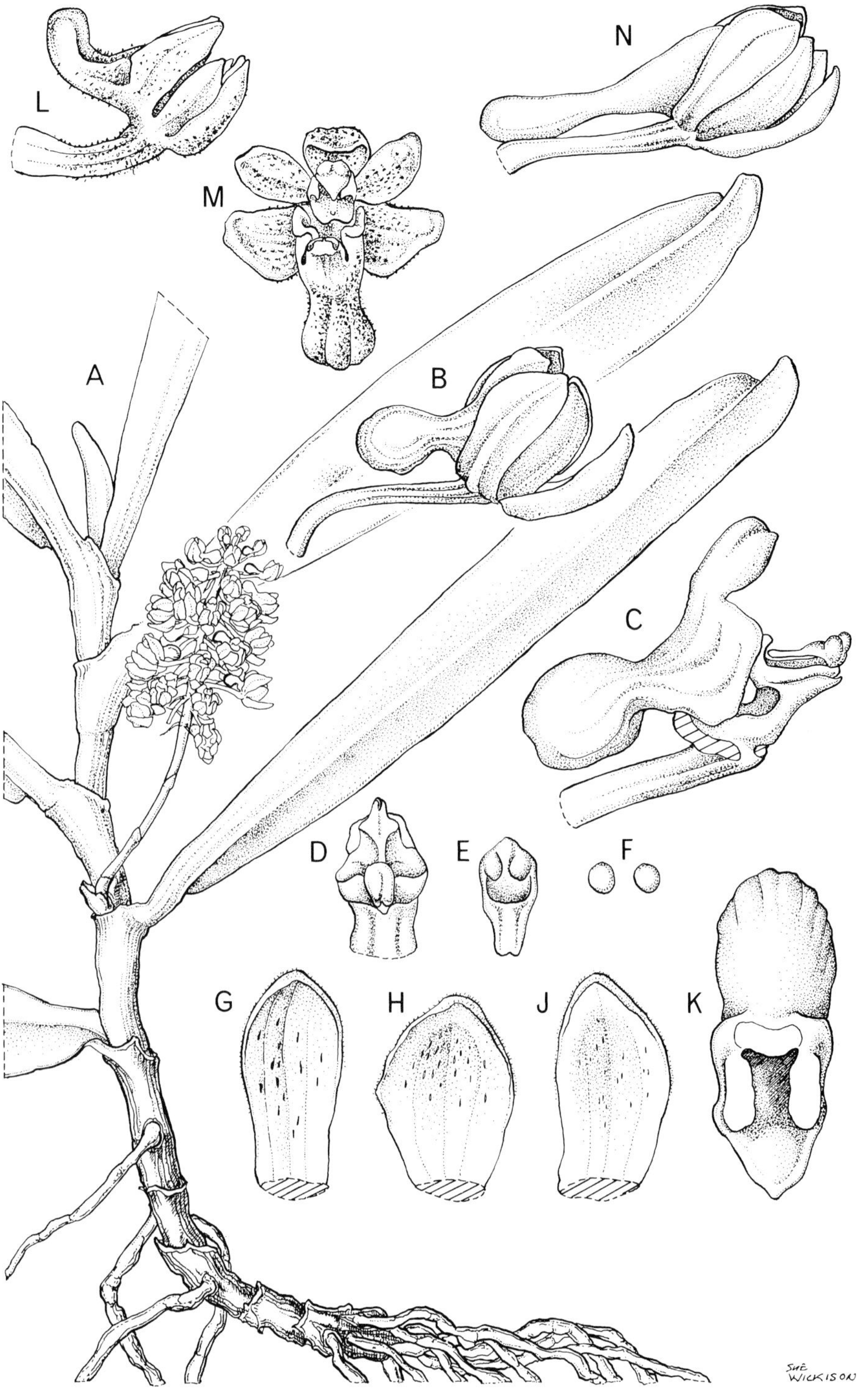
L
N
M
A
B
C
D
E
F
G
H
J
K
SUE WICKISON

***3. R. wassellii** *Dockr.*, Aust. Sarcanth.: 7 (1967). Type: Australia, *Wassell* s.n., cult. Melbourne Botanic Gardens (holotype MEL).

An epiphytic plant forming large straggly clumps. *Stems* flattened, pendulous, sparsely branched, up to 80 cm long, 1 cm wide, with thick, cord-like roots. *Leaves* coriaceous, distichous, 3–6 per stem, crowded, narrowly oblong, up to 14 cm long, 3 cm wide, unequally bilobed at the apex, dark green, sheathing at base. *Inflorescence* lateral, up to 15 cm long, stiffly pendulous, densely many-flowered; rhachis maroon. *Flowers* fleshy, maroon with a white lip with a maroon tip and a white column; sepals and petals incurved; dorsal sepal ovate, 4.5 mm long, 3 mm wide; lateral sepals ovate, 4.5 mm long, 3.5 mm wide; petals ovate, 4–4.5 mm long, 3.5–4 mm wide; lip trilobed, 3 mm long, 2 mm wide, the lateral lobes erect, oblong, fleshy, the mid-lobe projected forwards, triangular, fleshy, the spur slightly incurved, clavate, 10 mm long; column 2 mm long.

DISTRIBUTION: Introduced to Bougainville. Endemic to Australia.
HABITAT: Not known.
COLLECTION: *Mason* 2102 (K).

A single collection of this plant was made at Arawa in Bougainville where it was cultivated by McKillop. Although said to be native it could well have been introduced by him from Australia. Several other exotic orchids can be found growing on trees on his former estate at Arawa. (See Slade, 1965).

Unfortunately we have only seen the preserved flowers of this specimen and the description of the habit is from Australian material.

4. R. woodfordii *(Rolfe) Garay* in Bot. Mus. Leafl. Harv. Univ. 23: 197 (1972). Type: Solomon Islands, Rennell, *Woodford* s.n. (holotype K!).
Saccolabium woodfordii Rolfe in Bull. Misc. Inf., Kew 1907: 131 (1907).
Malleola woodfordii (Rolfe) P.F. Hunt in Kew Bull. 24: 99 (1970).

An epiphytic plant forming clumps. *Stems* sturdy, leafy. *Leaves* alternate, sturdy, coriaceous, oblong-elliptic, up to 19 cm long, 2.6 cm wide, unequally bilobed at the apex, blue-green, sheathing at base. *Inflorescence* lateral, up to 6 cm long, densely many-flowered; peduncle 1–2 cm long. *Flowers* pale orange with a deep green tip to the lip; sepals and petals incurved, not opening widely, fleshy; dorsal sepal ovate, 6 mm long, 2 mm wide; lateral sepals ovate, 6 mm long, 2.5 mm wide; petals ovate, 5 mm long, 3 mm wide; lip trilobed, 3 mm long, 2 mm wide, the lateral lobes erect, obliquely triangular, the mid-lobe projected forwards, triangular, obtuse, fleshy, the spur clavate-subglobular, 3 mm long; column 2 mm long. (See fig. 76 A–K).

DISTRIBUTION: Malaita, Rennell and Santa Isabel.
HABITAT: Lowland forest, c. 100 m.
COLLECTIONS: *Hunt* in RSS 2860 & 3034(K); *Woodford* s.n. (K).

77. TAENIOPHYLLUM Blume

Small leafless monopodial epiphytic plants with short stems bearing many long, spreading, green roots. *Roots* usually appressed to the bark of the supporting tree, dorsiventrally flattened or terete. *Inflorescences* lateral, with a short peduncle; rhachis slowly elongating, bearing flowers in succession, 1 or 2 at a time; bracts alternate, in 2 ranks. *Flowers* small; sepals and petals free and widely spreading or joined at the base and not opening widely; lip fixed to the base of the column, spurred; column short; pollinia 4.

In this genus the leaves are reduced to tiny brown scales and their function is carried out by the green roots.

A genus of about one hundred species from tropical Asia to Australia and the Pacific Islands, four species being recorded.

1. Roots terete, 1 mm wide or less 2
   Roots dorsiventrally flattened, 2 mm wide or more 3
2. Spur globose, shorter than the lip, 1–1.5 mm long .... **2. T. kompsopus**
   Spur clavate, as long as or longer than the lip, 3–4 mm long .... **3. T. malianum**
3. Inflorescence hairy; spur fusiform, 8–9 mm long .... **4. T. latipetalum**
   Inflorescence glabrous; spur cylindrical, obtuse, less than 2.5 mm long .... **1. T. fasciola**

1. T. fasciola *(Forst. f.) Reichb. f.* in Seem., Fl. Vit.: 296 (1868). Type: Tahiti, *G. Forster* 172 (holotype BM!, isotype P).

Epidendrum fasciola Forst. f., Fl. Ins. Austr. Prodr.: 60, n. 320 (1786).

Taeniophyllum seemannii Reichb. f. in Seem., Fl. Vit.: 296 (1868). Type: Fiji, *Seemann* 593 (holotype W, isotype L).

Taeniophyllum asperulum Reichb. f. in Otia Bot. Hamb. 1: 53 (1878). Type: Society Islands, *Wilkes* s.n. (holotype W).

Taeniophyllum decipiens Schltr. in Fedde, Repert. Spec. Nov. Regni Veg. 9: 112 (1911). Type: Samoa, *Vaupel* 278 (holotype B†, isotype K!, W).

Taeniophyllum parhamiae L.O. Williams in Bot. Mus. Leafl. Harv. Univ. 7: 148 (1939). Type: Fiji, *Parham* 3 (holotype AMES!).

An epiphytic herb. *Roots* dorsiventrally flattened, 5–15 cm long, 2–3 mm broad, fasciculate. *Inflorescences* lateral, short, up to 2 cm long; bracts distichous. *Flowers* greenish to pale yellow; dorsal sepal ovate-oblong, 3.5 mm long, 2 mm wide; lateral sepals ovate, 3.5 mm long, 2 mm wide; petals slightly falcate, 3–3.5 mm long, 1–1.5 mm wide; lip slipper-like, broadly obovate, acute, when flattened, 2 mm long, 2–3 mm wide, the spur forming a ± right-angle with the ovary, cylindrical, obtuse, 1.5–2 mm long, 1.5 mm wide; column 1.5 mm long. (See fig. 77 G–Q and plate 15C).

DISTRIBUTION: New Georgia and Rennell. Also in the Mariana Islands (Guam), the Horn Islands, the Cook Islands, Tahiti, Vanuatu, New Caledonia, Fiji, Samoa, and Tonga.

HABITAT: Strand forest, sea level to 30 m.

COLLECTIONS: *Chaplin* 820 (K): *Cribb* 6008 (K); *Cribb & Morrison* 1781 (K).

2. T. kompsopus *Schltr.* in K. Schum. & Laut., Nachtr. Fl. Deutsch. Südsee: 220 (1905). Type: New Guinea, *Schlechter* 13812 (holotype B†).

An epiphytic herb. *Roots* terete, 8–30 cm long, 1 mm broad. *Inflorescences* lateral, up to 5 cm long; bracts distichous. *Flowers* pale yellow-orange or white; dorsal sepal oblong, 3 mm long, 1 mm wide; lateral sepals lanceolate, 3 mm long, 1 mm wide; petals lanceolate, 3 mm long, 1 mm wide; lip obovate, acuminate, 2–2.5 mm long, 1.5–2 mm wide, the spur globose, 1–1.5 mm long, 1–1.5 mm wide; column 1–1.5 mm long. (See fig. 77 A–B and plate 15E).

DISTRIBUTION: Guadalcanal, Makira and Rendova. Also in New Guinea and Vanuatu.
HABITAT: Rain forest, 50–500 m.
COLLECTIONS: *Cribb* 6001 (K); *Cribb & Campbell* 5192 (K); *Cribb, Dennis & Campbell* 5001 (K); *Dennis* in RSS 4430 & 7928 (K); *Hunt* in RSS 2300 (K); *Whitmore* in BSIP 3812 (K); *Wickison* 83 (K).

M. Clements (pers. comm.) recently collected this species with G. Hermon Slade on Efate, Vanuatu, and this extends the known distribution of this species to the southeast.

3. T. latipetalum *Schltr.* in Fedde, Repert. Spec. Nov. Regni Veg., Beih. 1: 1034 (1913). Type: New Guinea, *Schlechter* 19693 (holotype B†).

An epiphytic herb. *Roots* dorsiventrally flattened, 8–37 cm long, 2–3 mm broad. *Inflorescences* lateral, up to 13 cm long, hairy; bracts distichous. *Flowers* greenish yellow; pedicel and ovary pubescent; dorsal sepal oblong, 4 mm long, 1.5–2 mm wide; lateral sepals ovate, 4 mm long, 1.5–2 mm wide; petals ovate, 3–3.5 mm long, 2 mm wide; lip obscurely trilobed, obovate, 3 mm long, 2–2.5 mm wide, with a fleshy callus at the base, the spur fusiform, 8–9 mm long, 1–1.5 mm wide. (See fig. 77 E–F).

DISTRIBUTION: Bougainville, Guadalcanal and Makira. Also in New Guinea.
HABITAT: Rain forest, up to 1200 m.
COLLECTIONS: *Cribb, Dennis & Campbell* 5030 & 5079 (K); *Hunt* in RSS 2210 & 2259 (K); *Lavarack & Ridsdale* in NGF 31183 (K); *Wickison* 83A (K).

4. T. malianum *Schltr.* in Fedde, Repert. Spec. Nov. Regni Veg., Beih. 1: 1022 (1913). Type: New Guinea, *Schlechter* 18364 (holotype B†).

An epiphytic herb. *Roots* terete, 6–20 cm long, 1 mm or less broad. *Inflorescences* lateral, up to 4 cm long; bracts distichous. *Flowers* translucent yellow; dorsal sepal oblong, 4 mm long, 1 mm wide; lateral sepals lanceolate, 4 mm long, 1–1.5 mm wide; petals lanceolate, 4 mm long, 1 mm wide; lip obovate, acuminate, 3–4 mm long, 2–2.5 mm wide, the spur clavate, as long as or longer than the lip, 3–4 mm long, 1.5 mm wide. (See fig. 77 C–D).

DISTRIBUTION: Guadalcanal. Also in New Guinea and Australia.
HABITAT: Rain forest, 300 m.
COLLECTION: *Cribb, Dennis & Campbell* 5000 (K).

78. TUBEROLABIUM Yamamoto

Small to medium-sized monopodial epiphytic herbs. *Stems* short, leafy. *Leaves* distichous, crowded, thick, sheathing at base. *Inflorescences* lateral, short, bearing a succession of flowers, or a few at a time. *Flowers* small; sepals and petals free, subsimilar; lip immovably joined to the apex of the column-foot, trilobed, deeply saccate, spurred at the base; column short; pollinia 2.

A genus of about eleven species from SE. Asia to the Pacific Islands, a single species being recorded.

T. papuanum *(Schltr.) J.J. Wood* in Lewis & Cribb, Orchids of Vanuatu: 150 (1989). Types: New Guinea, *Schlechter* 17166 & 18975 (syntypes B†).
Saccolabium papuanum Schltr. in Fedde, Repert. Spec. Nov. Regni Veg., Beih. 1: 978 (1913).
Sarcochilus societatis J.W. Moore in Bull. Bish. Mus. 102: 24 (1933). Type: Society Islands, *Moore* 411 (holotype BISH).
Saccolabium subluteum Rupp in N. Queensl. Nat., 21 (105): 1 (1953). Type: Australia, *Goessling-St. Cloud* 10/1952 (holotype NSW-lost), cult. ANBG ex *Wrigley & Telford* NQ 784A (neotype CBG).
Trachoma subluteum (Rupp) Garay in Bot. Mus. Leafl. Harv. Univ., 23: 208 (1972).
Trachoma societatis (J.W. Moore) N. Hallé in l'Orchidophile 40: 1481 (1980).
Trachoma papuanum (Schltr.) M. Clements, J.J. Wood & D. Jones in Clements, Aust. Orch. Res. 1: 145 (1989).

An epiphytic herb forming small, semi-pendulous clumps. *Stems* up to 5 cm long, with a few cord-like roots. *Leaves* 3–6, distichous, crowded, elliptic-ovate with an oblique point, up to 15 cm long, 2.5 cm wide, thick and fleshy, light green. *Inflorescence* lateral, few-flowered; peduncle up to 5 mm long; rhachis up to 1.5 cm long (elongating as subsequent flowers develop). *Flowers* dull pale yellow with fine brown-red speckles near the base on the interior surface, with a white lip and column, not opening widely; dorsal sepal ovate, 4–5 mm long, 2–2.5 mm wide; lateral sepals obliquely elliptic, 4–5 mm long, 2.5–3 mm wide; petals narrowly ovate, 3–4 mm long, 1 mm wide; lip trilobed, the lateral lobes short and fleshy, the mid-lobe short and fleshy, the spur subcylindrical, obtuse, 1.5 mm long; column 2 mm long.

DISTRIBUTION: Gizo. Also in New Guinea, Vanuatu, New Caledonia, Fiji, Tahiti, the Tubuai Islands and Australia.

HABITAT: No information from the Solomon Islands, in Vanuatu this species occurs in montane forest.

COLLECTION: *Im Thurn* 349 (K).

At present there is some discussion as to the use of the generic name *Tuberolabium* in preference to *Trachoma*. J.J. Wood (in press) has established that the genus *Trachoma* was separated from *Tuberolabium* primarily by the absence of a column-foot. However, the type species *Tuberolabium kotoense* Yamamoto, from Taiwan, lacks a true column-foot. Wood therefore concludes that *Trachoma* should be included in *Tuberolabium*, the earlier name.

79. **TRICHOGLOTTIS** Blume

Epiphytic monopodial herbs. *Stems* erect, ± climbing, drooping or pendulous, leafy in apical half. *Leaves* alternate, narrowly oblong or ovate, with sheathing bases. *Inflorescences* lateral, short, 1- to few-flowered. *Flowers* small to rather large; white to yellowish with red-brown markings; sepals broader than the petals, concave; lateral sepals joined at the base to a short column-foot; lip rigidly attached to the column-foot, entire or lobed, with a fleshy disc which may be hairy, saccate or spurred at the base, with a ± erect or valvate, ligulate or slender appendage on the posterior wall of the spur just below the column-foot; column short with a very short foot; pollinia 4.

A genus of about sixty species from tropical Asia to the Solomon Islands and Australia, two species being recorded.

Leaves 11–14 cm long, 1.2–1.5 cm wide; lip trilobed .. .. **1. T. papuana**
Leaves 5.5–7 cm long, 0.8–1.2 cm wide; lip apparently 5-lobed .. .. .. **2. T. sororia**

1. T. papuana *Schltr.* in Fedde, Repert. Spec. Nov. Regni Veg., Beih. 1: 993 (1913). Types: New Guinea, *Schlechter* 17297 (syntype B†, isosyntypes AMES, E, K!, NSW) & 19401 (syntype B†).
Trichoglottis flexuosa J.J. Smith in Nova Guinea 8: 125 (1909) non Rolfe (1905). Type: New Guinea, *Djibdja* 492 (holotype BO).

An epiphytic herb. *Stems* pendent, up to 50 cm long, woody towards the base, leafy in apical part, light green with a purple tinge around the inflorescence. *Leaves* fleshy, distichous, lanceolate, acute, 11–14 cm long, 1.2–1.5 cm wide, sheathing at base. *Inflorescence* lateral, arising opposite a leaf base, 2– to 4-flowered; peduncle 1.5–2 mm long. *Flowers* scented, the sepals and petals white with a red-brown blotch in the center, the lip and column white with purple markings; dorsal sepal oblong, 7 mm long, 2.7 mm wide; lateral sepals obliquely ovate, 7 mm long, 3.5 mm wide; petals obliquely spathulate, 6 mm long, 2.5 mm wide; lip trilobed, 5 mm long, 2 mm wide, the lateral lobes small, triangular, erect, the mid-lobe concave, ovate, acute, pubescent at base, the spur short and rounded; column short, 4 mm long. (See plate 15F).

DISTRIBUTION: Bougainville, Guadalcanal, Kolombangara, Malaita, Ranongga and Santa Isabel. Also in New Guinea.
VERNACULAR NAME: Marapauna (Bougainville).
HABITAT: Strand to lower montane forest, sea level to 800 m.
COLLECTIONS: *Comins* 347 (K); *Cribb & Campbell* 5114 (K); *Cribb, Dennis & Campbell* 5040 (K); *Guppy* 7/84 & 2/85 (K); *Hunt* in RSS 2776 (K); *Lavarack & Ridsdale* in NGF 30610 & 31037 (K); *Mauriasi* et al. BSIP 12474 & 16525 (K); *Mitchell* 79 (K); *Waterhouse* 707 (K); *Wickison* 206 & 218 (K).

Fig. 77. Taeniophyllum komsopus A-B. A, inflorescence x 6; **B**, lip x 10. **T. malianum C-D. C**, flower x 6; **D**, lip x 10. **T. latipetalum E-F. E**, flower x 6; **F**, lip x 10. **T. fasciola G-Q. G**, habit x 2/3; **H**, flower x 6; **J**, lip x 10; **K**, column, lip and spur x 8; **L**, dorsal and ventral view of anther-cap x 8; **M**, pollinia x 8; **N** column x 10; **O**, dorsal sepal x 8; **P**, petal x 8; **Q**, lateral sepal x 8. **A-B** drawn from *Wickison* 83; **C-D** from *Cribb, Dennis & Campbell* 5000; **E-F** from *Wickison* 83A; **G-Q** from *Cribb & Morrison* 1781. All drawn by Sue Wickison.

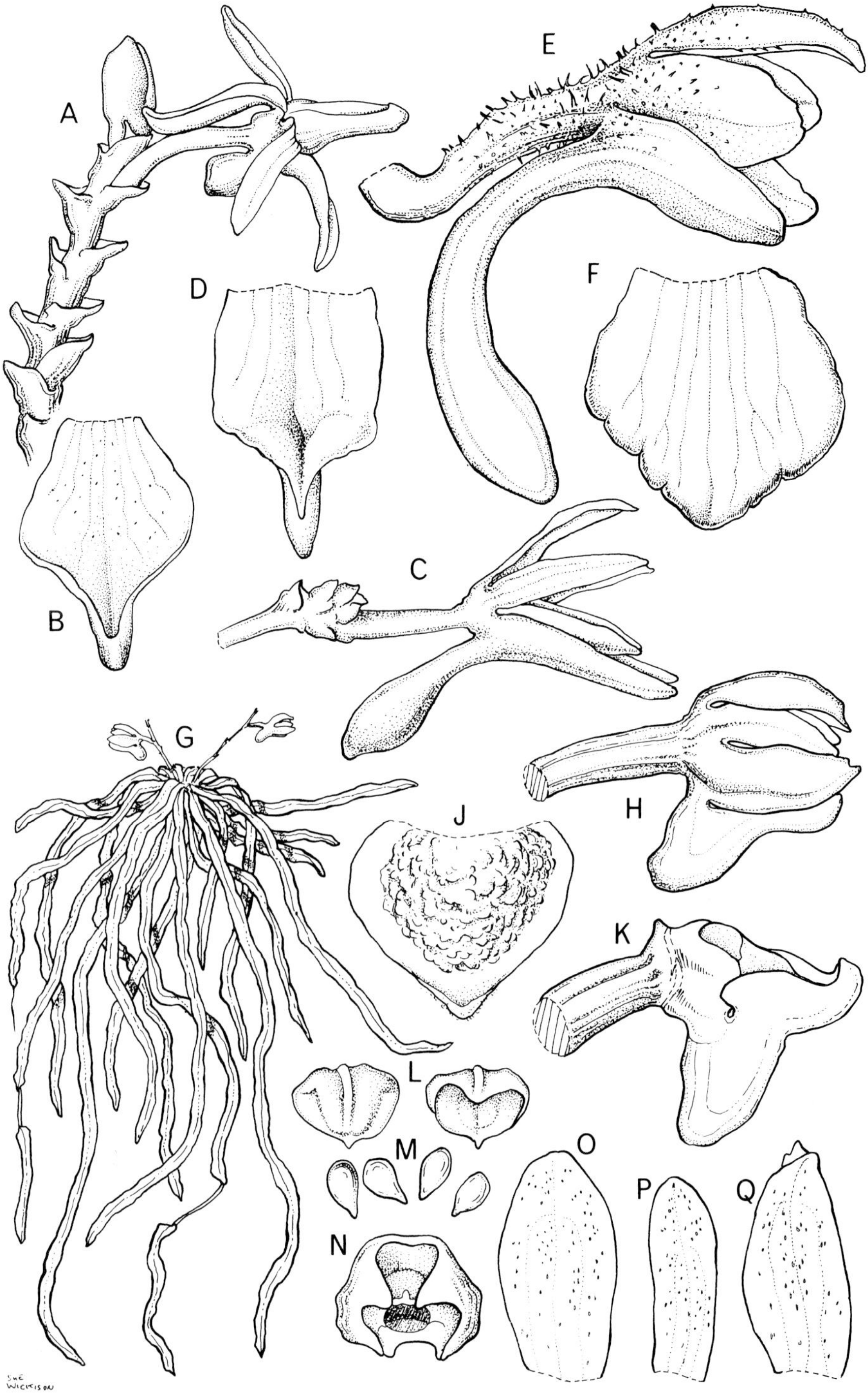
A
B
C
D
E
F
G
H
J
K
L
M
N
O
P
Q
SUE WICKISON

2. T. sororia *Schltr.* in Fedde, Repert. Spec. Nov. Regni Veg., Beih., 1: 993 (1913). Type: New Guinea, *Schlechter* 16576 (holotype B†, syntypes AMES, L, NSW).

An epiphytic herb. *Stems* pendent, up to 70 cm long, leafy in apical half. *Leaves* distichous, ovate-lanceolate, 5.5–7 cm long, 8–12 mm wide. *Inflorescence* lateral, arising opposite a leaf base, few-flowered; peduncle short, c. 2 mm long. *Flowers* pale yellow with red spots, the lip and column pale yellow with a cream-coloured spur; dorsal sepal spathulate, 7.5 mm long, 3.5 mm wide; lateral sepals ovate, 6 mm long, 4 mm wide; petals obliquely linear, 6.5 mm long, 1 mm wide; lip apparently 5-lobed, 4.5 mm long, 4.5 mm wide, the lateral lobes rounded and erect, the mid-lobe trilobed, with the side-lobules obliquely oblong with undulate margins, spreading, the mid-lobule convex, narrowly ovate, acute, fleshy, the spur globose, 1 mm long, 1 mm wide, upcurved; column stout, 3.5 mm long, 4 m wide.

DISTRIBUTION: Makira. Also in New Guinea.
HABITAT: Trees overhanging river, 50 m.
COLLECTION: *Hunt* in RSS 2255 (K).

80. CHRYSOGLOSSUM Blume

Terrestrial herbs with a creeping rhizome. *Pseudobulbs* of 1 internode, bearing a leaf, (pseudobulbs often indistinct and plants may appear to consist merely of a rhizome bearing a succession of stalked leaves). *Inflorescence* apparently arising from the rhizome, racemose, several-flowered. *Flowers* well-spaced; sepals and petals free and of equal length; lip trilobed, with a short spur; column with small wings on its front edge and with a conspicuous foot; pollinia 4.

A small genus of about six species from SE. Asia to the Pacific Islands, a single species being recorded.

C. vesicatum *Reichb. f.* in Seem., Fl. Vit.: 304 (1868). Type: Fiji, *Seemann* 611 (holotype W).
Collabium vesicatum (Reichb. f.) Schltr. in Fedde, Repert. Spec. Nov. Regni Veg. Beih. 1: 98 (1911).
Chrysoglossum aneityumense Ames in Journ. Arn. Arb. 14: 105 (1933). Type: Vanuatu, Anatom, *Morrison* s.n. (holotype AMES!, isotype K!).

A terrestrial herb. *Rhizome* slender, partly concealed by the fibrous remains of sheathing bracts. *Leaf* solitary, apparently arising from the rhizome, elliptic, acuminate, 9–12 cm long, 2.5–4 cm wide, dark green; petiole 1.5–3 cm long. *Inflorescence* racemose, apparently arising from the rhizome, up to 22 cm long, laxly 5- to 8 -flowered. *Flowers* yellowish-green with a yellowish lip and column; pedicel and ovary slender, 0.5–1 cm long; dorsal sepal oblong-lanceolate, acute, 9 mm long, 2.6 mm wide; lateral sepals lanceolate, falcate, acute, 10.5 mm long, 3–3.5 mm wide; petals lanceolate, falcate, 9.5–10 mm long, 2.5 mm wide; lip trilobed, 10.5 mm long, 8 mm wide, the lateral lobes small, triangular, acute, the mid-lobe broadly oblong, acute, the disc with 2 calli which extend to two-thirds the length of the lip, the spur conical, 2 mm long, 1 mm wide; column

4.5 mm long, with an irregularly dentate apex and with small wings on its front edge.

DISTRIBUTION: Makira. Also in Vanuatu, New Caledonia and Fiji.
HABITAT: Mossy ridge-top forest, 800 m.
COLLECTION: *Hunt* in RSS 2323 (K).

This identification is provisional as our specimen differs somewhat from the type in having a smaller lip with acute lateral lobes and a broadly oblong mid-lobe, and more material is needed to confirm the identification (see Kores, 1989). This species appears to be similar to *C. papuanum* J.J. Smith from New Guinea and New Ireland and the two should be critically compared.

Lewis & Cribb (1989) mistakenly cited *C. neocaledonicum* Schltr. and *C. gibbsiae* Rolfe as conspecific with *C. vesicatum*. They are in fact conspecific with *C. ornatum* Blume, a widespread species from SE. Asia to Vanuatu, New Caledonia, Fiji and Somoa, but as yet unrecorded from the Solomon Islands or New Guinea. *Chrysoglossum ornatum* differs from *C. vesicatum* in having a differently shaped lip with 3 callus ridges, the outer two extending to two-thirds the length of the lip and the central one extending to a third of the length (see Seidenfaden, 1983 and Kores, 1989). The description given by Lewis & Cribb (1989) under *C. vesicatum* refers, in fact, to *C. ornatum*.

81. PILOPHYLLUM Schlechter

Terrestrial herbs with a short, creeping rhizome. *Pseudobulbs* above ground bearing a solitary leaf or an erect inflorescence. *Leaf* erect, plicate, relatively large, hairy. *Inflorescence* racemose, laxly many-flowered. *Flowers* non-resupinate; sepals and petals free and spreading; lip trilobed; column with lateral appendages on its front edge, and with a distinct foot; pollinia 2.

A small genus of two species from the Malay Peninsula to the Solomon Islands, a single species being recorded.

P. villosum *(Blume) Schltr.* in Die Orchideen: 131 (1914). Type: Java, *Blume* s.n. (holotype L).
Chrysoglossum villosum Blume, Bijdr. Fl. Ned. Ind.: 338 (1825).

A terrestrial herb. *Pseudobulbs* clustered, 1–3 cm apart, mostly above ground level, sheathed by fibres and bracts, unifoliate at the apex. *Leaf* erect, ovate, 24–27 cm long, 9.5–12 cm wide, with 3 prominent veins, green with brown hairs; petiole 7.5 cm long, densely hairy. *Inflorescence* terminal from the apex of the pseudobulb, up to 70 cm long, red-purple with brown hairs; rhachis up to 30 cm long with 5 to 25 flowers; bracts hairy. *Flowers* non-resupinate, glabrous, yellow with dark red stripes and with a white lip with purple spots; pedicel and ovary hairy, 1.5 cm long; dorsal sepal narrowly lanceolate, concave at apex, apiculate, 1.2–1.7 cm long, 3–3.5 mm wide; lateral sepals falcate, apiculate, slightly shorter and broader than the dorsal sepal, 1.2–1.5 cm long, 3–4 mm wide; petals falcate-spathulate, apiculate; lip trilobed, 7 mm long, 7 mm wide, the lateral lobes oblong, spreading, 4 mm long, 1 mm wide, the mid-lobe narrowed from a broad base and expanding abruptly into a reniform apiculate apex, 1 mm long, 3 mm wide, the disc with 2 prominent keels extending to the reniform apex; column 9 mm long with 2 narrow, 3 mm long, lateral appendages in the middle, the column-foot 3 mm long.

DISTRIBUTION: Guadalcanal and Malaita. Also in the Malay Peninsula and Archipelago and New Guinea.
HABITAT: Montane forest, 650–900 m.
COLLECTIONS: *Corner* 3037 (K); *Cribb, Dennis & Campbell* 5061 (K).

Both collections from the Solomon Islands are sterile and the description of the flowers and inflorescence has been taken from collections from the Malay Archipelago.

82. EULOPHIA R. Brown

Terrestrial herbs with a short rhizome. *Stems* pseudobulbous, squat and with terminal leaves or slender, leafy and sheathed in cataphylls at the base (only the latter condition found in the Solomon Islands and Bougainville). *Leaves* lanceolate, plicate. *Inflorescence* lateral, arising from near the base of the stem, erect, racemose, few- to many-flowered. *Flowers* relatively small to large; sepals free, similar; petals smaller than the sepals; lip entire or trilobed, spurred at base; column slender; pollinia 4.

A pantropical genus of about two hundred species, a single species being recorded.

E. nuda *Lindley*, Gen. Sp. Orch. Pl.: 180 (1833). Type: Malaya, *Wallich* s.n. (holotype K!).
Eulophia squalida Lindley in Bot. Reg. 27: misc. 77 (1841). Type: Philippines, *Cuming* s.n. (holotype K!).
Eulophia macgregorii Ames in Philipp. Journ. Sc., Bot. 9: 12 (1914). Type: Mariana Islands, Guam, *R.C. McGregor* 631 (holotype AMES!).

For full synonymy see Seidenfaden (1983).

A terrestrial herb. *Stems* erect, up to 15 cm long, leafy, sheathed by cataphylls at base. *Leaves* 2 or more near the apex of the stem, elliptic-lanceolate, up to 30 cm long, 3.5 cm wide, narrowing at base, with 5–7 prominent veins. *Inflorescence* lateral, arising from near the base of the stem, up to 80 cm long, thick and fleshy, few- to many-flowered. *Flowers* greenish-white turning yellow; pedicel and ovary up to 4 cm long; dorsal sepal lanceolate, 2.2–2.5 cm long, 4 mm wide; lateral sepals obliquely lanceolate, 2.5 cm long, 5 mm wide; petals ovate, 1.9 cm long, 7 mm wide; lip obovate, entire, 1.9 cm long, 1 cm wide, the disc with 7 longitudinal callus ridges, the apical margin undulate, the spur broad, flattened, 3 mm long; column 1 cm long.

DISTRIBUTION: Guadalcanal. Widely distributed from Sri Lanka, Asia and the Malay Penisula to Sumatra, the Mariana Islands (Guam), Vanuatu, Fiji and Tonga.
HABITAT: Rain forest.
COLLECTION: *Dennis* s.n. (K).

83. GEODORUM G. Jackson

Terrestrial plants. *Stems* short, later swelling at base to form almost round, subterranean pseudobulbs. *Leaves* terminal, few, plicate, the uppermost the largest, broad, petiolate. *Inflorescence* separate from the leaf-bearing stem, erect; rhachis recurved, straightening prior to seed dispersal, densely several- to many-flowered. *Flowers* not opening widely, bell-shaped; sepals and petals similar; lip entire or obscurely trilobed, forming with the column-foot a short saccate base, not spurred; column short with a distinct foot; pollinia 4.

A small genus of about ten species in the Old World Tropics, mainly in Africa and Madagascar, a single species being recorded.

G. densiflorum *(Lam.) Schltr.* in Fedde, Repert. Spec. Nov. Regni Veg. 4: 259 (1919). Type: Malabar coast, [Anon] (lectotype Rheede's plate, Hort. Malab. 11: 35).
Limodorum densiflorum Lam., Encycl. Meth. Bot. 3: 516 (1792).
Cymbidium pictum R. Br., Prodr. Fl. Nov. Holl.: 331 (1810), non Link & Otto (1821). Type: Australia, *R. Brown* s.n. (lectotype BM!).
Geodorum pictum (R. Br.) Lindley, Gen. Sp. Orch. Pl.: 175 (1833).
Geodorum pacificum Rolfe in Bull. Misc. Inf., Kew 1908: 71 (1908). Types: Tonga Islands: Vavau, *Crosby* s.n.; Solomon Islands, *Woodford* s.n. (syntypes K!).
Geodorum tricarinatum Schltr., in Fedde, Repert. Spec. Nov. Regni Veg. 9: 101 (1911). Type: Samoa, *Vaupel* 285 (holotype B†).
Geodorum neocaledonicum Kraenzl. in Viertelj. Nat. Ges. Zür. 74: 82 (1929). Type: New Caledonia, *Daeniker* 1384 (holotype Z).

For full synonymy see Clements (1989).

A terrestrial herb. *Stems* up to 8 cm long, later swelling at the base to form a fleshy, ovoid pseudobulb. *Leaves* 2–3, from apical part of the stem, ovate-lanceolate, plicate, with the uppermost the largest, up to 30 cm long, 7 cm wide, with 5 or 7 principal veins; petiole up to 8 cm long. *Inflorescence* separate from the leaf-bearing stalk, with the peduncle up to 30 cm tall, rather fleshy; rhachis recurved when the flowers reach maturity, 3–5 cm long, bearing a succession of many flowers, after fertilization the peduncle straightens and both the peduncle and rhachis extend to as much as twice their former length. *Flowers* pink with purple markings on the lip; dorsal sepal ovate, 1.4 cm long, 3–4 mm wide; lateral sepals ovate, 1.4 cm long, 3–4 mm wide; petals ovate, 1.4 cm long, 3 mm wide; lip oblong to obscurely trilobed, 1.4–1.5 cm long, 7–8 mm wide, the lateral lobes erect, obliquely triangular, the mid-lobe broadly triangular, the disc with an obscure low broad keel; column 5 mm long, column-foot 3 mm long.

DISTRIBUTION: Santa Ana Island. Widespread from southern China, Burma, India and Sri Lanka to New Guinea, New Britain, Vanuatu, New Caledonia, Fiji, Samoa, Tonga, Niue and Australia.
HABITAT: Common in secondary scrub and old cultivation areas, sea level to 100 m.
COLLECTIONS: *Dennis* 4 (K); *Woodford* s.n. (K).

There has been considerable confusion surrounding the identity of this *Geodorum* species (as noted by Clements, 1989). In 1983 Seidenfaden proposed a new interpretation for the name *G. densiflorum* considering it to have a much narrower distribution than previously thought. Consequently Jones (1988) considered the Australian species to be referrable to *G. neocaledonicum* and Lewis & Cribb (1989) considered the Vanuatu plants to be referrable to *G. pacificum*. However, Clements (1989) and Kores (1989) have since established that the correct name for this species is *G. densiflorum* and both *G. neocaledonicum* and *G. pacificum* are conspecific with it.

84. OECEOCLADES Lindley

Terrestrial plants. *Pseudobulbs* short and one-noded to long, slender and few-noded. *Leaves* 1–3, terminal, conduplicate or plicate. *Inflorescence* lateral, arising from near the base of the pseudobulb, erect, racemose, many-flowered. *Flowers* with narrow, acute sepals; petals usually shorter and broader than sepals; lip trilobed or 4-lobed, with a short spur; column slender; pollinia 4.
A genus of about twenty species in the Old World Tropics, mainly in Africa and Madagascar, a single species being recorded.

O. pulchra *(Thouars) Cribb & M. Clements* in Clements, Aust. Orch. Res. 1:99 (1989). Type: Reunion, *Thouars* s.n. (holotype P).
Limodorum pulchrum Thou., Orch. Iles Austr. Afr.: t.43, 44 (1822).
Eulophia pulchra (Thouars) Lindley, Gen. Sp. Orch. Pl.: 182 (1833).
Eulophia macrostachya Lindley, Gen. Sp. Orch. Pl.: 183 (1833). Type: Sri Lanka, *Macrae* 27 (holotype K!).
Eulophia novaeebudae Kraenzl. in Bull. Soc. Bot. Fr. 76: 301 (1929). Type: Vanuatu, Efate, *Levat* s.n. (holotype MPU).

For full synonymy see Clements (1989).

A terrestrial herb. *Pseudobulbs* slender, up to 10 cm long, sheathed by cataphylls and the fibres of decayed cataphylls. *Leaves* apical, 2 (or 3), plicate, elliptic, up to 30 cm long, 7 cm wide, with 3 strongly marked veins; petiole up to 10 cm long. *Inflorescences* lateral, arising from near the base of the pseudobulb, up to 75 cm long, several- to many-flowered. *Flowers* green with a pale yellow lip with purple lines; pedicel and ovary 1.5–2 cm long; dorsal sepal ovate, acute, 1–1.2 cm long, 4–5 mm wide; lateral sepals obliquely ovate-falcate, acute, 1–1.2 cm long, 4–5 mm wide; petals broader than sepals, ovate, acute, 9–10 mm long, 5–6 mm wide; lip trilobed, 9–12 mm long, 10–12 mm wide, the lateral lobes erect, rounded, the mid-lobe much broader than long, broadly cleft, with a divided callus at its base, the spur short and sphaerical, 2 mm long, 2 mm wide; column 5 mm long.

DISTRIBUTION: Guadalcanal and Malaita. Widely distributed from tropical Africa, Madagascar and Asia to New Guinea, Vanuatu, the Mariana Islands, Fiji, Tonga, Niue and Australia.
HABITAT: Rain forest, c. 300 m.
COLLECTIONS: *Robinson* K169 (K); *Wickison* 122 & 205 (K).

85. GRAMMATOPHYLLUM Blume

Large epiphytic or rarely lithophytic herbs, with both elongate, spreading and erect, acuminate roots. *Pseudobulbs* ovoid to elongate, large, few- to many-leaved. *Leaves* distichous, plicate, large. *Inflorescences* lateral, unbranched, erect or drooping, many-flowered. *Flowers* large, yellow marked with brown or maroon; sepals and petals subsimilar, free, spreading; lip trilobed, much smaller than sepals, fused at base to base of the column, usually hairy, the callus of 2 or 3 ridges; column clavate; pollinia 4.
A genus of about ten species from SE. Asia to the Solomon Islands, two species being recorded.

Pseudobulbs ovoid, usually less than 25 cm long; leaves lanceolate, over 3.5 cm wide; without sterile basal flowers; sepals 2.5–3.8 cm long, up to 1.5 cm wide, brown with yellow margins or with large brown spots on a green or yellow background; lip rounded at base, with a subcircular mid-lobe .... **1. G. scriptum** var. **boweri**
Pseudobulbs elongate, usually more than 1 m long; leaves many, linear, less than 2.5 cm wide; basal 3–5 flowers of inflorescence often sterile and with 4 tepals, the remainder fertile and normal; sepals 4–5.5 cm long, 2 cm wide covered with many small maroon spots; lip cordate at base, with a triangular mid-lobe .... **2. G. speciosum**

1. G. scriptum *Blume* in Rumphia 4: 48 (1836). Type: Amboina, *Rumphius* s.n. (holotype L-lost, lectotype illustration in Rumphius, Herb. Amb. 6: 98, t. 42!).
Angraecum scriptum Rumph., Herb. Amb. 6: 95 (1750) nom. illeg. pre 1753.

var. **boweri** (F. Muell.) Schltr. in Fedde, Repert. Spec. Nov. Regni Veg., Beih. 1: 951 (1912). Type: Solomon Islands, *Goldfinch* (holotype MEL).
Grammatophyllum boweri F. Muell. in Wing's South Sci. Rec.: 1 (1883).
Grammatophyllum guiliemisecundi Kraenzl. in Gartenfl.: 114 (1894). Type: New Guinea, *Micholitz* (flower of holotype K!).

An epiphytic plant forming large compact clumps. *Pseudobulbs* ovoid, up to 25 cm long, green turning glossy yellow. *Leaves* distichous, lanceolate, petiolate, up to about 70 cm long, 3.5–8 cm wide. *Inflorescences* lateral, arcuate, many-flowered. *Flowers* large, yellow blotched with dull brown with a yellow or off-white lip veined with brown; dorsal sepal ovate, 2.5–3 cm long, 1.5–2 cm wide; lateral sepals ovate, 2.8–3.3 cm long, 1.5–1.7 cm wide; petals spathulate, 2.7–3 cm long, 1.2–1.4 cm wide; lip trilobed towards the apex, 1.9–2.2 cm long, 1.6–2 cm wide, hairy all over, the mid-lobe semicircular, the callus of 2 ridges; column 1.3–1.7 cm long. (See fig. 78 A–K and plate 15D).

DISTRIBUTION: Bougainville, Florida Group (Nggela), Guadalcanal, Makira, New Georgia, the Santa Cruz Islands and Santa Isabel. Also in New Guinea and New Britain.

HABITAT: Usually on trees by the sea, also on coconuts in plantations, sea level to 100 m.

COLLECTIONS: *Brown* 3302 (BM); *Chaplin* 786 & 788 (K); *Comins* 328 (K); *Crocker* 15 (K); *Dickson* AB8 (K); *Hunt* in RSS 2641 (K); *Im Thurn* 336 (K); *Foreman* in NGF 45710 (LAE); *Officers of HMS Penguin* s.n. (K); *Waterhouse* 630 (K); *Wickison* 66 (K); *Womersley & Whitmore* in BSIP 1186 (LAE).

Grammatophyllum scriptum is a variable species. Var. *boweri,* found in New Guinea, Bougainville and the Solomon Islands, is distinguished from the typical variety by the extensive dull brown blotching on its sepals and petals.

2. G. speciosum *Blume*, Bijdr. Fl. Ned. Ind.: 378 (1825). Type: Java, *Blume* s.n. (holotype L).

Grammatophyllum cominsii Rolfe in Ann. Bot. 5: 506 (1891). Type: Solomon Islands, *Comins* 57 (holotype K!).

A large epiphyte or lithophyte plant forming clumps. *Pseudobulbs* usually becoming pendent, fusiform to cylindrical, many-noded, up to 4 m long. *Leaves* many, distichous, linear, c. 1 m long, less than 2.5 cm wide. *Inflorescences* lateral, densely many-flowered, up to 2 m long, with the lowermost 3 to 5 flowers lacking a lip and column. *Flowers* yellow or greenish yellow spotted with maroon; sepals and petals with undulate margins; sepals ovate, 4.5–5.5 cm long, 2 cm wide; petals ovate, 4.5–5.5 cm long, 2 cm wide; lip trilobed, 1.7–2.2 cm long, 2–2.5 cm wide, pubescent, the mid-lobe narrowly triangular, the callus of 3 ridges; column 1–1.5 cm long. (See fig. 78 L–N and plate 15A).

DISTRIBUTION: Bougainville, Kolombangara and the Russell Islands. Widely distributed from SE. Asia, the Malay Archipelago and the Philippines to New Guinea.

HABITAT: Epiphyte high up in large trees in primary forest, also occasionally on rocks, up to 300 m.

COLLECTIONS: *Comins* 57 (K); *Lever* s.n. (K); *Waterhouse* 622 (K); *Wickison* 86 (K).

A widespread and variable species particularly in flower size and ground colour. *Grammatophyllum wallisii* Reichb. f., from the Philippines, is probably conspecific, differing only in the ivory ground colour of the flowers.

Fig. 78. Grammatophyllum scriptum var. **boweri A-K. A**, habit reduced; **B**, flower x 1; **C**, column x 3; **D**, dorsal sepal x 5/3; **E**, petal x 5/3; **F**, lateral sepal x 5/3; **G**, pollinia x 4; **H**, dorsal and ventral view of anther-cap x 4; **J**, side view of lip x 2; **K**, lip x 2. **G. speciosum L-N. L**, flower x 2/3; **M**, lip x 2; **N**, side view of lip x 2. **A-K** drawn from *Wickison* 66; **L-N** from *Wickison* 86. All drawn by Sue Wickison.

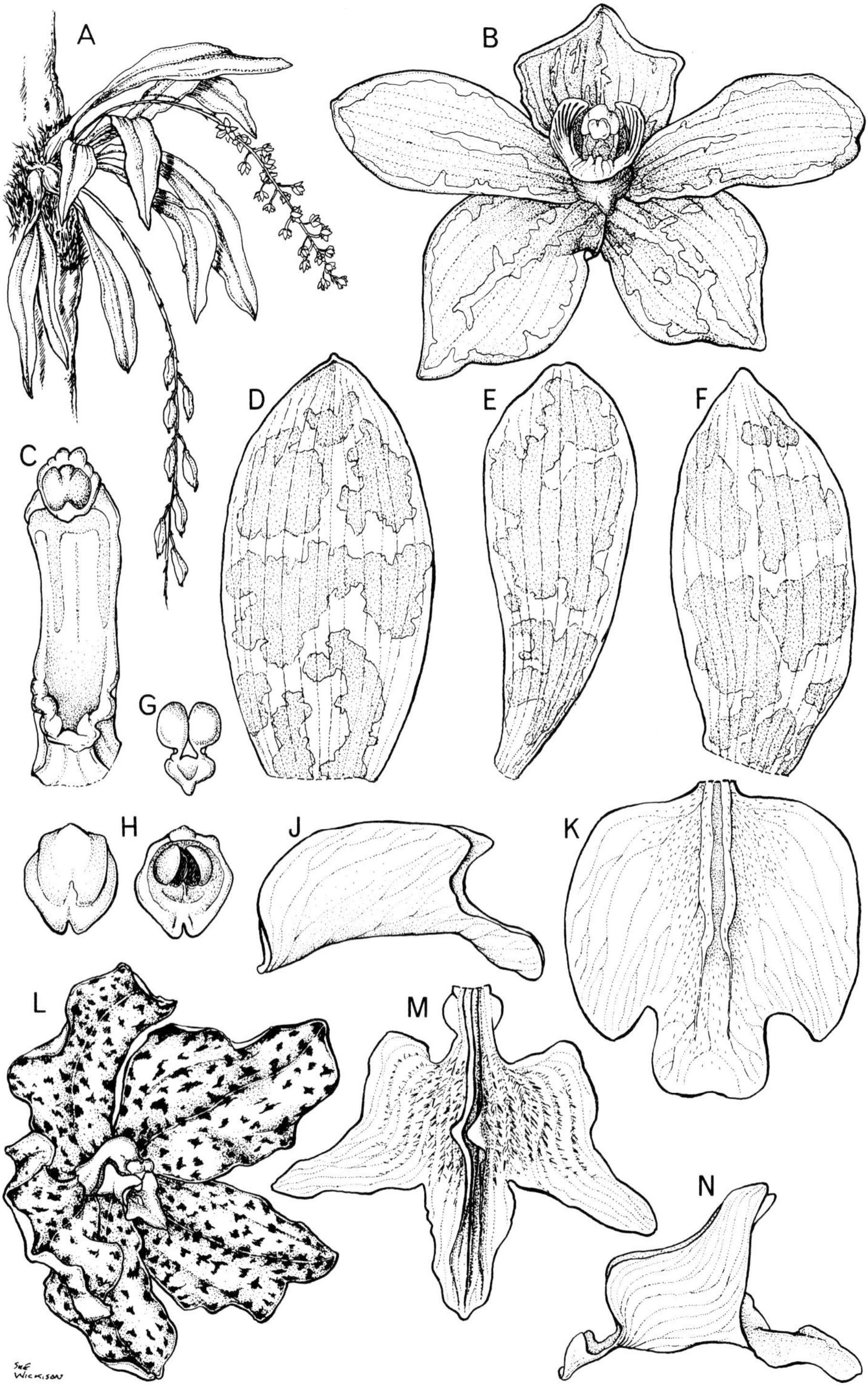
A
B
C
D
E
F
G
H
J
K
L
M
N

86. DIPODIUM R. Brown

Saprophytic or autotrophic, terrestrial or climbing herbs of diverse habit. *Roots* thick. *Stems* either climbing, continuing to grow indefinitely and rooting throughout, more or less erect-stemmed free-standing plants, or leafless saprophytes with only the inflorescence showing above the ground. (Only the first type is found in the Solomon Islands and Bougainville.) *Leaves* if present distichous, lanceolate, fleshy, with prominent veins. *Inflorescences* lateral or terminal, simple, more or less erect and containing from a few to 50 flowers which are of moderate size. *Flowers* widely spreading; sepals and petals free, similar, usually spotted; lip trilobed, close and parallel to the column, the lateral lobes slender and more or less embracing the column to form a tube, the mid-lobe much larger than the lateral lobes, ovate, convex, pubescent in part; column about half the length of the lip; pollinia 2.

A genus of about twenty species from tropical Asia to Australia and the Pacific Islands, a single species being recorded.

D. pictum *(Lindley) Reichb. f.* in Xenia Orch. 2: 15 (1862). Type: Java, *Lobb* s.n. (holotype K!).
Wailesia picta Lindley in Journ. Hort. Soc. 4: 262 (1849).
Dipodium pandanum Bailey in Queensl. Agric. Journ. 6: 287 (1900). Type: New Guinea, *Bailey* s.n. (holotype BRI).

A climbing leafy herb, up to 12 m high. *Roots* stout, up to 5 mm in diameter, produced along the stem. *Stems* at base of and spiralling up tree trunks, leafy. *Leaves* brittle, distichous, imbricate, lanceolate, acute or acutely bifid, up to 45 cm long, 6 cm wide, sheathing at base; leaf sheaths persistent. *Inflorescences* lateral, arising from the leaf axils in the upper part of each plant, suberect or erect, unbranched, laxly 5- to 25-flowered, up to 25 cm long; bracts ovate-elliptic, acute, up to 5 mm long, 2 mm wide. *Flowers* pale yellow spotted with red with a white lip veined with pink on the base of the mid-lobe, the column apex yellow; sepals and petals similar, spathulate or oblong-elliptic, obtuse or rounded, 1.6–1.8 cm long, 5–6 mm wide; lip trilobed at base, fused at base to column, 1.2–1.7 cm long, the lateral lobes small, appressed to the sides of the column, the mid-lobe convex, elliptic, obovate, acute, pubescent all over but with longest hairs towards the apex and on the mid-line; column 6–7 mm long. (See fig. 79 and plate 16D).

DISTRIBUTION: Guadalcanal, Kolombangara, Makira, Malaita, Florida Islands (Nggela), Santa Ana and Santa Isabel. Also in New Guinea, New Ireland and Australia.

HABITAT: Scrambling at base of trees or climbing up tree trunks in mossy ridge-top forest of *Pandanus*, *Casuarina* etc., and in secondary forest, 250–1470 m.

Fig. 79. Dipodium pictum. A, habit reduced; **B**, inflorescence x 2/3; **C**, flower x 1; **D**; lip x 2; **E**, dorsal and ventral view of anther-cap x 4; **F**, pollinia x 4; **G**, column x 3; **H**, side view of lip and column x 2; **J**, leaf apex x 2/3; **K**, dorsal sepal x 2; **L**, petal x 2; **M**, lateral sepal x 2. Drawn from *Chaplin* 846 by Sue Wickison.

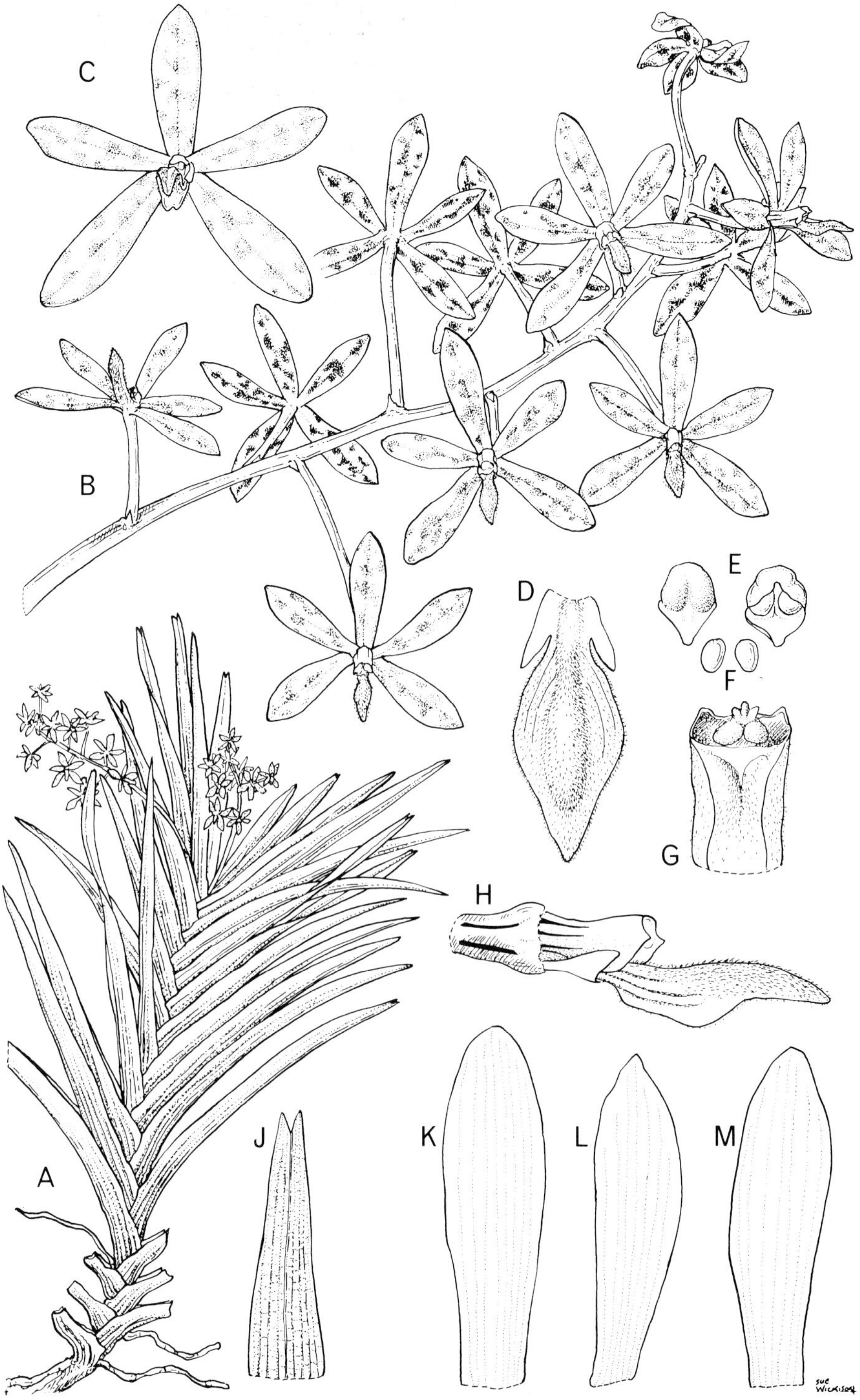
C
B
D
E
F
G
H
A
J
K
L
M
SUE WICKISON

COLLECTIONS: *Brass* 3210 (BM); *Chaplin* 846, 847 & 848 (K); *Cribb & Campbell* 5129 (K); *Dennis* in RSS 2275 & 2351 (K); *Robinson* K222 (K); *Sirute et al.* in BSIP 10089 (K); *Wickison* 244 (K).

This is an interesting species with an unusual growth habit. It is a true climber, with its stem extending up trees in an open spiral and being supported by wiry roots. The stems grow from the apex and become dormant and die out behind. It is common for pieces of the old, brittle rhizomes to break off and fall to the forest floor. Here a new shoot emerges from a dormant node and grows through the leaf litter until it finds a rock or tree on which to climb. Successful plants lose contact with the ground entirely and become completely epiphytic (Jones, 1988).

See also Dennis (1976) for a detailed discussion of its habitat.

This species is similar to *D. scandens* (Blume) J.J. Smith, from Java, and the two should be critically compared.

87. ACRIOPSIS Blume

Epiphytic herbs. *Rhizome* creeping, branched. *Roots* slender, branched, with tufts of erect, acuminate branches at base of pseudobulb. *Pseudobulbs* crowded, ovoid, covered at the base by thin, silvery bracts. *Leaves* apical, (1-)2–3(-4), oblong or linear, the midrib sunken above, prominent below, petiolate. *Inflorescence* arising from the base of the pseudobulb on a short, rooting rhizome, a raceme or panicle, usually many-flowered; peduncle relatively long, terete; floral bracts persistent. *Flowers* small, more or less twisted but not resupinate, widely open; sepals lanceolate, concave at the apex; petals spreading, oblong to obovate, about as long as the sepals; lip trilobed, pandurate or entire, the disc with 2 keels; column ± straight or slightly S-shaped, with 2 long parallel, porrect or decurved, elongate stelidia; pollinia 4.

A small genus of six species from Asia to the Malay Peninsula, Indonesia, New Guinea, the Pacific Islands and Australia, a single species being recorded.

A. javanica *Reinw. ex Blume*, Bijdr. Fl. Ned. Ind.: 377 (1825). Type: Java, *Blume* s.n. (lectotype L).

var. **javanica**

For full synonymy see Minderhoud & de Vogel (1986).

An epiphytic herb. *Pseudobulbs* crowded, up to 6 cm long, 2 cm wide. *Leaves* apical, (2-)3(-4), oblong-linear, 5–28 cm long, 3–20 mm wide; petiole 15–25 mm long. *Inflorescence* arising from the base of the pseudobulb, up to 90 cm long; rhachis branched, the branches up to 25 cm long, sometimes with secondary branches, many-flowered, 8–38 cm long; peduncle 8.5–55 cm long; bracts lanceolate, 1.5–3 mm long. *Flowers* with pale yellow sepals, off-white petals

---

Fig. 80. Acriopsis javanica var. **javanica. A**, habit x 2/3; **B**, flower x 4; **C**, side view of flower x 4; **D**, dorsal and ventral view of anther-cap x 10; **E**, pollinia x 10; **F**, lip x 6; **G**, column x 8; **H**, dorsal sepal x 6; **J**, petal x 6; **K**, lateral sepal x 6. Drawn from *Wickison* 80 by Sue Wickison.

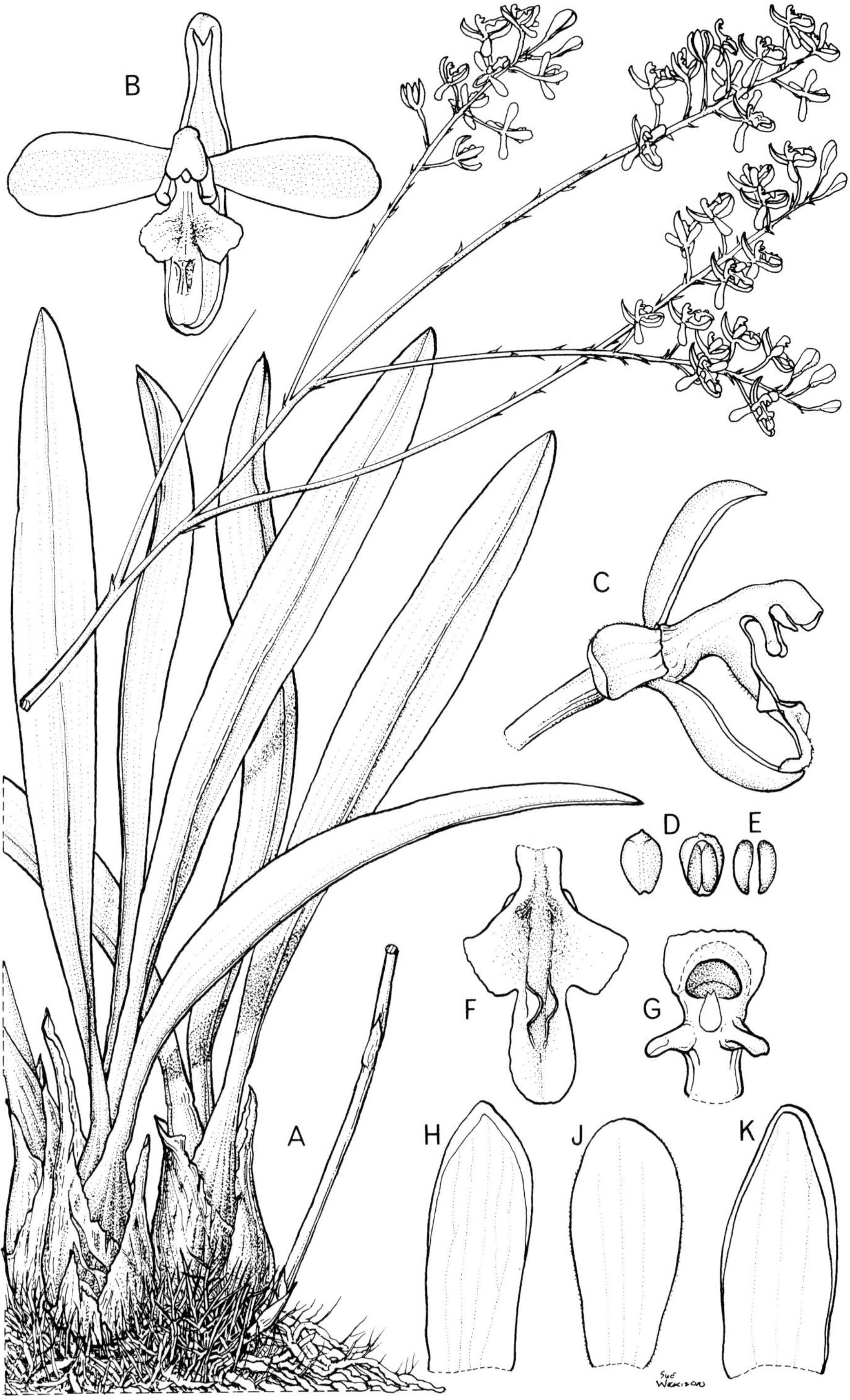
B
C
D
E
F
G
A
H
J
K

with a yellow stripe down the centre, a white lip and a light green column; pedicel and ovary 3–6.4 mm long; dorsal sepal ovate, 5–6 mm long, 1–2 mm wide; lateral sepals ovate, 5–6.5 mm long, 1–2.2 mm wide; petals elliptic-spathulate, 4–6 mm long, 1.3–3 mm wide, with a rounded apex; lip trilobed, 6 mm long, 4 mm wide when flattened, the lateral lobes obliquely oblong, the mid-lobe oblong with a rounded apex, the disc with 2 small, semi-circular raised calli at the base of the mid-lobe; column 3.5–4 mm long, with 1 mm long stelidia. (See fig. 80)

DISTRIBUTION: Bougainville, Choiseul, Guadalcanal, Florida group, Kolombangara, New Georgia and Santa Isabel. Widely distributed from Thailand and the Malay Peninsula and Archipelago to New Guinea and Australia.

HABITAT: Plantations, logging areas and rain forest, sea level to 650 m.

COLLECTIONS: *Chaplin* 769 (K); *Cribb & Campbell* 5101 (K); *Mitchell* 23 & 34 (K); *Schodde & Craven* 3959 (LAE); *Waterhouse* 535 & 668 (K); *Wickison* 80 (K); *Woodford* s.n. & 2 (K).

Minderhoud & de Vogel (1986) note that many collectors record that this orchid grows in ants' nests on tree trunks, and often has ant passages between the pseudobulbs. This is ascribed to the fact that the ants eat drops of fatty oil (elaeiosomes) which are located in the cells of the seed wing. The ants carry the seeds to their nest where many are able to germinate. Possibly the plants profit from nutrients in the waste material in the ants' nests.

APPENDIX I: NAME CHANGES IN THE SOLOMON ISLANDS

The country and islands have undergone changes in name which are often confusing. The first European to discover the Solomons was the Spanish adventurer, Alvaro de Mendaña, in 1567. He named the islands after King Solomon, Yslas su Llamen de Salomon (The Isles of Solomon), after his expedition found alluvial gold in the rivers on Guadalcanal. In 1893 the British took over the administration of the islands and renamed the territory the British Solomon Islands Protectorate. Finally, in 1975, when the islands became independent the name was formally changed to the Solomon Islands.

Many islands have had more than one name and confusion has often been compounded by variations in spelling.

Table 1: Names of Solomon islands in present usage and other common variations.

In present usage	Previously/Alternatives
Choiseul	Lauru
Florida Islands	Nggela Islands
Guadalcanal	Guadalcanar, Galekana, Solomon
Kolombangara	Nduke
Makira	San Cristobal
Malaita	Mala
Maramasike	Small Malaita
Ontong Java Atoll	Lord Howe Atoll*
Reef Islands	Swallow or Matema Islands
Santa Ana	Owa Rafa, Owa Raha
Santa Cruz	Nendo, Ndeni, Nitendi, Ndende
Santa Isabel	Santa Ysabel, Ysabel, Bughotu
Shortland Island	Alu
Vanikoro	Vanicoro, Vanikolo
Vella Lavella	Vella

* This should not be confused with Lord Howe Island along the southern shore of Nendo in Temotu Province nor the other Lord Howe Island, 700 km north-east of Sydney, Australia.

APPENDIX II: KWARA'AE VERNACULAR NAMES

"In my botanical excursions in these islands, I received the greatest assistance from the natives; and I was particularly struck with the familiar knowledge of their trees and plants which these islanders possessed. They have names not only for nearly all the trees, but for several of the grasses; and in the case of the former, when I was uncertain as to whether I had come upon any specimen before, they would obtain its flower, or fruit, or foliage, and point out to me its comparative characters."

H.B. Guppy, (1887a, p. 280)

Many of the vernacular names citied in this account are Kwara'ae names. The Kwara'ae area of Malaita extends from Auki on the west coast, north and eastwards over the highland bush interior, to the opposite coast and Kwai Island. Kwara'ae is also spoken by a large number of people in other communities. For a variety of reasons, in the early 1960's, Kwara'ae was adopted by botanists and foresters as the vernacular language for plant taxonomy in the Solomon Islands (Hancock & Henderson, 1988). Indeed, the Kwara'ae classification is diverse and comprehensive.

Several orchids have vernacular names especially common orchids, such as *Liparis condylobulbon* and *Spathoglottis plicata*. Useful orchids, such as *Diplocaulobium solomonense* and *Tropidia disticha* also have vernacular names. For a discussion on Kwara'ae names see Whitmore (1966) and Hancock & Henderson (1988). A detailed list of Kwara'ae words used in plant names is given in Whitmore (1966).

APPENDIX III: LIST OF HERBARIUM ABBREVIATIONS

AD	State Herbarium of South Australia, Australia.
AMES	Orchid Herbarium of Oakes Ames, Botanical Museum of Harvard University, Massachusetts, USA.
B	Botanischer Garten und Botanisches Museum Berlin-Dahlem, Germany.
BISH	Herbarium Pacificum, Department of Botany, Bernice P. Bishop Museum, Honolulu, Hawaii, USA.
BM	Herbarium, The Natural History Museum, London, U.K.
BO	Herbarium Bogoriense, Bogor, Indonesia.
BRI	Queensland Herbarium, Brisbane, Australia.
CAL	Central National Herbarium, Botanical Survey of India, Calcutta, India.
CANB	Australian National Herbarium, Canberra, Australia.
E	Herbarium, Royal Botanic Garden, Edinburgh, Scotland, U.K.
FI	Herbarium, Universitatis Florentinae, Museo Botanico, Firenze, Italy.
G	Herbarium, Conservatoire et Jardin botanique de la Ville de Genève, Switzerland.
K	The Herbarium, Royal Botanic Gardens, Kew, U.K.
L	Rijksherbarium, Leiden, Netherlands.
LAE	Herbarium, Division of Botany, Department of Forests, Lae, Papua New Guinea.
LINN	The Linnean Society of London, U.K.
MEL	National Herbarium of Victoria, Royal Botanic Gardens, Melbourne, Victoria, Australia.
MO	Herbarium, Missouri Botanic Garden, St Louis, U.S.A.
NSW	National Herbarium of New South Wales, Sydney, Australia.
NY	Herbarium, The New York Botanical Garden, New York, U.S.A.
P	Museum National d'Histoire Naturelle de Phanerogamie, Paris, France.
QRS	Herbarium, Queensland Regional Station, C.S.I.R.O., Division of Forestry Research, Australia.
S	Herbarium, Swedish Museum of Natural History (Natuurhistoriska Riksmuseet), Stockholm, Sweden.
UPS	Herbarium, Institute of Systematic Botany, University of Uppsala, Sweden.
Z	Institut fur systematische Botanik der Universität Zürich, Switzerland.

REFERENCES AND FURTHER READING

Ames, O. (1932). Contributions to the flora of the New Hebrides and Santa Cruz Islands; orchids collected by S.F. Kajewski in 1928 & 1929. Journ. Arn. Arb. 13: 101–112.

——— (1933). Additional notes on the orchids of the New Hebrides and Santa Cruz Islands. Journ. Arn. Arb. 14: 101–112.

Beaglehole, J.C. (1966). The exploration of the Pacific. A. & C. Black Ltd., London.

Brieger, F.G. (1981). Subtribus Dendrobiinae. Schlechter, Die Orchideen ed. 3, 1(11–12): 636–752.

Carr, C.E. (1934). On a collection of orchids from the Solomon Islands. Bull. Misc. Inf., Kew 1934: 375–383.

Christiansen, S. (1964). On Lake Tegano, Rennell Island, and some remarks on the problem of Rennell's origin. Geogr. Tidsskr. 63: 99–111.

Clements, M.A. (1989). Catalogue of Australian orchids. Australian Orchid Research 1: 1–160, Australian Orchid Foundation, Melbourne.

Coleman, P.J. (1966). The Solomon Islands as an island arc. Nature, 5055: 1249–1251.

——— (1970). Geology of the Solomon Islands and New Hebrides as part of the Melanesian re-entrant, south-west Pacific. Pacific Sci. 24: 289–314.

Comber, J. (1990). Orchids of Java. Royal Botanic Gardens, Kew.

Corner, E.J.H. (1969). A discussion on the results of the Royal Society Expedition to the British Solomon Islands Protectorate, 1965. Phil. Trans. Roy. Soc. Lond. B, 800(255): 185–631.

Cribb, P.J. (1983a). *Dendrobium* sect. *Ceratobium* (Orchidaceae) in the Pacific Islands. Kew Bull. 37(4): 577–590.

——— (1983b). A revision of *Dendrobium* section *Latouria*. Kew Bull. 38(2): 229–306.

——— (1986). A revision of *Dendrobium* section *Spatulata*. Kew Bull. 41(3): 615–692.

——— (1987a). The genus *Dendrobium* in New Guinea, Australia and the Pacific Islands. Proceedings of the 12th World Orchid Conference, 210–214.

——— (1987b). The genus *Paphiopedilum*. Collingridge & The Royal Botanic Gardens, Kew.

——— (1990a). Desert island discoveries: orchid hunting on a coral island. Orchid Review 98(1162): 243–248 & in Orchadian 10(1): 5–11.

——— (1990b). The orchids of the south-west Pacific Islands and their conservation. Proceedings of the 1st Australasian Native Orchid Conference and Show, 1990, 174–178. The Australasian Native Orchid Society.

Cribb, P.J., Campbell, J. & Dennis, G. (1985). *Paphiopedilum* in the Solomon Islands: the rediscovery of '*P. dennisii*'. Orch. Review 93 (1098): 130–131.

Cribb, P.J. & Lewis, B.A. (1989a). New orchids from Vanuatu in the south west Pacific, Part 1. Orch. Review 97 (1150): 251–153.

—— (1989b). New orchids from Vanuatu in the south west Pacific, Part 2. Orch. Review 97 (1152): 329–332.

—— (1989c). New orchids from Vanuatu in the south west Pacific, Part 3. Orch. Review 97(1153): 360–363.

—— (1989d). New orchids from Vanuatu in the south west Pacific, Part 4. Orch. Review 97(1154): 389–390.

—— (1990). Little known orchids of Espiritu Santo. Proceedings of the 1st Australasian Native Orchid Conference and Show, 1990, 124–128. The Australasian Native Orchid Society.

Cribb, P.J., Lewis, B.A. & Dickson, M. (1991). *Diplocaulobium* (Reichb. f.) Kraenzl. (Orchidaceae) in the Solomon Islands and Bougainville. Lindleyana 6(1): 27–35.

Cribb, P.J. & Robbins, S. (1991). The New Guinea orchid collections of Rudolf Schlechter in Herbarium Bogoriense. Lindleyana 5(4): 244–248.

Cribb, P.J. & Tang, C.Z. (1982). *Spathoglottis* in Australia and the Pacific Islands. Kew Bull. 36 (4): 721–729.

Cribb, P.J., Wood, J.J. & Stirton C. (1988). An index of Rudolf Schlechter's New Guinea orchid specimens in world herbaria. Lindleyana 3(2): 105–116.

Davies, R. (1972). The orchids of Honiara. South Pacific Bulletin 22(2): 26–28.

Dennis, G. (1976). Orchids of the Solomon Islands. Orchadian 5(6): 87–88.

Dockrill, A.W. (1969). Australian indigenous orchids. The Society for growing Australian plants, Sydney.

de Vogel, E.F. (1969). Monograph of the tribe *Apostasieae* (Orchidaceae). Blumea 17(2): 315–350.

—— (1988). Revisions in Coelogyninae (Orchidaceae). The genus *Pholidota.* Orchid Monographs, 3: 1–118. Rijksherbarium, Leiden.

Docters van Leeuwen, W.M. (1937). The biology of *Epipogium roseum* (D. Don) Lindl. Blumea, suppl. 1: 57–65.

Dressler, R.L. (1981). The orchids, natural history and classification. Harvard University Press.

Edridge, S. (1985). Solomon Islands Bibliography to 1980. Institute of Pacific Studies, The University of the South Pacific, Fiji, the Alexander Turnbull Library, New Zealand and The Solomon Islands National Library, Honiara.

Fowlie, J.A. (1968). A new species of *Paphiopedilum* from the island of Bougainville. Orchid Digest 32: 283–285.

Garay, L.A. & Kittredge, W. (1985). Notes from the Ames Orchid Herbarium. Bot. Mus. Leafl. Harv. Univ. 30(3): 47–59.

Garay, L.A. & Sweet, H.R. (1974). Orchids of the Southern Ryukyu Islands. Harvard University Press.

Glasser, T. (1987). Solomon Islands: paradise lost and found. The Courier 102: 44–51.

Guillaumin, A. (1965). Plantes nouvelles rares ou critiques des serres du Muséum. (Notules sur quelques Orchidées d'Indochine. 38). Bull. Mus. Nat. Hist. Paris, ser. 2, 37(1): 199–200.

Guppy, H.B. (1887a). The Solomon Islands and their natives. Swan Sonnenschein, Lowrey & Co., London.

—— (1887b). The Solomon Islands: their geology, general features and suitability for colonization. Swan Sonnenschein, Lowrey & Co., London.

Hallé, N. (1977). Flore de la Nouvelle-Calédonie et Dependances, 8: Orchidacées. Museum National d'Histoire Naturelle, Paris.

—— (1990). Additions aux *Orchidaceae* du Vanuatu. Adansonia, ser 4, 12, no. 2: 199–203.

Hancock, I.R. & Henderson, C.P. (1988). Flora of the Solomon Islands: Research Bulletin no.7. Dodo Creek Research Station, Ministry of Agriculture & Lands, Honiara, Solomon Islands.

Hansell, J.R.F. & Wall, J.R.D (1976). Land resources of the Solomon Islands, volume 1, introduction and recommendations, Land Resource Study 18. Land Resource Division, Ministry of Overseas Development, London.

Hansen, B. & Dissing, H. (1980). List of plants collected on Rennell and Bellona. Noona Dan Papers no. 138. Natural History of Rennell Island, British Solomon Islands 8: 15–25.

Henderson, C.P. & Hancock, I.R. (1988). A Guide to the useful plants of Solomon Islands. Research Department, Ministry of Agriculture and Lands, Honiara, Solomon Islands.

Holttum, R.E. (1964). Orchids of Malaya (third edition). Government Printing Office, Singapore.

Hunt, P.F. (1969). Orchids of the Solomon Islands. Phil. Trans. Roy. Soc. B, 255: 581–587.

Jones, D.L. (1988). Native Orchids of Australia. Reed Books Ltd., New South Wales.

Jones, K., Lim, K.Y. & Cribb, P.J. (1982). The chromosomes of Orchids VII. *Dendrobium*. Kew Bull. 37(2): 221–227.

Kajewski, S.F. (1930). A plant collector's notes on the New Hebrides and Santa Cruz Islands. Journ. Arn. Arb. 11: 172–180.

Kajewski, S.F. (1946). Plant collecting in the Solomon Islands. Journ. Arn. Arb. 27(3): 292–304.

Kent, A.H. (1900). Veitch's manual of the Coniferae, ed. 2: 543. James Veitch & Sons Ltd, Chelsea.

Kerr, R. (1966). The remarkable orchids of Arawa. Austr. Orch. Review 31: 85–89.

Kores, P.J. (1989). A precursory study of Fijian Orchids. Allertonia 5(1): 1–222.

Lever, R.J.A.W. (1973). The early history of agriculture in the Solomon Islands. S. Pacific Bull. 23 (1): 26–30.

Lewis, B.A. (1989). Orchids of Vanuatu. Orch. Review 97(1151): 278–285 & in Orchadian (in press).

—— (1990). A Bibliography for south-west Pacific Orchidaceae. Proceedings of the 1st Australasian Native Orchid Conference and Show, 1990: 130–134. The Australasian Native Orchid Society.

Lewis, B.A. & Cribb, P.J (1989). Orchids of Vanuatu. Royal Botanic Gardens, Kew.

—— (in press). *Coelogyne susanae*, a new species of Orchidaceae from the Solomon Islands and Bougainville. Kew Bull. 46(2): 317–319.

Merrill, E.D. (1946). Plant life of the Pacific world. Macmillan, New York.

Minderhoud, M.E. & de Vogel, E.F. (1986). A taxonomic revision of the genus *Acriopsis* Reinwardt ex Blume (Acriopsidinae, Orchidaceae). Orchid Monographs 1:1–86, Rijksherbarium, Leiden.

Millar, A. (1978). Orchids of Papua New Guinea, an introduction. Australian National University Press, Canberra.

Mitchell, R.B. (1989). Orchid hunting in Choiseul. Kew Mag. 6(3): 124–134.

Naval Intelligence Division (1944). Pacific Islands. Vol. 3: Western Pacific (Tonga to Solomon Islands) 511–606. Naval Intelligence Division.

Pendleton, R.C. (1949). The rain forest shadow effect on the plant formation of Guadalcanal. Ecol. Monogr. 19: 75–93.

Petherbridge, D.W. (1947). Orchids on Bougainville. Austr. Orch. Review, 12: 22–23.

Rassmussen, F.N. (1977). The genus *Corymborkis* Thou. (Orchidaceae). A taxonomic revision. Bot. Tidsskr. 71: 161–192.

Rauschert, S. (1983). Beitrag zur Nomenclature der Orchidaceae. Feddes, Repert. Spec. Nov. Regni Veg. 94(7–8): 433–471.

Reeve, T.M. (1990). New Guinea Orchidaceae: A synopsis of the genus *Dendrobium*. Proceedings of the 1st Australasian Native Orchid Conference and Show, 1990: 53–63. Australasian Native Orchid Society.

Reeve, T.M. & Woods, P.J.B. (1981). *Dendrobium delicatulum* Kraenzl. (section Oxyglosssum). Orchadian 7 (1): 18–21.

—— (1989). A revision of *Dendrobium* section *Oxyglossum* (Orchidaceae). Notes from the Royal Botanic Garden Edinburgh: 46(2): 1–305.

Schlechter, R. (1911–14). Die Orchidaceen von Deutsch-Neu-Guinea. Fedde, Repert. Spec. Nov. Regni Veg., Beih. 1: 1–1079. (Avaliable in English translation from the Australian Orchid Foundation, ed. D.F. Blaxell, 1982).

Schuiteman, A. (1990). A new combination in *Agrostophyllum* (Orchidaceae). Blumea 35: 165.

Schumann, K. & Lauterbach, K. (1905). Nachträge zur Flora der Deutschen Schutzgebiete in der Südsee. Leipzig.

Senghas, K. (1978). *Hapalochilus*, eine alte, neue Orchideengattung aus Neu-Guinea. Die Orchidee 29(6): 245–248 (1978).

Seidenfaden, G. (1973). Notes on *Cirrhopetalum*. Dansk Bot. Arkiv 29 (1): 1–260.

—— (1976). Orchid genera in Thailand 4: *Liparis*. Dansk. Bot. Arkiv 31 (1): 1–105.

—— (1978). Orchid genera in Thailand 6: *Neottioideae*. Dansk. Bot. Arkiv 32 (2): 1–195.

—— (1979). Orchid genera in Thailand 8: *Bulbophyllum*. Dansk. Bot. Arkiv 33 (3): 1–228.

—— (1980). Orchid genera in Thailand 9: *Flickingeria* & *Epigeneium*. Dansk. Bot. Arkiv 34 (1): 1–104.

—— (1982). Orchid genera in Thailand 10: *Trichotosia* and *Eria*. Opera Botanica 62: 1–157.

—— (1983). Orchid genera in Thailand 11: *Cymbidieae*. Opera Botanica 72: 1–124.

—— (1985). Orchid genera in Thailand 12: *Dendrobium*. Opera Botanica 83: 1–295.

—— (1988). Orchid genera in Thailand 14: fifty-nine vandoid genera. Opera Botanica 95: 1–398.

Seidenfaden, G. & Wood, J.J. (in press). The orchids of Peninsular Malaysia and Singapore. A revision of R.E. Holttum: Orchids of Malaya. Singapore National Printers.

Slade, G.H. (1965). A note on the orchids of Arawa plantation. Austr. Orch. Review, 30(1): 29.

—— (1980). *Dendrobium mohlianum* – A case of pollination by birds. Amer. Orch. Soc. Bull. 49: 869–870.

Smith, J.J. (1905). Orchids of Java. E.J. Brill, Leiden.

—— (1909–34). Nova Guinea. Résultats de l'expedition scientifique Néerlandaise à la Nouvelle-Guinée. E.J. Brill, Leiden.

van Steenis, C.G.G.J. (1950). Flora Malesiana 1: 1–639. Noordhoff-Kolff N.V. (Djakarta). Reprint 1985, Koeltz Scientific Books, Germany.

Tan, K. (1974). Taxonomy of *Arachnis*, *Armodorum*, *Esmeralda* and *Dimorphorchis* (Orchidaceae). University Microfilms, Ann Arbor, Michigan, U.S.A.

Taylor, G.R. (1973). Preliminary observations on the structural history of Rennell Island, South Solomon Sea. Bull. Geol. Soc. Amer. 84: 2795–2805.

Thorne, A. & Cribb, P.J. (1984). Orchids of the Solomon Islands and Bougainville, a preliminary checklist. Royal Botanic Gardens, Kew.

Turner, H. (in press). A revision of the orchid genera *Tainia*, *Ania*, *Mischobulbon* and *Hancockia*. Orchid Monographs, Rijksherbarium, Leiden.

van Balgooy, M.M.J. (1960). Preliminary plant geographical analysis of the Pacific. Blumea 10: 385–430.

van Royen, P. (1980). The orchids of the high mountains of New Guinea. J. Cramer, Germany.
—— (1983). The genus *Corybas* in its eastern areas. J. Cramer, Germany.

Walter, H. & Lieth, H. (1967). Kimadiagramm Weltaltas. VEB Gustav Fisher Verlag Jena.

Warren, R. (1990a). New Guinea orchids – I: *Dendrobium* section *Latouria*. Amer. Orch. Soc. Bull. 59(2): 128–134.
—— (1990b). New Guinea Orchids – II: The Antelope *Dendrobiums*. Amer. Orch. Soc. Bull. 59(3): 252–259.
—— (1990c). New Guinea *Dendrobiums* – III: section *Pedilonum*. Amer. Orch. Soc. Bull. 59(5): 473–479.
—— (1990d). New Guinea orchids – 4: *Dendrobium* section *Calyptrochilus*. Amer. Orch. Soc. Bull. 59(6): 590– 597.
Warren, R. & Woods, P (1990). New Guinea *Dendrobiums*–5: section *Oxyglossum*-'The jewels in the canopy'. Amer. Orch. Soc. Bull. 59(7): 704–711.

Whitmore, T.C. (1966). Guide to the Forests of the British Solomon Islands. Oxford University Press.
—— (1969). The vegetation of the Solomon Islands. Phil. Trans. Roy. Soc. B, 255: 259–270.
—— (1973). Plate tectonics and some aspects of Pacific plant geography. New Phytol. 72: 1185–1190.

Williams, L.O. (1940). Orchidaceae. In R. Fosberg, ed., Melanesian vascular plants. Lloydia 3(2): 111.

Wilson, G. (1924). *Dendrobium laevifolium*. Orch. Review 32: 280.

Womersley, J.S. (1967). The rediscovery of *Dendrobium laevifolium*. Austr. Orch. Review 32: 131–132.

Wood, J.J. (1978). Two *Dendrobium* species, *Dendrobium codonosepalum* J.J. Smith & *Dendrobium punamense* Schltr. Orch. Review 86(1020): 180–183.

—— (1983). Notes on Pacific *Bulbophyllum* (Orchidaceae). Kew Bull. 38 (1): 71–73.

—— (in press). Notes on *Trachoma, Tuberolabium* and *Parapteroceras* (Orchidaceae). Nord. J. Bot.

ACKNOWLEDGEMENTS

Without the tremendous help and encouragement of G. Hermon Slade, Gerald MacCraith and the Australian Orchid Foundation this book would not have been possible. We are also endebted to Geoffrey Dennis, long term resident in Honiara and the country's most knowledgable naturalist, who has encouraged us throughout and accompanied us on some delightful adventures. We are also beholden to him for the Foreward. We would also like to thank John Petersen, Graham Chaplin, John Campbell and Alasdair & Alexandra Morrison for their help and assistance in the Solomon Islands and Bougainville.

Mark Clements and Thomas Cope have kindly reviewed the manuscript.

The following people helped with comments and assistance on particular genera: J.J. Vermeulen with the genus *Bulbophyllum*, J. Renz with the genera *Peristylus* and *Habenaria*, E. Daucey, P. Woods, J.J. Wood, E. de Vogel and H. Turner. We would also like to thank J.J. Vermeulen for allowing us to use several of his illustrations of *Bulbophyllum* species.

Many people have helped with this project and we would especially like to thank the Solomon Islands Ministry of Forestry, Natural Resources and Mines, the late Pastor Soiaka, Newman Teghata, Jim Sevesi, Graham Chaplin, Robert Johns, Ben Wallace, Robert and Caron Mitchell, Margaret Dickson, Audrey Thorne, Sarah Robbins and Sandra Bell for their help; John McDonagh and Marianne Cribb for assistance and encouragement; Robert Allkin for help and support with the ALICE database and Jeff Wheatley for assistance with computer queries; Alan Radcliffe-Smith and Michael Lock for checking the latin diagnoses; Lady R. Sainsbury and Hewlett Packard (U.K.) for assistance with computer facilities; the Keeper of the Herbarium, the Royal Botanic Gardens, Kew; the Keeper of the Herbarium, Natural History Museum, London; the Director of the Conservatoire et Jardin Botanique, Geneva; the Director of the Laboratoire de Phanérogamie, Paris; the Director of Rijksherbarium, The Netherlands; the curator of the Herbarium, Department of Forestry, Lae; and the Keeper of the Orchid Herbarium of Oakes Ames, Botanical Museum of Harvard University.

The illustrations have been drawn by Sue Wickison who spent over two years in the Solomon Islands (1986–1988) sketching and collecting orchids. We are grateful to her for her enthusiasm and excellent work. Sue would especially like to thank the Winston Churchill Trust, who awarded her a travelling fellowship to go to the Solomon Islands, Rotring International, the Cannonbury Art Shop and Pro Arte Brushes. She would also like to thank George Anderson and the British High Commission, Christian Billy and his family (Provincial Member for West Malaita), Graham Chaplin, Danny Kennedy, John Lazarus and staff (Levers Pacific Timbers, Barora), Chris Turnbill (Forestry Dept., Honiara), staff and students at the Catholic Mission at Vanga Point, her parents, friends in the U.K. and especially Robert Barraclough, Geoffrey Dennis, Colin Morrow and Alison Worsnop.

Plate 1. **A** Rennell, Kanggava Bay

B Coastal forest, Munda, New Georgia, with *Dendrobium gouldii* in the foreground

C Montane forest, Guadalcanal, showing a landslip

D Copper mine, Panguna, Bougainville

E Logging, Barora, New Georgia

Plate 2. **A** *Paphiopedilum wentworthianum* **B** *Paphiopedilum bougainvilleanum*

C *Corymborkis* cf. *veratrifolia* **D** *Macodes cominsii*

E *Vrydagzynea argyrotaenia* **F** *Neuwiedia veratrifolia*

Plate 3. **A** *Corybas gemmatus* **B** *Acianthus vulcanicus*
C *Spathoglottis plicata* **D** *Ceratostylis subulata*
E *Goodyera viridiflora* **F** *Phaius amboinensis* var. *papuanus*

Plate 4. **A** *Calanthe ventilabrum* **B** *Calanthe hololeuca*

C *Calanthe triplicata* **D** *Appendicula polystachya*

E *Chilopogon oxysepalum* **F** *Trichotosia collina*

Plate 5. **A** *Coelogyne veitchii* **B** *Coelogyne asperata*

C *Coelogyne carinata* **D** *Coelogyne susanae*

E *Eria fitzalanii* **F** *Eria kingii*

Plate 6. **A** *Liparis condylobulbon* **B** *Liparis layardii*

C *Octarrhena angraecoides* **D** *Malaxis xanthochila*

E *Phreatia stenostachya* **F** *Phreatia matthewsii*

Plate 7. **A** *Pholidota imbricata* **B** *Mediocalcar paradoxum*

C *Glomera montana* **D** *Glossorhyncha macdonaldii*

E *Dendrobium bilobum* **F** *Dendrobium platygastrium*

Plate 8. **A** *Dendrobium gouldii,* form from the Florida Islands **B** *Dendrobium gouldii,* form from Guadalcanal

C *Dendrobium gouldii,* form from Bougainville **D** *Dendrobium rennellii*

E *Dendrobium sylvanum* **F** *Dendrobium conanthum*

Plate 9. **A** *Dendrobium mirbelianum* **B** *Dendrobium macranthum*

C *Dendrobium spectabile* **D** *Dendrobium magistratus*

E *Dendrobium ruginosum* **F** *Dendrobium johnsoniae*

Plate 10. **A** *Dendrobium goldfinchii* **B** *Dendrobium rhodostictum*
C *Dendrobium macrophyllum* **D** *Dendrobium polysema*
E *Dendrobium capituliflorum* **F** *Dendrobium purpureum*

Plate 11 **A** *Dendrobium mohlianum* **B** *Dendrobium puniceum*
C *Dendrobium laevifolium* **D** *Cadetia hispida*
E *Flickingeria comata* **F** *Flickingeria convexa*

Plate 12. **A** *Diplocaulobium magnilabre* **B** *Diplocaulobium meckynosepalum*

C *Diplocaulobium bicolor* **D** *Diplocaulobium subintegrum*

E *Bulbophyllum microrhombos* **F** *Bulbophyllum citrinilabre*

Plate 13. **A** *Bulbophyllum levatii* **B** *Bulbophyllum graciliscapum*

C *Bulbophyllum longiscapum* **D** *Bulbophyllum blumei*

E *Bulbophyllum orbiculare* **F** *Bulbophyllum hassallii*

Plate 14. **A** *Bulbophyllum trachyanthum* **B** *Bulbophyllum grandiflorum*

C *Gunnarella robertsii* **D** *Bulbophyllum dennisii*

E *Schoenorchis micrantha* **F** *Vanda hindsii*

Plate 15. **A** *Grammatophyllum speciosum* **B** *Malleola pallida*

C *Taeniophyllum fasciola* growing with *Dischidia* sp. **D** *Grammatophyllum scriptum* var. *boweri*

E *Taeniophyllum kompsopus* **F** *Trichoglottis papuana*

Plate 16. **A** *Pomatocalpa marsupiale* **B** *Rhinerrhiza moorei*
C *Robiquetia mooreana* **D** *Dipodium pictum*
E *Thrixspermum platystachys* **F** *Ascoglossum calopterum*

INDEX TO VERNACULAR NAMES

The vernacular names are cross-referenced to the scientific names. Numbers in bold refer to illustrations.

INDEX TO SCIENTIFIC NAMES

Names in bold capitals are accepted genera. Names in *italics* are synonyms. Species not recorded from the Solomon Islands and Bougainville but mentioned in the text are preceded by a short line. Numbers in bold refer to illustrations.